AF395811

Springer Asia Pacific Mathematics Series

Volume 9

The Springer Asia Pacific Mathematics Series promotes high-quality scientific research connected to Asia-Pacific countries. Monographs, edited volumes, lecture notes, and textbooks will cover topics related to pure and applied mathematics as well as interdisciplinary research.

All material will be rigorously peer-reviewed and should be related to mathematical research conducted at or with institutions in the APAC region. This series is open to subseries, which may be topical or exclusively supplied by a single research institution.

Zhi-Zhong Sun • Qifeng Zhang • Guang-hua Gao

Numerical Solutions to Partial Differential Equations with Finite Difference Methods

 Springer

Zhi-Zhong Sun
School of Mathematics
Southeast University
Nanjing, Jiangsu, China

Qifeng Zhang
Department of Mathematics
Zhejiang Sci-Tech University
Hangzhou, Zhejiang, China

Guang-hua Gao
College of Science
Nanjing University of Posts and
Telecommunications
Nanjing, Jiangsu, China

ISSN 3091-2555   ISSN 3091-2563   (electronic)
Springer Asia Pacific Mathematics Series
ISBN 978-981-95-5562-8   ISBN 978-981-95-5563-5   (eBook)
https://doi.org/10.1007/978-981-95-5563-5

Foreword

A large number of mathematical models in modern science, technology, and engineering can be described by differential equations. Many basic models themselves can be characterized by differential equations. The solutions to most of differential equations (especially partial differential equations) are difficult to express in explicit analytic forms.

In the process of science computerization, the computation in science and engineering has begun its new development. Numerical solutions of differential equations have also been developed in an unprecedented way.

Since the basic laws in science are mostly described by differential equations, one of the main tasks of scientific and engineering computation is to solve various differential equations. Therefore, the requirement to master and apply the numerical methods of differential equations nowadays is no longer limited to students in the department of mathematics. A large number of researchers engaged in mechanics, physics, astronomy, electronics, power, aviation, aerospace, civil engineering, geological exploration, oil field development, and other fields of engineering and technical personnel also regard this subject as an important research tool in their field.

This textbook introduces the numerical methods in depth for solving linear/nonlinear partial differential equations with boundary or initial-boundary value conditions, which include two-point boundary value problems, elliptic equations, parabolic equations, hyperbolic equations, high-dimensional evolution equations, fractional differential equations, the Schrödinger equation, the Burgers' equation, and the Korteweg-de Vries equation.

This textbook mainly concentrates on finite difference methods, which strives to achieve the following goals:

Featured Contents For each difference scheme, the following six aspects will be discussed. (1) Derivation of the difference scheme; (2) Existence and uniqueness of the difference solution; (3) Implementation of the difference scheme; (4) Numerical examples; (5) Priori estimate of the difference solution; (6) Convergence and stability of the difference solution. The first four aspects are fundamental. For the last two aspects, the priori estimate of the difference solution is essential, with which the convergence and the stability can be obtained immediately.

Scattered Difficulty We first provide a microcosm of the difference method by numerically solving the two-point boundary value problem and then apply this method to more complex problems one by one.

Emphasis on Practicability After the demonstration of the numerical examples, we would ask students to imitate and get the numerical results on the computer, then analyze the results, and finally achieve the purpose of mastering various numerical methods.

Last but not least, the summary and extension section in each chapter leaves more room for further study to whom with more ability.

Teachers can arrange teaching process according to the level of students and the amount of class hour. The first five chapters are basic, and the last four chapters are advanced. Sections 2.3, 3.7, 4.5, 5.2, 5.4, 6.5, 7.3, 8.3, and 9.3 are optional. Sections 5.2 and 5.4 are based on Sect. 2.3. Skipping Chap. 5 does not affect the reading of subsequent chapters, and skipping Chap. 6 also does not affect the reading of subsequent chapters.

The Chinese version of this book was drafted since 1995 in Southeast University and first published in January 2005. Since then, it has been selected as a textbook by many colleges and universities in China. After more than 20 years of teaching practice, the Chinese version of the textbook has come to its third edition at the beginning of 2022. In order to benefit more people in the whole world, we are very happy to share its English version. As one of the most influential publishing houses in the world, we have no hesitation and are very pleased to choose Springer Press.

During the process of publishing the Chinese versions of the textbooks, the graduate students of professor Zhi-Zhong Sun help to do some numerical computation and text editing. They are Ruilian Du, Fule Li, Xueling Li, Jianming Liu, Renjun Qi, Jincheng Ren, Hong Sun, Xuping Wang, Jingyu Wu, Zaibin Zhang, and Xuan Zhao. Many thanks to them.

We appreciate for the valuable publishing opportunity provided by Springer. We are also grateful to the editor Daniel Wang for his hard work and friendly cooperation.

Due to the authors' limited ability, mistakes are inevitable. We sincerely hope that experts and readers could provide valuable advice and suggestion.

<table>
<tr><td>Nanjing, China</td><td>Zhi-Zhong Sun</td></tr>
<tr><td>Hangzhou, China</td><td>Qifeng Zhang</td></tr>
<tr><td>Nanjing, China</td><td>Guang-hua Gao</td></tr>
<tr><td>October 15, 2025</td><td></td></tr>
</table>

Competing Interests The authors have no competing interests to declare that are relevant to the content of this manuscript.

Contents

About the Authors

Zhi-Zhong Sun, born in March 1963, receives his bachelor degree from the department of mathematics, Nanjing University in 1984; master degree from department of mathematics, Nanjing University in 1987; and PhD from Institute of Computational Mathematics and Scientific/Engineering Computing, Academy of Mathematics and Systems Science, Chinese Academy of Sciences in 1990. He is a faculty at the school of mathematics, Southeast University since 1990. He has been a full professor since April 1998 and a doctoral supervisor since July 2004 and retired in April 2023.

He is an academic leader of Jiangsu Province "Qinglan Project" and executive director of the computational mathematics society in Jiangsu Province. He majors in the computational mathematics and scientific engineering computing and is interested in the theory of difference methods in numerical solution of partial differential equations. He teaches computational methods, numerical analysis, numerical solutions of partial differential equations, and numerical methods of nonlinear evolution equations. He has trained 32 master students, 12 doctoral students, and 2 postdoctoral students. He has chaired five National Natural Science Foundation projects in China and one Natural Science Foundation project of Jiangsu province. He has published 8 monographs, 4 textbooks, 5 auxiliary textbooks, and more than 160 regular research papers. He is a highly cited scholar of Elsevier in 2020–2025, and on the list of top 2% scientists in 2022–2025 via the recent data from Elsevier and Standford University. The course of numerical analysis for engineering graduates was awarded as the outstanding graduate course of Innovation Project for Graduate Education in

Jiangsu Province. He has won the First Prize of Jiangsu Higher Education Teaching Achievement Award (Rank 6), Jiangsu Excellent Postgraduate Textbook Award, Jiangsu Science and Technology Award (Rank 2), and the title of National Excellent Coach in Mathematical Modeling.

Qifeng Zhang, born in September 1987, receives his PhD from the school of mathematics and statistics, Huazhong University of Science and Technology in 2014. He visits McGill University during 2013–2014 as a joint-training student and becomes a faculty at the department of mathematics, Zhejiang Sci-Tech University since 2014. He has been an associate professor since December 2017 and a master supervisor since June 2015. He is engaged in postdoctoral research under the supervision of Professor Zhi-Zhong Sun during 2018–2021.

He majors in computational mathematics and scientific engineering computing and teaches numerical analysis, numerical solutions of partial differential equations, and linear algebra. As a project leader, he has completed one National Natural Science Foundation project in China and three Natural Science Foundation projects of Zhejiang Province. Up to now, he has coauthored 1 monograph, 2 textbooks, and over 60 regular research papers.

Guang-hua Gao, born in November 1985, obtains her PhD from the School of mathematics, Southeast University in 2012, and is a faculty at the college of science, Nanjing University of Posts and Telecommunications since 2012. She has been an associate professor since September 2016 and a master supervisor since March 2014.

Her research interests are in the numerical solutions of partial differential equations, especially fractional differential equations in recent years. More than 50 academic papers have been published as an author or co-author and 3 monographs on numerical solutions of partial differential equations have been published as a co-author. As a director, she has completed the research work of two National Natural Science Foundation projects in China and two Natural Science Foundation projects of Jiangsu Province. Up to now, she has supervised eight graduate students.

Chapter 1
Finite Difference Methods for Two-Point Boundary Value Problems

The finite difference method is one of the most widely used numerical techniques for solving the problem of the differential equations. Its fundamental idea is to approximate the original differential equations and their corresponding boundary conditions with discrete difference equations involving a finite number of unknowns. Then the solution of these difference equations serves as an approximation to the solution of the original differential equations. The two-point boundary value problem of an ordinary differential equation can be interpreted as a boundary value problem of one-dimensional elliptic equation. In this chapter, we explore the finite difference solution to a model problem and introduce several key concepts in the numerical solution for differential equations, including the maximum principle, the energy method, and Richardson extrapolation.

1.1 The Dirichlet Boundary Value Problem

Consider the following two-point boundary value problem

$$
\begin{cases}
-u'' + q(x)u = f(x), & 0 < x < L, & (1.1a) \\
u(0) = \alpha, \quad u(L) = \beta, & & (1.1b)
\end{cases}
$$

where $q(x) \geqslant 0$, $f(x)$ is a known function, α and β are two given constants.

When $q(x) \equiv 0$, replacing x by s in (1.1a) and integrating once with respect to s from 0 to x, one has

$$
u'(x) = u'(0) - \int_0^x f(s)\mathrm{d}s.
$$

© Science Press 2026
Z.-Z. Sun et al., *Numerical Solutions to Partial Differential Equations with Finite Difference Methods*, Springer Asia Pacific Mathematics Series 9,
https://doi.org/10.1007/978-981-95-5563-5_1

Replacing x by ξ in the above equality, then integrating with respect to ξ from 0 to x once again, and noticing the left boundary value condition $u(0) = \alpha$, one gets

$$
\begin{aligned}
u(x) &= \alpha + u'(0)x - \int_0^x \left[\int_0^\xi f(s)\,ds \right] d\xi \\
&= \alpha + u'(0)x - \int_0^x \left[\int_s^x f(s)\,d\xi \right] ds \\
&= \alpha + u'(0)x - \int_0^x (x - s) f(s)\,ds.
\end{aligned}
$$

Utilizing the right boundary value condition $u(L) = \beta$ in the equality above produces

$$
u'(0) = \frac{\beta - \alpha + \displaystyle\int_0^L (L - s) f(s)\,ds}{L}.
$$

Therefore, the solution of (1.1) can be expressed by

$$
u(x) = \alpha + \left[\beta - \alpha + \int_0^L (L - s) f(s)\,ds \right] \frac{x}{L} - \int_0^x (x - s) f(s)\,ds.
$$

It is necessary to use numerical integration if one wishes to determine the value of the solution at a specific point. When $q(x) \not\equiv 0$, it is typically difficult, or even impossible, to obtain an exact expression for the solution in the same manner. Interested readers are encouraged to explore the case where $q(x) \equiv 1$.

Although finding an exact solution is challenging, we can attempt to provide a priori estimate of the solution. To achieve this, we introduce the following inequalities.

1.1.1 Fundamental Differential Inequalities

Throughout the book, $C^m[0, L]$ denotes the set of all functions that have continuous derivatives up to the m-th order on the closed interval $[0, L]$. When $m = 0$, m is omitted. Suppose $u \in C[0, L]$ and denote

$$
\|u\|_\infty = \max_{0 \leqslant x \leqslant L} |u(x)|, \qquad \|u\| = \sqrt{\int_0^L u^2(x)\,dx}.
$$

If $u \in C^1[0, L]$, then we further denote

$$|u|_1 = \sqrt{\int_0^L [u'(x)]^2 dx}, \quad \|u\|_1 = \sqrt{\|u\|^2 + |u|_1^2}.$$

Lemma 1.1

(I) *For any functions $u \in C^2[0, L]$ and $v \in C^1[0, L]$, it holds that*

$$-\int_0^L u''(x)v(x)dx = \int_0^L u'(x)v'(x)dx + u'(0)v(0) - u'(L)v(L). \quad (1.2)$$

(II) *For any function $v \in C^2[0, L]$ satisfying $v(0) = v(L) = 0$, it holds that*

$$-\int_0^L v''(x)v(x)dx = |v|_1^2. \quad (1.3)$$

(III) *For any function $v \in C^1[0, L]$ satisfying $v(0) = v(L) = 0$, it holds that*

$$\|v\|_\infty \leqslant \frac{\sqrt{L}}{2}|v|_1, \quad \|v\| \leqslant \frac{L}{\sqrt{6}}|v|_1. \quad (1.4)$$

(IV) *For any function $v \in C^1[0, L]$ satisfying $v(0) = v(L) = 0$, it holds that*

$$\|v\|_\infty^2 \leqslant \epsilon |v|_1^2 + \frac{1}{4\epsilon}\|v\|^2 \quad (1.5)$$

with ϵ any positive constant.
(V) *For any function $v \in C^1[0, L]$, it holds that*

$$\|v\|_\infty^2 \leqslant \epsilon |v|_1^2 + \left(\frac{1}{\epsilon} + \frac{1}{L}\right)\|v\|^2 \quad (1.6)$$

with ϵ any positive constant.

Proof

(I) The result (1.2) follows directly from integration by parts.
(II) It is easy to derive (1.3) from (1.2).
(III) For an arbitrary $x \in (0, L)$, noticing $v(0) = v(L) = 0$, we have

$$v(x) = \int_0^x v'(s)ds, \quad (1.7)$$

$$v(x) = -\int_x^L v'(s)ds. \quad (1.8)$$

Squaring on both sides of (1.7) and (1.8), respectively, then applying the Cauchy-Schwarz inequality, it yields

$$v^2(x) \leq \int_0^x ds \int_0^x [v'(s)]^2 ds = x \int_0^x [v'(s)]^2 ds, \tag{1.9}$$

$$v^2(x) \leq \int_x^L ds \int_x^L [v'(s)]^2 ds = (L-x) \int_x^L [v'(s)]^2 ds. \tag{1.10}$$

Multiplying (1.9) by $(L-x)$, and (1.10) by x, respectively, and adding up both results produce

$$L v^2(x) \leq x(L-x) \int_0^L [v'(s)]^2 ds = x(L-x)|v|_1^2. \tag{1.11}$$

Noticing when $x \in (0, L)$, it holds that

$$x(L-x) \leq \frac{L^2}{4},$$

and it follows easily from (1.11) that

$$L v^2(x) \leq \frac{L^2}{4} |v|_1^2, \quad x \in (0, L).$$

Taking square roots on both sides of the above inequality, it gives

$$|v(x)| \leq \frac{\sqrt{L}}{2} |v|_1, \quad x \in (0, L),$$

which implies

$$\|v\|_\infty \leq \frac{\sqrt{L}}{2} |v|_1.$$

Integrating on both sides of (1.11) with respect to x, it yields

$$L \int_0^L v^2(x) dx \leq |v|_1^2 \int_0^L x(L-x) dx = \frac{L^3}{6} |v|_1^2.$$

Taking square roots on both sides of the above inequality arrives at

$$\|v\| \leq \frac{L}{\sqrt{6}} |v|_1.$$

(IV) Noticing $v(0) = v(L) = 0$, we have

$$v^2(x) = \int_0^x \frac{\mathrm{d}}{\mathrm{d}s}[v^2(s)]\mathrm{d}s = 2\int_0^x v(s)v'(s)\mathrm{d}s,$$

$$v^2(x) = -\int_x^L \frac{\mathrm{d}}{\mathrm{d}s}[v^2(s)]\mathrm{d}s = -2\int_x^L v(s)v'(s)\mathrm{d}s.$$

Summing up the above two equalities and dividing the corresponding result by 2 produce

$$v^2(x) \leqslant \int_0^x \left|v(s)v'(s)\right|\mathrm{d}s + \int_x^L \left|v(s)v'(s)\right|\mathrm{d}s$$

$$= \int_0^L \left|v(s)v'(s)\right|\mathrm{d}s, \quad 0 \leqslant x \leqslant L.$$

For an arbitrary $\epsilon > 0$, we have

$$v^2(x) \leqslant \epsilon \int_0^L [v'(s)]^2\mathrm{d}s + \frac{1}{4\epsilon}\int_0^L [v(s)]^2\mathrm{d}s = \epsilon|v|_1^2 + \frac{1}{4\epsilon}\|v\|^2, \quad 0 \leqslant x \leqslant L.$$

Therefore, the inequality (1.5) holds.

(V) Suppose there is an $x^* \in [0, L]$ satisfying

$$|v(x^*)| = \|v\|_\infty.$$

When $x \in [0, x^*]$,

$$\|v\|_\infty^2 = v^2(x^*) = v^2(x) + \int_x^{x^*} \left[\frac{\mathrm{d}}{\mathrm{d}s}v^2(s)\right]\mathrm{d}s$$

$$= v^2(x) + 2\int_x^{x^*} v(s)v'(s)\mathrm{d}s \leqslant v^2(x) + 2\int_x^{x^*} \left|v(s)v'(s)\right|\mathrm{d}s$$

$$\leqslant v^2(x) + 2\int_0^L \left|v(s)v'(s)\right|\mathrm{d}s \leqslant v^2(x) + 2\|v\| \cdot |v|_1;$$

When $x \in [x^*, L]$,

$$\|v\|_\infty^2 = v^2(x^*) = v^2(x) - \int_{x^*}^x \left[\frac{\mathrm{d}}{\mathrm{d}s}v^2(s)\right]\mathrm{d}s$$

$$= v^2(x) - 2\int_{x^*}^x v(s)v'(s)\mathrm{d}s \leqslant v^2(x) + 2\int_{x^*}^x \left|v(s)v'(s)\right|\mathrm{d}s$$

$$\leqslant v^2(x) + 2\int_0^L \left|v(s)v'(s)\right|\mathrm{d}s \leqslant v^2(x) + 2\|v\| \cdot |v|_1.$$

It follows from adding up the above two inequalities that

$$\|v\|_\infty^2 \leqslant v^2(x) + 2\|v\| \cdot |v|_1, \quad x \in [0, L].$$

Integrating on both sides of the above inequality with respect to x over $[0, L]$, we get

$$L\|v\|_\infty^2 \leqslant \|v\|^2 + 2L\|v\| \cdot |v|_1.$$

Applying the Young's inequality to the cross term yields

$$\|v\|_\infty^2 \leqslant 2\|v\| \cdot |v|_1 + \frac{1}{L}\|v\|^2 \leqslant \epsilon|v|_1^2 + \frac{1}{\epsilon}\|v\|^2 + \frac{1}{L}\|v\|^2 = \epsilon|v|_1^2 + \left(\frac{1}{\epsilon} + \frac{1}{L}\right)\|v\|^2.$$

In other words, the inequality (1.6) is established. $\square$

1.1.2   *A Priori Estimate of the Solution*

First, a priori estimate of the solution is provided for the homogeneous boundary value problem.

Theorem 1.1 *Let* $v \in C^2[0, L]$ *be a solution of the two-point boundary value problem*

$$\begin{cases} -v'' + q(x)v = f(x), & 0 < x < L, & \text{(1.12a)} \\ v(0) = 0, & v(L) = 0 & \text{(1.12b)} \end{cases}$$

with $q(x) \geqslant 0$. *Then we have*

$$|v|_1 \leqslant \frac{L}{\sqrt{6}}\|f\|, \tag{1.13}$$

$$\|v\|_\infty \leqslant \frac{L^2}{2\sqrt{6}}\|f\|_\infty. \tag{1.14}$$

Proof

(I) Multiplying both sides of (1.12a) by $v(x)$ and integrating the result with respect to x on $(0, L)$, we have

$$-\int_0^L v''(x)v(x)dx + \int_0^L q(x)v^2(x)dx = \int_0^L f(x)v(x)dx. \tag{1.15}$$

Noticing (1.12b) and making use of Lemma 1.1, we have

$$-\int_0^L v''(x)v(x)\mathrm{d}x = |v|_1^2.$$

It follows from $q(x) \geqslant 0$ that

$$\int_0^L q(x)v^2(x)\mathrm{d}x \geqslant 0.$$

Furthermore, by using the Cauchy-Schwarz inequality, it yields

$$\int_0^L f(x)v(x)\mathrm{d}x \leqslant \|f\| \cdot \|v\|.$$

Substituting the above one equality and two inequalities into (1.15), one has

$$|v|_1^2 \leqslant \|f\| \cdot \|v\|.$$

Utilizing Lemma 1.1 again, we get

$$|v|_1^2 \leqslant \frac{L}{\sqrt{6}}\|f\| \cdot |v|_1,$$

which produces

$$|v|_1 \leqslant \frac{L}{\sqrt{6}}\|f\|.$$

(II) Noticing

$$\|f\| \leqslant \sqrt{L}\|f\|_\infty,$$

and combining (1.13) with Lemma 1.1, we arrive at

$$\|v\|_\infty \leqslant \frac{\sqrt{L}}{2}|v|_1 \leqslant \frac{\sqrt{L}}{2} \cdot \frac{L}{\sqrt{6}}\|f\| \leqslant \frac{L^2}{2\sqrt{6}}\|f\|_\infty.$$

$$\square$$

The inequalities (1.13) and (1.14) are referred to as **priori estimates** of the solution to the two-point boundary value problem (1.12).

1.2 Finite Difference Schemes

As discussed in the previous section, it is often difficult, or even impossible, to obtain an exact solution to the problem (1.1) for a general $q(x)$. This motivates researchers to seek an approximate solution, often through numerical methods.

We begin by presenting several commonly used for numerical differentiation formulas.

Lemma 1.2 *Let h and c be two given constants, and $h > 0$.*

(I) *If $g \in C^2[c - h, c + h]$, then we have*

$$g(c) = \frac{1}{2}[g(c - h) + g(c + h)] - \frac{h^2}{2}g''(\xi_0), \quad c - h < \xi_0 < c + h.$$

(II) *If $g \in C^2[c, c + h]$, then we have*

$$g'(c) = \frac{1}{h}[g(c + h) - g(c)] - \frac{h}{2}g''(\xi_1), \quad c < \xi_1 < c + h.$$

(III) *If $g \in C^2[c - h, c]$, then we have*

$$g'(c) = \frac{1}{h}[g(c) - g(c - h)] + \frac{h}{2}g''(\xi_2), \quad c - h < \xi_2 < c.$$

(IV) *If $g \in C^3[c - h, c + h]$, then we have*

$$g'(c) = \frac{1}{2h}[g(c + h) - g(c - h)] - \frac{h^2}{6}g'''(\xi_3), \quad c - h < \xi_3 < c + h.$$

(V) *If $g \in C^4[c - h, c + h]$, then we have*

$$g''(c) = \frac{1}{h^2}[g(c+h) - 2g(c) + g(c-h)] - \frac{h^2}{12}g^{(4)}(\xi_4), \quad c-h < \xi_4 < c+h.$$

(VI) *If $g \in C^3[c, c + h]$, then we have*

$$g''(c) = \frac{2}{h}\left[\frac{g(c + h) - g(c)}{h} - g'(c)\right] - \frac{h}{3}g'''(\xi_5), \quad c < \xi_5 < c + h.$$

If $g \in C^4[c, c + h]$, then we have

$$g''(c) = \frac{2}{h}\left[\frac{g(c + h) - g(c)}{h} - g'(c)\right] - \frac{h}{3}g'''(c) - \frac{h^2}{12}g^{(4)}(\xi_6),$$

$$c < \xi_6 < c + h.$$

(VII) *If $g \in C^3[c-h, c]$, then we have*

$$g''(c) = \frac{2}{h}\left[g'(c) - \frac{g(c) - g(c-h)}{h}\right] + \frac{h}{3}g'''(\xi_7), \quad c - h < \xi_7 < c.$$

If $g \in C^4[c-h, c]$, then we have

$$g''(c) = \frac{2}{h}\left[g'(c) - \frac{g(c) - g(c-h)}{h}\right] + \frac{h}{3}g'''(c) - \frac{h^2}{12}g^{(4)}(\xi_8),$$

$$c - h < \xi_8 < c.$$

(VIII) *If $g \in C^6[c-h, c+h]$, then we have*

$$\frac{1}{12}[g''(c-h) + 10g''(c) + g''(c+h)]$$

$$= \frac{1}{h^2}[g(c+h) - 2g(c) + g(c-h)] + \frac{h^4}{240}g^{(6)}(\xi_9), \quad c - h < \xi_9 < c + h.$$

Proof (I)–(VII) can be easily verified using Taylor expansions with differential remainders. In the following, we prove (VIII) by applying the Taylor expansion with an integral remainder.

By the Taylor expansion with an integral remainder, we have

$$g(c+h) = g(c) + hg'(c) + \frac{h^2}{2}g''(c) + \frac{h^3}{6}g'''(c) + \frac{h^4}{24}g^{(4)}(c)$$

$$+ \frac{h^5}{120}g^{(5)}(c) + \frac{h^6}{120}\int_0^1 g^{(6)}(c+sh)(1-s)^5 ds,$$

$$g(c-h) = g(c) - hg'(c) + \frac{h^2}{2}g''(c) - \frac{h^3}{6}g'''(c) + \frac{h^4}{24}g^{(4)}(c)$$

$$- \frac{h^5}{120}g^{(5)}(c) + \frac{h^6}{120}\int_0^1 g^{(6)}(c-sh)(1-s)^5 ds.$$

Adding the above two equalities together produces

$$\frac{1}{h^2}[g(c+h) - 2g(c) + g(c-h)]$$

$$= g''(c) + \frac{h^2}{12}g^{(4)}(c) + \frac{h^4}{120}\int_0^1 \left[g^{(6)}(c+sh) + g^{(6)}(c-sh)\right](1-s)^5 ds.$$

$$(1.16)$$

Analogously, using the Taylor expansion again with an integral remainder yields

$$g''(c+h) = g''(c) + hg'''(c) + \frac{h^2}{2}g^{(4)}(c) + \frac{h^3}{6}g^{(5)}(c)$$
$$+ \frac{h^4}{6}\int_0^1 g^{(6)}(c+sh)(1-s)^3 ds,$$

$$g''(c-h) = g''(c) - hg'''(c) + \frac{h^2}{2}g^{(4)}(c) - \frac{h^3}{6}g^{(5)}(c)$$
$$+ \frac{h^4}{6}\int_0^1 g^{(6)}(c-sh)(1-s)^3 ds,$$

which generates

$$\frac{1}{12}\left[g''(c+h) + 10g''(c) + g''(c-h)\right]$$
$$= g''(c) + \frac{h^2}{12}g^{(4)}(c) + \frac{h^4}{72}\int_0^1 \left[g^{(6)}(c+sh) + g^{(6)}(c-sh)\right](1-s)^3 ds.$$

$$(1.17)$$

Subtracting (1.16) from (1.17) and applying the mean value theorem of integrals give

$$\frac{1}{12}\left[g''(c+h) + 10g''(c) + g''(c-h)\right] - \frac{1}{h^2}\left[g(c+h) - 2g(c) + g(c-h)\right]$$
$$= \frac{h^4}{360}\int_0^1 \left[g^{(6)}(c+sh) + g^{(6)}(c-sh)\right](1-s)^3\left[5 - 3(1-s)^2\right]ds$$
$$= \frac{h^4}{360}\left[g^{(6)}(c+\hat{s}h) + g^{(6)}(c-\hat{s}h)\right]\int_0^1 (1-s)^3\left[5 - 3(1-s)^2\right]ds$$
$$= \frac{h^4}{480}\left[g^{(6)}(c+\hat{s}h) + g^{(6)}(c-\hat{s}h)\right]$$
$$= \frac{h^4}{240}g^{(6)}(\xi_9), \quad \hat{s} \in (0,1), \quad \xi_9 \in (c-h, c+h),$$

which completes the proof. $\square$

1.2.1 *Derivation of the Difference Scheme*

Without any special assumptions, throughout the book, we assume that the solution to the differential equation under consideration possesses the necessary regularity.

The first step in solving the two-point boundary value problem using the finite difference method is to perform a **grid partition** (or subdivision) of the interval $[0, L]$, dividing it into m equidistant subintervals. Let $h = \frac{L}{m}$, and define the node points as $x_i = a + ih$ for $0 \leqslant i \leqslant m$. The set of node points is denoted by $\Omega_h = \{x_i \mid 0 \leqslant i \leqslant m\}$. Here, h is the **step size**, x_i is the **node point**, and Ω_h is the **grid set**. A function defined on Ω_h is called a **grid function**.

Suppose $v = \{v_i \mid 0 \leqslant i \leqslant m\}$ represents a grid function on Ω_h. Denote

$$v_{i-\frac{1}{2}} = \frac{1}{2}\left(v_i + v_{i-1}\right), \quad \delta_x v_{i-\frac{1}{2}} = \frac{1}{h}\left(v_i - v_{i-1}\right), \quad \delta_x^2 v_i = \frac{1}{h}\left(\delta_x v_{i+\frac{1}{2}} - \delta_x v_{i-\frac{1}{2}}\right).$$

Define the grid function $U = \{U_i \mid 0 \leqslant i \leqslant m\}$, where

$$U_i = u(x_i), \quad 0 \leqslant i \leqslant m.$$

Considering the problem (1.1) at the node point x_i, one has

$$\begin{cases} -u''(x_i) + q(x_i)u(x_i) = f(x_i), & 1 \leqslant i \leqslant m - 1, & (1.18\text{a}) \\ u(x_0) = \alpha, \quad u(x_m) = \beta. & & (1.18\text{b}) \end{cases}$$

According to Lemma 1.2, we have

$$u''(x_i) = \frac{1}{h^2}\left[u(x_{i-1}) - 2u(x_i) + u(x_{i+1})\right] - \frac{h^2}{12}u^{(4)}(\xi_i)$$

$$= \delta_x^2 U_i - \frac{h^2}{12}u^{(4)}(\xi_i), \quad x_{i-1} < \xi_i < x_{i+1}.$$

Substituting the above equality into (1.18a) and noticing (1.18b), we arrive at

$$\begin{cases} -\delta_x^2 U_i + q(x_i)U_i = f(x_i) - \frac{h^2}{12}u^{(4)}(\xi_i), & 1 \leqslant i \leqslant m - 1, & (1.19\text{a}) \\ U_0 = \alpha, \quad U_m = \beta. & & (1.19\text{b}) \end{cases}$$

We call (1.19a) the **discretization** of (1.18a) and $-\frac{h^2}{12}u^{(4)}(\xi_i)$ the dicretization error.

Omitting the small term $-\frac{h^2}{12}u^{(4)}(\xi_i)$, one gets the following approximate equality:

$$\begin{cases} -\delta_x^2 U_i + q(x_i)U_i \approx f(x_i), & 1 \leqslant i \leqslant m - 1, & (1.20\text{a}) \\ U_0 = \alpha, \quad U_m = \beta, & & (1.20\text{b}) \end{cases}$$

where (1.20) can be viewed as an approximation of the problem (1.18).

By replacing U_i with u_i, and the approximation symbol " $\approx$ " with equality " $=$ " in (1.20), we obtain the following system of equations:

$$\begin{cases} -\delta_x^2 u_i + q(x_i)u_i = f(x_i), & 1 \leqslant i \leqslant m - 1, & \text{(1.21a)} \\ u_0 = \alpha, \quad u_m = \beta. & & \text{(1.21b)} \end{cases}$$

We refer to the system (1.21) as the **difference scheme** for solving the problem (1.1). This system consists of $m + 1$ equations and $m + 1$ unknowns, namely, $u_0, u_1, \ldots, u_m$.

The goal is to determine the value of $(u_0, u_1, \ldots, u_m)$ and interpret u_i as the approximation to U_i. Noting the boundary conditions in (1.21b), the system (1.21a) can be viewed as an $(m - 1)$-th order system of linear equations in the unknown $(u_1, u_2, \ldots, u_{m-1})$.

Conversely, if the exact solution U_i is substituted for u_i into (1.21a), we would obtain the approximate equation (1.20a). The difference

$$R_i \equiv -\delta_x^2 U_i + q(x_i)U_i - f(x_i)$$

between the left-hand side and the right-hand side of (1.20a) is called the **local truncation error** of the difference scheme (1.21).

In general, R_i is closely related to h, i.e., $R_i = R_i(h)$. From (1.19a), it is clear that

$$R_i = -\frac{h^2}{12} u^{(4)}(\xi_i), \tag{1.22}$$

which is a small term that has been omitted in the derivation of the difference scheme. This term reflects the degree of approximation of the difference equation (1.21a) to the differential equation (1.18a).

When

$$\max_{1 \leqslant i \leqslant m-1} |R_i| \to 0 \quad (h \to 0),$$

the difference scheme (1.21) is said to be **consistent** with the original problem (1.1).

Once the difference scheme is established, the following natural questions arise: Does the difference scheme have a solution? Which method can be used to obtain the solution of the difference scheme? Can the difference solution be used as an approximate solution to the differential equation problem (i.e., **convergence**)? How does the error introduced during the computation affect the difference solution (i.e., **stability**)?

1.2.2   Existence of the Difference Solution

Theorem 1.2 *The difference scheme (1.21) is uniquely solvable.*

Proof   The difference scheme (1.21) is linear. Therefore, it suffices to consider its homogeneous one

$$
\begin{cases}
-\delta_x^2 u_i + q(x_i)u_i = 0, & 1 \leqslant i \leqslant m-1, & (1.23a) \\[2mm]
u_0 = 0, \quad u_m = 0. & & (1.23b)
\end{cases}
$$

Let $\displaystyle\max_{0 \leqslant i \leqslant m} |u_i| = M$. Now we suppose $M > 0$. Then in the light of (1.23b), we know that there is a k $(1 \leqslant k \leqslant m-1)$ satisfying $|u_k| = M$, and at least one of $|u_{k-1}|$ and $|u_{k+1}|$ is strictly less than M. Without loss of generality, we consider the difference equation with $i = k$ in (1.23a)

$$
-u_{k-1} + \left[2 + h^2 q(x_k)\right] u_k - u_{k+1} = 0,
$$

namely,

$$
\left[2 + h^2 q(x_k)\right] u_k = u_{k-1} + u_{k+1}.
$$

Taking the absolute value of both sides of the above equality, we obtain

$$
2M \leqslant \left[2 + h^2 q(x_k)\right] \cdot |u_k| = |u_{k-1} + u_{k+1}| \leqslant |u_{k-1}| + |u_{k+1}| < M + M = 2M,
$$

which leads to a contradiction with the assumption that $M > 0$. Thus, it forces $M = 0$, which implies that (1.23) admits only the trivial solution

$$
u_i = 0, \quad 0 \leqslant i \leqslant m.
$$

Therefore, we conclude that the difference scheme (1.21) is uniquely solvable. $\square$

1.2.3   Implementation of the Difference Scheme and Numerical Examples

Multiplying both sides of (1.21a) by h^2, one has

$$
-u_{i-1} + \left[2 + h^2 q(x_i)\right] u_i - u_{i+1} = h^2 f(x_i), \quad 1 \leqslant i \leqslant m-1.
$$

Thus, (1.21) can be written as

$$
\begin{pmatrix}
1 & 0 & & & \\
-1 & 2+h^2q(x_1) & -1 & & \\
& \ddots & \ddots & \ddots & \\
& & -1 & 2+h^2q(x_{m-1}) & -1 \\
& & & 0 & 1
\end{pmatrix}
\begin{pmatrix}
u_0 \\ u_1 \\ \vdots \\ u_{m-1} \\ u_m
\end{pmatrix}
=
\begin{pmatrix}
\alpha \\ h^2 f(x_1) \\ \vdots \\ h^2 f(x_{m-1}) \\ \beta
\end{pmatrix},
$$

or

$$
\begin{pmatrix}
2+h^2q(x_1) & -1 & & & \\
-1 & 2+h^2q(x_2) & -1 & & \\
& \ddots & \ddots & \ddots & \\
& & -1 & 2+h^2q(x_{m-2}) & -1 \\
& & & -1 & 2+h^2q(x_{m-1})
\end{pmatrix}
\begin{pmatrix}
u_1 \\ u_2 \\ \vdots \\ u_{m-2} \\ u_{m-1}
\end{pmatrix}
$$

$$
=
\begin{pmatrix}
h^2 f(x_1)+\alpha \\
h^2 f(x_2) \\
\vdots \\
h^2 f(x_{m-2}) \\
h^2 f(x_{m-1})+\beta
\end{pmatrix}.
$$

The coefficient matrix of the above system of linear equations is tridiagonal and is thus referred to as a **tridiagonal system of linear equations**.

For solving such a system, the **"chasing-catching" method** (also known as the **double sweep method** or **Thomas algorithm**) can be used. Given a tridiagonal system of linear equations,

$$
\begin{cases}
\beta_l u_l + \gamma_l u_{l+1} & = d_l, \\
\alpha_{l+1} u_l + \beta_{l+1} u_{l+1} + \gamma_{l+1} u_{l+2} & = d_{l+1}, \\
\quad\quad \cdots\cdots & \\
\alpha_{n-1} u_{n-2} + \beta_{n-1} u_{n-1} + \gamma_{n-1} u_n & = d_{n-1}, \\
\alpha_n u_{n-1} + \beta_n u_n & = d_n.
\end{cases}
\tag{1.24}
$$

Solving u_l from the first equation in (1.24), we have

$$
u_l = \frac{d_l}{\beta_l} - \frac{\gamma_l}{\beta_l} u_{l+1}.
$$

Denoting

$$
g_l = \frac{d_l}{\beta_l}, \quad w_l = \frac{\gamma_l}{\beta_l},
$$

we have

$$u_l = g_l - w_l u_{l+1}. \tag{1.25}$$

Substituting the above equality into the second one in (1.24), we arrive at

$$\alpha_{l+1}(g_l - w_l u_{l+1}) + \beta_{l+1} u_{l+1} + \gamma_{l+1} u_{l+2} = d_{l+1},$$

namely,

$$u_{l+1} = g_{l+1} - w_{l+1} u_{l+2},$$

where

$$g_{l+1} = \frac{d_{l+1} - \alpha_{l+1} g_l}{\beta_{l+1} - \alpha_{l+1} w_l}, \qquad w_{l+1} = \frac{\gamma_{l+1}}{\beta_{l+1} - \alpha_{l+1} w_l}.$$

In exactly the same way, one has

$$u_i = g_i - w_i u_{i+1}, \quad l+1 \leqslant i \leqslant n-1, \tag{1.26}$$

where

$$g_i = \frac{d_i - \alpha_i g_{i-1}}{\beta_i - \alpha_i w_{i-1}}, \qquad w_i = \frac{\gamma_i}{\beta_i - \alpha_i w_{i-1}}.$$

Plugging the related expression

$$u_{n-1} = g_{n-1} - w_{n-1} u_n$$

into the last one in (1.24), we obtain

$$\alpha_n(g_{n-1} - w_{n-1} u_n) + \beta_n u_n = d_n.$$

Therefore,

$$u_n = g_n,$$

where

$$g_n = \frac{d_n - \alpha_n g_{n-1}}{\beta_n - \alpha_n w_{n-1}}.$$

Suppose u_n has been figured out. By combining (1.26) with (1.25), we can solve $u_{n-1}, u_{n-2}, \ldots, u_{l+1}, u_l$ in sequence. The whole process can be divided into two steps:

Step one: Determine g_l, w_l; g_{l+1}, w_{l+1}; g_{l+2}, w_{l+2}; $\ldots$; g_{n-1}, w_{n-1}; g_n in turn.

Step two: Determine $u_n, u_{n-1}, u_{n-2}, \ldots, u_l$ in reverse order.

The calculation formulas are as follows:

(I) $g_l = \dfrac{d_l}{\beta_l}, \quad w_l = \dfrac{\gamma_l}{\beta_l},$

$$g_i = \frac{d_i - \alpha_i g_{i-1}}{\beta_i - \alpha_i w_{i-1}}, \quad w_i = \frac{\gamma_i}{\beta_i - \alpha_i w_{i-1}}, \quad i = l+1, l+2, \ldots, n-1,$$

$$g_n = \frac{d_n - \alpha_n g_{n-1}}{\beta_n - \alpha_n w_{n-1}}.$$

(II) $u_n = g_n,$

$$u_i = g_i - w_i u_{i+1}, \quad i = n-1, n-2, \ldots, l.$$

Generally, in **step one**, the index progresses from small to large, commonly termed the **"chasing"** process. In **step two**, the index progresses from large to small, known as the **"catching"** process. The entire process is called the **double sweep method**, **Thomas algorithm**, or the **"chasing-catching" method**.

When $l = 1$ and $n = N$, the total computational effort of the **"chasing-catching" method** involves $(5N - 4)$ multiplications and divisions, and $(3N - 3)$ additions and subtractions.

Example 1.1 Apply the difference scheme (1.21) to compute the following two-point boundary value problem:

$$\begin{cases} -u'' + u = e^x(\sin x - 2\cos x), & 0 < x < \pi, \\ u(0) = 0, \quad u(\pi) = 0. \end{cases} \tag{1.27}$$

The exact solution of this problem is $u(x) = e^x \sin x$.

Divide $[0, \pi]$ into m equal subintervals and denote $h = \pi/m$, $x_i = ih$, $0 \leqslant i \leqslant m$. The difference scheme for solving (1.27) reads

$$\begin{cases} -\delta_x^2 u_i + u_i = e^{x_i}(\sin x_i - 2\cos x_i), & 1 \leqslant i \leqslant m-1, \\ u_0 = 0, \quad u_m = 0. \end{cases}$$

Table 1.1 lists the exact solutions (ES) at four different points and corresponding numerical solutions (NS) with varying step sizes. Table 1.2 gives the absolute values of numerical errors at four different node points, as well as the maximum errors of

Table 1.1 (Example 1.1) The exact and numerical solutions at four node points with different step sizes

h	x			
	$\pi/5$	$2\pi/5$	$3\pi/5$	$4\pi/5$
$\pi/10$	1.064007	3.266548	6.166191	7.178725
$\pi/20$	1.092311	3.322830	6.239355	7.237039
$\pi/40$	1.099410	3.336920	6.257628	7.251546
$\pi/80$	1.101186	3.340444	6.262195	7.255169
$\pi/160$	1.101630	3.341320	6.263337	7.256074
ES	1.101778	3.341619	6.263717	7.256376

Table 1.2 (Example 1.1) The absolute values of the numerical errors and maximum numerical errors at four node points with different step sizes

h	x				$E_\infty(h)$	$E_\infty(2h)/E_\infty(h)$
	$\pi/5$	$2\pi/5$	$3\pi/5$	$4\pi/5$		
$\pi/10$	3.777e$-$2	7.507e$-$2	9.753e$-$2	7.765e$-$2	9.753e$-$2	
$\pi/20$	9.467e$-$3	1.879e$-$2	2.436e$-$2	1.934e$-$2	2.439e$-$2	3.999
$\pi/40$	2.368e$-$3	4.698e$-$3	6.089e$-$3	4.829e$-$3	6.115e$-$3	3.999
$\pi/80$	5.921e$-$4	1.175e$-$3	1.522e$-$3	1.207e$-$3	1.529e$-$3	3.999
$\pi/160$	1.480e$-$4	2.937e$-$4	3.806e$-$4	3.017e$-$4	3.822e$-$4	4.001

Fig. 1.1 (Example 1.1) Curves of the exact and numerical solutions with $h = \pi/10$

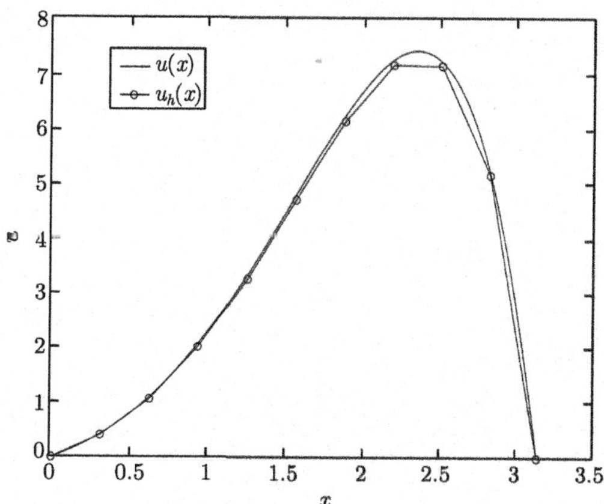

numerical solutions defined by

$$E_\infty(h) = \max_{0 \leqslant i \leqslant m} |u(x_i) - u_i|$$

with different step sizes.

As observed from Table 1.2, reducing the step size h by a factor of two yields a corresponding reduction in the maximum numerical error by a factor of four. Figure 1.1 shows the curves of the exact solution (ES) $u(x)$ and the numerical solution (NS) $u_h(x)$ with $h = \pi/10$. Figure 1.2 illustrates the curves of the numerical errors $|u(x) - u_h(x)|$.

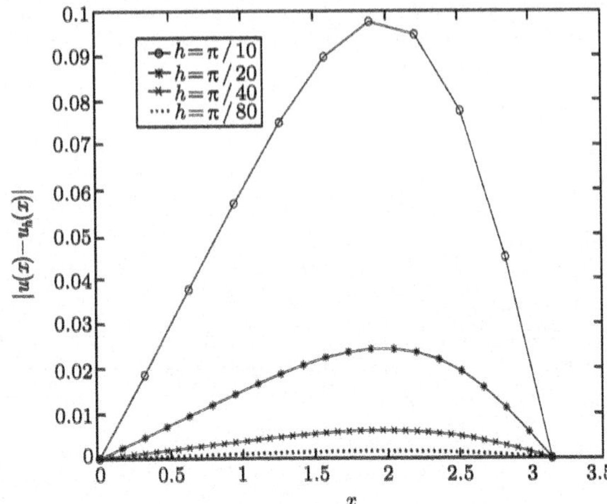

Fig. 1.2 (Example 1.1) The error curves of numerical solutions with varying step sizes

1.2.4   A Priori Estimate of the Difference Solution

The priori estimate of the difference solution plays a key role in the numerical analysis.

Denote

$$\mathcal{U}_h = \{v \mid v = \{v_i \mid 0 \leqslant i \leqslant m\} \text{ is the grid function on } \Omega_h\},$$

$$\overset{\circ}{\mathcal{U}}_h = \{v \mid v = \{v_i \mid 0 \leqslant i \leqslant m\} \in \mathcal{U}_h \text{ and } v_0 = v_m = 0\}.$$

Let $u, v \in \mathcal{U}_h$ and introduce the following notation:

$$(u, v) = h\left(\frac{1}{2}u_0 v_0 + \sum_{i=1}^{m-1} u_i v_i + \frac{1}{2}u_m v_m\right), \quad \|u\| = \sqrt{(u, u)},$$

$$(\delta_x u, \delta_x v) = h \sum_{i=1}^{m} (\delta_x u_{i-\frac{1}{2}})(\delta_x v_{i-\frac{1}{2}}), \quad \|\delta_x u\| = \sqrt{(\delta_x u, \delta_x u)},$$

$$(\delta_x^2 u, \delta_x^2 v) = h \sum_{i=1}^{m-1} (\delta_x^2 u_i)(\delta_x^2 v_i), \quad \|\delta_x^2 u\| = \sqrt{(\delta_x^2 u, \delta_x^2 u)},$$

$$\|u\|_\infty = \max_{0 \leqslant i \leqslant m} |u_i|, \quad |u|_1 = \|\delta_x u\|, \quad |u|_2 = \|\delta_x^2 u\|,$$

$$\|u\|_1 = \sqrt{\|u\|^2 + |u|_1^2}, \quad \|u\|_2 = \sqrt{\|u\|^2 + |u|_1^2 + |u|_2^2}.$$

It is easy to prove:

(1) (u, v) is an inner product on $\mathcal{U}_h$. $(\delta_x u, \delta_x v)$ and $(\delta_x^2 u, \delta_x^2 v)$ are inner products on $\overset{\circ}{\mathcal{U}}_h$.

(2) $\|u\|$, $\|u\|_1$, $\|u\|_2$, $\|u\|_\infty$ are norms on $\mathcal{U}_h$. They are, respectively, referred to as (discrete) 2-norm (average norm), H^1-norm, H^2-norm, and infinite norm (maximum/uniform norm).

(3) $|u|_1$ and $|u|_2$ are, respectively, semi-norms on $\mathcal{U}_h$.

(4) $|u|_1$ and $|u|_2$ are, respectively, norms on $\overset{\circ}{\mathcal{U}}_h$. $\|\delta_x u\|$ is called 2-norm of the difference quotient. $|u|_1$ is called H^1-norm.

(5) $|u|_1$ and $\|u\|_1$ are two equivalent norms on $\overset{\circ}{\mathcal{U}}_h$. $|u|_2$ and $\|u\|_2$ are also two equivalent norms on $\overset{\circ}{\mathcal{U}}_h$.

A priori estimate of the difference scheme (1.21) can be derived by either the maximum principle or the energy method.

Maximum Principle

Let $v = (v_0, v_1, \ldots, v_{m-1}, v_m) \in \mathcal{U}_h$ and denote

$$L_h v_i = -\delta_x^2 v_i + q(x_i) v_i, \quad 1 \leqslant i \leqslant m - 1.$$

Lemma 1.3 *Suppose*

$$L_h v_i \leqslant 0, \quad 1 \leqslant i \leqslant m - 1. \tag{1.28}$$

If $M = \max\limits_{0 \leqslant i \leqslant m} v_i \geqslant 0$, *then we have*

$$\max\limits_{1 \leqslant i \leqslant m-1} v_i \leqslant \max\limits_{i=0,m} v_i.$$

Proof The proof is carried out by contradiction. Suppose

$$\max\limits_{1 \leqslant i \leqslant m-1} v_i > \max\limits_{i=0,m} v_i.$$

Then there is an $i_0 \in \{1, 2, \ldots, m - 1\}$ such that $v_{i_0} = M$, and at least one of v_{i_0-1} and v_{i_0+1} is strictly less than M. Therefore,

$$L_h v_{i_0} = -\frac{1}{h^2}(v_{i_0-1} - 2v_{i_0} + v_{i_0+1}) + q(x_i) v_{i_0} > 0,$$

which contradicts the condition in (1.28). $\qquad\square$

Theorem 1.3 *Let* $v = \{v_i \mid 0 \leqslant i \leqslant m\}$ *be the solution of the difference scheme*

$$\begin{cases} -\delta_x^2 v_i + q(x_i) v_i = f_i, & 1 \leqslant i \leqslant m - 1, \\ v_0 = \alpha, \quad v_m = \beta. \end{cases}$$

Then we have

$$\|v\|_\infty \leqslant \max\{|\alpha|, |\beta|\} + \frac{L^2}{8} \max_{1 \leqslant i \leqslant m-1} |f_i|. \qquad (1.29)$$

Proof Define

$$C = \max_{1 \leqslant i \leqslant m-1} |f_i|, \quad P(x) = x(L-x), \quad P_i = P(x_i), \quad w_i = \frac{1}{2} C P_i, \quad 0 \leqslant i \leqslant m.$$

Then we have

$$L_h(\pm v_i - w_i) = \pm L_h v_i - L_h w_i = \pm f_i - \frac{1}{2} C L_h P_i$$

$$= \pm f_i - \frac{1}{2} C(2 + q(x_i)P_i) = \pm f_i - C - \frac{1}{2} C q(x_i) P_i \leqslant 0, \quad 1 \leqslant i \leqslant m-1.$$

(I) If $\max\limits_{0 \leqslant i \leqslant m} \{\pm v_i - w_i\} \geqslant 0$, it follows from Lemma 1.3 that

$$\max_{1 \leqslant i \leqslant m-1} \{\pm v_i - w_i\} \leqslant \max_{i=0,m} \{\pm v_i - w_i\} = \max_{i=0,m} \{\pm v_i\} \leqslant \max\{|\alpha|, |\beta|\}.$$

Thus,

$$\max_{1 \leqslant i \leqslant m-1} \{\pm v_i\} = \max_{1 \leqslant i \leqslant m-1} \{\pm v_i - w_i + w_i\}$$

$$\leqslant \max_{1 \leqslant i \leqslant m-1} \{\pm v_i - w_i\} + \max_{1 \leqslant i \leqslant m-1} w_i$$

$$\leqslant \max\{|\alpha|, |\beta|\} + \frac{1}{2} C \max_{1 \leqslant i \leqslant m-1} P_i$$

$$\leqslant \max\{|\alpha|, |\beta|\} + \frac{L^2}{8} \max_{1 \leqslant i \leqslant m-1} |f_i|,$$

which implies

$$\|v\|_\infty \leqslant \max\{|\alpha|, |\beta|\} + \frac{L^2}{8} \max_{1 \leqslant i \leqslant m-1} |f_i|.$$

(II) If $\max\limits_{0 \leqslant i \leqslant m} \{\pm v_i - w_i\} < 0$, then we have

$$\max_{0 \leqslant i \leqslant m} \{\pm v_i\} = \max_{0 \leqslant i \leqslant m} \{\pm v_i - w_i + w_i\}$$

$$\leqslant \max_{0 \leqslant i \leqslant m} \{\pm v_i - w_i\} + \max_{0 \leqslant i \leqslant m} w_i$$

$$\leqslant \frac{1}{2}C \max_{1\leqslant i\leqslant m-1} P_i$$

$$\leqslant \frac{L^2}{8} \max_{1\leqslant i\leqslant m-1} |f_i|.$$

Therefore,

$$\|v\|_\infty \leqslant \frac{L^2}{8} \max_{1\leqslant i\leqslant m-1} |f_i|,$$

which implies that (1.29) also holds.

$\square$

Energy Method

We begin with the following lemma, which presents several frequently used identities and inequalities.

Lemma 1.4

(I) *For any functions $u = \{u_i \mid 0 \leqslant i \leqslant m\} \in \mathcal{U}_h$ and $v = \{v_i \mid 0 \leqslant i \leqslant m\} \in \mathcal{U}_h$, we have*

$$- (\delta_x^2 u, v) = (\delta_x u, \delta_x v) + (\delta_x u_{\frac{1}{2}})v_0 - (\delta_x u_{m-\frac{1}{2}})v_m. \tag{1.30}$$

(II) *For any $v \in \overset{\circ}{\mathcal{U}}_h$, we have*

$$-(\delta_x^2 v, v) = |v|_1^2, \tag{1.31}$$

$$\|v\|_\infty \leqslant \frac{\sqrt{L}}{2}|v|_1, \tag{1.32}$$

$$\|v\| \leqslant \frac{L}{\sqrt{6}}|v|_1. \tag{1.33}$$

(III) *For any $v \in \mathcal{U}_h$, we have*

$$|v|_1^2 \leqslant \frac{4}{h^2}\|v\|^2. \tag{1.34}$$

(IV) *For any $v \in \overset{\circ}{\mathcal{U}}_h$, we have*

$$\|v\|_\infty^2 \leqslant \epsilon |v|_1^2 + \frac{1}{4\epsilon}\|v\|^2 \tag{1.35}$$

with ϵ an arbitrary positive constant.

(V) *For any $v \in \mathcal{U}_h$, we have*

$$\|v\|_\infty^2 \leqslant \epsilon \, |v|_1^2 + \left(\frac{1}{\epsilon} + \frac{1}{L}\right) \|v\|^2 \tag{1.36}$$

with ϵ an arbitrary positive constant.

Proof

(I) Direct calculations give

$$-(\delta_x^2 u, v) = -h \sum_{i=1}^{m-1} (\delta_x^2 u_i) v_i$$

$$= -\sum_{i=1}^{m-1} \left(\delta_x u_{i+\frac{1}{2}} - \delta_x u_{i-\frac{1}{2}}\right) v_i$$

$$= \sum_{i=1}^{m-1} \left(\delta_x u_{i-\frac{1}{2}}\right) v_i - \sum_{i=1}^{m-1} \left(\delta_x u_{i+\frac{1}{2}}\right) v_i$$

$$= \sum_{i=1}^{m-1} \left(\delta_x u_{i-\frac{1}{2}}\right) v_i - \sum_{i=2}^{m} \left(\delta_x u_{i-\frac{1}{2}}\right) v_{i-1}$$

$$= \sum_{i=1}^{m} \left(\delta_x u_{i-\frac{1}{2}}\right) (v_i - v_{i-1}) + \left(\delta_x u_{\frac{1}{2}}\right) v_0 - \left(\delta_x u_{m-\frac{1}{2}}\right) v_m$$

$$= (\delta_x u, \delta_x v) + (\delta_x u_{\frac{1}{2}}) v_0 - (\delta_x u_{m-\frac{1}{2}}) v_m.$$

Therefore, (1.30) is proved.

(II) It easily follows from (1.30) that (1.31) holds.

For $1 \leqslant i \leqslant m - 1$, one has

$$v_i = \sum_{j=1}^{i} (v_j - v_{j-1}) = h \sum_{j=1}^{i} \delta_x v_{j-\frac{1}{2}},$$

$$v_i = -\sum_{j=i+1}^{m} (v_j - v_{j-1}) = -h \sum_{j=i+1}^{m} \delta_x v_{j-\frac{1}{2}}.$$

Squaring on both sides of the above two equalities and then applying the Cauchy-Schwarz inequality, we have

$$v_i^2 \leqslant \left(h \sum_{j=1}^{i} 1^2\right) h \sum_{j=1}^{i} (\delta_x v_{j-\frac{1}{2}})^2 = x_i h \sum_{j=1}^{i} (\delta_x v_{j-\frac{1}{2}})^2, \tag{1.37}$$

$$v_i^2 \leqslant \left(h \sum_{j=i+1}^{m} 1^2 \right) h \sum_{j=i+1}^{m} (\delta_x v_{j-\frac{1}{2}})^2 = (L - x_i) h \sum_{j=i+1}^{m} (\delta_x v_{j-\frac{1}{2}})^2.$$

$$(1.38)$$

Multiplying (1.37) by $(L - x_i)$ and (1.38) by x_i and then adding the results together produces

$$Lv_i^2 \leqslant x_i(L - x_i) h \sum_{j=1}^{m} (\delta_x v_{j-\frac{1}{2}})^2 = x_i(L - x_i)|v|_1^2, \quad 1 \leqslant i \leqslant m - 1.$$

$$(1.39)$$

By means of the inequality

$$x_i(L - x_i) \leqslant \frac{L^2}{4},$$

we have

$$Lv_i^2 \leqslant \frac{L^2}{4}|v|_1^2,$$

which reduces to

$$|v_i| \leqslant \frac{\sqrt{L}}{2}|v|_1, \quad 1 \leqslant i \leqslant m - 1.$$

Noticing $v_0 = 0$ and $v_m = 0$, we conclude

$$\|v\|_\infty \leqslant \frac{\sqrt{L}}{2}|v|_1.$$

Multiplying both sides of (1.39) by h and summing over i produce

$$L\|v\|^2 \leqslant h \sum_{i=1}^{m-1} x_i(L - x_i)|v|_1^2 = h \sum_{i=1}^{m-1} (ih) \times \big((m - i)h\big)|v|_1^2$$

$$= h^3 \left(m \sum_{i=1}^{m-1} i - \sum_{i=1}^{m-1} i^2 \right) |v|_1^2 = \frac{1}{6}m(m^2 - 1)h^3|v|_1^2$$

$$\leqslant \frac{1}{6}(mh)^3|v|_1^2 = \frac{1}{6}L^3|v|_1^2,$$

which means

$$\|v\| \leqslant \frac{L}{\sqrt{6}}|v|_1.$$

(III)   Note that (1.34) directly follows from

$$|v|_1^2 = h \sum_{i=1}^{m} \left(\delta_x v_{i-\frac{1}{2}}\right)^2 = \frac{1}{h^2} h \sum_{i=1}^{m} (v_i - v_{i-1})^2$$

$$\leqslant \frac{2}{h^2} h \sum_{i=1}^{m} \left(v_i^2 + v_{i-1}^2\right) = \frac{4}{h^2} h \left(\frac{1}{2} v_0^2 + \sum_{i=1}^{m-1} v_i^2 + \frac{1}{2} v_m^2\right) = \frac{4}{h^2} \|v\|^2.$$

(IV)   For $1 \leqslant i \leqslant m - 1$, one has

$$v_i^2 = \sum_{l=0}^{i-1} (v_{l+1}^2 - v_l^2) = 2h \sum_{l=0}^{i-1} \left(v_{l+\frac{1}{2}}\right)\left(\delta_x v_{l+\frac{1}{2}}\right),$$

$$v_i^2 = - \sum_{l=i}^{m-1} (v_{l+1}^2 - v_l^2) = -2h \sum_{l=i}^{m-1} \left(v_{l+\frac{1}{2}}\right)\left(\delta_x v_{l+\frac{1}{2}}\right).$$

Adding the above two equalities together, we get

$$v_i^2 \leqslant h \sum_{l=0}^{m-1} |v_{l+\frac{1}{2}}| \cdot |\delta_x v_{l+\frac{1}{2}}|$$

$$\leqslant \epsilon h \sum_{l=0}^{m-1} \left(\delta_x v_{l+\frac{1}{2}}\right)^2 + \frac{1}{4\epsilon} h \sum_{l=0}^{m-1} \left(v_{l+\frac{1}{2}}\right)^2 \leqslant \epsilon |v|_1^2 + \frac{1}{4\epsilon} \|v\|^2.$$

Therefore, (1.35) is proved.

(V)   Suppose

$$|v_l| = \|v\|_\infty, \quad 0 \leqslant l \leqslant m.$$

When $0 \leqslant i \leqslant l$,

$$\|v\|_\infty^2 = v_l^2 = v_i^2 + \sum_{j=i+1}^{l} (v_j^2 - v_{j-1}^2)$$

$$= v_i^2 + 2h \sum_{j=i+1}^{l} v_{j-\frac{1}{2}} \delta_x v_{j-\frac{1}{2}} \leqslant v_i^2 + 2h \sum_{j=i+1}^{l} |v_{j-\frac{1}{2}} \delta_x v_{j-\frac{1}{2}}|$$

$$\leqslant v_i^2 + 2h \sum_{j=1}^{m} |v_{j-\frac{1}{2}} \delta_x v_{j-\frac{1}{2}}| \leqslant v_i^2 + 2\|v\| \cdot |v|_1.$$

When $l \leqslant i \leqslant m$,

$$\|v\|_\infty^2 = v_l^2 = v_i^2 - \sum_{j=l+1}^{i} (v_j^2 - v_{j-1}^2)$$

$$= v_i^2 - 2h \sum_{j=l+1}^{i} v_{j-\frac{1}{2}} \delta_x v_{j-\frac{1}{2}} \leqslant v_i^2 + 2h \sum_{j=l+1}^{i} |v_{j-\frac{1}{2}} \delta_x v_{j-\frac{1}{2}}|$$

$$\leqslant v_i^2 + 2h \sum_{j=1}^{m} |v_{j-\frac{1}{2}} \delta_x v_{j-\frac{1}{2}}| \leqslant v_i^2 + 2\|v\| \cdot |v|_1.$$

Combining the above two inequalities yields

$$\|v\|_\infty^2 \leqslant v_i^2 + 2\|v\| \cdot |v|_1, \quad 0 \leqslant i \leqslant m.$$

Multiplying the above inequality with $i = 0$ and $i = m$, respectively, by $\frac{1}{2}h$ and the above other inequalities with $1 \leqslant i \leqslant m - 1$ by h, and then summing up the results, we arrive at

$$L\|v\|_\infty^2 \leqslant \|v\|^2 + 2L\|v\| \cdot |v|_1.$$

Therefore,

$$\|v\|_\infty^2 \leqslant 2\|v\| \cdot |v|_1 + \frac{1}{L}\|v\|^2 \leqslant \epsilon |v|_1^2 + \left(\frac{1}{\epsilon} + \frac{1}{L}\right) \|v\|^2.$$

$\square$

Equalities (1.30) and (1.31) are conventionally termed **summation by parts**, while (1.32)–(1.33) and (1.35)–(1.36) are referred to as **embedding inequalities**. Inequality (1.34) is commonly known as an **inverse estimate**.

Theorem 1.4 *Let $v = \{v_i \mid 0 \leqslant i \leqslant m\}$ be the solution of the difference scheme*

$$\begin{cases} -\delta_x^2 v_i + q(x_i)v_i = f_i, & 1 \leqslant i \leqslant m - 1, & \text{(1.40a)} \\ v_0 = 0, \quad v_m = 0. & & \text{(1.40b)} \end{cases}$$

Then we have

$$|v|_1 \leqslant \frac{L}{\sqrt{6}}\|f\|, \quad \|v\|_\infty \leqslant \frac{L^2}{2\sqrt{6}} \|f\|_\infty,$$

where

$$\|f\|_\infty = \max_{1\leqslant i\leqslant m-1} |f_i|, \quad \|f\| = \sqrt{h\sum_{i=1}^{m-1} f_i^2}.$$

Proof Multiplying both sides of (1.40a) by hv_i and summing over i, one has

$$h\sum_{i=1}^{m-1}(-\delta_x^2 v_i)v_i + h\sum_{i=1}^{m-1} q(x_i)v_i^2 = h\sum_{i=1}^{m-1} f_i v_i. \tag{1.41}$$

It follows from Lemma 1.4 that

$$h\sum_{i=1}^{m-1}(-\delta_x^2 v_i)v_i = |v|_1^2.$$

Due to the non-negativity of $q(x)$, it yields

$$h\sum_{i=1}^{m-1} q(x_i)v_i^2 \geqslant 0.$$

With the help of the Cauchy-Schwarz inequality, it yields

$$h\sum_{i=1}^{m-1} f_i v_i \leqslant \sqrt{h\sum_{i=1}^{m-1} f_i^2} \cdot \sqrt{h\sum_{i=1}^{m-1} v_i^2} = \|f\| \cdot \|v\|.$$

Inserting the above one equality and two inequalities into (1.41) generates

$$|v|_1^2 \leqslant \|f\| \cdot \|v\|.$$

The embedding inequality (1.33) further implies that

$$|v|_1^2 \leqslant \|f\| \cdot \frac{L}{\sqrt{6}}|v|_1,$$

which is

$$|v|_1 \leqslant \frac{L}{\sqrt{6}}\|f\|.$$

By means of Lemma 1.4 again, we have

$$\|v\|_\infty \leqslant \frac{\sqrt{L}}{2}|v|_1 \leqslant \frac{\sqrt{L}}{2}\cdot\frac{L}{\sqrt{6}}\|f\| \leqslant \frac{L^2}{2\sqrt{6}}\|f\|_\infty,$$

which completes the proof. $\square$

The estimate of the solution to the nonhomogeneous boundary value problem can be obtained through the homogenization method of the boundary conditions.

Theorem 1.5 *Let* $v = \{v_i \mid 0 \leqslant i \leqslant m\}$ *be the solution of the difference scheme*

$$\begin{cases} -\delta_x^2 v_i + q(x_i)v_i = f_i, & 1 \leqslant i \leqslant m-1, \\ v_0 = \alpha, \quad v_m = \beta. \end{cases}$$

Then we have

$$\|v\|_\infty \leqslant \frac{L^2}{2\sqrt{6}}\|f\|_\infty + \left(\frac{L^2}{2\sqrt{6}}\kappa + 1\right)\max\{|\alpha|, |\beta|\}, \tag{1.42}$$

where $\kappa = \max\limits_{0\leqslant x\leqslant L}|q(x)|$ *and* $\|f\|_\infty = \max\limits_{1\leqslant i\leqslant m-1}|f_i|.$

Proof Denote

$$p(x) = \frac{L-x}{L}\alpha + \frac{x}{L}\beta, \quad q_i = q(x_i), \quad p_i = p(x_i), \quad w_i = v_i - p_i, \quad 0 \leqslant i \leqslant m.$$

It is evident that $w = (w_0, w_1, \ldots, w_{m-1}, w_m)$ satisfies the following system of difference equations

$$\begin{cases} -\delta_x^2 w_i + q(x_i)w_i = f_i - q_i p_i, & 1 \leqslant i \leqslant m-1, \\ w_0 = 0, \quad w_m = 0. \end{cases}$$

An application of Theorem 1.4 yields

$$\begin{aligned} \|w\|_\infty &\leqslant \frac{L^2}{2\sqrt{6}}\|f - qp\|_\infty \\ &\leqslant \frac{L^2}{2\sqrt{6}}\left(\|f\|_\infty + \kappa\|p\|_\infty\right) \\ &\leqslant \frac{L^2}{2\sqrt{6}}\left(\|f\|_\infty + \kappa\max\{|\alpha|, |\beta|\}\right). \end{aligned}$$

Therefore,

$$\|v\|_\infty = \|w + p\|_\infty \leqslant \|w\|_\infty + \|p\|_\infty$$

$$\leqslant \frac{L^2}{2\sqrt{6}} \left(\|f\|_\infty + \kappa \max\{|\alpha|, |\beta|\}\right) + \max\{|\alpha|, |\beta|\}$$

$$= \frac{L^2}{2\sqrt{6}} \|f\|_\infty + \left(\frac{L^2}{2\sqrt{6}}\kappa + 1\right) \max\{|\alpha|, |\beta|\}.$$

$\square$

A Brief Comparison Between the Maximum Principle and the Energy Method
By comparing the priori estimates (1.29) and (1.42), we observe that they are in the same form while differing only in the coefficients of their right-hand terms.

The energy method and the maximum principle method are two essential techniques for analyzing difference schemes. Some difference schemes satisfy the maximum principle, making it preferable to use in this case. However, for difference schemes where the maximum principle fails to apply, the energy method often provides a viable alternative.

1.2.5 Convergence and Stability of the Difference Solution

Convergence

Theorem 1.6 (Convergence) *Let* $\{u(x) \,|\, 0 \leqslant x \leqslant L\}$ *be the solution of the problem (1.1) and* $\{u_i \,|\, 0 \leqslant i \leqslant m\}$ *be the solution of the difference scheme (1.21). Denote*

$$e_i = u(x_i) - u_i, \quad 0 \leqslant i \leqslant m.$$

Then it holds that

$$\|e\|_\infty \leqslant \frac{M_4 L^2}{24\sqrt{6}} h^2,$$

where

$$M_4 = \max_{0 \leqslant x \leqslant L} |u^{(4)}(x)|.$$

Proof Subtracting (1.21) from (1.19) leads to the system of error equations

$$\begin{cases} -\delta_x^2 e_i + q(x_i)e_i = R_i, & 1 \leqslant i \leqslant m - 1, \\ e_0 = 0, \quad e_m = 0. \end{cases}$$

With the help of Theorem 1.4 and noticing (1.22), we have

$$\|e\|_\infty \leqslant \frac{L^2}{2\sqrt{6}} \max_{1\leqslant i\leqslant m-1} |R_i| \leqslant \frac{M_4 L^2}{24\sqrt{6}} h^2.$$

$\square$

Definition 1.1 If $\|e\| = O(h^p)$, the difference scheme is convergent with the **convergence order** p in the norm $\|\cdot\|$.

According to Definition 1.1 and Theorem 1.6, the difference scheme (1.21) is convergent with the convergence order two in the maximum norm $\|\cdot\|_\infty$.

Stability

In practical calculations, errors are inevitable. For instance, there may be a small error g_i in computing $f(x_i)$. Let $v = (v_0, v_1, \ldots, v_m)$ be the solution of the difference scheme

$$\begin{cases} -\delta_x^2 v_i + q(x_i)v_i = f(x_i) + g_i, & 1 \leqslant i \leqslant m-1, \\ v_0 = \alpha, \quad v_m = \beta. \end{cases} \tag{1.43}$$

Denote

$$\varepsilon_i = v_i - u_i, \quad 0 \leqslant i \leqslant m.$$

Subtracting (1.21) from (1.43) gives

$$\begin{cases} -\delta_x^2 \varepsilon_i + q(x_i)\varepsilon_i = g_i, & 1 \leqslant i \leqslant m-1, \\ \varepsilon_0 = 0, \quad \varepsilon_m = 0, \end{cases}$$

which is referred to as the **system of perturbation equations**, which is similar to (1.21) formally. Using Theorem 1.4, one has

$$\|\varepsilon\|_\infty \leqslant \frac{L^2}{2\sqrt{6}} \max_{1\leqslant i\leqslant m-1} |g_i|.$$

When $\max\limits_{1\leqslant i\leqslant m-1} |g_i|$ is small enough, $\|\varepsilon\|_\infty$ is also very small. Consequently, we have the following theorem.

Theorem 1.7 (Stability) *The solution of the difference scheme (1.21) is stable with respect to the right-hand side function in the following sense: Let $\{u_i \mid 0 \leqslant i \leqslant m\}$ be the solution of the difference scheme*

$$\begin{cases} -\delta_x^2 u_i + q(x_i)u_i = f_i, & 1 \leqslant i \leqslant m-1, \\ u_0 = 0, \quad u_m = 0. \end{cases}$$

Then it holds that

$$\|u\|_\infty \leqslant \frac{L^2}{2\sqrt{6}} \max_{1 \leqslant i \leqslant m-1} |f_i|.$$

1.2.6   The Richardson Extrapolation Method

Let $p_0(h)$ be an approximation to the unknown quality p. Suppose when $h \to 0$, the following relation holds:

$$p_0(h) = p + \alpha h^2 + O(h^4), \tag{1.44}$$

where α is a nonzero constant independent of h. The equality (1.44) is called an **asymptotic expansion** of $p_0(h)$. Since

$$p_0(h) - p = \alpha h^2 + O(h^4),$$

we conclude that $p_0(h)$ is an approximation to p of order two.

Replacing h with $h/2$ in (1.44) produces

$$p_0\left(\frac{h}{2}\right) = p + \alpha \left(\frac{h}{2}\right)^2 + O\left(\left(\frac{h}{2}\right)^4\right). \tag{1.45}$$

It is easy to know that $p_0(h/2)$ is also an approximation to p of order two. Next, multiplying both sides of (1.45) by 4/3 and (1.44) by 1/3 and then subtracting the corresponding results, we arrive at

$$\frac{4}{3} p_0\left(\frac{h}{2}\right) - \frac{1}{3} p_0(h) = p + O(h^4).$$

Denote

$$p_1(h) = \frac{4}{3} p_0\left(\frac{h}{2}\right) - \frac{1}{3} p_0(h), \tag{1.46}$$

then we have

$$p_1(h) = p + O(h^4).$$

If we consider $p_1(h)$ as an alternative approximation to p, it demonstrates superiority compared to both $p_0(h)$ and $p_0(h/2)$, owing to its higher-order accuracy.

The equality (1.46) is called the **Richardson extrapolation formula**. The method of obtaining high-accuracy approximation by a linear combination of low-accuracy approximations $p_0(h)$ and $p_0(h/2)$ is called **Richardson extrapolation**. This extrapolation is highly effective in the difference method. Denote the solution of (1.21) by $u_i(h)$ with the step size h, $0 \leqslant i \leqslant m$. The following result holds.

Theorem 1.8 *Suppose the two-point boundary value problem*

$$\begin{cases} -w''(x) + q(x)w(x) = \dfrac{1}{12}u^{(4)}(x), & 0 < x < L, \\ w(0) = 0, \quad w(L) = 0 \end{cases} \tag{1.47}$$

has a smooth solution, then we have

$$\max_{0 \leqslant i \leqslant m} \left| u(x_i) - \left[\frac{4}{3}u_{2i}\left(\frac{h}{2}\right) - \frac{1}{3}u_i(h) \right] \right| = O(h^4),$$

where $h = L/m$, $x_i = ih$, $0 \leqslant i \leqslant m$.

Proof　With the help of Taylor expansion, one has

$$u''(x_i) = \delta_x^2 U_i - \frac{h^2}{12}u^{(4)}(x_i) - \frac{h^4}{360}u^{(6)}(\eta_i), \quad \eta_i \in (x_{i-1}, x_{i+1}).$$

Thus, (1.19) can be written as

$$\begin{cases} -\delta_x^2 U_i + q(x_i)U_i = f(x_i) - \dfrac{h^2}{12}u^{(4)}(x_i) - \dfrac{h^4}{360}u^{(6)}(\eta_i), & 1 \leqslant i \leqslant m - 1, \\ U_0 = \alpha, \quad U_m = \beta. \end{cases}$$

$$\tag{1.48}$$

Subtracting (1.21) from (1.48), the system of error equations reads

$$\begin{cases} -\delta_x^2 e_i + q(x_i)e_i = -\dfrac{h^2}{12}u^{(4)}(x_i) - \dfrac{h^4}{360}u^{(6)}(\eta_i), & 1 \leqslant i \leqslant m - 1, \\ e_0 = 0, \quad e_m = 0. \end{cases}$$

$$\tag{1.49}$$

Denote

$$W_i = w(x_i), \quad 0 \leqslant i \leqslant m.$$

Discretizing (1.47) generates

$$\begin{cases} -\delta_x^2 W_i + q(x_i)W_i = \dfrac{1}{12}u^{(4)}(x_i) - \dfrac{h^2}{12}w^{(4)}(\tilde{\eta}_i), & 1 \leqslant i \leqslant m - 1, \\ w_0 = 0, \quad w_m = 0, \end{cases}$$

$$\tag{1.50}$$

where $\tilde{\eta}_i \in (x_{i-1}, x_{i+1})$.

Define

$$r_i = e_i + h^2 W_i, \quad 0 \leqslant i \leqslant m.$$

Multiplying both sides of (1.50) by h^2, and then adding the result with (1.49), one gets

$$\begin{cases} -\delta_x^2 r_i + q(x_i) r_i = -\dfrac{h^4}{360} u^{(6)}(\eta_i) - \dfrac{h^4}{12} w^{(4)}(\tilde{\eta}_i), & 1 \leqslant i \leqslant m - 1, \\ r_0 = 0, \quad r_m = 0. \end{cases}$$

Borrowing Theorem 1.4, we have

$$\max_{0 \leqslant i \leqslant m} |r_i| \leqslant \frac{L^2}{2\sqrt{6}} \left[\frac{h^4}{360} \max_{0 \leqslant x \leqslant L} |u^{(6)}(x)| + \frac{h^4}{12} \max_{0 \leqslant x \leqslant L} |w^{(4)}(x)| \right],$$

which implies

$$u(x_i) - u_i(h) + h^2 W_i = O(h^4), \quad 0 \leqslant i \leqslant m.$$

Rearranging the above equality yields

$$u_i(h) = u(x_i) + h^2 w(x_i) + O(h^4), \quad 0 \leqslant i \leqslant m.$$

Similarly, one has

$$u_{2i}\left(\frac{h}{2}\right) = u(x_i) + \left(\frac{h}{2}\right)^2 w(x_i) + O(h^4), \quad 0 \leqslant i \leqslant m.$$

In combination of the above two equalities, we have

$$\frac{4}{3} u_{2i}\left(\frac{h}{2}\right) - \frac{1}{3} u_i(h) = u(x_i) + O(h^4), \quad 0 \leqslant i \leqslant m,$$

which completes the proof. $\qquad\qquad\qquad\qquad\qquad\qquad\qquad\qquad\qquad\quad\square$

Example 1.2 Compute the two-point boundary value problem (1.27) by the Richardson extrapolation method.

Table 1.3 lists numerical errors in the maximum norm defined by

$$\hat{E}_\infty(h) = \max_{1 \leqslant i \leqslant m-1} \left| u(x_i) - \left[\frac{4}{3} u_{2i}\left(\frac{h}{2}\right) - \frac{1}{3} u_i(h) \right] \right|.$$

By comparing the numerical results in Tables 1.3 and 1.2, it is evident that the extrapolation method significantly enhances the numerical accuracy.

Table 1.3 (Example 1.2) Maximum errors of numerical solutions with different step sizes

h	$\hat{E}_\infty(h)$	$\hat{E}_\infty(2h)/\hat{E}_\infty(h)$
$\pi/10$	1.009e−4	
$\pi/20$	6.846e−6	14.74
$\pi/40$	4.308e−7	15.89
$\pi/80$	2.697e−8	15.97
$\pi/160$	1.689e−9	15.97
$\pi/320$	9.588e−11	17.62

1.2.7   The Compact Difference Scheme

Let $w = \{w_i \mid 0 \leqslant i \leqslant m\} \in \mathcal{U}_h$. Define an operator

$$(\mathcal{A}w)_i = \begin{cases} \dfrac{1}{12}(w_{i-1} + 10w_i + w_{i+1}), & 1 \leqslant i \leqslant m - 1, \\ w_i, & i = 0, m. \end{cases}$$

Considering Eq. (1.1a) at the point x_i, one has

$$-u''(x_i) + q(x_i)u(x_i) = f(x_i), \quad 0 \leqslant i \leqslant m.$$

Performing the operator $\mathcal{A}$ on both sides of the above equality, we have

$$-\mathcal{A}u''(x_i) + \mathcal{A}[q(x_i)u(x_i)] = \mathcal{A}f(x_i), \quad 1 \leqslant i \leqslant m - 1. \tag{1.51}$$

It follows from Lemma 1.2 that

$$\mathcal{A}u''(x_i) = \delta_x^2 u(x_i) + \frac{h^4}{240}u^{(6)}(\xi_i),$$

where $\xi_i \in (x_{i-1}, x_{i+1})$.

Substituting the above formula into (1.51), we have

$$\begin{cases} -\delta_x^2 U_i + \mathcal{A}[q(x_i)U_i] = \mathcal{A}f(x_i) + \frac{h^4}{240}u^{(6)}(\xi_i), & 1 \leqslant i \leqslant m - 1, \\ U_0 = \alpha, \quad U_m = \beta. \end{cases}$$

Omitting the small term $\frac{1}{240}u^{(6)}(\xi_i)h^4$ and replacing U_i by u_i, another difference scheme reads

$$\begin{cases} -\delta_x^2 u_i + \dfrac{1}{12}[q(x_{i-1})u_{i-1} + 10q(x_i)u_i + q(x_{i+1})u_{i+1}] \\ \qquad = \dfrac{1}{12}[f(x_{i-1}) + 10f(x_i) + f(x_{i+1})], \quad 1 \leqslant i \leqslant m - 1, \\ u_0 = \alpha, \quad u_m = \beta. \end{cases} \tag{1.52}$$

It can be proved that the difference scheme (1.52) is uniquely solvable, stable and convergent with the convergence order four in the maximum norm.

The difference scheme (1.52) achieves the highest order of four based on a three-point (x_{i-1}, x_i, x_{i+1}) stencil without requiring the derivatives of the right-hand side function $f(x)$. This type of difference scheme is referred to as a **compact difference scheme**.

Example 1.3 Apply the difference scheme (1.52) to solve the two-point boundary value problem (1.27) in Example 1.1.

Divide the interval $[0, \pi]$ into m equal subintervals. Denote $h = \pi/m$, $x_i = ih$, $0 \leqslant i \leqslant m$. The compact difference scheme for solving (1.27) reads

$$
\begin{cases}
-\dfrac{1}{h^2}(u_{i-1} - 2u_i + u_{i+1}) + \dfrac{1}{12}(u_{i-1} + 10u_i + u_{i+1}) \\
\quad = \dfrac{1}{12}[e^{x_{i-1}}(\sin x_{i-1} - 2\cos x_{i-1}) + 10e^{x_i}(\sin x_i - 2\cos x_i) \\
\quad\quad + e^{x_{i+1}}(\sin x_{i+1} - 2\cos x_{i+1})], \quad 1 \leqslant i \leqslant m-1, \\
u_0 = 0, \quad u_m = 0.
\end{cases}
$$

Table 1.4 presents the exact and numerical solutions at four node points for various step sizes. Table 1.5 shows the absolute values of the numerical errors at the same four node points, along with the numerical errors in the maximum norm, defined as

$$
E_\infty(h) = \max_{0 \leqslant i \leqslant m} |u(x_i) - u_i|
$$

for different step sizes.

As observed in Table 1.5, when the step size h is halved, the maximum error decreases to 1/16 of its original value. Figure 1.3 displays the curves of the exact solution and the numerical solution with $h = \pi/10$. Since the relative error of the numerical solution is approximately 1/10000, the curves for the numerical and exact solutions are nearly indistinguishable to the naked eye. Figure 1.4 illustrates the error curves of the numerical solutions for various step sizes.

Table 1.4 (Example 1.3) The exact and numerical solutions at four node points with different step sizes

	x			
h	$\pi/5$	$2\pi/5$	$3\pi/5$	$4\pi/5$
$\pi/10$	1.101789	3.341453	6.263188	7.255585
$\pi/20$	1.101778	3.341608	6.263684	7.256326
$\pi/40$	1.101778	3.341618	6.263715	7.256373
$\pi/80$	1.101778	3.341618	6.263717	7.256376
$\pi/160$	1.101778	3.341619	6.263717	7.256376
ES	1.101778	3.341619	6.263717	7.256376

Table 1.5 (Example 1.3) The absolute values of numerical errors and maximum errors of numerical solutions at four node points with different step sizes

h	x				$E_\infty(h)$	$E_\infty(2h)/E_\infty(h)$
	$\pi/5$	$2\pi/5$	$3\pi/5$	$4\pi/5$		
$\pi/10$	1.105e−5	1.658e−4	5.289e−4	7.910e−4	7.910e−4	
$\pi/20$	5.547e−7	1.063e−5	3.339e−5	4.971e−5	4.971e−5	15.91
$\pi/40$	3.253e−8	6.685e−7	2.093e−6	3.111e−6	3.111e−6	15.98
$\pi/80$	2.000e−9	4.185e−8	1.309e−7	1.945e−7	1.946e−7	15.99
$\pi/160$	1.246e−10	2.616e−9	8.180e−9	1.216e−8	1.216e−8	16.00

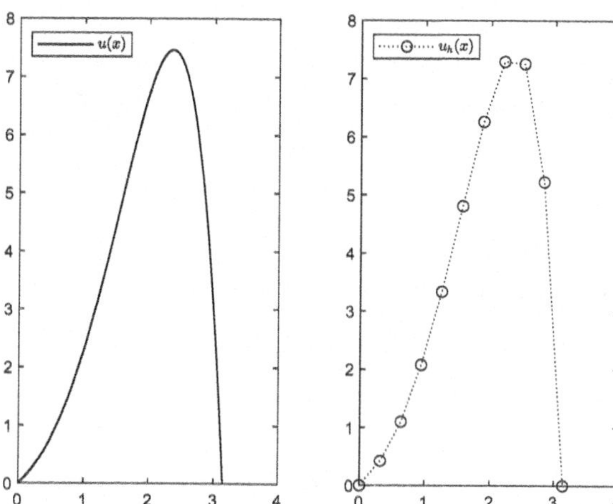

Fig. 1.3 (Example 1.3) The curves of the exact and numerical solutions plotted with $h = \pi/10$

1.3   The Derivative Boundary Value Problem

Consider the following derivative boundary value problem

$$\begin{cases} -u'' + q(x)u = f(x), & 0 < x < L, \\ -u'(0) + \lambda_1 u(0) = \alpha, & u'(L) + \lambda_2 u(L) = \beta, \end{cases}$$

$$(1.53a)$$
$$(1.53b)$$

where $q(x) \geqslant 0$ and $f(x)$ are given continuous functions, $\lambda_1 \geqslant 0$, $\lambda_2 \geqslant 0$, α and β are known constants. Additionally, it is assumed that $\lambda_1 + \lambda_2 = 0$ and $q(x) \equiv 0$ do not hold simultaneously.

Let $v = (v_0, v_1, \ldots, v_m) \in \mathcal{U}_h$. Denote

$$D_+ v_i = \frac{1}{h}(v_{i+1} - v_i), \quad D_- v_i = \frac{1}{h}(v_i - v_{i-1}).$$

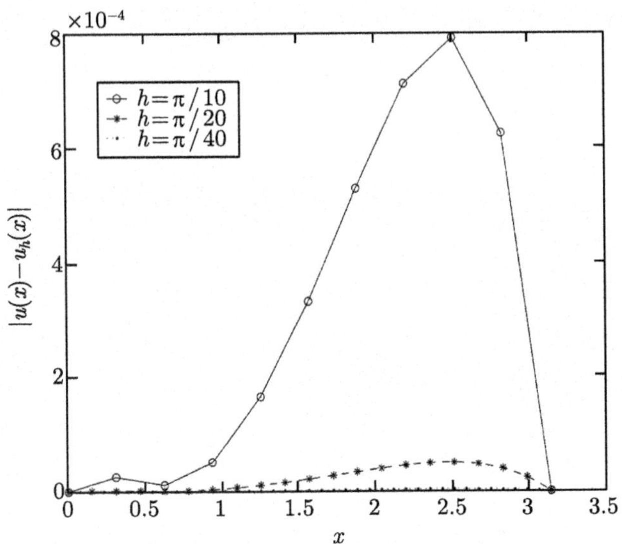

Fig. 1.4 (Example 1.3) The error curves of numerical solutions with different step sizes

1.3.1   *Derivation of the Difference Scheme*

Considering (1.53) at the node point x_i, one has

$$
\begin{cases}
-u''(x_i) + q(x_i)u(x_i) = f(x_i), & 1 \leqslant i \leqslant m-1, & \text{(1.54a)} \\
-u'(x_0) + \lambda_1 u(x_0) = \alpha, & u'(x_m) + \lambda_2 u(x_m) = \beta. & \text{(1.54b)}
\end{cases}
$$

Applying

$$
u''(x_i) = \delta_x^2 U_i - \frac{h^2}{12} u^{(4)}(\xi_i), \quad \xi_i \in (x_{i-1}, x_{i+1}), \quad 1 \leqslant i \leqslant m-1
$$

to (1.54a) and

$$
\begin{cases}
u'(x_0) = \dfrac{u(x_1)-u(x_0)}{h} - \dfrac{h}{2}u''(\xi_0) = D_+U_0 - \dfrac{h}{2}u''(\xi_0), & \xi_0 \in (x_0, x_1), \\[2ex]
u'(x_m) = \dfrac{u(x_m)-u(x_{m-1})}{h} + \dfrac{h}{2}u''(\xi_m) = D_-U_m + \dfrac{h}{2}u''(\xi_m), & \xi_m \in (x_{m-1}, x_m)
\end{cases}
$$

to (1.54b), respectively, we arrive at

$$
\begin{cases}
-\delta_x^2 U_i + q(x_i)U_i = f(x_i) - \dfrac{h^2}{12}u^{(4)}(\xi_i), & 1 \leqslant i \leqslant m-1, \quad (1.55\text{a}) \\[2ex]
-D_+U_0 + \lambda_1 U_0 = \alpha - \dfrac{h}{2}u''(\xi_0), & (1.55\text{b}) \\[2ex]
D_-U_m + \lambda_2 U_m = \beta - \dfrac{h}{2}u''(\xi_m). & (1.55\text{c})
\end{cases}
$$

Ignoring small terms in the above equalities and replacing U_i with u_i, a difference scheme reads

$$
\begin{cases}
-\delta_x^2 u_i + q(x_i)u_i = f(x_i), & 1 \leqslant i \leqslant m-1, \quad (1.56\text{a}) \\[2ex]
-D_+u_0 + \lambda_1 u_0 = \alpha, \quad D_-u_m + \lambda_2 u_m = \beta. & (1.56\text{b})
\end{cases}
$$

The local truncation error of (1.56a) is $O(h^2)$ and that of (1.56b) is $O(h)$.

To improve the approximation accuracy of the derivative boundary value conditions, by noticing (1.53a), one has

$$
u''(x_0) = q(x_0)u(x_0) - f(x_0), \quad u''(x_m) = q(x_m)u(x_m) - f(x_m).
$$

From the Taylor expansion, it derives to

$$
\begin{aligned}
u'(x_0) &= D_+U_0 - \frac{h}{2}u''(x_0) - \frac{h^2}{6}u'''(\bar{\xi}_0) \\[1ex]
&= D_+U_0 - \frac{h}{2}[q(x_0)u(x_0) - f(x_0)] - \frac{h^2}{6}u'''(\bar{\xi}_0), \quad x_0 < \bar{\xi}_0 < x_1, \\[2ex]
u'(x_m) &= D_-U_m + \frac{h}{2}u''(x_m) - \frac{h^2}{6}u'''(\bar{\xi}_m) \\[1ex]
&= D_-U_m + \frac{h}{2}[q(x_m)u(x_m) - f(x_m)] - \frac{h^2}{6}u'''(\bar{\xi}_m), \quad x_{m-1} < \bar{\xi}_m < x_m.
\end{aligned}
$$

Inserting the above two equalities into (1.54b) and noticing (1.55a), we have

$$
\begin{cases}
-\delta_x^2 U_i + q(x_i)U_i = f(x_i) - \dfrac{h^2}{12}u^{(4)}(\xi_i), & 1 \leqslant i \leqslant m-1, \\[2ex]
-D_+U_0 + \dfrac{h}{2}[q(x_0)U_0 - f(x_0)] + \lambda_1 U_0 = \alpha - \dfrac{h^2}{6}u'''(\bar{\xi}_0), \\[2ex]
D_-U_m + \dfrac{h}{2}[q(x_m)U_m - f(x_m)] + \lambda_2 U_m = \beta + \dfrac{h^2}{6}u'''(\bar{\xi}_m).
\end{cases}
$$

Omitting the small terms and replacing U_i by u_i, another finite difference scheme for solving the problem (1.53) reads

$$
\begin{cases}
-\delta_x^2 u_i + q(x_i)u_i = f(x_i), & 1 \leqslant i \leqslant m-1, & (1.57a) \\[2mm]
-D_+ u_0 + \dfrac{h}{2} q(x_0)u_0 + \lambda_1 u_0 = \alpha + \dfrac{h}{2} f(x_0), & & (1.57b) \\[2mm]
D_- u_m + \dfrac{h}{2} q(x_m)u_m + \lambda_2 u_m = \beta + \dfrac{h}{2} f(x_m). & & (1.57c)
\end{cases}
$$

The local truncation errors of the difference equation (1.57a) for (1.53a), and of (1.57b)-(1.57c) for the boundary value condition (1.53b), are both $O(h^2)$.

Equations (1.57b) and (1.57c) can be written as

$$
-\frac{2}{h}\left[D_+ u_0 - \left(\lambda_1 u_0 - \alpha\right)\right] + q(x_0)u_0 = f(x_0), \tag{1.58}
$$

$$
-\frac{2}{h}\left[\left(\beta - \lambda_2 u_m\right) - D_- u_m\right] + q(x_m)u_m = f(x_m). \tag{1.59}
$$

Consequently, Eqs. (1.58) and (1.59) can be viewed as the discretization of

$$
-u''(x_0) + q(x_0)u(x_0) = f(x_0)
$$

and

$$
-u''(x_m) + q(x_m)u(x_m) = f(x_m),
$$

respectively, which, in fact, are explained by

$$
-\frac{2}{h}\left[D_+ U_0 - \left(\lambda_1 U_0 - \alpha\right)\right] + q(x_0)U_0 = f(x_0) + O(h),
$$

$$
-\frac{2}{h}\left[\left(\beta - \lambda_2 U_m\right) - D_- U_m\right] + q(x_m)U_m = f(x_m) + O(h).
$$

1.3.2 Implementation of the Difference Scheme and Numerical Examples

The difference scheme (1.56) can be written as the following system of linear equations:

$$
\begin{pmatrix}
1+\lambda_1 h & -1 & & & & \\
-1 & 2+h^2 q(x_1) & -1 & & & \\
& \ddots & \ddots & \ddots & & \\
& & -1 & 2+h^2 q(x_{m-1}) & -1 \\
& & & -1 & 1+\lambda_2 h
\end{pmatrix}
\begin{pmatrix}
u_0 \\ u_1 \\ \vdots \\ u_{m-1} \\ u_m
\end{pmatrix}
$$

$$
=
\begin{pmatrix}
h\alpha \\
h^2 f(x_1) \\
\vdots \\
h^2 f(x_{m-1}) \\
h\beta
\end{pmatrix},
$$

which can be solved by the double sweep method introduced in Sect. 1.2.3.

Example 1.4   Apply the difference scheme (1.56) to solve the following two-point boundary value problem:

$$
\begin{cases}
-u'' + u = e^x(\sin x - 2\cos x), & 0 < x < \pi, \\
-u'(0) = -1, \quad u'(\pi) = -e^\pi.
\end{cases}
\tag{1.60}
$$

The exact solution of the problem is $u(x) = e^x \sin x$.

Divide the interval $[0, \pi]$ into m equal subintervals and denote $h = \pi/m$, $x_i = ih$, $0 \leqslant i \leqslant m$. The difference scheme for solving (1.60) reads

$$
\begin{cases}
-\dfrac{u_{i-1} - 2u_i + u_{i+1}}{h^2} + u_i = e^{x_i}(\sin x_i - 2\cos x_i), & 1 \leqslant i \leqslant m-1, \\
-\dfrac{u_1 - u_0}{h} = -1, \quad \dfrac{u_m - u_{m-1}}{h} = -e^\pi.
\end{cases}
$$

Table 1.6 lists the exact and numerical solutions at four points with varying step sizes. Table 1.7 shows the absolute values of the numerical errors $|u(x_i) - u_i|$ with different step sizes and the maximum error defined by

$$
E_\infty(h) = \max_{0 \leqslant i \leqslant m} |u(x_i) - u_i|.
$$

Table 1.6 (Example 1.4) The exact and numerical solutions at four node points with varying step sizes

		x			
	h	$\pi/5$	$2\pi/5$	$3\pi/5$	$4\pi/5$
	$\pi/160$	1.064444	3.271905	6.133363	7.012527
	$\pi/320$	1.083244	3.306943	6.198800	7.134814
	$\pi/640$	1.092544	3.324326	6.231323	7.195686
	$\pi/1280$	1.097169	3.332984	6.247536	7.226053
	ES	1.101778	3.341619	6.263717	7.256376

Table 1.7 (Example 1.4) The absolute values of numerical errors at four node points and maximum errors with varying step sizes

h	x				$E_\infty(h)$	$E_\infty(2h)/E_\infty(h)$
	$\pi/5$	$2\pi/5$	$3\pi/5$	$4\pi/5$		
$\pi/160$	3.733e$-$2	6.971e$-$2	1.304e$-$1	2.438e$-$1	4.565e$-$1	
$\pi/320$	1.853e$-$2	3.468e$-$2	6.492e$-$2	1.216e$-$1	2.277e$-$1	2.005
$\pi/640$	9.234e$-$3	1.729e$-$2	3.239e$-$2	6.069e$-$2	1.137e$-$1	2.003
$\pi/1280$	4.609e$-$3	8.635e$-$3	1.618e$-$2	3.032e$-$2	5.683e$-$2	2.001

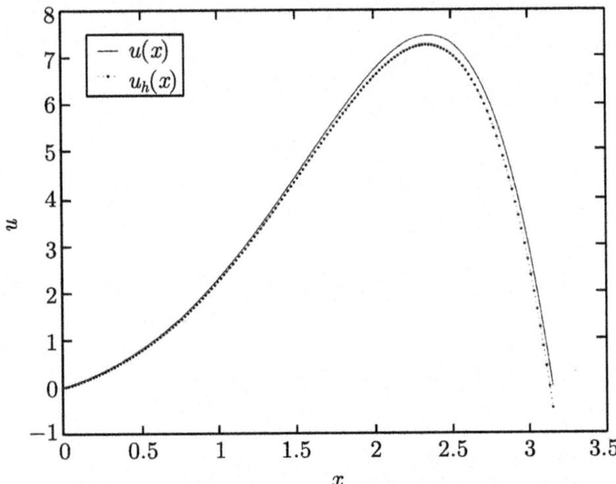

Fig. 1.5 (Example 1.4) The curves of the exact and numerical solutions plotted with the step size $h = \pi/160$

As we see from Table 1.7, when the step size is reduced by half, the maximum error is reduced by half. It displays that $E_\infty(h) \approx ch$. Figure 1.5 shows the curves of the exact and numerical solutions with the step size $h = \pi/160$. Figure 1.6 illustrates the numerical error curves with different step sizes.

The difference scheme (1.57) can be written as the following matrix-vector form:

$$
\begin{pmatrix}
1 + h\lambda_1 + \dfrac{h^2}{2}q(x_0) & -1 & & & \\
-1 & 2 + h^2 q(x_1) & -1 & & \\
& \ddots & \ddots & \ddots & \\
& & -1 & 2 + h^2 q(x_{m-1}) & -1 \\
& & & -1 & 1 + h\lambda_2 + \dfrac{h^2}{2}q(x_m)
\end{pmatrix}
$$

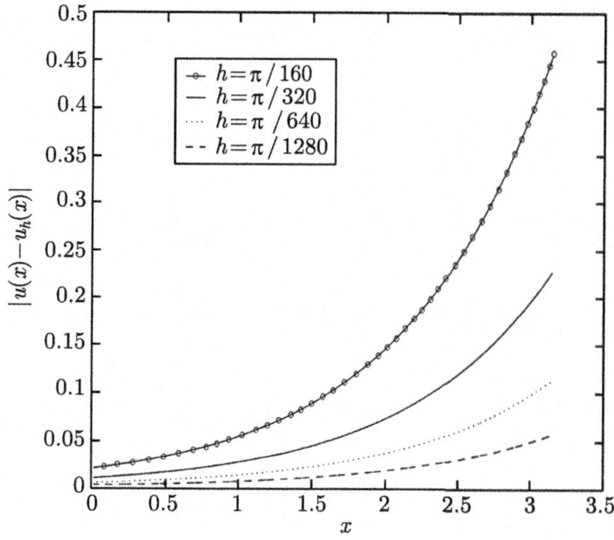

Fig. 1.6 (Example 1.4) The numerical error curves plotted with different step sizes

$$
\cdot
\begin{pmatrix}
u_0 \\
u_1 \\
\vdots \\
u_{m-1} \\
u_m
\end{pmatrix}
=
\begin{pmatrix}
h\alpha + \dfrac{h^2}{2} f(x_0) \\
h^2 f(x_1) \\
\vdots \\
h^2 f(x_{m-1}) \\
h\beta + \dfrac{h^2}{2} f(x_m)
\end{pmatrix},
$$

which can be solved by the double sweep method.

Example 1.5 Apply the difference scheme (1.57) to solve the two-point boundary value problem (1.60) in Example 1.4.

Divide $[0, \pi]$ into m equal subintervals and denote $h = \pi/m$, $x_i = ih$, $0 \leqslant i \leqslant m$. The difference scheme reads

$$
\begin{cases}
-\dfrac{u_{i-1} - 2u_i + u_{i+1}}{h^2} + u_i = \mathrm{e}^{x_i}(\sin x_i - 2\cos x_i), \quad 1 \leqslant i \leqslant m-1, \\[2mm]
-\dfrac{u_1 - u_0}{h} + \dfrac{h}{2} u_0 = -1 - h, \quad \dfrac{u_m - u_{m-1}}{h} + \dfrac{h}{2} u_m = (h-1)\mathrm{e}^{\pi}.
\end{cases}
$$

Table 1.8 (Example 1.5) The exact and numerical solutions at four node points with different step sizes

h	x			
	$\pi/5$	$2\pi/5$	$3\pi/5$	$4\pi/5$
$\pi/10$	1.113391	3.358955	6.339003	7.501848
$\pi/20$	1.104682	3.346009	6.282778	7.318382
$\pi/40$	1.102504	3.342720	6.268497	7.271918
$\pi/80$	1.101959	3.341894	6.264913	7.260264
$\pi/160$	1.101778	3.341687	6.264016	7.257348
ES	1.101778	3.341619	6.263717	7.256376

Table 1.9 (Example 1.5) The absolute values of numerical errors at four node points and maximum errors of numerical solutions with different step sizes

h	x				$E_\infty(h)$	$E_\infty(2h)/E_\infty(h)$
	$\pi/5$	$2\pi/5$	$3\pi/5$	$4\pi/5$		
$\pi/10$	1.161e−2	1.734e−2	7.529e−2	2.455e−1	6.041e−1	
$\pi/20$	2.905e−3	4.391e−3	1.906e−2	6.201e−2	1.524e−1	3.964
$\pi/40$	7.263e−4	1.101e−3	4.780e−3	1.554e−2	3.818e−2	3.992
$\pi/80$	1.816e−4	2.755e−4	1.196e−3	3.888e−3	9.550e−3	3.998
$\pi/160$	4.539e−5	6.890e−5	2.991e−4	9.722e−4	2.388e−3	3.999

Table 1.8 gives the exact and numerical solutions at four points with different step sizes. Table 1.9 lists the absolute values of numerical errors $|u(x_i) - u_i|$ and the maximum errors defined by

$$E_\infty(h) = \max_{0 \leqslant i \leqslant m} |u(x_i) - u_i|$$

with different step sizes.

As we see from Table 1.9, when the step size is reduced by half, the maximum error is reduced to a quarter of the original. This indicates that $E_\infty(h) \approx ch^2$. Figure 1.7 shows the curves of the exact solutions and the numerical solutions with the step size $h = \pi/10$. Figure 1.8 presents the numerical error curves with different step sizes.

1.4   Summary and Extension

In Sect. 1.1, the closed-form solution to the two-point boundary value problem is derived by integrating twice. Then, several fundamental equalities and inequalities are introduced. The energy method is employed to analyze a priori estimate of the solution for the two-point homogeneous boundary value problem of ordinary differential equations. Additionally, the energy method serves as a fundamental tool for analyzing the numerical methods for solving these differential equations.

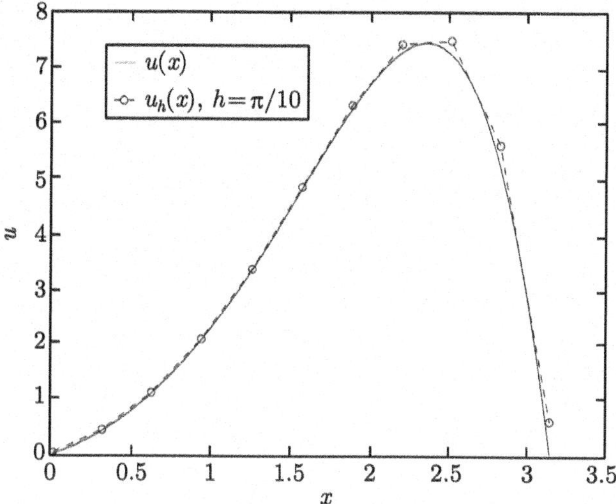

Fig. 1.7 (Example 1.5) The curves of the exact and numerical solutions with the step size $h = \pi/10$

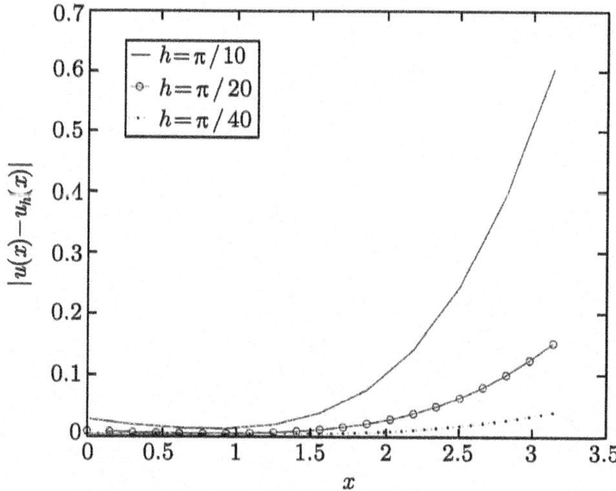

Fig. 1.8 (Example 1.5) The numerical error curves with different step sizes

In Sect. 1.2, the difference scheme (1.21) for solving the two-point boundary value problem (1.1) is developed. The uniqueness of the solution is proved, and numerical analyses along with numerical examples are provided. The maximum principle and the energy method are employed to establish the convergence and stability of the difference scheme (1.21). The two equalities and five inequalities introduced in Lemma 1.4 are essential for the subsequent energy analysis. Most of the results in Lemma 1.4 are adapted from [1].

Grid partition, grid functions, local truncation error, convergence, and stability are fundamental concepts in the difference method. To improve the accuracy of numerical solutions, there are two primary approaches: one is directly constructing a high-order accurate difference scheme to reduce the truncation error and the other is applying Richardson extrapolation procedure.

For the derivative boundary value problem, we have only derived the difference scheme. Readers can prove the uniqueness, convergence, and stability of the difference scheme by applying the methods described in Sect. 1.2. From the numerical results in Example 1.4, we can infer that the difference scheme (1.56) is convergent with convergence order one, and (1.57) is convergent with convergence order two. Interested readers are encouraged to work out the proofs.

For a variable-coefficient differential equation

$$-u'' + p(x)u' + q(x)u = f(x),$$

the corresponding difference scheme can be written as

$$-\delta_x^2 u_i + p(x_i) \cdot \frac{u_{i+1} - u_{i-1}}{2h} + q(x_i)u_i = f(x_i).$$

For the differential equation in a conservative form

$$-(a(x)u')' + q(x)u = f(x),$$

the difference scheme can be established as

$$-\frac{1}{h}\left[a(x_{i+\frac{1}{2}})\delta_x u_{i+\frac{1}{2}} - a(x_{i-\frac{1}{2}})\delta_x u_{i-\frac{1}{2}}\right] + q(x_i)u_i = f(x_i),$$

where $x_{i+\frac{1}{2}} = \frac{1}{2}(x_i + x_{i+1})$.

1.5   Exercise

1.1 Let $f \in C^3[c-h, c+h]$. Prove

$$\frac{f(c+h) - f(c-h)}{2h} = f'(c) + \frac{h^2}{6}f'''(\xi), \quad c-h < \xi < c+h;$$

$$\frac{f(c+h) + f(c-h)}{2} = f(c) + \frac{h^2}{2}\int_0^1 \left[f''(c+sh) + f''(c-sh)\right](1-s)ds;$$

$$\frac{f(c+h) - f(c-h)}{2h} = f'(c) + \frac{h^2}{4}\int_0^1 \left[f'''(c+sh) + f'''(c-sh)\right](1-s)^2 ds;$$

$$f''(c) = \frac{2}{h}\left[\frac{f(c+h) - f(c)}{h} - f'(c)\right] - h\int_0^1 f'''(c+sh)(1-s)^2 \mathrm{d}s;$$

$$\frac{f(c+h) - f(c-h)}{2h} = \frac{1}{2}\int_0^1 \left[f'(c+sh) + f'(c-sh)\right]\mathrm{d}s;$$

$$\frac{1}{h^2}\left[f(c+h) - 2f(c) + f(c-h)\right] = \int_0^1 \left[f''(c+sh) + f''(c-sh)\right](1-s)\mathrm{d}s.$$

1.2 Compute the two-point boundary value problem

$$\begin{cases} -u'' = 6x, & 0 < x < 1, \\ u(0) = 0, & u(1) = 1 \end{cases}$$

by the difference scheme (1.21). Take $h = 1/4$, calculate the numerical solutions at the points $1/4, 1/2, 3/4$, and compare them with the exact solution. Explain the observed phenomenon. The exact solution is $u(x) = x(2 - x^2)$.

1.3 Let $u = \{u_i \mid 0 \leqslant i \leqslant m\} \in \mathcal{U}_h$ and $v = \{v_i \mid 0 \leqslant i \leqslant m\} \in \mathcal{U}_h$. Try to prove the Cauchy-Schwarz inequality

$$|(u, v)| \leqslant \|u\| \cdot \|v\|.$$

1.4 Let $u = \{u_i \mid 0 \leqslant i \leqslant m\} \in \mathcal{U}_h$ and $u_0 = 0$. Prove

$$\|u\|_\infty \leqslant \sqrt{L}\,|u|_1, \qquad \|u\| \leqslant \frac{L}{\sqrt{2}}\,|u|_1.$$

1.5 Let $u = \{u_i \mid 0 \leqslant i \leqslant m\} \in \mathcal{U}_h$. Prove

$$\|u\|_\infty^2 \leqslant 2\max\{u_0^2, u_m^2\} + \frac{L}{2}|u|_1^2.$$

1.6 Let $u = \{u_i \mid 0 \leqslant i \leqslant m\} \in \mathcal{U}_h$. Denote

$$\|\delta_x v\|_\infty = \max_{0 \leqslant i \leqslant m-1} |\delta_x v_{i+\frac{1}{2}}|, \qquad \|\delta_x v\| = \sqrt{h\sum_{i=0}^{m-1}(\delta_x v_{i+\frac{1}{2}})^2},$$

$$\|\delta_x^2 v\| = \sqrt{h\sum_{i=1}^{m-1}(\delta_x^2 v_i)^2}.$$

For an arbitrary $\epsilon > 0$, prove

$$\|\delta_x v\|_\infty^2 \leqslant \epsilon \|\delta_x^2 v\|^2 + \left(\frac{1}{L} + \frac{1}{\epsilon} \right) \|\delta_x v\|^2.$$

1.7   Compute the two-point boundary value problem

$$\begin{cases} -u'' + u = \left(\dfrac{1}{20} x^4 - 6 \right) x, & 0 < x < 1, \\[2mm] u(0) = 0, \quad u(1) = \dfrac{21}{20} \end{cases}$$

by the compact difference scheme (1.52). Take $h = 1/4$, calculate the numerical solutions at the points $1/4, 1/2, 3/4$, and compare them with the exact solution. Explain the observed phenomenon. The exact solution is $u(x) = \frac{1}{20} x^5 + x^3$.

1.8   Compute the two-point boundary value problem

$$\begin{cases} -u'' = 0, & 0 < x < 1, \\[2mm] -u'(0) + u(0) = -1, & u'(1) = 4. \end{cases}$$

by the difference scheme (1.56). Take $h = 1/4$, calculate numerical solutions at the points $1/4, 1/2, 3/4$, and compare them with the exact solution. Explain the observed phenomenon. The exact solution is $u(x) = 4x + 3$.

1.9   Compute the two-point boundary value problem

$$\begin{cases} -u'' + \left(x - \dfrac{1}{2} \right)^2 u = \left(x^2 - x + \dfrac{5}{4} \right) \sin x, & 0 < x < \dfrac{\pi}{2}, \\[2mm] u(0) = 0, \quad u\left(\dfrac{\pi}{2} \right) = 1 \end{cases}$$

by the difference scheme (1.21). Fill in Tables 1.10 and 1.11, and draw the curves of the exact solution, the numerical solution, and the numerical error. The exact solution is $u(x) = \sin x$.

1.10   Compute the two-point boundary value problem

$$\begin{cases} -u'' + (1 + \sin x)u = e^x \sin x, & 0 < x < 1, \\[2mm] -u'(0) + u(0) = 0, \quad u'(1) + 2u(1) = 3e \end{cases}$$

by the difference schemes (1.56) and (1.57). Fill in Tables 1.12 and 1.13, and draw the curves of the exact solution, the numerical solution, and the numerical error. The exact solution is $u(x) = e^x$.

Table 1.10 (Exercise 1.9) The exact and numerical solutions at part of points with different step sizes

h	x						
	$\pi/16$	$2\pi/16$	$3\pi/16$	$4\pi/16$	$5\pi/16$	$6\pi/16$	$7\pi/16$
$\pi/16$							
$\pi/32$							
$\pi/64$							
$\pi/128$							
ES							

Table 1.11 (Exercise 1.9) The absolute values of numerical errors and the maximum errors of numerical solutions at part of points with different step sizes

h	x							$E_\infty(h)$	$\frac{E_\infty(2h)}{E_\infty(h)}$
	$\pi/16$	$2\pi/16$	$3\pi/16$	$4\pi/16$	$5\pi/16$	$6\pi/16$	$7\pi/16$		
$\pi/16$									
$\pi/32$									
$\pi/64$									
$\pi/128$									

Table 1.12 (Exercise 1.10) The exact and numerical solutions at part of points

h	x		
	1/4	2/4	3/4
1/4			
1/8			
1/16			
1/32			
1/64			
ES			

Table 1.13 (Exercise 1.10) The absolute values of numerical errors and the maximum errors of numerical solutions at part of points with different step sizes

h	x			$E_\infty(h)$	$E_\infty(2h)/E_\infty(h)$
	1/4	2/4	3/4		
1/4					
1/8					
1/16					
1/32					
1/64					

Reference

1. Samarskiĭ, A.A., Andreev, V.B.: Difference Methods for Elliptic Equations. Nauka, Moscow (1976)

Chapter 2
Finite Difference Methods for Elliptic Equations

The description of various steady-state physical processes often leads to elliptic partial differential equations, such as those encountered in stationary heat conduction, diffusion problems, current distribution in conductors, electrostatics, magnetostatics, elastic theory, and percolation theory, among others.

Analytical solutions to elliptic boundary value problems can be obtained only under certain special conditions. Even when analytical solutions are possible, the calculations are typically quite complex. As a result, numerical methods are commonly employed to solve these problems.

Two representative examples of elliptic differential equations include the two-dimensional Poisson equation

$$-\left(\frac{\partial^2 u}{\partial x^2} + \frac{\partial^2 u}{\partial y^2}\right) = f(x, y), \quad (x, y) \in \Omega$$

and the Laplace equation

$$-\left(\frac{\partial^2 u}{\partial x^2} + \frac{\partial^2 u}{\partial y^2}\right) = 0, \quad (x, y) \in \Omega,$$

where $\Omega \subset \mathbf{R}^2$ is a bounded domain.

The boundary value conditions commonly fall into one of the three categories:

(1) The first kind of the boundary value condition (Dirichlet boundary value condition) $u\big|_{\Gamma} = \varphi(x, y)$;
(2) The second kind of the boundary value condition (Neumann boundary value condition) $\frac{\partial u}{\partial n}\big|_{\Gamma} = \varphi_1(x, y)$;

© Science Press 2026

Z.-Z. Sun et al., *Numerical Solutions to Partial Differential Equations with Finite Difference Methods*, Springer Asia Pacific Mathematics Series 9,
https://doi.org/10.1007/978-981-95-5563-5_2

(3) The third kind of the boundary value condition (Robin boundary value condition) $\left[\frac{\partial u}{\partial n} + \lambda(x, y)u\right]\Big|_{\Gamma} = \psi(x, y),$

where Γ denotes the boundary of Ω, and $\boldsymbol{n}$ is the unit outward normal vector of Γ and $\lambda(x, y)\big|_{\Gamma} \not\equiv 0$. The second and third kind of the boundary value conditions are collectively referred to as the **derivative boundary** value conditions.

In this chapter, we focus on the two-dimensional Poisson equation and explore several finite difference methods for its numerical solution. The discussion centers around the following three key aspects:

- How to choose an appropriate grid to discretize the differential equation into a system of difference equations?
- The existence and uniqueness of the difference solution, as well as the methods for solving the resulting system of difference equations;
- Whether the difference solution converges to the exact solution of the differential equation as the grid step sizes in each direction tend to zero.

2.1　The Dirichlet Boundary Value Problem

Consider the Dirichlet boundary value problem for the two-dimensional Poisson equation:

$$\begin{cases} -\Delta u = f(x, y), & (x, y) \in \Omega, & (2.1a) \\ u = \varphi(x, y), & (x, y) \in \Gamma, & (2.1b) \end{cases}$$

where $\Delta = \frac{\partial^2}{\partial x^2} + \frac{\partial^2}{\partial y^2}$ is the Laplace operator. For brevity, we restrict our attention to a rectangular domain $\Omega = (0, L_1) \times (0, L_2)$ with Γ the boundary. We begin by introducing a priori estimate for the solution of the homogeneous Dirichlet boundary value problem (i.e., a boundary value problem of the first kind with homogeneous boundary conditions).

Denote

$$\kappa = \left(\frac{6}{L_1^2} + \frac{6}{L_2^2}\right)^{-\frac{1}{2}}. \tag{2.2}$$

It is easy to see

$$\kappa \leqslant \frac{1}{6}\sqrt{3L_1 L_2}. \tag{2.3}$$

Lemma 2.1 *Denote $\Omega = (0, L_1) \times (0, L_2)$ with Γ the boundary of Ω. Suppose $w \in C^1(\bar{\Omega})$ and when $(x, y) \in \Gamma$, $w(x, y) = 0$. Then we have*

$$\|w\| \leqslant \kappa |w|_1, \tag{2.4}$$

where

$$\|w\|^2 = \iint_\Omega w^2(x, y)\mathrm{d}x\mathrm{d}y, \quad |w|_1^2 = \iint_\Omega \left[w_x^2(x, y) + w_y^2(x, y) \right]\mathrm{d}x\mathrm{d}y.$$

Proof By means of Lemma 1.1, it yields

$$\int_0^{L_1} w^2(x, y)\mathrm{d}x \leqslant \frac{L_1^2}{6} \int_0^{L_1} w_x^2(x, y)\mathrm{d}x, \quad y \in (0, L_2),$$

$$\int_0^{L_2} w^2(x, y)\mathrm{d}y \leqslant \frac{L_2^2}{6} \int_0^{L_2} w_y^2(x, y)\mathrm{d}y, \quad x \in (0, L_1),$$

or, to put them in another way,

$$\frac{6}{L_1^2} \int_0^{L_1} w^2(x, y)\mathrm{d}x \leqslant \int_0^{L_1} w_x^2(x, y)\mathrm{d}x, \quad y \in (0, L_2), \tag{2.5}$$

$$\frac{6}{L_2^2} \int_0^{L_2} w^2(x, y)\mathrm{d}y \leqslant \int_0^{L_2} w_y^2(x, y)\mathrm{d}y, \quad x \in (0, L_1). \tag{2.6}$$

Integrating both sides of (2.5) with respect to y on $(0, L_2)$, integrating both sides of (2.6) with respect to x on $(0, L_1)$, and then summing up the results produce (2.4). $\qquad\qquad\square$

Lemma 2.2 *Denote $\Omega = (0, L_1) \times (0, L_2)$ with Γ the boundary of Ω. Suppose $w \in C^2(\Omega)$ and when $(x, y) \in \Gamma$, $w(x, y) = 0$. Then we have*

$$\|w\|_\infty \leqslant \frac{1}{12} \sqrt{3(\sqrt{2} + 1)L_1 L_2} \, \|\Delta w\|.$$

Proof Let

$$|w(x^*, y^*)| = \|w\|_\infty.$$

In combination of Lemma 1.1 and noticing $w(0, y) = w(L_1, y) = 0$, $w(x, 0) = w(x, L_2) = 0$ and $w_x(x, 0) = w_x(x, L_2) = 0$, we have

$$\|w\|_\infty^2 = w^2(x^*, y^*)$$

$$\leqslant \epsilon_1 \int_0^{L_1} w_x^2(x, y^*)\mathrm{d}x + \frac{1}{4\epsilon_1} \int_0^{L_1} w^2(x, y^*)\mathrm{d}x$$

$$\leqslant \epsilon_1 \int_0^{L_1} \left[\epsilon_2 \int_0^{L_2} w_{xy}^2(x, y)\mathrm{d}y + \frac{1}{4\epsilon_2} \int_0^{L_2} w_x^2(x, y)\mathrm{d}y \right] \mathrm{d}x$$

$$+ \frac{1}{4\epsilon_1} \int_0^{L_1} \left[\epsilon_2 \int_0^{L_2} w_y^2(x, y)\mathrm{d}y + \frac{1}{4\epsilon_2} \int_0^{L_2} w^2(x, y)\mathrm{d}y \right] \mathrm{d}x$$

$$= \epsilon_1\epsilon_2 \|w_{xy}\|^2 + \frac{\epsilon_1}{4\epsilon_2} \|w_x\|^2 + \frac{\epsilon_2}{4\epsilon_1} \|w_y\|^2 + \frac{1}{16\epsilon_1\epsilon_2} \|w\|^2.$$

In addition, it is easy to know

$$\|\Delta w\|^2 = \|w_{xx}\|^2 + \|w_{yy}\|^2 + 2\|w_{xy}\|^2,$$

which implies

$$\|w_{xy}\|^2 \leqslant \frac{1}{2} \|\Delta w\|^2.$$

Thus,

$$\|w\|_\infty^2 \leqslant \frac{1}{2}\epsilon_1\epsilon_2 \|\Delta w\|^2 + \frac{\epsilon_1}{4\epsilon_2} \|w_x\|^2 + \frac{\epsilon_2}{4\epsilon_1} \|w_y\|^2 + \frac{1}{16\epsilon_1\epsilon_2} \|w\|^2.$$

Taking $\epsilon_1 = \epsilon_2 = \sqrt{2\epsilon}$, we have

$$\|w\|_\infty^2 \leqslant \epsilon \|\Delta w\|^2 + \frac{1}{4} |w|_1^2 + \frac{1}{32\epsilon} \|w\|^2. \tag{2.7}$$

With the help of Lemma 2.1, it yields

$$\|w\| \leqslant \kappa |w|_1.$$

Further, based on

$$|w|_1^2 = -(\Delta w, w) \leqslant \|\Delta w\| \cdot \|w\|,$$

we have

$$|w|_1 \leqslant \kappa \|\Delta w\| \tag{2.8}$$

and

$$\|w\| \leqslant \kappa^2 \|\Delta w\|. \tag{2.9}$$

Substituting (2.8) and (2.9) into (2.7), one has

$$\|w\|_\infty^2 \leqslant \epsilon \|\Delta w\|^2 + \frac{1}{4}\kappa^2 \|\Delta w\|^2 + \frac{1}{32\epsilon}\kappa^4 \|\Delta w\|^2.$$

Taking $\epsilon = \frac{\sqrt{2}}{8}\kappa^2$, it yields

$$\|w\|_\infty^2 \leqslant \left(\frac{\sqrt{2}}{4} + \frac{1}{4}\right)\kappa^2 \|\Delta w\|^2.$$

Therefore,

$$\|w\|_\infty \leqslant \frac{\kappa}{2}\sqrt{\sqrt{2}+1}\,\|\Delta w\| \leqslant \frac{1}{12}\sqrt{3(\sqrt{2}+1)L_1 L_2}\,\|\Delta w\|.$$

$\square$

Theorem 2.1 *Let $\{v(x, y) \mid (x, y) \in \bar{\Omega}\}$ be the solution of the Dirichlet boundary value problem of the elliptic equation*

$$\begin{cases} -\Delta v = f(x, y), & (x, y) \in \Omega, & \text{(2.10a)} \\ v(x, y) = 0, & (x, y) \in \Gamma, & \text{(2.10b)} \end{cases}$$

where $\Omega = (0, L_1) \times (0, L_2)$ and Γ is the boundary of Ω. Then we have

$$\|v\|_\infty \leqslant \frac{1}{12}\sqrt{3(\sqrt{2}+1)L_1 L_2}\,\|f\|.$$

Proof Taking the inner product of (2.10a) with $-\Delta v$, we get

$$\|\Delta v\|^2 = -(\Delta v, f) \leqslant \|\Delta v\| \cdot \|f\|,$$

Hence,

$$\|\Delta v\| \leqslant \|f\|.$$

By means of Lemma 2.2, we have

$$\|v\|_\infty \leqslant \frac{1}{12}\sqrt{3(\sqrt{2}+1)L_1 L_2}\,\|\Delta v\| \leqslant \frac{1}{12}\sqrt{3(\sqrt{2}+1)L_1 L_2}\,\|f\|.$$

$\square$

2.2 The Five-Point Difference Scheme

2.2.1 Derivation of the Difference Scheme

Divide $[0, L_1]$ into m_1 equal subintervals and denote $h_1 = L_1/m_1$, $x_i = ih_1$, $0 \leqslant i \leqslant m_1$. Divide $[0, L_2]$ into m_2 equal subintervals and denote $h_2 = L_2/m_2$, $y_j = jh_2$, $0 \leqslant j \leqslant m_2$. h_1 and h_2 denote the step sizes in x and y directions, respectively. Divide the region $\bar{\Omega}$ into mn small rectangles with two clusters of parallel lines

$$x = x_i, \quad 0 \leqslant i \leqslant m_1;$$

$$y = y_j, \quad 0 \leqslant j \leqslant m_2.$$

The intersection points (x_i, y_j) of two clusters of lines are called grid points as shown in Fig. 2.1.

Denote

$$\Omega_h = \left\{ (x_i, y_j) \mid 0 \leqslant i \leqslant m_1, \ 0 \leqslant j \leqslant m_2 \right\}.$$

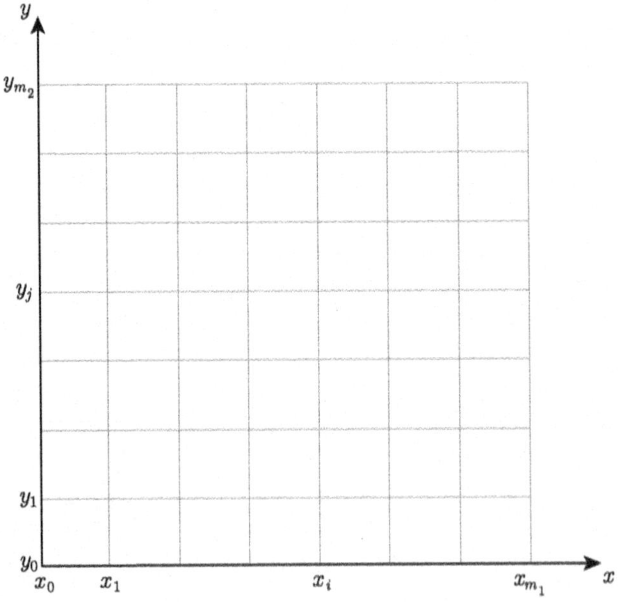

Fig. 2.1 Rectangular grid subdivision

Next, define the set of interior grid points by

$$\overset{\circ}{\Omega}_h = \left\{ (x_i, y_j) \mid 1 \leqslant i \leqslant m_1 - 1, \ 1 \leqslant j \leqslant m_2 - 1 \right\}.$$

The set of boundary grid points is denoted by

$$\Gamma_h = \Omega_h \setminus \overset{\circ}{\Omega}_h.$$

The boundary grid points can be further classified into two categories: a point located at one of the four vertices of Ω is called a corner point, while the remaining points are called interior grid points on the boundary. Clearly, we have the relationship $\Omega_h = \overset{\circ}{\Omega}_h \cup \Gamma_h$. For convenience, we define the following sets:

$$\omega = \left\{ (i, j) \mid (x_i, y_j) \in \overset{\circ}{\Omega}_h \right\}, \quad \gamma = \left\{ (i, j) \mid (x_i, y_j) \in \Gamma_h \right\}, \quad \bar{\omega} = \omega \cup \gamma.$$

Denote the following sets:

$$\mathcal{V}_h = \left\{ v \mid v = \{v_{ij} \mid (i, j) \in \bar{\omega}\} \text{ is the grid function on } \Omega_h \right\},$$

$$\overset{\circ}{\mathcal{V}}_h = \left\{ v \mid v \in \mathcal{V}_h; \text{ when } (i, j) \in \gamma, \ v_{ij} = 0 \right\}.$$

Let $v = \{v_{ij} \mid 0 \leqslant i \leqslant m_1, \ 0 \leqslant j \leqslant m_2\} \in \mathcal{V}_h$ and introduce the following notation:

$$\delta_x v_{i-\frac{1}{2},j} = \frac{1}{h_1}(v_{ij} - v_{i-1,j}), \quad \delta_x^2 v_{ij} = \frac{1}{h_1}\left(\delta_x v_{i+\frac{1}{2},j} - \delta_x v_{i-\frac{1}{2},j}\right),$$

$$\delta_y v_{i,j-\frac{1}{2}} = \frac{1}{h_2}(v_{ij} - v_{i,j-1}), \quad \delta_y^2 v_{ij} = \frac{1}{h_2}\left(\delta_y v_{i,j+\frac{1}{2}} - \delta_y v_{i,j-\frac{1}{2}}\right),$$

$$\Delta_h v_{ij} = \delta_x^2 v_{ij} + \delta_y^2 v_{ij}, \quad \|v\|_\infty = \max_{0 \leqslant i \leqslant m_1, 0 \leqslant j \leqslant m_2} |v_{ij}|,$$

where $\|v\|_\infty$ is called the maximum norm of v.

Considering the boundary value problem (2.1) at the grid point (x_i, y_j), we have

$$\begin{cases} -\left[\dfrac{\partial^2 u}{\partial x^2}(x_i, y_j) + \dfrac{\partial^2 u}{\partial y^2}(x_i, y_j)\right] = f(x_i, y_j), & (i, j) \in \omega, \quad (2.11a) \\[2mm] u(x_i, y_j) = \varphi(x_i, y_j), & (i, j) \in \gamma. \quad\quad\quad (2.11b) \end{cases}$$

Define the grid function on Ω_h by

$$U = \left\{ U_{ij} \mid (i, j) \in \bar{\omega} \right\},$$

where

$$U_{ij} = u(x_i, y_j), \quad (i, j) \in \bar{\omega}.$$

By using Lemma 1.2, we have

$$\frac{\partial^2 u}{\partial x^2}(x_i, y_j) = \frac{1}{h_1^2}\left[u(x_{i-1}, y_j) - 2u(x_i, y_j) + u(x_{i+1}, y_j) \right] - \frac{h_1^2}{12}\frac{\partial^4 u(\xi_{ij}, y_j)}{\partial x^4}$$

$$= \delta_x^2 U_{ij} - \frac{h_1^2}{12}\frac{\partial^4 u(\xi_{ij}, y_j)}{\partial x^4}, \quad x_{i-1} < \xi_{ij} < x_{i+1};$$

$$\frac{\partial^2 u}{\partial y^2}(x_i, y_j) = \frac{1}{h_2^2}\left[u(x_i, y_{j-1}) - 2u(x_i, y_j) + u(x_i, y_{j+1}) \right] - \frac{h_2^2}{12}\frac{\partial^4 u(x_i, \eta_{ij})}{\partial y^4}$$

$$= \delta_y^2 U_{ij} - \frac{h_2^2}{12}\frac{\partial^4 u(x_i, \eta_{ij})}{\partial y^4}, \quad y_{j-1} < \eta_{ij} < y_{j+1}.$$

Plugging the above two equalities into (2.11a) and noticing (2.11b), we arrive at

$$\begin{cases} -\Delta_h U_{ij} = f(x_i, y_j) - \frac{h_1^2}{12}\frac{\partial^4 u(\xi_{ij}, y_j)}{\partial x^4} - \frac{h_2^2}{12}\frac{\partial^4 u(x_i, \eta_{ij})}{\partial y^4}, \quad (i, j) \in \omega, \\ U_{ij} = \varphi(x_i, y_j), \quad (i, j) \in \gamma. \end{cases} \tag{2.12}$$

Omitting the small term

$$(R_1)_{ij} = -\frac{h_1^2}{12}\frac{\partial^4 u(\xi_{ij}, y_j)}{\partial x^4} - \frac{h_2^2}{12}\frac{\partial^4 u(x_i, \eta_{ij})}{\partial y^4} \tag{2.13}$$

in (2.12) and replacing U_{ij} with u_{ij}, a difference scheme for solving (2.1) reads

$$\begin{cases} -\Delta_h u_{ij} = f(x_i, y_j), \quad (i, j) \in \omega, & \tag{2.14a} \\ u_{ij} = \varphi(x_i, y_j), \quad (i, j) \in \gamma. & \tag{2.14b} \end{cases}$$

Here, $(R_1)_{ij}$ represents the local truncation error of the difference scheme (2.14), which indicates the level of approximation of the difference equation (2.14a) to the differential equation (2.11a). In other words, $(R_1)_{ij}$ is the difference between both sides of (2.14a) when the approximate solution u_{ij} is replaced by the exact solution U_{ij}, i.e.,

$$(R_1)_{ij} = -\Delta_h U_{ij} - f(x_i, y_j).$$

Denote

$$M_4 = \max \left\{ \max_{(x,y)\in \bar{\Omega}} \left| \frac{\partial^4 u(x,\,y)}{\partial x^4} \right|, \ \max_{(x,y)\in \bar{\Omega}} \left| \frac{\partial^4 u(x,\,y)}{\partial y^4} \right| \right\}. \tag{2.15}$$

It is easy to know from (2.13) and (2.15) that

$$\left| (R_1)_{ij} \right| \leqslant \frac{M_4}{12}(h_1^2 + h_2^2), \quad (i,\,j) \in \omega. \tag{2.16}$$

2.2.2   Existence of the Difference Solution

Theorem 2.2 *There is a unique solution to the difference scheme (2.14).*

Proof   Since the difference scheme (2.14) is linear, it suffices to consider its homogeneous one

$$\begin{cases} -\Delta_h u_{ij} = 0, & (i,\,j) \in \omega, & (2.17\text{a}) \\[2mm] u_{ij} = 0, & (i,\,j) \in \gamma. & (2.17\text{b}) \end{cases}$$

Assume $\|u\|_\infty = M > 0$. It follows from (2.17b) that there is an $(i_0,\,j_0) \in \omega$ satisfying $|u_{i_0,j_0}| = M$, and at least one of $|u_{i_0-1,j_0}|$, $|u_{i_0+1,j_0}|$, $|u_{i_0,j_0-1}|$, and $|u_{i_0,j_0+1}|$ is strictly less than M. Considering the difference equation (2.17a) with $(i,\,j) = (i_0,\,j_0)$, we have

$$\left(\frac{2}{h_1^2} + \frac{2}{h_2^2} \right) u_{i_0,j_0} = \frac{1}{h_1^2}(u_{i_0-1,j_0} + u_{i_0+1,j_0}) + \frac{1}{h_2^2}(u_{i_0,j_0-1} + u_{i_0,j_0+1}).$$

Taking the absolute values on both sides of the above inequality, we obtain

$$\left(\frac{2}{h_1^2} + \frac{2}{h_2^2} \right) M \leqslant \frac{1}{h_1^2}(|u_{i_0-1,j_0}| + |u_{i_0+1,j_0}|) + \frac{1}{h_2^2}(|u_{i_0,j_0-1}| + |u_{i_0,j_0+1}|)$$

$$< \left(\frac{2}{h_1^2} + \frac{2}{h_2^2} \right) M,$$

which contradicts the assumption condition $M > 0$. Thus, $M = 0$, which implies that the difference scheme (2.14) is uniquely solvable. $\qquad\square$

2.2.3　Implementation of the Difference Scheme and Numerical Examples

The difference scheme (2.14) is a system of linear equations in the unknown $\left\{u_{ij}\mid 1\leqslant i\leqslant m_1-1,\ 1\leqslant j\leqslant m_2-1\right\}$. The difference equation (2.14a) can be rewritten as

$$-\frac{1}{h_2^2}u_{i,j-1}-\frac{1}{h_1^2}u_{i-1,j}+2\left(\frac{1}{h_1^2}+\frac{1}{h_2^2}\right)u_{ij}-\frac{1}{h_1^2}u_{i+1,j}-\frac{1}{h_2^2}u_{i,j+1}$$

$$=f(x_i,y_j),\quad (i,j)\in\omega. \tag{2.18}$$

Denote

$$\boldsymbol{u}_j=\begin{pmatrix} u_{1j}\\ u_{2j}\\ \vdots\\ u_{m_1-1,j}\end{pmatrix},\quad 0\leqslant j\leqslant m_2.$$

Noticing (2.14b), Eq. (2.18) can be further rewritten as

$$\boldsymbol{D}\boldsymbol{u}_{j-1}+\boldsymbol{C}\boldsymbol{u}_j+\boldsymbol{D}\boldsymbol{u}_{j+1}=\boldsymbol{f}_j,\quad 1\leqslant j\leqslant m_2-1, \tag{2.19}$$

where

$$\boldsymbol{C}=\begin{pmatrix} 2\left(\frac{1}{h_1^2}+\frac{1}{h_2^2}\right) & -\frac{1}{h_1^2} & & & \\ -\frac{1}{h_1^2} & 2\left(\frac{1}{h_1^2}+\frac{1}{h_2^2}\right) & -\frac{1}{h_1^2} & & \\ & \ddots & \ddots & \ddots & \\ & & -\frac{1}{h_1^2} & 2\left(\frac{1}{h_1^2}+\frac{1}{h_2^2}\right) & -\frac{1}{h_1^2} \\ & & & -\frac{1}{h_1^2} & 2\left(\frac{1}{h_1^2}+\frac{1}{h_2^2}\right) \end{pmatrix},$$

$$\boldsymbol{D}=\begin{pmatrix} -\frac{1}{h_2^2} & & & \\ & -\frac{1}{h_2^2} & & \\ & & \ddots & \\ & & & -\frac{1}{h_2^2} \\ & & & & -\frac{1}{h_2^2} \end{pmatrix},\quad \boldsymbol{f}_j=\begin{pmatrix} f(x_1,y_j)+\frac{1}{h_1^2}\varphi(x_0,y_j)\\ f(x_2,y_j)\\ \vdots\\ f(x_{m_1-2},y_j)\\ f(x_{m_1-1},y_j)+\frac{1}{h_1^2}\varphi(x_{m_1},y_j)\end{pmatrix}.$$

Equation (2.19) can be further written as

$$
\begin{pmatrix}
C & D & & & \\
D & C & D & & \\
 & \ddots & \ddots & \ddots & \\
 & & D & C & D \\
 & & & D & C
\end{pmatrix}
\begin{pmatrix}
u_1 \\
u_2 \\
\vdots \\
u_{m_2-2} \\
u_{m_2-1}
\end{pmatrix}
=
\begin{pmatrix}
f_1 - D u_0 \\
f_2 \\
\vdots \\
f_{m_2-2} \\
f_{m_2-1} - D u_{m_2}
\end{pmatrix}.
\tag{2.20}
$$

The coefficient matrix of the above system of linear equations is a tridiagonal block matrix, with no more than five nonzero elements in each row. A matrix that contains mostly zero elements is commonly referred to as a **sparse matrix**. Iterative methods are frequently employed to solve systems of linear equations with large, sparse coefficient matrices. It can be shown that the coefficient matrix of (2.20) is symmetric and positive definite. We now consider the following two iterative methods as examples.

Jacobi Iteration Method For $k = 0, 1, 2, \ldots$, calculate

$$
u_{ij}^{(k+1)} = \left[f(x_i, y_j) + \frac{1}{h_2^2} u_{i,j-1}^{(k)} + \frac{1}{h_1^2} u_{i-1,j}^{(k)} + \frac{1}{h_1^2} u_{i+1,j}^{(k)} + \frac{1}{h_2^2} u_{i,j+1}^{(k)} \right] \Big/ \left[2\Big(\frac{1}{h_1^2} + \frac{1}{h_2^2}\Big) \right],
$$

$$
i = 1, 2, \ldots, m_1 - 1; \quad j = 1, 2, \ldots, m_2 - 1.
$$

Gauss-Seidel Iteration Method For $k = 0, 1, 2, \ldots$, calculate

$$
u_{ij}^{(k+1)} = \left[f(x_i, y_j) + \frac{1}{h_2^2} u_{i,j-1}^{(k+1)} + \frac{1}{h_1^2} u_{i-1,j}^{(k+1)} + \frac{1}{h_1^2} u_{i+1,j}^{(k)} + \frac{1}{h_2^2} u_{i,j+1}^{(k)} \right] \Big/ \left[2\Big(\frac{1}{h_1^2} + \frac{1}{h_2^2}\Big) \right],
$$

$$
i = 1, 2, \ldots, m_1 - 1; \quad j = 1, 2, \ldots, m_2 - 1.
$$

Example 2.1 Apply the difference scheme (2.14) to compute the following problem:

$$
\begin{cases}
-\Delta u = (\pi^2 - 1)e^x \sin(\pi y), & 0 < x < 2, \quad 0 < y < 1, \\
u(0, y) = \sin(\pi y), \quad u(2, y) = e^2 \sin(\pi y), & 0 \leqslant y \leqslant 1, \\
u(x, 0) = 0, \quad u(x, 1) = 0, & 0 < x < 2.
\end{cases}
\tag{2.21}
$$

The exact solution is $u(x, y) = e^x \sin(\pi y)$.

We divide the interval $[0, 2]$ into m_1 equal subintervals and the interval $[0, 1]$ into m_2 equal subintervals. The Gauss-Seidel iterative method is used to solve the difference scheme (2.14) with an accuracy of $\|u^{(l+1)} - u^{(l)}\|_\infty \leqslant \frac{1}{2} \times 10^{-10}$.

Table 2.1 presents the exact and numerical solutions at five grid points, computed with different step sizes. Table 2.2 shows the absolute values of the differences

Table 2.1 (Example 2.1) The exact and numerical solutions at some grid points with different step sizes

(h_1, h_2)	(x, y)				
	$(1/2, 1/4)$	$(1, 1/4)$	$(3/2, 1/4)$	$(1/2, 1/2)$	$(1, 1/2)$
$(1/8, 1/8)$	1.179943	1.946264	3.198998	1.668692	2.752434
$(1/16, 1/16)$	1.169343	1.928138	3.176531	1.653700	2.726799
$(1/32, 1/32)$	1.166702	1.923620	3.170908	1.649965	2.720410
$(1/64, 1/64)$	1.166042	1.922492	3.169502	1.649032	2.718814
ES	1.165822	1.922116	3.169033	1.648721	2.718282

Table 2.2 (Example 2.1) The absolute values of numerical errors at some grid points with different step sizes

(h_1, h_2)	(x, y)				
	$(1/2, 1/4)$	$(1, 1/4)$	$(3/2, 1/4)$	$(1/2, 1/2)$	$(1, 1/2)$
$(1/8, 1/8)$	1.412e−2	2.415e−2	2.997e−2	1.997e−2	3.415e−2
$(1/16, 1/16)$	3.521e−3	6.023e−3	7.499e−3	4.979e−3	8.517e−3
$(1/32, 1/32)$	8.796e−4	1.505e−3	1.875e−3	1.244e−3	2.128e−3
$(1/64, 1/64)$	2.198e−4	3.761e−4	4.688e−4	3.109e−4	5.319e−4

Table 2.3 (Example 2.1) The maximum errors of numerical solutions with different step sizes

(h_1, h_2)	$E_\infty(h_1, h_2)$	$E_\infty(2h_1, 2h_2)/E_\infty(h_1, h_2)$
$(1/8, 1/8)$	4.238e−2	
$(1/16, 1/16)$	1.061e−2	3.994
$(1/32, 1/32)$	2.656e−3	3.995
$(1/64, 1/64)$	6.640e−4	4.000

between the numerical and exact solutions for various step sizes. Table 2.3 provides the maximum errors

$$E_\infty(h_1, h_2) = \max_{(i,j)\in\omega} \left| u(x_i, y_j) - u_{ij} \right|$$

of the numerical solutions for different step sizes.

It can be observed from Table 2.3 that the maximum errors decrease to about one-fourth of the original when both step sizes h_1 and h_2 are halved. The surfaces of the numerical solutions and the exact solution are shown in Figs. 2.2 and 2.3, respectively. Figure 2.4 displays the numerical error surfaces for different step sizes.

Fig. 2.2 (Example 2.1) The numerical solution surface ($h_1 = h_2 = 1/8$)

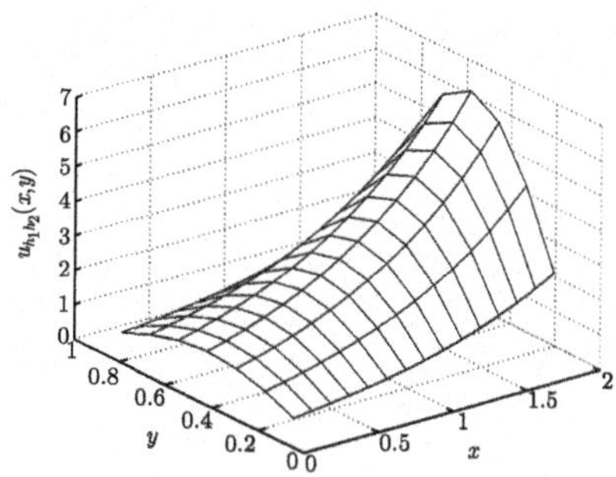

Fig. 2.3 (Example 2.1) The exact solution surface

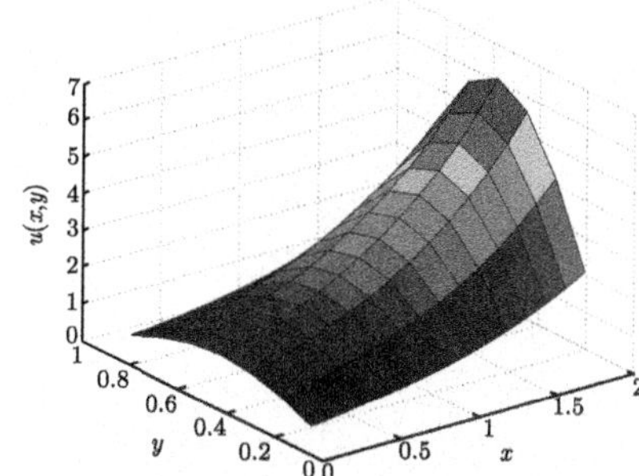

Fig. 2.4 (Example 2.1) The numerical error surfaces with different step sizes

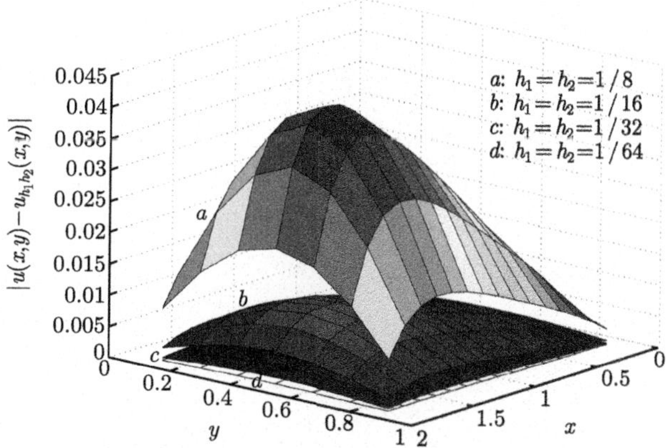

2.2.4   A Priori Estimate of the Difference Solution

In the section, we apply the maximum principle to give a priori estimate of the difference scheme

$$\begin{cases} -\Delta_h v_{ij} = g_{ij}, & (i, j) \in \omega, \\ v_{ij} = \varphi_{ij}, & (i, j) \in \gamma. \end{cases} \tag{2.22}$$

Let $v = \{v_{ij} \mid (i, j) \in \bar{\omega}\}$ be the grid function on Ω_h. Denote

$$(L_h v)_{ij} = -\left(\delta_x^2 v_{ij} + \delta_y^2 v_{ij}\right), \quad (i, j) \in \omega.$$

Lemma 2.3 (Maximum Principle) *Let $v = \{v_{ij} \mid (i, j) \in \bar{\omega}\}$ be the grid function on Ω_h. If*

$$(L_h v)_{ij} \leqslant 0, \quad (i, j) \in \omega,$$

then we have

$$\max_{(i,j)\in\omega} v_{ij} \leqslant \max_{(i,j)\in\gamma} v_{ij}.$$

Proof The proof is by contradiction. Assume

$$\max_{(i,j)\in\omega} v_{ij} > \max_{(i,j)\in\gamma} v_{ij}$$

and $\max\limits_{(i,j)\in\omega} v_{ij} = M$. Then there must be an $(i_0, j_0) \in \omega$ satisfying $v_{i_0, j_0} = M$, and at least one of v_{i_0-1, j_0}, v_{i_0+1, j_0}, v_{i_0, j_0-1}, and v_{i_0, j_0+1} is less than M. Thus,

$$(L_h v)_{i_0, j_0} = 2\left(\frac{1}{h_1^2} + \frac{1}{h_2^2}\right) v_{i_0, j_0} - \frac{1}{h_1^2}\left(v_{i_0-1, j_0} + v_{i_0+1, j_0}\right) - \frac{1}{h_2^2}\left(v_{i_0, j_0-1} + v_{i_0, j_0+1}\right)$$

$$> 2\left(\frac{1}{h_1^2} + \frac{1}{h_2^2}\right) M - \frac{1}{h_1^2}\left(M + M\right) - \frac{1}{h_2^2}\left(M + M\right) = 0.$$

This contradicts the assumption. □

Theorem 2.3 *Let $\{v_{ij} \mid (i, j) \in \bar{\omega}\}$ be the solution of the difference scheme (2.22). Then we have*

$$\max_{(i,j)\in\omega} |v_{ij}| \leqslant \max_{(i,j)\in\gamma} |\varphi_{ij}| + \frac{1}{16}\left(L_1^2 + L_2^2\right) \max_{(i,j)\in\omega} |g_{ij}|.$$

Proof Denote

$$C = \max_{(i,j)\in\omega} |g_{ij}|, \quad P(x, y) = x(L_1 - x) + y(L_2 - y),$$

and define the grid function on Ω_h by

$$w_{ij} = \frac{1}{4} C P(x_i, y_j), \quad (i, j) \in \bar{\omega},$$

then we have

$$w_{ij} \geqslant 0, \quad (i, j) \in \bar{\omega},$$

$$(L_h w)_{ij} = C, \quad (i, j) \in \omega.$$

Therefore,

$$L_h(\pm v - w)_{ij} = \pm(L_h v)_{ij} - (L_h w)_{ij} = \pm g_{ij} - C \leqslant 0, \quad (i, j) \in \omega.$$

It follows from Lemma 2.3 that

$$\max_{(i,j)\in\omega} (\pm v - w)_{ij} \leqslant \max_{(i,j)\in\gamma} (\pm v - w)_{ij} \leqslant \max_{(i,j)\in\gamma} |\pm v_{ij}| + \max_{(i,j)\in\gamma} (-w_{ij}) \leqslant \max_{(i,j)\in\gamma} |v_{ij}|.$$

Thus,

$$\max_{(i,j)\in\omega} (\pm v)_{ij} = \max_{(i,j)\in\omega} (\pm v - w + w)_{ij}$$

$$\leqslant \max_{(i,j)\in\omega} (\pm v - w)_{ij} + \max_{(i,j)\in\omega} w_{ij}$$

$$\leqslant \max_{(i,j)\in\gamma} |v_{ij}| + \max_{(i,j)\in\omega} w_{ij}$$

$$\leqslant \max_{(i,j)\in\gamma} |\varphi_{ij}| + \frac{1}{16}\left(L_1^2 + L_2^2\right) \max_{(i,j)\in\omega} |g_{ij}|,$$

which implies

$$\max_{(i,j)\in\omega} |v_{ij}| \leqslant \max_{(i,j)\in\gamma} |\varphi_{ij}| + \frac{1}{16}\left(L_1^2 + L_2^2\right) \max_{(i,j)\in\omega} |g_{ij}|.$$

$$\square$$

2.2.5   Convergence and Stability of the Difference Solution

Convergence

Theorem 2.4 *Let $\{u(x, y) \mid (x, y) \in \bar{\Omega}\}$ be the solution of the problem (2.1) and $\{u_{ij} \mid (i, j) \in \bar{\omega}\}$ be the solution of the difference scheme (2.14). Then we have*

$$\max_{(i,j)\in\omega} |u(x_i, y_j) - u_{ij}| \leqslant \frac{1}{192} M_4 \left(L_1^2 + L_2^2\right)(h_1^2 + h_2^2),$$

where M_4 is defined by (2.15).

Proof Denote

$$e_{ij} = u(x_i, y_j) - u_{ij}, \quad (i, j) \in \bar{\omega}.$$

Subtracting (2.14) from (2.12), we have the system of error equations

$$\begin{cases} -\Delta_h e_{ij} = (R_1)_{ij}, & (i, j) \in \omega, \\ e_{ij} = 0, & (i, j) \in \gamma, \end{cases} \tag{2.23}$$

where $(R_1)_{ij}$ is defined by (2.13). It follows from (2.16) that

$$\max_{(i,j)\in\omega} |(R_1)_{ij}| \leqslant \frac{1}{12} M_4 (h_1^2 + h_2^2).$$

Applying Theorem 2.3, it derives to

$$\max_{(i,j)\in\omega} |e_{ij}| \leqslant \frac{1}{16}(L_1^2 + L_2^2) \max_{(i,j)\in\omega} |(R_1)_{ij}| \leqslant \frac{1}{192} M_4 \left(L_1^2 + L_2^2\right)(h_1^2 + h_2^2).$$

$$\square$$

Stability

Suppose there are perturbation errors f_{ij} in calculating the right-hand side term $f(x_i, y_j)$ and φ_{ij} in calculating the boundary values $\varphi(x_i, y_j)$ when the difference scheme (2.14) is applied. Let $\{v_{ij} \mid (i, j) \in \bar{\omega}\}$ be the solution of the difference scheme

$$\begin{cases} -\Delta_h v_{ij} = f(x_i, y_j) + f_{ij}, & (i, j) \in \omega, \\ v_{ij} = \varphi(x_i, y_j) + \varphi_{ij}, & (i, j) \in \gamma. \end{cases} \tag{2.24}$$

Denote

$$\varepsilon_{ij} = v_{ij} - u_{ij}, \quad (i, j) \in \omega \cup \gamma.$$

Subtracting (2.14) from (2.24) produces

$$\begin{cases} -\Delta_h \varepsilon_{ij} = f_{ij}, & (i, j) \in \omega, \\ \varepsilon_{ij} = \varphi_{ij}, & (i, j) \in \gamma. \end{cases} \tag{2.25}$$

By means of Theorem 2.3, one has

$$\max_{(i,j)\in\omega} |\varepsilon_{ij}| \leqslant \max_{(i,j)\in\gamma} |\varphi_{ij}| + \frac{1}{16}\left(L_1^2 + L_2^2\right) \max_{(i,j)\in\omega} |f_{ij}|.$$

From the above inequality, we can see that when both $\max\limits_{(i,j)\in\gamma} |\varphi_{ij}|$ and $\max\limits_{(i,j)\in\omega} |f_{ij}|$ are small, $\max\limits_{(i,j)\in\omega} |\varepsilon_{ij}|$ is also small. Therefore, the solution of the difference scheme (2.14) is stable with respect to the boundary values and the right-hand side term.

Indeed, (2.25) is referred to as the **system of perturbation equations**, which has the same form as the difference scheme (2.14). Consequently, we can state the following theorem.

Theorem 2.5 *The solution of the difference scheme (2.14) is stable with respect to the boundary values and the right-hand side term in the following sense: Let $\{u_{ij}\}$ be the solution of system*

$$\begin{cases} -\Delta_h u_{ij} = f_{ij}, & (i, j) \in \omega, \\ u_{ij} = \varphi_{ij}, & (i, j) \in \gamma. \end{cases}$$

Then we have

$$\max_{(i,j)\in\omega} |u_{ij}| \leqslant \max_{(i,j)\in\gamma} |\varphi_{ij}| + \frac{1}{16}\left(L_1^2 + L_2^2\right) \max_{(i,j)\in\omega} |f_{ij}|.$$

2.2.6 The Richardson Extrapolation Method

Denote the solution of the difference scheme (2.14) by $u_{ij}(h_1, h_2)$.

Theorem 2.6 *Suppose there are smooth solutions to the following two problems:*

$$\begin{cases} -\Delta v = \frac{1}{12}\frac{\partial^4 u(x,y)}{\partial x^4}, & (x, y) \in \Omega, \\ v = 0, & (x, y) \in \Gamma \end{cases} \tag{2.26}$$

and

$$\begin{cases} -\Delta w = \frac{1}{12}\frac{\partial^4 u(x,y)}{\partial y^4}, & (x, y) \in \Omega, \\ w = 0, & (x, y) \in \Gamma, \end{cases} \tag{2.27}$$

then we have

$$\max_{(i,j)\in\omega} \left|u(x_i, y_j) - \left[\frac{4}{3}u_{2i,2j}\left(\frac{h_1}{2}, \frac{h_2}{2}\right) - \frac{1}{3}u_{ij}(h_1, h_2)\right]\right| = O(h_1^4 + h_2^4),$$

where $h_1 = L_1/m_1$ and $h_2 = L_2/m_2$.

Proof Note that (2.12) can be rewritten as

$$\begin{cases} -\Delta_h U_{ij} = f(x_i, y_j) - \dfrac{h_1^2}{12}\dfrac{\partial^4 u(x_i,y_j)}{\partial x^4} - \dfrac{h_2^2}{12}\dfrac{\partial^4 u(x_i,y_j)}{\partial y^4} \\[2mm] \qquad\qquad - \dfrac{h_1^4}{360}\dfrac{\partial^6 u(\overline{\xi}_{ij},y_j)}{\partial x^6} - \dfrac{h_2^4}{360}\dfrac{\partial^6 u(x_i,\overline{\eta}_{ij})}{\partial y^6}, \quad (i,j) \in \omega, \\[2mm] u(x_i, y_j) = \varphi(x_i, y_j), \quad (i,j) \in \gamma, \end{cases}$$

where $\overline{\xi}_{ij} \in (x_{i-1}, x_{i+1})$ and $\overline{\eta}_{ij} \in (y_{j-1}, y_{j+1})$. The system of error equations (2.23) can be rewritten as

$$\begin{cases} -\Delta_h e_{ij} = -\dfrac{h_1^2}{12}\dfrac{\partial^4 u(x_i, y_j)}{\partial x^4} - \dfrac{h_2^2}{12}\dfrac{\partial^4 u(x_i, y_j)}{\partial y^4} \\[3mm] \qquad\qquad - \dfrac{h_1^4}{360}\dfrac{\partial^6 u(\overline{\xi}_{ij}, y_j)}{\partial x^6} - \dfrac{h_2^4}{360}\dfrac{\partial^6 u(x_i, \overline{\eta}_{ij})}{\partial y^6}, \quad (i,j) \in \omega, \quad (2.28) \\[3mm] e_{ij} = 0, \quad (i,j) \in \gamma. \end{cases}$$

Let

$$V_{ij} = v(x_i, y_j), \quad W_{ij} = w(x_i, y_j), \quad (i,j) \in \bar{\omega}.$$

Discretizing (2.26) and (2.27), respectively, we have

$$\begin{cases} -\Delta_h V_{ij} = \dfrac{1}{12}\dfrac{\partial^4 u(x_i,y_j)}{\partial x^4} - \dfrac{h_1^2}{12}\dfrac{\partial^4 v(x_{ij}^{(1)},y_j)}{\partial x^4} - \dfrac{h_2^2}{12}\dfrac{\partial^4 v(x_i,y_{ij}^{(1)})}{\partial y^4}, \quad (i,j) \in \omega, \\[2mm] V_{ij} = 0, \quad (i,j) \in \gamma \end{cases}$$

$$(2.29)$$

and

$$\begin{cases} -\Delta_h W_{ij} = \dfrac{1}{12}\dfrac{\partial^4 u(x_i,y_j)}{\partial y^4} - \dfrac{h_1^2}{12}\dfrac{\partial^4 w(x_{ij}^{(2)},y_j)}{\partial x^4} - \dfrac{h_2^2}{12}\dfrac{\partial^4 w(x_i,y_{ij}^{(2)})}{\partial y^4}, \quad (i,j) \in \omega, \\[2mm] W_{ij} = 0, \quad (i,j) \in \gamma, \end{cases}$$

$$(2.30)$$

where $x_{ij}^{(1)}$, $x_{ij}^{(2)} \in (x_{i-1}, x_{i+1})$ and $y_{ij}^{(1)}$, $y_{ij}^{(2)} \in (y_{j-1}, y_{j+1})$.
Denote

$$r_{ij} = e_{ij} + h_1^2 V_{ij} + h_2^2 W_{ij}.$$

Multiplying both sides of (2.29) by h_1^2 and (2.30) by h_2^2, respectively, and then summing up the results with (2.28), we arrive at

$$
\begin{cases}
-\Delta_h r_{ij} = -\dfrac{h_1^4}{360}\dfrac{\partial^6 u(\bar{\xi}_{ij},y_j)}{\partial x^6} - \dfrac{h_2^4}{360}\dfrac{\partial^6 u(x_i,\bar{\eta}_{ij})}{\partial y^6} \\[2mm]
\qquad -h_1^2\left[\dfrac{h_1^2}{12}\dfrac{\partial^4 v(x_{ij}^{(1)},y_j)}{\partial x^4} + \dfrac{h_2^2}{12}\dfrac{\partial^4 v(x_i,y_{ij}^{(1)})}{\partial y^4}\right] \\[2mm]
\qquad -h_2^2\left[\dfrac{h_1^2}{12}\dfrac{\partial^4 w(x_{ij}^{(2)},y_j)}{\partial x^4} + \dfrac{h_2^2}{12}\dfrac{\partial^4 w(x_i,y_{ij}^{(2)})}{\partial y^4}\right], \quad (i,j)\in\omega, \\[2mm]
r_{ij} = 0, \quad (i,j)\in\gamma.
\end{cases}
$$

It follows from Theorem 2.3 that

$$
r_{ij} = O(h_1^4 + h_2^4), \quad (i,j)\in\omega;
$$

in other words,

$$
u_{ij}(h_1,h_2) = u(x_i,y_j) + h_1^2 v(x_i,y_j) + h_2^2 w(x_i,y_j) + O(h_1^4 + h_2^4), \quad (i,j)\in\omega.
\tag{2.31}
$$

By the same token, we obtain

$$
u_{2i,2j}\left(\frac{h_1}{2},\frac{h_2}{2}\right) = u(x_i,y_j) + \left(\frac{h_1}{2}\right)^2 v(x_i,y_j) + \left(\frac{h_2}{2}\right)^2 w(x_i,y_j)
$$
$$
+ O\left(\left(\frac{h_1}{2}\right)^4 + \left(\frac{h_2}{2}\right)^4\right), \quad (i,j)\in\omega.
\tag{2.32}
$$

Multiplying both sides of (2.32) by 4/3 and (2.31) by 1/3 and then subtracting the results, we have

$$
\frac{4}{3}u_{2i,2j}\left(\frac{h_1}{2},\frac{h_2}{2}\right) - \frac{1}{3}u_{ij}(h_1,h_2) = u(x_i,y_j) + O(h_1^4 + h_2^4), \quad (i,j)\in\omega.
$$

This completes the proof. $\qquad\qquad\qquad\qquad\qquad\qquad\qquad\qquad\square$

Example 2.2 Compute the problem (2.21) in Example 2.1 by the Richardson extrapolation method.

Table 2.4 lists the maximum numerical errors

$$
\tilde{E}_\infty(h_1,h_2) = \max_{(i,j)\in\omega}\left|u(x_i,y_j) - \left[\frac{4}{3}u_{2i,2j}\left(\frac{h_1}{2},\frac{h_2}{2}\right) - \frac{1}{3}u_{ij}(h_1,h_2)\right]\right|
$$

calculated using the extrapolation method with different step sizes.

It can be seen from Table 2.4 that when h_1 and h_2 are both reduced by half, the maximum numerical errors decrease to 1/16 of the original. Comparing Table 2.4

Table 2.4 (Example 2.2) The maximum numerical errors calculated by the extrapolation method once with different step sizes

(h_1, h_2)	$\widetilde{E}_\infty(h_1, h_2)$	$\widetilde{E}_\infty(2h_1, 2h_2)/\widetilde{E}_\infty(h_1, h_2)$
(1/8,1/8)	2.109e$-$4	
(1/16,1/16)	1.402e$-$5	15.04
(1/32,1/32)	8.717e$-$7	16.08
(1/64,1/64)	4.809e$-$8	18.13

with Table 2.3, we see that the extrapolation method significantly improves the accuracy of the numerical solution.

2.3 The Compact Difference Scheme

This section derives a difference scheme with the accuracy $O(h_1^4 + h_2^4)$ for the problem (2.1).

For $v = \{v_{ij} \mid (i, j) \in \bar{\omega}\} \in \mathcal{V}_h$, define the operators

$$(\mathcal{A}v)_{ij} = \begin{cases} \frac{1}{12}(v_{i-1,j} + 10v_{ij} + v_{i+1,j}), & 1 \leqslant i \leqslant m_1 - 1, \quad 0 \leqslant j \leqslant m_2, \\ v_{ij}, & i = 0, m_1, \quad 0 \leqslant j \leqslant m_2, \end{cases}$$

$$(\mathcal{B}v)_{ij} = \begin{cases} \frac{1}{12}(v_{i,j-1} + 10v_{ij} + v_{i,j+1}), & 1 \leqslant j \leqslant m_2 - 1, \quad 0 \leqslant i \leqslant m_1, \\ v_{ij}, & j = 0, m_2, \quad 0 \leqslant i \leqslant m_1. \end{cases}$$

It is easy to know that

$$(\mathcal{A}v)_{ij} = \left(\mathcal{I} + \frac{h_1^2}{12}\delta_x^2\right)v_{ij}, \quad (\mathcal{B}v)_{ij} = \left(\mathcal{I} + \frac{h_2^2}{12}\delta_y^2\right)v_{ij}, \quad (i, j) \in \omega,$$

where $\mathcal{I}$ is an identity operator, namely, $\mathcal{I}v_{ij} = v_{ij}$.

Without causing confusion, they are abbreviated as $\mathcal{A}v_{ij}$ and $\mathcal{B}v_{ij}$, respectively. For $v, w \in \overset{\circ}{\mathcal{V}}_h$, introduce the following inner products and norms:

$$(v, w) = h_1 h_2 \sum_{i=1}^{m_1-1} \sum_{j=1}^{m_2-1} v_{ij} w_{ij}, \quad \|v\| = \sqrt{(v, v)},$$

$$(\delta_x v, \delta_x w) = h_1 h_2 \sum_{i=1}^{m_1} \sum_{j=1}^{m_2-1} (\delta_x v_{i-\frac{1}{2},j})(\delta_x w_{i-\frac{1}{2},j}), \quad \|\delta_x v\| = \sqrt{(\delta_x v, \delta_x v)},$$

$$(\delta_y v, \delta_y w) = h_1 h_2 \sum_{i=1}^{m_1-1} \sum_{j=1}^{m_2} (\delta_y v_{i,j-\frac{1}{2}})(\delta_y w_{i,j-\frac{1}{2}}), \quad \|\delta_y v\| = \sqrt{(\delta_y v, \delta_y v)},$$

$$(\Delta_h v, \Delta_h w) = h_1 h_2 \sum_{i=1}^{m_1-1} \sum_{j=1}^{m_2-1} (\Delta_h v_{ij})(\Delta_h w_{ij}), \quad \|\Delta_h v\| = \sqrt{(\Delta_h v, \Delta_h v)},$$

$$(\delta_x \delta_y v, \delta_x \delta_y w) = h_1 h_2 \sum_{i=1}^{m_1} \sum_{j=1}^{m_2} (\delta_x \delta_y v_{i-\frac{1}{2}, j-\frac{1}{2}})(\delta_x \delta_y w_{i-\frac{1}{2}, j-\frac{1}{2}}),$$

$$\|\delta_x \delta_y v\| = \sqrt{(\delta_x \delta_y v, \delta_x \delta_y v)}, \quad |v|_1 = \sqrt{\|\delta_x v\|^2 + \|\delta_y v\|^2}, \quad \|v\|_1 = \sqrt{\|v\|^2 + |v|_1^2}.$$

Generally, $\|v\|$, $\|v\|_1$, and $|v|_1$ are called the 2-norm (average norm) of v, the H^1-norm of v, and the 2-norm of the difference quotient of v, respectively.

2.3.1   Derivation of the Difference Scheme

Considering Eq. (2.1a) at the grid point (x_i, y_j), we have

$$-\frac{\partial^2 u}{\partial x^2}(x_i, y_j) - \frac{\partial^2 u}{\partial y^2}(x_i, y_j) = f(x_i, y_j), \quad (i, j) \in \bar{\omega}.$$

Performing $\mathcal{AB}$ on both sides of the above equality, one easily has

$$-\mathcal{AB}\frac{\partial^2 u}{\partial x^2}(x_i, y_j) - \mathcal{AB}\frac{\partial^2 u}{\partial y^2}(x_i, y_j) = \mathcal{AB}f(x_i, y_j), \quad (i, j) \in \omega,$$

namely,

$$-\mathcal{B}\left(\mathcal{A}\frac{\partial^2 u}{\partial x^2}(x_i, y_j)\right) - \mathcal{A}\left(\mathcal{B}\frac{\partial^2 u}{\partial y^2}(x_i, y_j)\right) = \mathcal{AB}f(x_i, y_j), \quad (i, j) \in \omega.$$

$$(2.33)$$

It follows from Lemma 1.2 that

$$\mathcal{A}\frac{\partial^2 u}{\partial x^2}(x_i, y_j) = \delta_x^2 U_{ij} + \frac{h_1^4}{240}\frac{\partial^6 u}{\partial x^6}(\xi_{ij}, y_j), \quad 1 \leqslant i \leqslant m_1 - 1, \quad 0 \leqslant j \leqslant m_2,$$

$$(2.34)$$

$$\mathcal{B}\frac{\partial^2 u}{\partial y^2}(x_i, y_j) = \delta_y^2 U_{ij} + \frac{h_2^4}{240}\frac{\partial^6 u}{\partial y^6}(x_i, \eta_{ij}), \quad 0 \leqslant i \leqslant m_1, \quad 1 \leqslant j \leqslant m_2 - 1,$$

$$(2.35)$$

where $\xi_{ij} \in (x_{i-1}, x_{i+1})$ and $\eta_{ij} \in (y_{j-1}, y_{j+1})$.

Denote

$$P_{ij} = \frac{h_1^4}{240} \frac{\partial^6 u}{\partial x^6} (\xi_{ij}, y_j), \quad 1 \leqslant i \leqslant m_1 - 1, \quad 0 \leqslant j \leqslant m_2, \tag{2.36}$$

$$Q_{ij} = \frac{h_2^4}{240} \frac{\partial^6 u}{\partial y^6} (x_i, \eta_{ij}), \quad 0 \leqslant i \leqslant m_1, \quad 1 \leqslant j \leqslant m_2 - 1. \tag{2.37}$$

Then (2.34) and (2.35) read

$$A\frac{\partial^2 u}{\partial x^2}(x_i, y_j) = \delta_x^2 U_{ij} + P_{ij}, \quad 1 \leqslant i \leqslant m_1 - 1, \quad 0 \leqslant j \leqslant m_2,$$

$$B\frac{\partial^2 u}{\partial y^2}(x_i, y_j) = \delta_y^2 U_{ij} + Q_{ij}, \quad 0 \leqslant i \leqslant m_1, \quad 1 \leqslant j \leqslant m_2 - 1.$$

Denote $f_{ij} = f(x_i, y_j), (i, j) \in \bar{\omega}$. Substituting the above two equalities into (2.33), we get

$$-\left[B(\delta_x^2 U_{ij} + P_{ij}) + A(\delta_y^2 U_{ij} + Q_{ij}) \right] = ABf_{ij}, \quad (i, j) \in \omega;$$

in other words,

$$-\left(B\delta_x^2 U_{ij} + A\delta_y^2 U_{ij} \right) = ABf_{ij} + (R_2)_{ij}, \quad (i, j) \in \omega, \tag{2.38}$$

where

$$(R_2)_{ij} = BP_{ij} + AQ_{ij}, \quad (i, j) \in \omega. \tag{2.39}$$

Noticing the boundary value condition (2.1b), we have

$$U_{ij} = \varphi(x_i, y_j), \quad (i, j) \in \gamma. \tag{2.40}$$

Omitting the small term $(R_2)_{ij}$ in (2.38), noticing (2.40), and replacing U_{ij} with u_{ij}, another difference scheme for solving the problem (2.1) reads

$$\begin{cases} -\left(B\delta_x^2 u_{ij} + A\delta_y^2 u_{ij} \right) = ABf_{ij}, & (i, j) \in \omega, \tag{2.41a} \\ u_{ij} = \varphi(x_i, y_j), & (i, j) \in \gamma. \tag{2.41b} \end{cases}$$

Denote

$$M_6 = \max\left\{ \max_{(x,y)\in \bar{\Omega}} \left| \frac{\partial^6 u(x, y)}{\partial x^6} \right|, \max_{(x,y)\in \bar{\Omega}} \left| \frac{\partial^6 u(x, y)}{\partial y^6} \right| \right\}. \tag{2.42}$$

From (2.36), (2.37), and (2.39), it yields

$$\left|(R_2)_{ij}\right| \leqslant \frac{1}{240} M_6(h_1^4 + h_2^4), \quad (i, j) \in \omega. \tag{2.43}$$

The difference scheme (2.41) is derived using the nine-point stencil $\{(x_{i+l}, y_{j+m}) \mid -1 \leqslant l, m \leqslant 1\}$, which only requires the values of $f(x, y)$ at these points. Despite this, it achieves the highest order four. This type of difference scheme is called a **compact difference scheme**.

2.3.2   Existence of the Difference Solution

Lemma 2.4 *For $v \in \overset{\circ}{\mathcal{V}}_h$, we have*

$$\|\delta_x v\|^2 \leqslant \frac{4}{h_1^2}\|v\|^2, \quad \|\delta_y v\|^2 \leqslant \frac{4}{h_2^2}\|v\|^2, \tag{2.44}$$

$$-\left(\mathcal{B}\delta_x^2 v + \mathcal{A}\delta_y^2 v, v\right) \geqslant \frac{2}{3}|v|_1^2. \tag{2.45}$$

Proof

(I) It follows from

$$
\begin{aligned}
\|\delta_x v\|^2 &= h_1 h_2 \sum_{i=1}^{m_1} \sum_{j=1}^{m_2-1} \left(\frac{v_{ij} - v_{i-1,j}}{h_1}\right)^2 \\
&\leqslant h_1 h_2 \sum_{i=1}^{m_1} \sum_{j=1}^{m_2-1} \frac{2(v_{ij})^2 + 2(v_{i-1,j})^2}{h_1^2} \leqslant \frac{4}{h_1^2}\|v\|^2
\end{aligned}
$$

and

$$
\begin{aligned}
\|\delta_y v\|^2 &= h_1 h_2 \sum_{i=1}^{m_1-1} \sum_{j=1}^{m_2} \left(\frac{v_{ij} - v_{i,j-1}}{h_2}\right)^2 \\
&\leqslant h_1 h_2 \sum_{i=1}^{m_1-1} \sum_{j=1}^{m_2} \frac{2(v_{ij})^2 + 2(v_{i,j-1})^2}{h_2^2} \leqslant \frac{4}{h_2^2}\|v\|^2
\end{aligned}
$$

that (2.44) holds.

(II) Combining the summation by parts with (2.44) and direct calculation, we have

$$
- \left(\mathcal{B} \delta_x^2 v + \mathcal{A} \delta_y^2 v, \, v \right)
$$

$$
= - \left[\left(\mathcal{I} + \frac{h_2^2}{12} \delta_y^2 \right) \delta_x^2 v + \left(\mathcal{I} + \frac{h_1^2}{12} \delta_x^2 \right) \delta_y^2 v, \, v \right]
$$

$$
= - \left(\delta_x^2 v, \, v \right) - \frac{h_2^2}{12} \left(\delta_y^2 \delta_x^2 v, \, v \right) - \left(\delta_y^2 v, \, v \right) - \frac{h_1^2}{12} \left(\delta_x^2 \delta_y^2 v, \, v \right)
$$

$$
= \| \delta_x v \|^2 + \frac{h_2^2}{12} \left(\delta_y \delta_x^2 v, \, \delta_y v \right) + \| \delta_y v \|^2 + \frac{h_1^2}{12} \left(\delta_x \delta_y^2 v, \, \delta_x v \right)
$$

$$
= \| \delta_x v \|^2 + \frac{h_2^2}{12} \left(\delta_x^2 \delta_y v, \, \delta_y v \right) + \| \delta_y v \|^2 + \frac{h_1^2}{12} \left(\delta_y^2 \delta_x v, \, \delta_x v \right)
$$

$$
= \| \delta_x v \|^2 - \frac{h_2^2}{12} \| \delta_x \delta_y v \|^2 + \| \delta_y v \|^2 - \frac{h_1^2}{12} \| \delta_y \delta_x v \|^2
$$

$$
= \| \delta_x v \|^2 - \frac{h_2^2}{12} \| \delta_y \delta_x v \|^2 + \| \delta_y v \|^2 - \frac{h_1^2}{12} \| \delta_x \delta_y v \|^2
$$

$$
\geqslant \| \delta_x v \|^2 - \frac{h_2^2}{12} \cdot \frac{4}{h_2^2} \| \delta_x v \|^2 + \| \delta_y v \|^2 - \frac{h_1^2}{12} \cdot \frac{4}{h_1^2} \| \delta_y v \|^2
$$

$$
= \frac{2}{3} \left(\| \delta_x v \|^2 + \| \delta_y v \|^2 \right)
$$

$$
= \frac{2}{3} |v|_1^2,
$$

which implies (2.45).

□

Theorem 2.7 *There is a unique solution to the difference scheme (2.41).*

Proof Since the difference scheme (2.41) is linear, it suffices to consider its homogeneous one

$$
\begin{cases}
- \left(\mathcal{B} \delta_x^2 u_{ij} + \mathcal{A} \delta_y^2 u_{ij} \right) = 0, & (i, j) \in \omega, & \text{(2.46a)} \\[2mm]
u_{ij} = 0, & (i, j) \in \gamma. & \text{(2.46b)}
\end{cases}
$$

Taking an inner product of (2.46a) with u, we have

$$
- \left(\mathcal{B} \delta_x^2 u + \mathcal{A} \delta_y^2 u, \, u \right) = 0.
$$

It follows from Lemma 2.4 that

$$\frac{2}{3}|u|_1^2 \leqslant 0.$$

Noticing (2.46b), we have

$$u_{ij} = 0, \quad 0 \leqslant i \leqslant m_1, \quad 0 \leqslant j \leqslant m_2.$$

Therefore, the difference scheme (2.41) is uniquely solvable. $\square$

2.3.3   *Implementation of the Difference Scheme and Numerical Examples*

The difference scheme (2.41) is a system of linear equations in the unknown $\{u_{ij} \mid (i, j) \in \bar{\omega}\}$. Note that (2.41a) can be written as

$$-\frac{1}{12}\left(\frac{1}{h_1^2} + \frac{1}{h_2^2}\right)u_{i-1,j-1} - \frac{1}{6}\left(\frac{5}{h_2^2} - \frac{1}{h_1^2}\right)u_{i,j-1} - \frac{1}{12}\left(\frac{1}{h_1^2} + \frac{1}{h_2^2}\right)u_{i+1,j-1}$$

$$-\frac{1}{6}\left(\frac{5}{h_1^2} - \frac{1}{h_2^2}\right)u_{i-1,j} + \frac{5}{3}\left(\frac{1}{h_1^2} + \frac{1}{h_2^2}\right)u_{ij} - \frac{1}{6}\left(\frac{5}{h_1^2} - \frac{1}{h_2^2}\right)u_{i+1,j}$$

$$-\frac{1}{12}\left(\frac{1}{h_1^2} + \frac{1}{h_2^2}\right)u_{i-1,j+1} - \frac{1}{6}\left(\frac{5}{h_2^2} - \frac{1}{h_1^2}\right)u_{i,j+1} - \frac{1}{12}\left(\frac{1}{h_1^2} + \frac{1}{h_2^2}\right)u_{i+1,j+1}$$

$$= \mathcal{A}\mathcal{B}f_{ij}, \quad (i, j) \in \omega. \tag{2.47}$$

Denote

$$\boldsymbol{u}_j = \begin{pmatrix} u_{1j} \\ u_{2j} \\ \vdots \\ u_{m_1-1,j} \end{pmatrix}, \quad 0 \leqslant j \leqslant m_2.$$

Noticing (2.41b), one further rewrites (2.47) as

$$\boldsymbol{D}\boldsymbol{u}_{j-1} + \boldsymbol{C}\boldsymbol{u}_j + \boldsymbol{D}\boldsymbol{u}_{j+1} = \boldsymbol{f}_j, \quad 1 \leqslant j \leqslant m_2 - 1, \tag{2.48}$$

where

$$
C = \begin{pmatrix} c_1 & c_2 & & & \\ c_2 & c_1 & c_2 & & \\ & \ddots & \ddots & \ddots & \\ & & c_2 & c_1 & c_2 \\ & & & c_2 & c_1 \end{pmatrix}, \qquad
D = \begin{pmatrix} c_3 & c_4 & & & \\ c_4 & c_3 & c_4 & & \\ & \ddots & \ddots & \ddots & \\ & & c_4 & c_3 & c_4 \\ & & & c_4 & c_3 \end{pmatrix},
$$

$$
c_1 = \frac{5}{3}\left(\frac{1}{h_1^2} + \frac{1}{h_2^2}\right), \qquad
c_2 = -\frac{1}{6}\left(\frac{5}{h_1^2} - \frac{1}{h_2^2}\right),
$$

$$
c_3 = -\frac{1}{6}\left(\frac{5}{h_2^2} - \frac{1}{h_1^2}\right), \qquad
c_4 = -\frac{1}{12}\left(\frac{1}{h_1^2} + \frac{1}{h_2^2}\right),
$$

$$
f_j = \begin{pmatrix} \mathcal{AB}f_{1,j} - c_4 u_{0,j-1} - c_2 u_{0j} - c_4 u_{0,j+1} \\ \mathcal{AB}f_{2j} \\ \vdots \\ \mathcal{AB}f_{m_1-2,j} \\ \mathcal{AB}f_{m_1-1,j} - c_4 u_{m_1,j-1} - c_2 u_{m_1,j} - c_4 u_{m_1,j+1} \end{pmatrix}.
$$

Indeed, (2.48) can be further written as

$$
\begin{pmatrix} C & D & & & \\ D & C & D & & \\ & \ddots & \ddots & \ddots & \\ & & D & C & D \\ & & & D & C \end{pmatrix}
\begin{pmatrix} u_1 \\ u_2 \\ \vdots \\ u_{m_2-2} \\ u_{m_2-1} \end{pmatrix}
=
\begin{pmatrix} f_1 - D u_0 \\ f_2 \\ \vdots \\ f_{m_2-2} \\ f_{m_2-1} - D u_{m_2} \end{pmatrix}. \tag{2.49}
$$

The coefficient matrix of the above system of linear equations is a tridiagonal block matrix with no more than nine nonzero elements in each row.

It can be proved that the coefficient matrix of (2.49) is symmetric and positive definite. In fact, for

$$
\begin{pmatrix} u_1 \\ u_2 \\ \vdots \\ u_{m_2-2} \\ u_{m_2-1} \end{pmatrix} \neq 0,
$$

and when $(i, j) \in \gamma$, $u_{ij} = 0$, with the help of Lemma 2.4, we have

$$h_1 h_2 \left(u_1^{\mathrm{T}}, u_2^{\mathrm{T}}, \ldots, u_{m_2-2}^{\mathrm{T}}, u_{m_2-1}^{\mathrm{T}} \right) \begin{pmatrix} C & D & & & \\ D & C & D & & \\ & \ddots & \ddots & \ddots & \\ & & D & C & D \\ & & & D & C \end{pmatrix} \begin{pmatrix} u_1 \\ u_2 \\ \vdots \\ u_{m_2-2} \\ u_{m_2-1} \end{pmatrix}$$

$$= -\left(\mathcal{B}\delta_x^2 u + \mathcal{A}\delta_y^2 u, u \right) \geqslant \tfrac{2}{3} |u|_1^2 > 0.$$

An iterative method is commonly employed to solve the system of linear equations (2.49) with a large and sparse coefficient matrix.

Example 2.3 Apply the compact difference scheme (2.41) to solve the problem (2.21) in Example 2.1.

Divide $[0, 2]$ into m_1 equal subintervals and $[0, 1]$ into m_2 equal subintervals. The Gauss-Seidel iteration method is applied to solve the system of difference equations (2.41) with an accuracy $\|u^{(l+1)} - u^{(l)}\|_\infty \leqslant \tfrac{1}{2} \times 10^{-10}$.

Table 2.5 lists the exact and numerical solutions at five grid points with different step sizes. Table 2.6 gives the absolute values of the errors $|u(x_i, y_j) - u_{ij}|$ between the numerical and exact solutions with different step sizes at these grid points.

Table 2.5 (Example 2.3) The exact and numerical solutions at some grid points with different step sizes

(h_1, h_2)	(x, y)				
	$(1/2, 1/4)$	$(1, 1/4)$	$(3/2, 1/4)$	$(1/2, 1/2)$	$(1, 1/2)$
$(1/8, 1/8)$	1.165930	1.922300	3.169263	1.648874	2.718543
$(1/16, 1/16)$	1.165829	1.922127	3.169047	1.648731	2.718298
$(1/32, 1/32)$	1.165822	1.922116	3.169034	1.648722	2.718283
$(1/64, 1/64)$	1.165822	1.922116	3.169033	1.648721	2.718282
ES	1.165822	1.922116	3.169033	1.648721	2.718282

Table 2.6 (Example 2.3) The absolute values of numerical errors at some grid points with different step sizes

(h_1, h_2)	(x, y)				
	$(1/2, 1/4)$	$(1, 1/4)$	$(3/2, 1/4)$	$(1/2, 1/2)$	$(1, 1/2)$
$(1/8, 1/8)$	1.079e−4	1.874e−4	2.302e−4	1.526e−4	2.612e−4
$(1/16, 1/16)$	6.713e−6	1.149e−5	1.432e−5	9.494e−6	1.624e−5
$(1/32, 1/32)$	4.119e−7	7.072e−7	8.872e−7	5.828e−7	1.001e−6
$(1/64, 1/64)$	2.342e−9	5.227e−9	2.840e−8	2.733e−9	8.196e−9

Table 2.7 (Example 2.3) The maximum numerical errors with different step sizes

(h_1, h_2)	$E_\infty(h_1, h_2)$	$E_\infty(2h_1, 2h_2)/E_\infty(h_1, h_2)$
(1/8,1/8)	3.255e$-$4	
(1/16,1/16)	2.025e$-$5	16.07
(1/32,1/32)	1.257e$-$6	16.11
(1/64,1/64)	4.564e$-$8	27.54

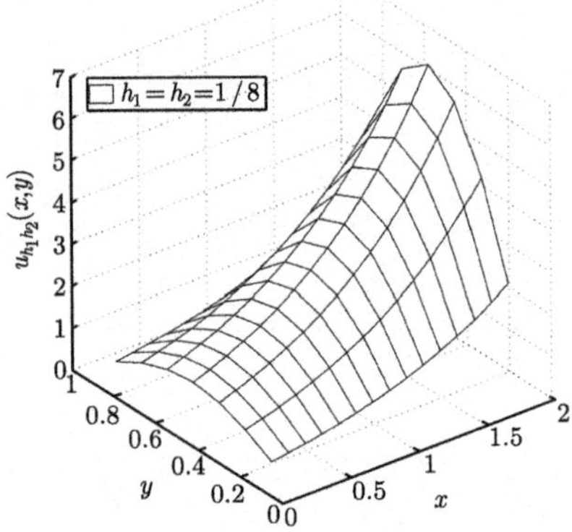

Fig. 2.5 (Example 2.3) The numerical solution surface

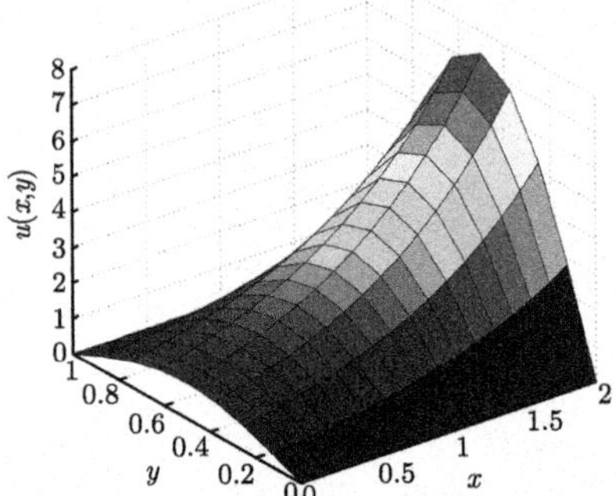

Fig. 2.6 (Example 2.3) The exact solution surface

Table 2.7 displays the maximum numerical errors

$$E_\infty(h_1, h_2) = \max_{(i,j)\in\omega} \left| u(x_i, y_j) - u_{ij} \right|$$

with different step sizes.

It can be seen from Table 2.7 that when the step sizes h_1 and h_2 are both reduced by half, the maximum numerical error is reduced to approximately 1/16 of the original.

Figure 2.5 shows the numerical solution surface with the step sizes $h_1 = h_2 = 1/8$. Figure 2.6 illustrates the exact solution surface and Figure 2.7 displays the numerical error surfaces with different step sizes.

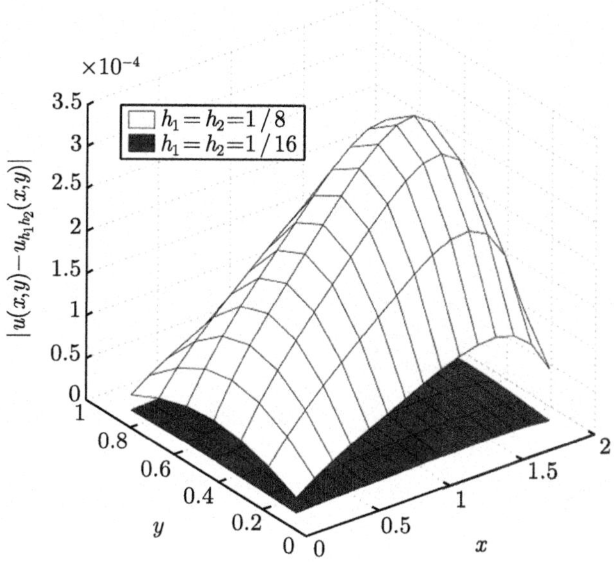

Fig. 2.7 (Example 2.3) Error surfaces

2.3.4   A Priori Estimate of the Difference Solution

Four lemmas are introduced before giving a priori estimate of the difference solution.

Lemma 2.5 *For* $v \in \overset{\circ}{\mathcal{V}}_h$, *we have*

$$\|\Delta_h v\|^2 = \|\delta_x^2 v\|^2 + 2\|\delta_x \delta_y v\|^2 + \|\delta_y^2 v\|^2.$$

Proof It can be calculated directly that

$$\begin{aligned}
\|\Delta_h v\|^2 &= (\delta_x^2 v + \delta_y^2 v, \, \delta_x^2 v + \delta_y^2 v) \\
&= (\delta_x^2 v, \delta_x^2 v) + 2(\delta_x^2 v, \delta_y^2 v) + (\delta_y^2 v, \delta_y^2 v) \\
&= \|\delta_x^2 v\|^2 - 2(\delta_x v, \delta_x \delta_y^2 v) + \|\delta_y^2 v\|^2 \\
&= \|\delta_x^2 v\|^2 - 2(\delta_x v, \delta_y^2 \delta_x v) + \|\delta_y^2 v\|^2 \\
&= \|\delta_x^2 v\|^2 + 2(\delta_y \delta_x v, \delta_y \delta_x v) + \|\delta_y^2 v\|^2 \\
&= \|\delta_x^2 v\|^2 + 2\|\delta_x \delta_y v\|^2 + \|\delta_y^2 v\|^2.
\end{aligned}$$

$\square$

Lemma 2.6 *For $v \in \overset{\circ}{\mathcal{V}}_h$, we have*

$$\frac{2}{3}\|\Delta_h v\|^2 \leq \left(\mathcal{B}\delta_x^2 v + \mathcal{A}\delta_y^2 v, \Delta_h v\right) \leq \|\Delta_h v\|^2.$$

Proof   It follows from the summation by parts and the inverse estimate (2.44) that

$$\left(\mathcal{B}\delta_x^2 v + \mathcal{A}\delta_y^2 v, \Delta_h v\right)$$

$$= \left(\left(\mathcal{I} + \frac{h_2^2}{12}\delta_y^2\right)\delta_x^2 v + \left(\mathcal{I} + \frac{h_1^2}{12}\delta_x^2\right)\delta_y^2 v, \delta_x^2 v + \delta_y^2 v\right)$$

$$= (\delta_x^2 v + \delta_y^2 v, \delta_x^2 v + \delta_y^2 v) + \frac{h_2^2}{12}(\delta_y^2\delta_x^2 v, \delta_x^2 v + \delta_y^2 v) + \frac{h_1^2}{12}(\delta_x^2\delta_y^2 v, \delta_x^2 v + \delta_y^2 v)$$

$$= \|\Delta_h v\|^2 - \frac{h_2^2}{12}\left(\|\delta_x^2\delta_y v\|^2 + \|\delta_x\delta_y^2 v\|^2\right) - \frac{h_1^2}{12}\left(\|\delta_y\delta_x^2 v\|^2 + \|\delta_x\delta_y^2 v\|^2\right)$$

$$= \|\Delta_h v\|^2 - \frac{h_2^2}{12}\left(\|\delta_y\delta_x^2 v\|^2 + \|\delta_y^2\delta_x v\|^2\right) - \frac{h_1^2}{12}\left(\|\delta_x^2\delta_y v\|^2 + \|\delta_x\delta_y^2 v\|^2\right)$$

$$\geq \|\Delta_h v\|^2 - \frac{h_2^2}{12}\left(\frac{4}{h_2^2}\|\delta_x^2 v\|^2 + \frac{4}{h_2^2}\|\delta_y\delta_x v\|^2\right) - \frac{h_1^2}{12}\left(\frac{4}{h_1^2}\|\delta_x\delta_y v\|^2 + \frac{4}{h_1^2}\|\delta_y^2 v\|^2\right)$$

$$= \|\Delta_h v\|^2 - \frac{1}{3}\left(\|\delta_x^2 v\|^2 + \|\delta_y\delta_x v\|^2 + \|\delta_x\delta_y v\|^2 + \|\delta_y^2 v\|^2\right)$$

$$= \frac{2}{3}\|\Delta_h v\|^2,$$

in which Lemma 2.5 is employed in the last equality.

In addition, noticing the third equality above, it is easy to know that

$$\left(\mathcal{B}\delta_x^2 v + \mathcal{A}\delta_y^2 v, \Delta_h v\right) \leq \|\Delta_h v\|^2,$$

which completes the proof. $\square$

Lemma 2.7 *For $v \in \overset{\circ}{\mathcal{V}}_h$, we have*

$$\|v\| \leq \kappa |v|_1,$$

where κ is defined by (2.2).

Proof Noticing $v_{0j} = v_{m_1,j} = 0$, it follows from Lemma 1.4 that

$$h_1 \sum_{i=1}^{m_1-1} (v_{ij})^2 \leq \frac{L_1^2}{6} h_1 \sum_{i=0}^{m_1-1} (\delta_x v_{i+\frac{1}{2},j})^2, \quad 1 \leq j \leq m_2 - 1.$$

Multiplying both sides of the above inequality by h_2 and summing over j from 1 to $m_2 - 1$, we have

$$\|v\|^2 \leqslant \frac{L_1^2}{6} \|\delta_x v\|^2.$$

The same argument implies

$$\|v\|^2 \leqslant \frac{L_2^2}{6} \|\delta_y v\|^2.$$

It follows from the above two inequalities that

$$\frac{6}{L_1^2} \|v\|^2 \leqslant \|\delta_x v\|^2, \qquad \frac{6}{L_2^2} \|v\|^2 \leqslant \|\delta_y v\|^2.$$

Then adding the above two inequalities together, we have

$$\left(\frac{6}{L_1^2} + \frac{6}{L_2^2} \right) \|v\|^2 \leqslant |v|_1^2,$$

namely,

$$\frac{1}{\kappa^2} \|v\|^2 \leqslant |v|_1^2.$$

Taking the square root on both sides of the above inequality produces

$$\|v\| \leqslant \kappa |v|_1.$$

$$\square$$

Remark 2.1 By Lemma 2.7, for any grid function $v \in \overset{\circ}{\mathcal{V}}_h$, we have

$$|v|_1^2 \leqslant \|v\|_1^2 \leqslant \left(1 + \kappa^2 \right) |v|_1^2.$$

Thus, the norms $| \cdot |_1$ and $\| \cdot \|_1$ are equivalent in the space $\overset{\circ}{\mathcal{V}}_h$.

Lemma 2.8 *For $v \in \overset{\circ}{\mathcal{V}}_h$, we have*

$$\|v\|_\infty \leqslant \frac{1}{12} \sqrt{3(\sqrt{2}+1)L_1 L_2} \, \|\Delta_h v\|. \tag{2.50}$$

Proof Let

$$|v_{i_0,j_0}| = \|v\|_\infty.$$

It follows from Lemma 1.4 that

$$\|v\|_\infty^2 = v_{i_0,j_0}^2 \leqslant \epsilon_1 h_1 \sum_{i=1}^{m_1} (\delta_x v_{i-\frac{1}{2},j_0})^2 + \frac{1}{4\epsilon_1} h_1 \sum_{i=1}^{m_1-1} (v_{i,j_0})^2$$

$$\leqslant \epsilon_1 h_1 \sum_{i=1}^{m_1} \left[\epsilon_2 h_2 \sum_{j=1}^{m_2} (\delta_y \delta_x v_{i-\frac{1}{2},j-\frac{1}{2}})^2 + \frac{1}{4\epsilon_2} h_2 \sum_{j=1}^{m_2-1} (\delta_x v_{i-\frac{1}{2},j})^2 \right]$$

$$+ \frac{1}{4\epsilon_1} h_1 \sum_{i=1}^{m_1-1} \left[\epsilon_2 h_2 \sum_{j=1}^{m_2} (\delta_y v_{i,j-\frac{1}{2}})^2 + \frac{1}{4\epsilon_2} h_2 \sum_{j=1}^{m_2-1} (v_{ij})^2 \right]$$

$$= \epsilon_1 \epsilon_2 \|\delta_x \delta_y v\|^2 + \frac{\epsilon_1}{4\epsilon_2} \|\delta_x v\|^2 + \frac{\epsilon_2}{4\epsilon_1} \|\delta_y v\|^2 + \frac{1}{16\epsilon_1\epsilon_2} \|v\|^2. \qquad (2.51)$$

By means of Lemma 2.5, we have

$$\|\delta_x \delta_y v\|^2 \leqslant \frac{1}{2} \|\Delta_h v\|^2. \qquad (2.52)$$

Inserting (2.52) into (2.51) and taking $\epsilon_1 = \epsilon_2 = \sqrt{2\epsilon}$, it immediately yields

$$\|v\|_\infty^2 \leqslant \epsilon \|\Delta_h v\|^2 + \frac{1}{4} |v|_1^2 + \frac{1}{32\epsilon} \|v\|^2. \qquad (2.53)$$

It is also easy to know that

$$|v|_1^2 = -(\Delta_h v, v) \leqslant \|v\| \cdot \|\Delta_h v\|.$$

In combination of Lemma 2.7, we have

$$|v|_1^2 \leqslant \kappa |v|_1 \cdot \|\Delta_h v\|.$$

Therefore,

$$|v|_1 \leqslant \kappa \|\Delta_h v\|.$$

Applying Lemma 2.7 again, we have

$$\|v\| \leqslant \kappa |v|_1 \leqslant \kappa^2 \|\Delta_h v\|.$$

Substituting the above inequalities into (2.53) and taking $\epsilon = \dfrac{\sqrt{2}}{8}\kappa^2$, one obtains

$$\|v\|_\infty^2 \leqslant \epsilon\|\Delta_h v\|^2 + \frac{1}{4}\kappa^2\|\Delta_h v\|^2 + \frac{1}{32\epsilon}\kappa^4\|\Delta_h v\|^2 = \frac{\sqrt{2}+1}{4}\kappa^2\|\Delta_h v\|^2.$$

Taking the square root of both sides of the above inequality and noting (2.3), the conclusion (2.50) follows immediately. $\qquad\square$

Theorem 2.8 *Let* $\{v_{ij} \mid (i,j) \in \bar{\omega}\}$ *be the solution of*

$$\begin{cases} -\left(\mathcal{B}\delta_x^2 v_{ij} + \mathcal{A}\delta_y^2 v_{ij}\right) = g_{ij}, & (i,j) \in \omega, & \text{(2.54a)} \\ v_{ij} = 0, & (i,j) \in \gamma. & \text{(2.54b)} \end{cases}$$

Then we have

$$\|v\|_\infty \leqslant \frac{1}{8}\sqrt{3(\sqrt{2}+1)L_1 L_2}\,\|g\|,$$

where

$$\|g\|^2 = h_1 h_2 \sum_{i=1}^{m_1-1}\sum_{j=1}^{m_2-1}(g_{ij})^2.$$

Proof Taking the inner product of (2.54a) with $-\Delta_h v$, we have

$$\left(\mathcal{B}\delta_x^2 v + \mathcal{A}\delta_y^2 v,\ \Delta_h v\right) = -(g,\ \Delta_h v).$$

It follows from Lemma 2.6 that

$$\frac{2}{3}\|\Delta_h v\|^2 \leqslant -(g,\ \Delta_h v) \leqslant \|g\| \cdot \|\Delta_h v\|,$$

Thus,

$$\|\Delta_h v\| \leqslant \frac{3}{2}\|g\|.$$

Applying Lemma 2.8, we have

$$\|v\|_\infty \leqslant \frac{1}{12}\sqrt{3(\sqrt{2}+1)L_1 L_2}\,\|\Delta_h v\| \leqslant \frac{1}{8}\sqrt{3(\sqrt{2}+1)L_1 L_2}\,\|g\|.$$

$\qquad\square$

2.3.5   Convergence and Stability of the Difference Solution

Convergence

Theorem 2.9 *Let $\{u(x, y) \mid (x, y) \in \bar{\Omega}\}$ be the solution of the problem (2.1) and $\{u_{ij} \mid (i, j) \in \bar{\omega}\}$ be the solution of the difference scheme (2.41). Denote*

$$e_{ij} = u(x_i, y_j) - u_{ij}, \quad 0 \leqslant i \leqslant m_1, \quad 0 \leqslant j \leqslant m_2,$$

then we have

$$\|e\|_\infty \leqslant \frac{1}{1920} \sqrt{3(\sqrt{2}+1)} \, L_1 L_2 M_6 (h_1^4 + h_2^4),$$

where M_6 is defined by (2.42).

Proof Subtracting (2.41) from (2.38) and (2.40), we have the system of error equations

$$\begin{cases} -\left(\mathcal{B}\delta_x^2 e_{ij} + \mathcal{A}\delta_y^2 e_{ij} \right) = (R_2)_{ij}, \quad (i, j) \in \omega, \\ e_{ij} = 0, \quad (i, j) \in \gamma. \end{cases}$$

Applying Theorem 2.8 and noticing (2.43), we have

$$\|e\|_\infty \leqslant \frac{1}{8} \sqrt{3(\sqrt{2}+1)L_1 L_2} \, \|R_2\|$$

$$\leqslant \frac{1}{1920} \sqrt{3(\sqrt{2}+1)} \, L_1 L_2 M_6 (h_1^4 + h_2^4),$$

which completes the proof. $\square$

Stability

From Theorem 2.8, one can easily get the following stability result.

Theorem 2.10 *The solution of the difference scheme (2.41) is stable with respect to the right-hand side term in the following sense: Let $\{u_{ij} \mid (i, j) \in \bar{\omega}\}$ be the solution of*

$$\begin{cases} -\left(\mathcal{B}\delta_x^2 u_{ij} + \mathcal{A}\delta_y^2 u_{ij} \right) = f_{ij}, \quad (i, j) \in \omega, \\ u_{ij} = 0, \quad (i, j) \in \gamma. \end{cases}$$

Then we have

$$\|u\|_\infty \leqslant \frac{1}{8} \sqrt{3(\sqrt{2}+1)L_1 L_2} \, \|f\|.$$

2.4   The Derivative Boundary Value Problem

Consider the derivative boundary value problem

$$
\begin{cases}
-\left(\dfrac{\partial^2 u}{\partial x^2} + \dfrac{\partial^2 u}{\partial y^2}\right) = f(x, y), \quad (x, y) \in \Omega, & (2.55a) \\[2ex]
\left[-\dfrac{\partial u}{\partial x} + \lambda_1(y)u\right]\Big|_{x=0} = \psi_1(y), \quad 0 \leqslant y \leqslant L_2, & (2.55b) \\[2ex]
\left[\dfrac{\partial u}{\partial x} + \lambda_2(y)u\right]\Big|_{x=L_1} = \psi_2(y), \quad 0 \leqslant y \leqslant L_2, & (2.55c) \\[2ex]
\left[-\dfrac{\partial u}{\partial y} + \lambda_3(x)u\right]\Big|_{y=0} = \psi_3(x), \quad 0 \leqslant x \leqslant L_1, & (2.55d) \\[2ex]
\left[\dfrac{\partial u}{\partial y} + \lambda_4(x)u\right]\Big|_{y=L_2} = \psi_4(x), \quad 0 \leqslant x \leqslant L_1, & (2.55e)
\end{cases}
$$

where $\Omega = (0, L_1) \times (0, L_2)$, $\lambda_1(y)$ and $\lambda_2(y)$ are nonnegative continuous functions on $[0, L_2]$, and $\lambda_3(x)$ and $\lambda_4(x)$ are nonnegative continuous functions on $[0, L_1]$, and not always equal to 0.

2.4.1   Derivation of the Difference Scheme

Denote

$$
D_x v_{ij} = \frac{1}{h_1}(v_{i+1,j} - v_{ij}), \quad D_{\bar{x}} v_{ij} = \frac{1}{h_1}(v_{ij} - v_{i-1,j}),
$$

$$
D_y v_{ij} = \frac{1}{h_2}(v_{i,j+1} - v_{ij}), \quad D_{\bar{y}} v_{ij} = \frac{1}{h_2}(v_{ij} - v_{i,j-1}).
$$

Considering Eq. (2.55a) at the grid point (x_i, y_j), we have

$$
-\left[\frac{\partial^2 u(x_i, y_j)}{\partial x^2} + \frac{\partial^2 u(x_i, y_j)}{\partial y^2}\right] = f(x_i, y_j), \quad 0 \leqslant i \leqslant m_1, \quad 0 \leqslant j \leqslant m_2.
$$

$$(2.56)$$

From the Taylor expansion and (2.55b)–(2.55e), we have

$$
\frac{\partial^2 u(x_i, y_j)}{\partial x^2} = \delta_x^2 U_{ij} - \frac{h_1^2}{12}\frac{\partial^4 u(\xi_{ij}, y_j)}{\partial x^4}, \quad 1 \leqslant i \leqslant m_1 - 1, \quad 0 \leqslant j \leqslant m_2, \quad (2.57)
$$

$$
\frac{\partial^2 u(x_i, y_j)}{\partial y^2} = \delta_y^2 U_{ij} - \frac{h_2^2}{12}\frac{\partial^4 u(x_i, \eta_{ij})}{\partial y^4}, \quad 0 \leqslant i \leqslant m_1, \quad 1 \leqslant j \leqslant m_2 - 1, \quad (2.58)
$$

$$\frac{\partial^2 u(x_0, y_j)}{\partial x^2}$$

$$= \frac{2}{h_1}\left[\frac{u(x_1, y_j) - u(x_0, y_j)}{h_1} - \frac{\partial u(x_0, y_j)}{\partial x}\right] - \frac{h_1}{3}\frac{\partial^3 u(\xi_{0j}, y_j)}{\partial x^3}$$

$$= \frac{2}{h_1}\left[D_x U_{0j} + \psi_1(y_j) - \lambda_1(y_j)u(x_0, y_j)\right] - \frac{h_1}{3}\frac{\partial^3 u(\xi_{0j}, y_j)}{\partial x^3},$$

$$0 \leqslant j \leqslant m_2, \qquad (2.59)$$

$$\frac{\partial^2 u(x_{m_1}, y_j)}{\partial x^2}$$

$$= \frac{2}{h_1}\left[\frac{\partial u(x_{m_1}, y_j)}{\partial x} - \frac{u(x_{m_1}, y_j) - u(x_{m_1-1}, y_j)}{h_1}\right] + \frac{h_1}{3}\frac{\partial^3 u(\xi_{m_1,j}, y_j)}{\partial x^3}$$

$$= \frac{2}{h_1}\left[\psi_2(y_j) - \lambda_2(y_j)u(x_{m_1}, y_j) - D_{\bar{x}} U_{m_1,j}\right] + \frac{h_1}{3}\frac{\partial^3 u(\xi_{m_1,j}, y_j)}{\partial x^3},$$

$$0 \leqslant j \leqslant m_2, \qquad (2.60)$$

$$\frac{\partial^2 u(x_i, y_0)}{\partial y^2}$$

$$= \frac{2}{h_2}\left[\frac{u(x_i, y_1) - u(x_i, y_0)}{h_2} - \frac{\partial u(x_i, y_0)}{\partial y}\right] - \frac{h_2}{3}\frac{\partial^3 u(x_i, \eta_{i0})}{\partial y^3}$$

$$= \frac{2}{h_2}\left[D_y U_{i0} + \psi_3(x_i) - \lambda_3(x_i)u(x_i, y_0)\right] - \frac{h_2}{3}\frac{\partial^3 u(x_i, \eta_{i0})}{\partial y^3},$$

$$0 \leqslant i \leqslant m_1, \qquad (2.61)$$

$$\frac{\partial^2 u(x_i, y_{m_2})}{\partial y^2}$$

$$= \frac{2}{h_2}\left[\frac{\partial u(x_i, y_{m_2})}{\partial y} - \frac{u(x_i, y_{m_2}) - u(x_i, y_{m_2-1})}{h_2}\right] + \frac{h_2}{3}\frac{\partial^3 u(x_i, \eta_{i,m_2})}{\partial y^3}$$

$$= \frac{2}{h_2}\left[\psi_4(x_i) - \lambda_4(x_i)u(x_i, y_{m_2}) - D_{\bar{y}} U_{i,m_2}\right] + \frac{h_2}{3}\frac{\partial^3 u(x_i, \eta_{i,m_2})}{\partial x^3},$$

$$0 \leqslant i \leqslant m_1, \qquad (2.62)$$

where $\xi_{0j} \in (x_0, x_1)$, $0 \leqslant j \leqslant m_2$ and $\eta_{i0} \in (y_0, y_1)$, $0 \leqslant i \leqslant m_1$;

$$\xi_{m_1,j} \in (x_{m_1-1}, x_{m_1}), \quad \xi_{ij} \in (x_{i-1}, x_{i+1}), \quad 1 \leqslant i \leqslant m_1 - 1, \quad 0 \leqslant j \leqslant m_2;$$

$$\eta_{i,m_2} \in (y_{m_2-1}, y_{m_2}), \quad \eta_{ij} \in (y_{j-1}, y_{j+1}), \quad 1 \leqslant j \leqslant m_2 - 1, \quad 0 \leqslant i \leqslant m_1.$$

Substituting (2.57)–(2.62) into (2.56), discarding the small terms, and replacing U_{ij} by u_{ij}, a difference scheme for the problem (2.55) reads

$$-\left(\delta_x^2 u_{ij} + \delta_y^2 u_{ij}\right) = f(x_i, y_j), \quad (i, j) \in \omega, \tag{2.63a}$$

$$-\frac{2}{h_1}\left[D_x u_{0j} + \psi_1(y_j) - \lambda_1(y_j)u_{0j}\right] - \delta_y^2 u_{0j} = f(x_0, y_j),$$
$$1 \leqslant j \leqslant m_2 - 1, \tag{2.63b}$$

$$-\frac{2}{h_1}\left[\psi_2(y_j) - \lambda_2(y_j)u_{m_1,j} - D_{\bar{x}}u_{m_1,j}\right] - \delta_y^2 u_{m_1,j} = f(x_{m_1}, y_j),$$
$$1 \leqslant j \leqslant m_2 - 1, \tag{2.63c}$$

$$-\delta_x^2 u_{i0} - \frac{2}{h_2}\left[D_y u_{i0} + \psi_3(x_i) - \lambda_3(x_i)u_{i0}\right] = f(x_i, y_0),$$
$$1 \leqslant i \leqslant m_1 - 1, \tag{2.63d}$$

$$-\delta_x^2 u_{i,m_2} - \frac{2}{h_2}\left[\psi_4(x_i) - \lambda_4(x_i)u_{i,m_2} - D_{\bar{y}}u_{i,m_2}\right] = f(x_i, y_{m_2}),$$
$$1 \leqslant i \leqslant m_1 - 1, \tag{2.63e}$$

$$-\frac{2}{h_1}\left[D_x u_{00} + \psi_1(y_0) - \lambda_1(y_0)u_{00}\right]$$
$$-\frac{2}{h_2}\left[D_y u_{00} + \psi_3(x_0) - \lambda_3(x_0)u_{00}\right] = f(x_0, y_0), \tag{2.63f}$$

$$-\frac{2}{h_1}\left[\psi_2(y_0) - \lambda_2(y_0)u_{m_1,0} - D_{\bar{x}}u_{m_1,0}\right]$$
$$-\frac{2}{h_2}\left[D_y u_{m_1,0} + \psi_3(x_{m_1}) - \lambda_3(x_{m_1})u_{m_1,0}\right] = f(x_{m_1}, y_0), \tag{2.63g}$$

$$-\frac{2}{h_1}\left[D_x u_{0,m_2} + \psi_1(y_{m_2}) - \lambda_1(y_{m_2})u_{0,m_2}\right]$$
$$-\frac{2}{h_2}\left[\psi_4(x_0) - \lambda_4(x_0)u_{0,m_2} - D_{\bar{y}}u_{0,m_2}\right] = f(x_0, y_{m_2}), \tag{2.63h}$$

$$-\frac{2}{h_1}\left[\psi_2(y_{m_2}) - \lambda_2(y_{m_2})u_{m_1,m_2} - D_{\bar{x}}u_{m_1,m_2}\right]$$
$$-\frac{2}{h_2}\left[\psi_4(x_{m_1}) - \lambda_4(x_{m_1})u_{m_1,m_2} - D_{\bar{y}}u_{m_1,m_2}\right] = f(x_{m_1}, y_{m_2}). \tag{2.63i}$$

Equation (2.63a) represents the difference equation at the inner points, Eqs. (2.63b)–(2.63e) correspond to the difference equations at the inner grid points on the boundary, and Eqs. (2.63f)–(2.63i) represent the difference equations at the four corner points.

2.4.2 Implementation of the Difference Scheme and Numerical Examples

The difference scheme (2.63) is a system of linear equations in $\{u_{ij} \mid (i, j) \in \bar{\omega}\}$. The Jacobi iterative method or Gauss-Seidel iterative method can be used to solve it.

Example 2.4 Apply the difference scheme (2.63) to compute the following problem:

$$
\begin{cases}
-\left(\dfrac{\partial^2 u}{\partial x^2} + \dfrac{\partial^2 u}{\partial y^2}\right) = (\pi^2 - 1)e^x \sin(\pi y), & 0 < x < 2, \quad 0 < y < 1, & \text{(2.64a)} \\[2ex]
\left(-\dfrac{\partial u}{\partial x} + u\right)\Big|_{x=0} = 0, \quad 0 \leqslant y \leqslant 1, & & \text{(2.64b)} \\[2ex]
\left(\dfrac{\partial u}{\partial x} + 2yu\right)\Big|_{x=2} = e^2(1 + 2y)\sin(\pi y), \quad 0 \leqslant y \leqslant 1, & & \text{(2.64c)} \\[2ex]
\left(-\dfrac{\partial u}{\partial y} + 2xu\right)\Big|_{y=0} = -\pi e^x, \quad 0 \leqslant x \leqslant 2, & & \text{(2.64d)} \\[2ex]
\left(\dfrac{\partial u}{\partial y} + x^2 u\right)\Big|_{y=1} = -\pi e^x, \quad 0 \leqslant x \leqslant 2. & & \text{(2.64e)}
\end{cases}
$$

The exact solution of the problem is $u(x, y) = e^x \sin(\pi y)$.

Divide $[0, 2]$ into m_1 equal subintervals and $[0, 1]$ into m_2 equal subintervals. The Gauss-Seidel iterative method is used to solve the system (2.63) with an accuracy $\|u^{(l+1)} - u^{(l)}\|_\infty \leqslant \frac{1}{2} \times 10^{-10}$.

Table 2.8 lists the exact and numerical solutions at five grid points with different step sizes. Table 2.9 gives the absolute value $|u(x_i, y_j) - u_{ij}|$ of the difference between the numerical solutions and exact solutions with different step sizes. Table 2.10 shows the maximum numerical errors

$$
E_\infty(h_1, h_2) = \max_{(i,j)\in\bar{\omega}} \left|u(x_i, y_j) - u_{ij}\right|
$$

with different step sizes.

As we see from Table 2.10, when h_1 and h_2 are both reduced by half, the maximum errors decrease to $1/4$ of the original.

Figure 2.8 shows the numerical solution surface with $h_1 = h_2 = 1/8$. Figure 2.9 displays the exact solution surface. Figure 2.10 illustrates the numerical error surfaces with different step sizes.

Table 2.8 (Example 2.4) The exact and numerical solutions at some grid points with different step sizes

(h_1, h_2)	(x, y)				
	$(0, 1/2)$	$(1/2, 1/2)$	$(1, 1/2)$	$(3/2, 1/2)$	$(2, 1/2)$
(1/8,1/8)	0.970647	1.612039	2.689368	4.469304	7.388177
(1/16,1/16)	0.992662	1.639547	2.711032	4.478549	7.388702
(1/32,1/32)	0.998160	1.646421	2.716462	4.480898	7.388957
(1/64,1/64)	0.999516	1.648118	2.717804	4.481476	7.389022
ES	1.000000	1.648721	2.718282	4.481689	7.389056

Table 2.9 (Example 2.4) The absolute values of numerical errors at some grid points with different step sizes

(h_1, h_2)	(x, y)				
	$(0, 1/2)$	$(1/2, 1/2)$	$(1, 1/2)$	$(3/2, 1/2)$	$(2, 1/2)$
(1/8,1/8)	2.935e−2	3.668e−2	2.891e−2	1.238e−2	8.789e−4
(1/16,1/16)	7.338e−3	9.174e−3	7.250e−3	3.140e−3	3.542e−4
(1/32,1/32)	1.840e−3	2.301e−3	1.819e−3	7.912e−4	9.938e−5
(1/64,1/64)	4.836e−4	6.037e−4	4.780e−4	2.127e−4	3.411e−5

Table 2.10 (Example 2.4) The maximum numerical errors with different step sizes

(h_1, h_2)	$E_\infty(h_1, h_2)$	$E_\infty(2h_1, 2h_2)/E_\infty(h_1, h_2)$
(1/8,1/8)	7.850e−2	
(1/16,1/16)	1.954e−2	4.017
(1/32,1/32)	4.879e−3	4.005
(1/64,1/64)	1.225e−3	3.983

Fig. 2.8 (Example 2.4) the numerical solution surface

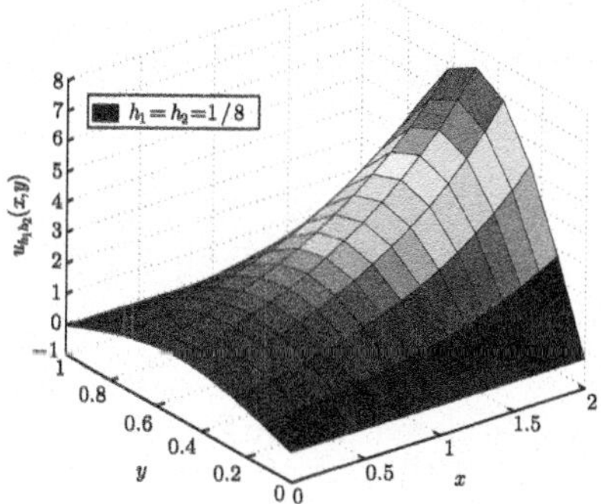

Fig. 2.9 (Example 2.4) the
exact solution surface

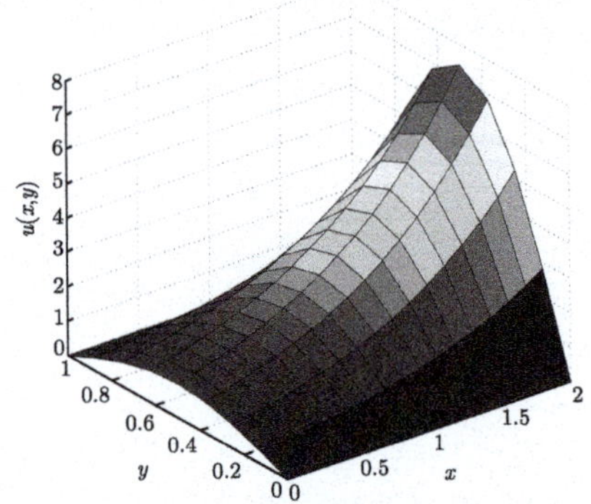

Fig. 2.10 (Example 2.4)
Error surfaces

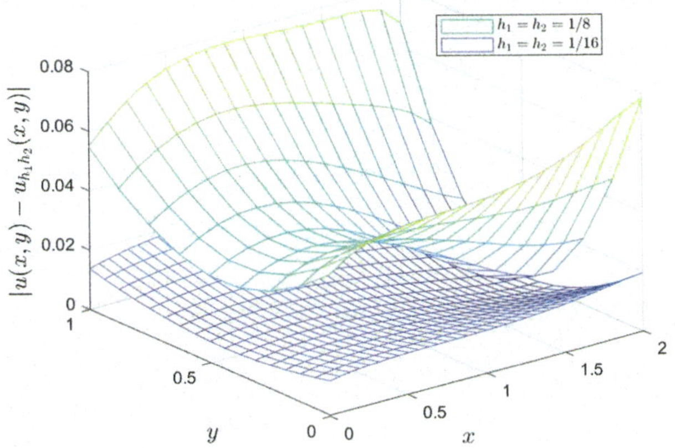

2.5   The Biharmonic Boundary Value Problem

As one of the examples of high-order elliptic equations, a biharmonic boundary
value problem

$$
\begin{cases}
\Delta^2 u \equiv \dfrac{\partial^4 u}{\partial x^4} + 2\dfrac{\partial^4 u}{\partial x^2 \partial y^2} + \dfrac{\partial^4 u}{\partial y^4} = f(x, y), & (x, y) \in \Omega, \\[2mm]
u(x, y) = \varphi(x, y), \quad \Delta u(x, y) = \psi(x, y), & (x, y) \in \Gamma
\end{cases}
\tag{2.65}
$$

is considered, where Ω is the rectangular region $(0, L_1) \times (0, L_2)$ with Γ its
boundary, and $\Delta u = \dfrac{\partial^2 u}{\partial x^2} + \dfrac{\partial^2 u}{\partial y^2}$.

Let

$$
v = \Delta u.
$$

Then (2.65) can be written as the following equivalent system of differential equations:

$$
\begin{cases}
\Delta v = f(x, y), & (x, y) \in \Omega, \\
\Delta u = v(x, y), & (x, y) \in \Omega, \\
u(x, y) = \varphi(x, y), & (x, y) \in \Gamma, \\
v(x, y) = \psi(x, y), & (x, y) \in \Gamma.
\end{cases}
\tag{2.66}
$$

The difference scheme for (2.66) can be derived in the form of

$$
\begin{cases}
\delta_x^2 v_{ij} + \delta_y^2 v_{ij} = f_{ij}, & (i, j) \in \omega, & \text{(2.67a)} \\
\delta_x^2 u_{ij} + \delta_y^2 u_{ij} = v_{ij}, & (i, j) \in \omega, & \text{(2.67b)} \\
u_{ij} = \varphi_{ij}, & (i, j) \in \gamma, & \text{(2.67c)} \\
v_{ij} = \psi_{ij}, & (i, j) \in \gamma, & \text{(2.67d)}
\end{cases}
$$

where $f_{ij} = f(x_i, y_j)$, $\varphi_{ij} = \varphi(x_i, y_j)$, and $\psi_{ij} = \psi(x_i, y_j)$. First, solving (2.67a) and (2.67d), one has $\{v_{ij} \mid (i, j) \in \omega\}$. Then solving (2.67b) and (2.67c), one has $\{u_{ij} \mid (i, j) \in \omega\}$. It can be proved that (2.67) is uniquely solvable and convergent with the convergence order two.

One can also establish a compact difference scheme for (2.66) as follows:

$$
\begin{cases}
\mathcal{B}\delta_x^2 v_{ij} + \mathcal{A}\delta_y^2 v_{ij} = \mathcal{A}\mathcal{B} f_{ij}, & (i, j) \in \omega, & \text{(2.68a)} \\
\mathcal{B}\delta_x^2 u_{ij} + \mathcal{A}\delta_y^2 u_{ij} = \mathcal{A}\mathcal{B} v_{ij}, & (i, j) \in \omega, & \text{(2.68b)} \\
u_{ij} = \varphi_{ij}, & (i, j) \in \gamma, & \text{(2.68c)} \\
v_{ij} = \psi_{ij}, & (i, j) \in \gamma. & \text{(2.68d)}
\end{cases}
$$

First, by solving (2.68a) and (2.68d), one obtains $\{v_{ij} \mid (i, j) \in \omega\}$. Then, by solving (2.68b) and (2.68c), one obtains $\{u_{ij} \mid (i, j) \in \omega\}$. It can be proved that (2.68) is also uniquely solvable and convergent, with the convergence order four.

2.6 Summary and Extension

In this chapter, finite difference methods for elliptic equations are discussed. For the Dirichlet boundary value problem of the Poisson equation on a rectangular domain, a five-point difference scheme is developed, and iterative methods can be used to solve it. The maximum principle is applied to prove that the difference scheme is second-order convergent under the infinity norm and stable with respect to boundary values and right-hand side terms. Richardson extrapolation is then used to increase the accuracy to order four. Interested readers can derive the corresponding results using the energy method.

A nine-point compact difference scheme is also established with the fourth-order accuracy in space by introducing two average operators. The energy method (H^2-analysis) is borrowed to prove that the difference scheme is uniquely solvable for arbitrary step sizes h_1 and h_2 and fourth-order convergent with respect to h_1 and h_2 in the maximum norm. It can be proved that the compact difference scheme satisfies the maximum principle when the step sizes are restricted to $1/\sqrt{5} \leqslant h_1/h_2 \leqslant \sqrt{5}$ [1]. In [2], the energy method is utilized to prove the unconditional stability and convergence of the solution of the compact difference scheme in the H^1-norm.

For the derivative boundary value problem and the biharmonic boundary value problem, only the difference scheme is presented. Readers are encouraged to apply the energy method or the maximum principle to prove uniqueness, convergence, and stability.

Assume that the region Ω under consideration is curvilinear as shown in Fig. 2.11. Take a point in the region Ω and denote it by (x_0, y_0). Take a fixed step size h and utilize two clusters of parallel lines

$$x = x_i \equiv x_0 + ih, \qquad i = 0, \pm 1, \pm 2, \ldots,$$

$$y = y_j \equiv y_0 + jh, \qquad j = 0, \pm 1, \pm 2, \ldots$$

to divide Ω into a finite number of squares. The intersections of two clusters of parallel lines are called grid points. Only the grid points located in $\bar{\Omega}$ are considered. These grid points can be further classified into three categories.

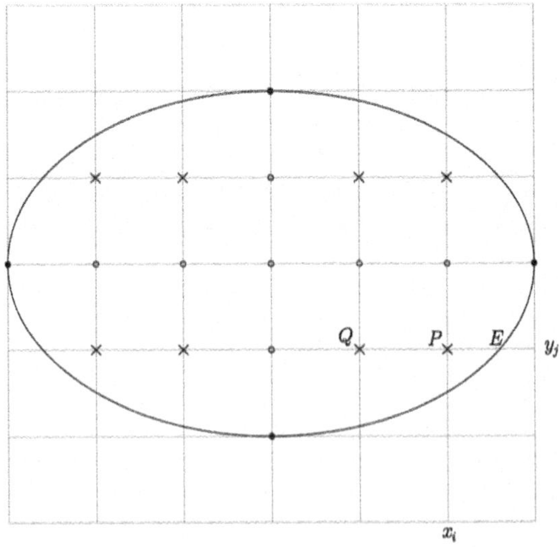

Fig. 2.11 Division of a curvilinear region

- A grid point is called the regular interior point if its four adjacent grid points are all in $\bar{\Omega}$. The set of all the regular interior points is denoted by $\overset{\circ}{\Omega}_h$, which is denoted by "o" in Fig. 2.11.
- A grid point is called the non-regular interior point if it belongs to Ω but does not belong to $\overset{\circ}{\Omega}_h$. It is denoted by "×" in Fig. 2.11.
- A grid point is called the boundary grid point if it is located on Γ, which is denoted by "•" as shown in Fig. 2.11.

For the Dirichlet boundary value problem, the value of u is known at the boundary grid points. A five-point difference scheme can be established at regular interior points, as described in the previous section, and a difference scheme can also be formulated based on interpolation at non-regular interior points. For example, let P be a non-regular interior point, as shown in Fig. 2.11, with coordinates (x_i, y_j). To the right of P, the line $y = y_j$ intersects the boundary of Ω at E, with coordinates $(x_i + h_E, y_j)$. The point Q, located to the left of P, is an interior point with coordinates (x_{i-1}, y_j). By performing linear interpolation for u based on the points Q and E, the value of u at P can be determined as follows:

$$u(x_i, y_j) = \frac{h_E}{h + h_E} u(x_{i-1}, y_j) + \frac{h}{h + h_E} u(E) - \frac{h h_E}{2} \frac{\partial^2 u(\xi_i, y_j)}{\partial x^2}.$$

Omitting the small term

$$-\frac{h h_E}{2} \frac{\partial^2 u(\xi_i, y_j)}{\partial x^2}$$

and replacing $u(x_i, y_j)$ with u_{ij}, the difference equation at P reads

$$u_{ij} - \frac{h_E}{h + h_E} u_{i-1,j} = \frac{h}{h + h_E} u(E).$$

If the differential equation under consideration is a variable-coefficient one, for example,

$$-\left[a(x, y)\frac{\partial^2 u}{\partial x^2} + b(x, y)\frac{\partial^2 u}{\partial y^2} \right] + c(x, y)\frac{\partial u}{\partial x} + d(x, y)\frac{\partial u}{\partial y} + e(x, y)u = f(x, y),$$

we can establish a difference scheme

$$-\left[a(x_i, y_j)\delta_x^2 u_{ij} + b(x_i, y_j)\delta_y^2 u_{ij} \right] + c(x_i, y_j)\frac{u_{i+1,j} - u_{i-1,j}}{2h_1}$$

$$+d(x_i, y_j)\frac{u_{i,j+1} - u_{i,j-1}}{2h_2} + e(x_i, y_j)u_{ij} = f(x_i, y_j).$$

2.7 Exercise

2.1 Use the difference scheme (2.14) to compute the following problem:

$$\begin{cases} -\left(\dfrac{\partial^2 u}{\partial x^2} + \dfrac{\partial^2 u}{\partial y^2}\right) = -6(x+y), & 0 < x < 1, \quad 0 < y < 1, \\[2mm] u(x,0) = x^3, \quad u(x,1) = 1+x^3, & 0 \leqslant x \leqslant 1, \\[2mm] u(0,y) = y^3, \quad u(1,y) = 1+y^3, & 0 \leqslant y \leqslant 1. \end{cases}$$

Take $h = 1/3$ and compute the numerical solutions at four points $(1/3, 1/3)$, $(2/3, 1/3)$, $(1/3, 2/3)$, $(2/3, 2/3)$. Compare them with the exact solutions and explain the observed phenomenon. The exact solution is $u(x,y) = x^3 + y^3$.

2.2 Define $\Omega_h = \{(x_i, y_j) \mid (i,j) \in \bar{\omega}\}$. Let $v = \{v_{ij} \mid (i,j) \in \bar{\omega}\}$ be a grid function on Ω_h and

$$(L_h v)_{ij} \equiv -\left(\delta_x^2 v_{ij} + \delta_y^2 v_{ij}\right) \geqslant 0, \quad (i,j) \in \omega.$$

Prove

$$\min_{(i,j)\in\omega} v_{ij} \geqslant \min_{(i,j)\in\gamma} v_{ij}.$$

2.3 Consider the problem

$$\begin{cases} -\left(\dfrac{\partial^2 u}{\partial x^2} + \dfrac{\partial^2 u}{\partial y^2}\right) = f(x,y), & (x,y) \in \Omega, \\[2mm] u = \varphi(x,y), & (x,y) \in \Gamma, \end{cases}$$

where $\Omega = (0,1) \times (0,1)$ and Γ is the boundary of Ω. Take a square grid partition with $h_1 = h_2 = 1/m$. Analyze the local truncation error of the difference scheme as follows:

$$\begin{cases} -\dfrac{1}{2h^2}(u_{i-1,j-1} + u_{i+1,j-1} + u_{i-1,j+1} + u_{i+1,j+1} - 4u_{ij}) \quad (2.69) \\[2mm] \quad = f(x_i, y_j), \qquad\qquad (i,j) \in \omega, \\[2mm] u_{ij} = \varphi(x_i, y_j), \qquad (i,j) \in \gamma. \end{cases}$$

Show that the maximum principle is valid and the difference scheme is convergent with the convergence order two.

2.4 Define $\Omega_h = \{(x_i, y_j) \mid (i, j) \in \bar{\omega}\}$. Let $v = \{v_{ij} \mid (i, j) \in \bar{\omega}\}$ be a function on Ω_h, and

$$(\hat{L}_h v)_{ij} \equiv -\left(\mathcal{B}\delta_x^2 v_{ij} + \mathcal{A}\delta_y^2 v_{ij}\right) \leqslant 0, \quad (i, j) \in \omega.$$

Prove

(1) When the step sizes h_1 and h_2 satisfy $1/\sqrt{5} \leqslant h_1/h_2 \leqslant \sqrt{5}$, it holds that
$$\max_{(i,j)\in\omega} v_{ij} \leqslant \max_{(i,j)\in\gamma} v_{ij}.$$
(2) Under the constraint of the above step ratio, the difference scheme (2.41) is fourth-order convergent in the maximum norm.

2.5 Assume $v \in \overset{\circ}{\mathcal{V}}_h$ and prove

$$-\left(\Delta_h v, v\right) = |v|_1^2.$$

2.6 Consider the difference scheme (2.63) and denote $U_j = (u_{0j}, u_{1j}, \ldots, u_{m_1,j})^{\mathrm{T}}$. If (2.63) is written as

$$\left\{\begin{array}{l}
A_0 U_0 \quad + \quad C_0 U_1 \hspace{6.5cm} = F_0, \\
B_1 U_0 \quad + \quad A_1 U_1 \quad + \quad C_1 U_2 \hspace{4.3cm} = F_1, \\
\hspace{1.7cm} B_2 U_1 \quad + \quad A_2 U_2 \quad + \quad C_2 U_3 \hspace{2.5cm} = F_2, \\
\hspace{4cm} \cdots \cdots \\
\hspace{1.2cm} B_{m_2-2} U_{m_2-3} + A_{m_2-2} U_{m_2-2} + C_{m_2-2} U_{m_2-1} \hspace{0.8cm} = F_{m_2-2}, \\
\hspace{2.6cm} B_{m_2-1} U_{m_2-2} + A_{m_2-1} U_{m_2-1} \quad + \quad C_{m_2-1} U_{m_2} = F_{m_2-1}, \\
\hspace{4.3cm} B_{m_2} U_{m_2-1} \quad + \quad A_{m_2} U_{m_2} = F_{m_2},
\end{array}\right.$$

then what are the specific expressions of A_j, B_j, C_j, and F_j, $j = 0, 1, \cdots, m_2$?

2.7 Use the difference scheme (2.14) to solve the following problem:

$$\left\{\begin{array}{l}
-\left(\dfrac{\partial^2 u}{\partial x^2} + \dfrac{\partial^2 u}{\partial y^2}\right) = 0, \quad 0 < x < 1, \ 0 < y < 1, \\[2mm]
u(0, y) = \sin y + \cos y, \quad u(1, y) = e(\sin y + \cos y), \quad 0 \leqslant y \leqslant 1, \\[2mm]
u(x, 0) = e^x, \quad u(x, 1) = e^x(\sin 1 + \cos 1), \quad 0 < x < 1.
\end{array}\right.$$

The exact solution of the problem is $u(x, y) = e^x(\sin y + \cos y)$. Take $(h_1, h_2) = \left(\frac{1}{4}, \frac{1}{4}\right), \left(\frac{1}{8}, \frac{1}{8}\right), \left(\frac{1}{16}, \frac{1}{16}\right), \left(\frac{1}{32}, \frac{1}{32}\right), \left(\frac{1}{64}, \frac{1}{64}\right)$, respectively, and fill data in Table 2.11. Draw the exact solution surface, the numerical solution surface, and the error surface.

Table 2.11 (Exercise 2.7) The exact and numerical solutions at some grid points with different step sizes

(h_1, h_2)	$\left(\frac{1}{4}, \frac{1}{4}\right)$	$\left(\frac{1}{2}, \frac{1}{4}\right)$	$\left(\frac{3}{4}, \frac{1}{4}\right)$	$\left(\frac{1}{4}, \frac{3}{4}\right)$	$\left(\frac{1}{2}, \frac{3}{4}\right)$	$\left(\frac{3}{4}, \frac{1}{4}\right)$
$\left(\frac{1}{4}, \frac{1}{4}\right)$						
$\left(\frac{1}{8}, \frac{1}{8}\right)$						
$\left(\frac{1}{16}, \frac{1}{16}\right)$						
$\left(\frac{1}{32}, \frac{1}{32}\right)$						
$\left(\frac{1}{64}, \frac{1}{64}\right)$						
ES						

The column group is headed (x, y).

References

1. Samarskiĭ, A.A., Andreev, V.B.: Difference Methods for Elliptic Equations. Nauka, Moscow (1976)
2. Sun, Z.Z.: Numerical Methods for Partial Differential Equations. 2nd edn. Science Press, Beijing (2012)

Chapter 3
Finite Difference Methods for Parabolic Equations

Partial differential equations of the parabolic type are commonly encountered in the study of heat conduction, gas expansion, and electromagnetic field propagation. One of the independent variables in these problems is time, typically denoted by t. Thus, parabolic equations often describe physical processes that evolve over time, known as unsteady physical processes. There are three main types of well-posed problems for parabolic equations: the pure initial value problem, the initial-boundary value problem on a semi-unbounded domain, and the initial-boundary value problem on a bounded domain.

In this chapter, we primarily focus on finite difference methods for the Dirichlet initial-boundary value problem of parabolic equations on a bounded domain. Specifically, we discuss how to establish a difference scheme, how to solve it, and how to analyze the convergence and stability of the scheme.

3.1 The Dirichlet Initial-Boundary Value Problem

We consider a finite difference scheme for the one-dimensional inhomogeneous heat conduction equation with Dirichlet initial-boundary value conditions (i.e., the first boundary value problem)

$$
\begin{cases}
\dfrac{\partial u}{\partial t} - a\dfrac{\partial^2 u}{\partial x^2} = f(x,t), & 0 < x < L, \quad 0 < t \leqslant T, & \text{(3.1a)} \\[2mm]
u(x,0) = \varphi(x), & 0 \leqslant x \leqslant L, & \text{(3.1b)} \\[2mm]
u(0,t) = \alpha(t), \quad u(L,t) = \beta(t), & 0 < t \leqslant T, & \text{(3.1c)}
\end{cases}
$$

© Science Press 2026

Z.-Z. Sun et al., *Numerical Solutions to Partial Differential Equations with Finite Difference Methods*, Springer Asia Pacific Mathematics Series 9, https://doi.org/10.1007/978-981-95-5563-5_3

where a is a positive constant, $f(x, t), \varphi(x), \alpha(t), \beta(t)$ are all given functions, the conditions $\varphi(0) = \alpha(0)$ and $\varphi(L) = \beta(0)$ ensure consistency, (3.1b) is the initial value condition, and (3.1c) is the boundary value condition.

Before introducing the difference scheme, we first give a priori estimate for the solution to the homogeneous boundary value problem by using the energy method.

Theorem 3.1 *Let $v(x, t)$ be the solution of the homogeneous* Dirichlet *boundary value problem of the parabolic type*

$$
\begin{cases}
\dfrac{\partial v}{\partial t} - a\dfrac{\partial^2 v}{\partial x^2} = f(x, t), & 0 < x < L, \quad 0 < t \leqslant T, & \text{(3.2a)} \\[2mm]
v(x, 0) = \varphi(x), & 0 \leqslant x \leqslant L, & \text{(3.2b)} \\[2mm]
v(0, t) = 0, \quad v(L, t) = 0, & 0 < t \leqslant T, & \text{(3.2c)}
\end{cases}
$$

and $\varphi(0) = \varphi(L) = 0$. Then we have

$$
\int_0^L v^2(x, t)\mathrm{d}x \leqslant \int_0^L \varphi^2(x)\mathrm{d}x + \frac{L^2}{12a} \int_0^t \left[\int_0^L f^2(x, s)\mathrm{d}x \right] \mathrm{d}s, \quad 0 \leqslant t \leqslant T;
$$

$$\text{(3.3)}$$

$$
\int_0^L \left[\frac{\partial v(x, t)}{\partial x} \right]^2 \mathrm{d}x \leqslant \int_0^L \left[\varphi'(x) \right]^2 \mathrm{d}x + \frac{1}{2a} \int_0^t \left[\int_0^L f^2(x, s)\mathrm{d}x \right] \mathrm{d}s,
$$

$$0 \leqslant t \leqslant T;$$
$$\text{(3.4)}$$

$$
\left(\max_{0 \leqslant x \leqslant L} |v(x, t)| \right)^2 \leqslant \frac{L}{4} \left(\int_0^L \left[\varphi'(x) \right]^2 \mathrm{d}x + \frac{1}{2a} \int_0^t \left[\int_0^L f^2(x, s)\mathrm{d}x \right] \mathrm{d}s \right),
$$

$$0 \leqslant t \leqslant T.$$
$$\text{(3.5)}$$

Proof

(I) Multiplying both sides of (3.2a) by $2v$ and integrating the result with respect to x on $(0, L)$, we have

$$
2 \int_0^L \frac{\partial v(x, t)}{\partial t} v(x, t)\mathrm{d}x - 2a \int_0^L \frac{\partial^2 v(x, t)}{\partial x^2} v(x, t)\mathrm{d}x
$$

$$
= 2 \int_0^L f(x, t)v(x, t)\mathrm{d}x.
$$

Utilizing the integration by parts and noticing (3.2c), we have

$$\frac{d}{dt} \int_0^L v^2(x, t)dx + 2a \int_0^L \left[\frac{\partial v(x, t)}{\partial x}\right]^2 dx$$

$$= 2 \int_0^L f(x, t)v(x, t)dx$$

$$\leqslant 2\sqrt{\int_0^L f^2(x, t)dx} \sqrt{\int_0^L v^2(x, t)dx}$$

$$\leqslant \frac{12a}{L^2} \int_0^L v^2(x, t)dx + \frac{L^2}{12a} \int_0^L f^2(x, t)dx.$$

With the application of Lemma 1.1, it yields

$$\int_0^L v^2(x, t)dx \leqslant \frac{L^2}{6} \int_0^L \left[\frac{\partial v(x, t)}{\partial x}\right]^2 dx,$$

which implies

$$\frac{d}{dt} \int_0^L v^2(x, t)dx \leqslant \frac{L^2}{12a} \int_0^L f^2(x, t)dx.$$

Integrating both sides of the above inequality with respect to t produces

$$\int_0^L v^2(x, t)dx \leqslant \int_0^L v^2(x, 0)dx + \frac{L^2}{12a} \int_0^t \left[\int_0^L f^2(x, s)dx\right] ds$$

$$= \int_0^L \varphi^2(x)dx + \frac{L^2}{12a} \int_0^t \left[\int_0^L f^2(x, s)dx\right] ds,$$

which is exactly (3.3).

(II) Multiplying both sides of (3.2a) by $\frac{\partial v}{\partial t}$ and integrating the result with respect to x on $(0, L)$, we have

$$\int_0^L \left[\frac{\partial v(x, t)}{\partial t}\right]^2 dx - a \int_0^L \frac{\partial^2 v(x, t)}{\partial x^2} \frac{\partial v(x, t)}{\partial t}dx$$

$$= \int_0^L f(x, t)\frac{\partial v(x, t)}{\partial t}dx$$

$$\leqslant \int_0^L \left[\frac{\partial v(x, t)}{\partial t}\right]^2 dx + \frac{1}{4} \int_0^L f^2(x, t)dx. \tag{3.6}$$

Noticing (3.2c), the second term in the above inequality becomes

$$-\int_0^L \frac{\partial^2 v(x,t)}{\partial x^2} \frac{\partial v(x,t)}{\partial t} dx$$

$$= -\frac{\partial v(x,t)}{\partial x} \frac{\partial v(x,t)}{\partial t}\bigg|_{x=0}^L + \int_0^L \frac{\partial v(x,t)}{\partial x} \frac{\partial^2 v(x,t)}{\partial x \partial t} dx$$

$$= \frac{1}{2} \frac{\mathrm{d}}{\mathrm{d}t} \int_0^L \left[\frac{\partial v(x,t)}{\partial x}\right]^2 dx. \tag{3.7}$$

Substituting (3.7) into (3.6) yields

$$\frac{a}{2} \cdot \frac{\mathrm{d}}{\mathrm{d}t} \int_0^L \left[\frac{\partial v(x,t)}{\partial x}\right]^2 dx \leqslant \frac{1}{4} \int_0^L f^2(x,t)dx,$$

namely,

$$\frac{\mathrm{d}}{\mathrm{d}t} \int_0^L \left[\frac{\partial v(x,t)}{\partial x}\right]^2 dx \leqslant \frac{1}{2a} \int_0^L f^2(x,t)dx, \quad 0 < t \leqslant T.$$

Integrating the above inequality with respect to t, we have

$$\int_0^L \left[\frac{\partial v(x,t)}{\partial x}\right]^2 dx \leqslant \int_0^L \left[\frac{\partial v(x,0)}{\partial x}\right]^2 dx + \frac{1}{2a} \int_0^t \left[\int_0^L f^2(x,s)dx\right] ds$$

$$= \int_0^L [\varphi'(x)]^2 dx + \frac{1}{2a} \int_0^t \left[\int_0^L f^2(x,s)dx\right] ds,$$

which is exactly (3.4).

(III) It follows from (3.4) and Lemma 1.1 that (3.5) holds.

$$\square$$

3.2 The Forward Euler Scheme

To solve problem (3.1) using the finite difference method, we first divide the computational domain

$$D = \{(x,t) \mid 0 < x < L,\ 0 < t \leqslant T\}.$$

Let m and n be two positive integers. Divide the spatial interval $[0, L]$ into m equal subintervals and temporal interval $[0, T]$ into n equal subintervals. Denote $h = L/m$, $\tau = T/n$, $x_i = ih$, $0 \leqslant i \leqslant m$; $t_k = k\tau$, $0 \leqslant k \leqslant n$. Here h

Fig. 3.1 Grid division

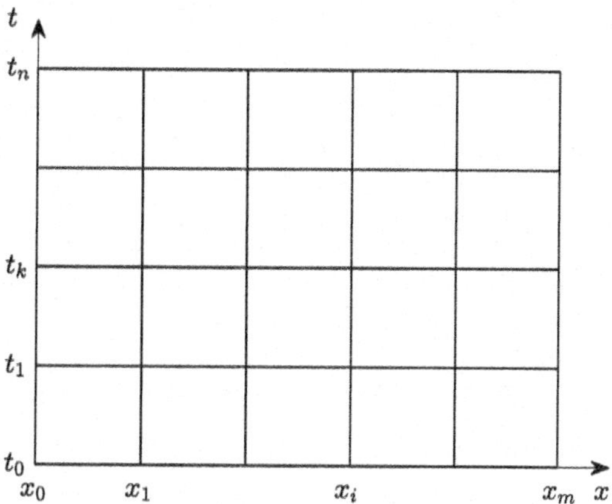

and τ are referred to as the spatial step size and the temporal step size, respectively. Denote **step ratio** as $r = a\tau/h^2$. In addition, denote $x_{i+\frac{1}{2}} = \frac{1}{2}(x_i + x_{i+1})$, $t_{k+\frac{1}{2}} = \frac{1}{2}(t_k + t_{k+1})$.

We draw two families of parallel lines:

$$x = x_i, \quad 0 \leqslant i \leqslant m; \quad t = t_k, \quad 0 \leqslant k \leqslant n$$

to divide the domain $\bar{D}$ into rectangular grids, as illustrated in Fig. 3.1. Define the following sets: $\Omega_h = \{x_i \mid 0 \leqslant i \leqslant m\}$, $\Omega_\tau = \{t_k \mid 0 \leqslant k \leqslant n\}$, $\Omega_{h\tau} = \Omega_h \times \Omega_\tau$. Each point (x_i, t_k) is called a **node point**; the node points on $t = 0$, $x = 0$, and $x = L$ are called **boundary node points**, while all other node points are called **inner node points**; furthermore, all the node points $\{(x_i, t_k) \mid 0 \leqslant i \leqslant m\}$ on the line $t = t_k$ are called the **k-th time level node points**.

Let $\{v_i^k \mid 0 \leqslant i \leqslant m, 0 \leqslant k \leqslant n\}$ be a grid function on $\Omega_{h\tau}$. Introduce the following notation:

$$v_i^{k+\frac{1}{2}} = \frac{1}{2}(v_i^k + v_i^{k+1}), \quad \delta_t v_i^{k+\frac{1}{2}} = \frac{1}{\tau}(v_i^{k+1} - v_i^k),$$

$$D_t v_i^k = \frac{1}{\tau}(v_i^{k+1} - v_i^k), \quad D_{\bar{t}} v_i^k = \frac{1}{\tau}(v_i^k - v_i^{k-1}),$$

$$D_x v_i^k = \frac{1}{h}(v_{i+1}^k - v_i^k), \quad D_{\bar{x}} v_i^k = \frac{1}{h}(v_i^k - v_{i-1}^k),$$

$$\delta_x v_{i+\frac{1}{2}}^k = \frac{1}{h}(v_{i+1}^k - v_i^k), \quad \delta_x^2 v_i^k = \frac{1}{h^2}(v_{i-1}^k - 2v_i^k + v_{i+1}^k),$$

$$v_i^{\bar{k}} = \frac{1}{2}(v_i^{k+1} + v_i^{k-1}), \quad \Delta_t v_i^k = \frac{1}{2\tau}(v_i^{k+1} - v_i^{k-1}).$$

Let $v^k = (v_0^k, v_1^k, \ldots, v_m^k)$. Then v^k is a grid function on Ω_h.

Introduce the same notation as those in Chap. 1:

$$\mathcal{U}_h = \{v \mid v = \{v_i \mid 0 \leqslant i \leqslant m\} \text{ be the grid function on } \Omega_h\},$$

$$\mathring{\mathcal{U}}_h = \{v \mid v = \{v_i \mid 0 \leqslant i \leqslant m\} \in \mathcal{U}_h, \text{ and } v_0 = v_m = 0\}.$$

Let $v \in \mathcal{U}_h$ and introduce the following norms or seminorm:

$$\|v\|_\infty = \max_{0 \leqslant i \leqslant m} |v_i|, \quad \|v\| = \sqrt{h\left(\frac{1}{2}v_0^2 + \sum_{i=1}^{m-1} v_i^2 + \frac{1}{2}v_m^2\right)},$$

$$|v|_1 = \sqrt{h\sum_{i=1}^{m}\left(\frac{v_i - v_{i-1}}{h}\right)^2}, \quad \|\delta_x v\| = |v|_1, \quad \|v\|_1 = \sqrt{\|v\|^2 + |v|_1^2}.$$

Then, $\|v\|_\infty$, $\|v\|$, and $\|v\|_1$ are all norms of v on $\mathcal{U}_h$, which are, respectively, called the maximum norm (infinity/uniform norm), 2-norm (also known as the L^2-norm or average norm), and H^1-norm. Moreover, $|v|_1$ is a seminorm on $\mathcal{U}_h$, and it is also a norm on $\mathring{\mathcal{U}}_h$, referred to as the 2-norm of the difference quotient.

3.2.1   Derivation of the Difference Scheme

Define the grid function

$$U = \{U_i^k \mid 0 \leqslant i \leqslant m, \ 0 \leqslant k \leqslant n\}$$

on $\Omega_{h\tau}$, where

$$U_i^k = u(x_i, t_k), \quad 0 \leqslant i \leqslant m, \quad 0 \leqslant k \leqslant n.$$

Considering Eq. (3.1a) at the node point (x_i, t_k), we have

$$\frac{\partial u}{\partial t}(x_i, t_k) - a\frac{\partial^2 u}{\partial x^2}(x_i, t_k) = f(x_i, t_k), \quad 1 \leqslant i \leqslant m - 1, \quad 0 \leqslant k \leqslant n - 1. \tag{3.8}$$

Substituting

$$\frac{\partial^2 u}{\partial x^2}(x_i, t_k) = \frac{1}{h^2}[u(x_{i-1}, t_k) - 2u(x_i, t_k) + u(x_{i+1}, t_k)] - \frac{h^2}{12}\frac{\partial^4 u}{\partial x^4}(\xi_{ik}, t_k)$$

$$= \delta_x^2 U_i^k - \frac{h^2}{12}\frac{\partial^4 u}{\partial x^4}(\xi_{ik}, t_k), \quad x_{i-1} < \xi_{ik} < x_{i+1}$$

and

$$\frac{\partial u}{\partial t}(x_i, t_k) = \frac{1}{\tau}[u(x_i, t_{k+1}) - u(x_i, t_k)] - \frac{\tau}{2}\frac{\partial^2 u}{\partial t^2}(x_i, \eta_{ik})$$

$$= D_t U_i^k - \frac{\tau}{2}\frac{\partial^2 u}{\partial t^2}(x_i, \eta_{ik}), \quad t_k < \eta_{ik} < t_{k+1}$$

into (3.8) arrives at

$$D_t U_i^k - a\delta_x^2 U_i^k = f(x_i, t_k) + \frac{\tau}{2}\frac{\partial^2 u}{\partial t^2}(x_i, \eta_{ik}) - \frac{ah^2}{12}\frac{\partial^4 u}{\partial x^4}(\xi_{ik}, t_k),$$

$$1 \leqslant i \leqslant m - 1, \quad 0 \leqslant k \leqslant n - 1. \tag{3.9}$$

In light of the initial-boundary value conditions (3.1b)–(3.1c), we have

$$\begin{cases} U_i^0 = \varphi(x_i), & 0 \leqslant i \leqslant m, \\ U_0^k = \alpha(t_k), & U_m^k = \beta(t_k), \quad 1 \leqslant k \leqslant n. \end{cases} \tag{3.10}$$

By neglecting the small term

$$(R_1)_i^k = \frac{\tau}{2}\frac{\partial^2 u}{\partial t^2}(x_i, \eta_{ik}) - \frac{ah^2}{12}\frac{\partial^4 u}{\partial x^4}(\xi_{ik}, t_k)$$

in (3.9) and replacing U_i^k with u_i^k, a difference scheme for solving (3.1) reads

$$\begin{cases} D_t u_i^k - a\delta_x^2 u_i^k = f(x_i, t_k), & 1 \leqslant i \leqslant m - 1, \quad 0 \leqslant k \leqslant n - 1, & (3.11a) \\ u_i^0 = \varphi(x_i), & 0 \leqslant i \leqslant m, & (3.11b) \\ u_0^k = \alpha(t_k), & u_m^k = \beta(t_k), \quad 1 \leqslant k \leqslant n. & (3.11c) \end{cases}$$

The difference scheme (3.11) is referred to as the **forward Euler scheme**, and $(R_1)_i^k$ represents the **local truncation error** of (3.11a).

Denote

$$c_1 = \max\left\{\frac{1}{2}\max_{(x,t)\in\bar{D}}\left|\frac{\partial^2 u}{\partial t^2}(x, t)\right|, \frac{a}{12}\max_{(x,t)\in\bar{D}}\left|\frac{\partial^4 u}{\partial x^4}(x, t)\right|\right\}, \tag{3.12}$$

and then we have

$$|(R_1)_i^k| \leqslant c_1(\tau + h^2), \quad 1 \leqslant i \leqslant m - 1, \quad 0 \leqslant k \leqslant n - 1. \tag{3.13}$$

The stencil of the difference scheme (3.11) is illustrated in Fig. 3.2, where the symbol "×" indicates the point at which the difference scheme is established, and "o" indicates the node points used in the difference scheme.

Fig. 3.2 The stencil of the
forward Euler scheme (3.11)

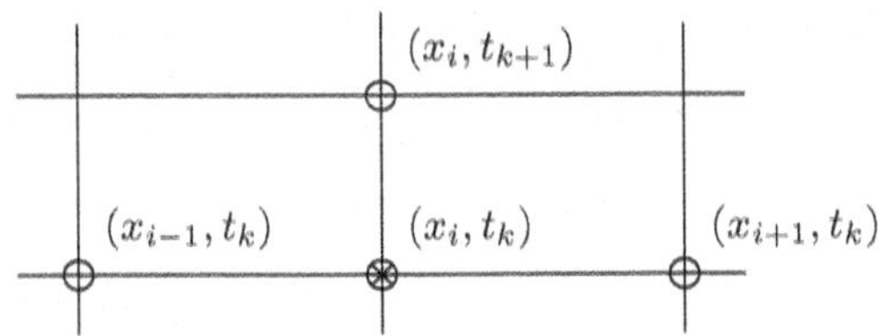

3.2.2   Existence of the Difference Solution

The difference equation (3.11a) can be rewritten as

$$u_i^{k+1} = (1-2r)u_i^k + r(u_{i-1}^k + u_{i+1}^k) + \tau f(x_i, t_k), \quad 1 \leqslant i \leqslant m-1, \quad 0 \leqslant k \leqslant n-1,$$

where r denotes the step ratio. The above equation indicates that numerical solutions
at the $(k+1)$-th time level can be expressed directly by the known values at the k-th
time level. If the value $\{u_i^k \mid 0 \leqslant i \leqslant m\}$ at the k-th time level is available, then
the value $\{u_i^{k+1} \mid 0 \leqslant i \leqslant m\}$ at the $(k+1)$-th time level can be directly obtained
from the above formula and (3.11c). Therefore, the difference scheme is uniquely
solvable for any value of the step ratio r.

The difference scheme (3.11), being explicit in form, is sometimes termed the
classical explicit scheme.

3.2.3   Implementation of the Difference Scheme and Numerical Examples

The classical explicit scheme (3.11) can be reformulated in matrix form as

$$\begin{pmatrix} u_1^{k+1} \\ u_2^{k+1} \\ \vdots \\ u_{m-2}^{k+1} \\ u_{m-1}^{k+1} \end{pmatrix} = \begin{pmatrix} 1-2r & r & & & \\ r & 1-2r & r & & \\ & \ddots & \ddots & \ddots & \\ & & r & 1-2r & r \\ & & & r & 1-2r \end{pmatrix} \begin{pmatrix} u_1^k \\ u_2^k \\ \vdots \\ u_{m-2}^k \\ u_{m-1}^k \end{pmatrix}$$

$$+ \begin{pmatrix} \tau f(x_1, t_k) + r u_0^k \\ \tau f(x_2, t_k) \\ \vdots \\ \tau f(x_{m-2}, t_k) \\ \tau f(x_{m-1}, t_k) + r u_m^k \end{pmatrix}.$$

Example 3.1   Apply the forward *Euler* scheme (3.11) to compute the problem

$$\begin{cases} \dfrac{\partial u}{\partial t} - \dfrac{\partial^2 u}{\partial x^2} = 0, & 0 < x < 1, \quad 0 < t \leqslant 1, \\ u(x, 0) = e^x, & 0 \leqslant x \leqslant 1, \\ u(0, t) = e^t, & u(1, t) = e^{1+t}, \quad 0 < t \leqslant 1. \end{cases}$$

The exact solution of the above initial-boundary value problem is $u(x, t) = e^{x+t}$.
Table 3.1 lists a subset of the numerical results with the step sizes $h = 1/10$ and $\tau = 1/200$ (the step ratio $r = 1/2$), which demonstrate that the numerical solution provides a good approximation to the exact solution. Table 3.2 gives some numerical results with the step sizes $h = 1/10$ and $\tau = 1/100$ (the step ratio $r = 1$). As the number of time steps increases, the numerical error accumulates and the solution deviates significantly from the exact values. Table 3.3 gives the maximum errors

$$E_\infty(h, \tau) = \max_{1 \leqslant i \leqslant m-1, 1 \leqslant k \leqslant n} |u(x_i, t_k) - u_i^k|$$

of numerical solutions by taking different step sizes with the fixed step ratio $r = 1/2$.

As shown in Table 3.3, halving the spatial step size and reducing the temporal step size to one-quarter of its original value lead to an approximate fourfold reduction in the maximum error. Figure 3.3 displays the exact and numerical solution profiles using step sizes $h = 1/10$, $\tau = 1/200$ at $t = 1$. Here, $u_{h\tau}(x, t)$ represents the numerical solution at the point (x, t) computed with step sizes h and τ. Given that the relative error is around two in ten thousand, the difference between both profiles is visually indistinguishable. Figure 3.4 shows the numerical error curves at $t = 1$, and Fig. 3.5 presents the numerical error surfaces for various step sizes.

Table 3.1 (Example 3.1)
The numerical solutions,
exact solutions, and absolute
values of errors at some node
points
$(h = 1/10, \tau = 1/200)$

k	(x, t)	NS	ES	$|\text{ES}-\text{NS}|$
20	(0.5, 0.1)	1.821889	1.822119	2.301e−4
40	(0.5, 0.2)	2.013409	2.013753	3.436e−4
60	(0.5, 0.3)	2.225128	2.225541	4.125e−4
80	(0.5, 0.4)	2.459135	2.459603	4.679e−4
100	(0.5, 0.5)	2.717760	2.718282	5.215e−4
120	(0.5, 0.6)	3.003588	3.004166	5.779e−4
140	(0.5, 0.7)	3.319478	3.320117	6.393e−4
160	(0.5, 0.8)	3.668590	3.669297	7.068e−4
180	(0.5, 0.9)	4.054419	4.055200	7.812e−4
200	(0.5, 1.0)	4.480826	4.481689	8.634e−4

Table 3.2 (Example 3.1) Numerical solutions, exact solutions, and absolute values of errors at some node points ($h = 1/10$, $\tau = 1/100$)

k	(x, t)	NS	ES	$|\text{ES}-\text{NS}|$
1	$(0.5, 0.01)$	1.665222	1.665291	6.897e−5
2	$(0.5, 0.02)$	1.681888	1.682028	1.393e−4
3	$(0.5, 0.03)$	1.698721	1.698932	2.111e−4
4	$(0.5, 0.04)$	1.715723	1.716007	2.843e−4
5	$(0.5, 0.05)$	1.732894	1.733253	3.589e−4
6	$(0.5, 0.06)$	1.750393	1.750673	2.794e−4
7	$(0.5, 0.07)$	1.767291	1.768267	9.761e−4
8	$(0.5, 0.08)$	1.787463	1.786038	1.424e−3
9	$(0.5, 0.09)$	1.796973	1.803988	7.015e−3
10	$(0.5, 0.10)$	1.842269	1.822119	2.015e−2
11	$(0.5, 0.11)$	1.775371	1.840431	6.506e−2
12	$(0.5, 0.12)$	2.054576	1.858928	1.956e−1
13	$(0.5, 0.13)$	1.286980	1.877611	5.906e−1
14	$(0.5, 0.14)$	3.651896	1.896481	1.755e+0
15	$(0.5, 0.15)$	− 3.277574	1.915541	5.193e+0
16	$(0.5, 0.16)$	1.721143e+1	1.934792	1.528e+1
17	$(0.5, 0.17)$	− 4.283715e + 1	1.954237	4.479e+1
18	$(0.5, 0.18)$	1.329477e+2	1.973878	1.310e+2

Table 3.3 (Example 3.1) The maximum errors with different step sizes ($r = 1/2$)

h	τ	$E_\infty(h, \tau)$	$E_\infty(2h, 4\tau)/E_\infty(h, \tau)$
1/10	1/200	8.634e−4	
1/20	1/800	2.175e−4	3.970
1/40	1/3200	5.437e−5	4.000
1/80	1/12800	1.359e−5	4.001

3.2.4   A Priori Estimate of the Difference Solution

We will use the maximum principle to give a priori estimate of the difference solution.

Let $v = \{v_i^k \mid 0 \leqslant i \leqslant m, 0 \leqslant k \leqslant n\}$ be a grid function on $\Omega_{h\tau}$. Denote

$$L_{\tau h} v_i^k = D_t v_i^k - a\delta_x^2 v_i^k, \quad 1 \leqslant i \leqslant m - 1, \quad 0 \leqslant k \leqslant n - 1.$$

Lemma 3.1 (Maximum Principle) *As shown in Fig. 3.6, denote*

$$\omega_k = \{(i, s) \mid 1 \leqslant i \leqslant m - 1, 1 \leqslant s \leqslant k\},$$

$$\gamma_k = \{(i, 0) \mid 0 \leqslant i \leqslant m\} \cup \{(0, s) \mid 1 \leqslant s \leqslant k\} \cup \{(m, s) \mid 1 \leqslant s \leqslant k\}.$$

If

$$(L_{\tau h} v)_i^s = D_t v_i^s - a\delta_x^2 v_i^s \leqslant 0, \quad 1 \leqslant i \leqslant m - 1, \quad 0 \leqslant s \leqslant k - 1, \qquad (3.14)$$

Fig. 3.3 (Example 3.1) The numerical and exact solution profiles at $t = 1$ ($h = 1/10$, $\tau = 1/200$)

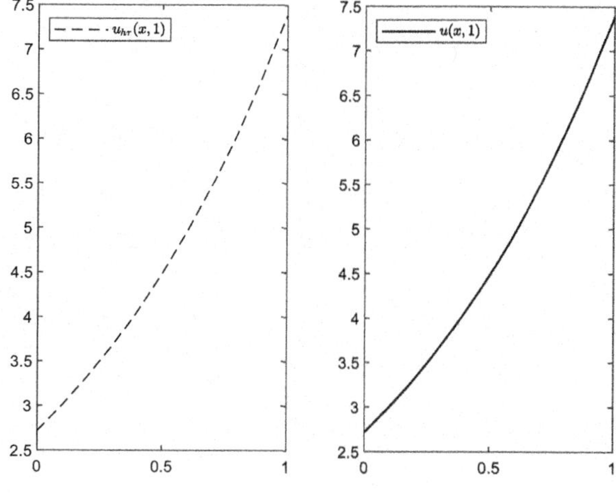

Fig. 3.4 (Example 3.1) Error curves of numerical solutions with different step sizes at $t = 1$ ($r = 1/2$)

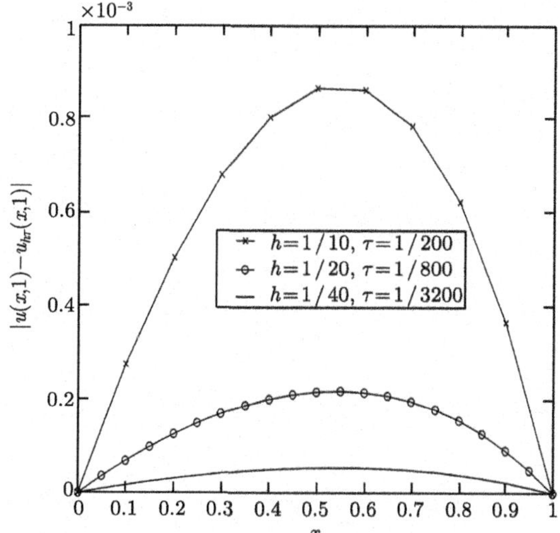

then when $r \leqslant \frac{1}{2}$, we have

$$\max_{(i,s)\in\omega_k} v_i^s \leqslant \max_{(i,s)\in\gamma_k} v_i^s.$$

Proof The proof is by contradiction. Suppose

$$\max_{(i,s)\in\omega_k} v_i^s > \max_{(i,s)\in\gamma_k} v_i^s$$

Fig. 3.5 (Example 3.1) Numerical error surfaces with different step sizes ($r = 1/2$)

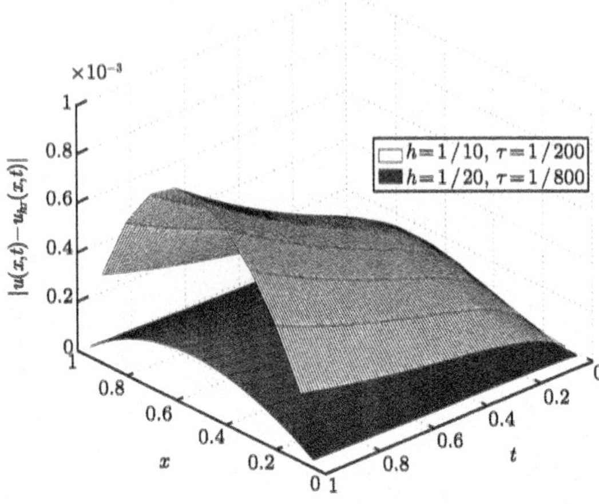

Fig. 3.6 Auxiliary figure for the proof of Lemma 3.1, Theorem 3.2, Lemma 3.2, and Theorem 3.8: "∘" denotes γ_k and "•" denotes ω_k

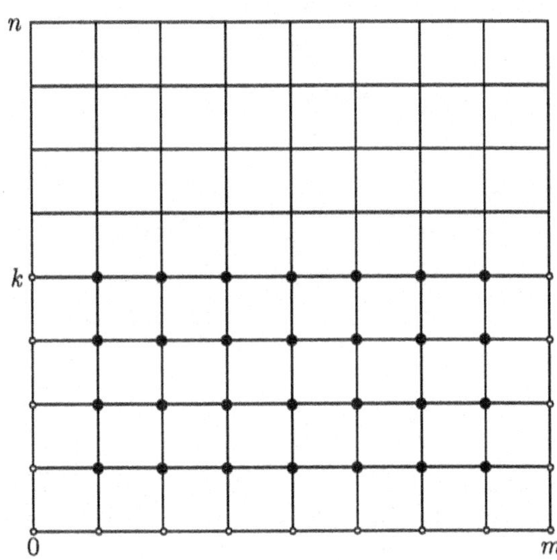

and denote

$$M = \max_{(i,s)\in\omega_k} v_i^s.$$

(I) Let $r < \frac{1}{2}$. At this time, we have

$$(L_{\tau h}v)_i^{s-1} = \frac{1}{\tau}\left\{v_i^s - \left[rv_{i-1}^{s-1} + (1-2r)v_i^{s-1} + rv_{i+1}^{s-1}\right]\right\}.$$

There are $i_0 \in \{1, 2, \ldots, m - 1\}$ and $s_0 \in \{1, 2, \ldots, k\}$ satisfying $v_{i_0}^{s_0} = M$, and at least one of $v_{i_0}^{s_0-1}$, $v_{i_0-1}^{s_0-1}$, and $v_{i_0+1}^{s_0-1}$ is strictly less than M. Hence,

$$(L_{\tau h} v)_{i_0}^{s_0-1} = \frac{1}{\tau}\left\{ v_{i_0}^{s_0} - \left[r v_{i_0-1}^{s_0-1} + (1 - 2r) v_{i_0}^{s_0-1} + r v_{i_0+1}^{s_0-1} \right] \right\} > 0,$$

which contradicts the condition (3.14).

(II) Let $r = \frac{1}{2}$. At this time, we have

$$(L_{\tau h} v)_i^{s-1} = \frac{1}{\tau}\left[v_i^s - \frac{1}{2}\left(v_{i-1}^{s-1} + v_{i+1}^{s-1} \right) \right].$$

There are $i_0 \in \{1, 2, \ldots, m - 1\}$ and $s_0 \in \{1, 2, \ldots, k\}$ satisfying $v_{i_0}^{s_0} = M$, and at least one of $v_{i_0-1}^{s_0-1}$ and $v_{i_0+1}^{s_0-1}$ is strictly less than M. Thereby,

$$(L_{\tau h} v)_{i_0}^{s_0-1} = \frac{1}{\tau}\left[v_{i_0}^{s_0} - \frac{1}{2}\left(v_{i_0-1}^{s_0-1} + v_{i_0+1}^{s_0-1} \right) \right] > 0,$$

which contradicts the condition (3.14).

$\square$

Theorem 3.2 *Let $v = \{v_i^k \mid 0 \leqslant i \leqslant m, \ 0 \leqslant k \leqslant n\}$ be the solution of the difference scheme*

$$\begin{cases} D_t v_i^k - a \delta_x^2 v_i^k = g_i^k, & 1 \leqslant i \leqslant m - 1, \quad 0 \leqslant k \leqslant n - 1, \\ v_i^0 = \varphi_i, & 0 \leqslant i \leqslant m, \\ v_0^k = \alpha^k, \quad v_m^k = \beta^k, & 1 \leqslant k \leqslant n. \end{cases}$$

Then when the step ratio $r \leqslant \frac{1}{2}$, we have

$$\max_{1 \leqslant i \leqslant m-1} |v_i^k| \leqslant \max\Big\{ \max_{0 \leqslant i \leqslant m} |\varphi_i|, \ \max_{1 \leqslant s \leqslant k} |\alpha^s|, \ \max_{1 \leqslant s \leqslant k} |\beta^s| \Big\}$$

$$+ \frac{L^2}{8a} \max_{0 \leqslant s \leqslant k-1} \max_{1 \leqslant i \leqslant m-1} |g_i^s|, \quad 0 \leqslant k \leqslant n.$$

Proof As shown in Fig. 3.6, for an arbitrary $k \in \{1, 2, \ldots, n\}$, define

$$\omega_k = \{(i, s) \mid 1 \leqslant i \leqslant m - 1, 1 \leqslant s \leqslant k\},$$

$$\gamma_k = \{(i, 0) \mid 0 \leqslant i \leqslant m\} \cup \{(0, s) \mid 1 \leqslant s \leqslant k\} \cup \{(m, s) \mid 1 \leqslant s \leqslant k\}.$$

Denote

$$C_k = \max_{0 \leqslant s \leqslant k-1} \max_{1 \leqslant i \leqslant m-1} |g_i^s|, \quad P_k(x) = \frac{C_k}{2a} x(L - x), \quad P_i^s = P_k(x_i), 0 \leqslant s \leqslant k - 1.$$

Calculation gives

$$L_{\tau h}(\pm v - P)_i^s = \pm g_i^s - C_k \leqslant 0, \quad 1 \leqslant i \leqslant m - 1, \quad 0 \leqslant s \leqslant k - 1.$$

It follows from Lemma 3.1 that

$$\max_{(i,s)\in\omega_k} (\pm v_i^s - P_i^s) \leqslant \max_{(i,s)\in\gamma_k} (\pm v_i^s - P_i^s)$$

$$\leqslant \max_{(i,s)\in\gamma_k} (\pm v_i^s) + \max_{(i,s)\in\gamma_k} (-P_i^s) \leqslant \max_{(i,s)\in\gamma_k} (\pm v_i^s).$$

Furthermore, we have

$$\max_{(i,s)\in\omega_k} (\pm v_i^s) = \max_{(i,s)\in\omega_k} (\pm v - P + P)_i^s$$

$$\leqslant \max_{(i,s)\in\omega_k} (\pm v - P)_i^s + \max_{(i,s)\in\omega_k} P_i^s \leqslant \max_{(i,s)\in\gamma_k} (\pm v_i^s) + \max_{(i,s)\in\omega_k} P_i^s$$

$$\leqslant \max\{ \max_{0\leqslant i\leqslant m} |\varphi_i|, \max_{1\leqslant s\leqslant k} |\alpha^s|, \max_{1\leqslant s\leqslant k} |\beta^s| \}$$

$$+ \frac{L^2}{8a} \max_{0\leqslant s\leqslant k-1} \max_{1\leqslant i\leqslant m-1} |g_i^s|.$$

Therefore,

$$\max_{(i,s)\in\omega_k} |v_i^s| \leqslant \max\{ \max_{0\leqslant i\leqslant m} |\varphi_i|, \max_{1\leqslant s\leqslant k} |\alpha^s|, \max_{1\leqslant s\leqslant k} |\beta^s| \} + \frac{L^2}{8a} \max_{0\leqslant s\leqslant k-1} \max_{1\leqslant i\leqslant m-1} |g_i^s|.$$

Particularly,

$$\max_{1\leqslant i\leqslant m-1} |v_i^k| \leqslant \max\{ \max_{0\leqslant i\leqslant m} |\varphi_i|, \max_{1\leqslant s\leqslant k} |\alpha^s|, \max_{1\leqslant s\leqslant k} |\beta^s| \}$$

$$+ \frac{L^2}{8a} \max_{0\leqslant s\leqslant k-1} \max_{1\leqslant i\leqslant m-1} |g_i^s|.$$

$$\square$$

In the case of vanishing boundary values, a simpler approach can be employed to derive a priori estimate.

Theorem 3.3 *Let* $v = \{v_i^k \mid 0 \leqslant i \leqslant m, \ 0 \leqslant k \leqslant n\}$ *be the solution of the difference scheme*

$$\begin{cases} D_t v_i^k - a\delta_x^2 v_i^k = g_i^k, & 1 \leqslant i \leqslant m - 1, \quad 0 \leqslant k \leqslant n - 1, & \text{(3.15a)} \\ v_i^0 = \varphi_i, & 0 \leqslant i \leqslant m, & \text{(3.15b)} \\ v_0^k = 0, \quad v_m^k = 0, & 1 \leqslant k \leqslant n. & \text{(3.15c)} \end{cases}$$

Then when the step ratio $r \leqslant 1/2$, we have

$$\|v^k\|_\infty \leqslant \|\varphi\|_\infty + \tau \sum_{l=0}^{k-1} \|g^l\|_\infty, \quad 0 \leqslant k \leqslant n,$$

where $\|g^l\|_\infty = \max\limits_{1 \leqslant i \leqslant m-1} |g_i^l|$.

Proof Rewriting (3.15a) as

$$v_i^{k+1} = (1 - 2r)v_i^k + r(v_{i-1}^k + v_{i+1}^k) + \tau g_i^k, \quad 1 \leqslant i \leqslant m - 1, \quad 0 \leqslant k \leqslant n - 1,$$

then when $r \leqslant \frac{1}{2}$, we have

$$|v_i^{k+1}| \leqslant (1 - 2r)\|v^k\|_\infty + r(\|v^k\|_\infty + \|v^k\|_\infty) + \tau\|g^k\|_\infty$$

$$= \|v^k\|_\infty + \tau\|g^k\|_\infty, \quad 1 \leqslant i \leqslant m - 1, \quad 0 \leqslant k \leqslant n - 1.$$

As a result,

$$\|v^{k+1}\|_\infty \leqslant \|v^k\|_\infty + \tau\|g^k\|_\infty, \quad 0 \leqslant k \leqslant n - 1.$$

It follows from the recursion that

$$\|v^k\|_\infty \leqslant \|v^0\|_\infty + \tau \sum_{l=0}^{k-1} \|g^l\|_\infty, \quad 0 \leqslant k \leqslant n,$$

which completes the proof. $\qquad\square$

3.2.5 Convergence and Stability of the Difference Solution

Convergence

Theorem 3.4 *Let $\{u(x, t) \mid (x, t) \in \bar{D}\}$ be the solution of the problem (3.1) and $\{u_i^k \mid 0 \leqslant i \leqslant m, 0 \leqslant k \leqslant n\}$ be the solution of the difference scheme (3.11). When $r \leqslant \frac{1}{2}$, we have*

$$\max_{0 \leqslant i \leqslant m} |u(x_i, t_k) - u_i^k| \leqslant c_1 T(\tau + h^2), \quad 0 \leqslant k \leqslant n,$$

where c_1 is defined by (3.12).

Proof Denote

$$e_i^k = u(x_i, t_k) - u_i^k, \quad 0 \leqslant i \leqslant m, \quad 0 \leqslant k \leqslant n.$$

Subtracting (3.11) from (3.9) and (3.10), the system of error equations is produced as

$$\begin{cases} D_t e_i^k - a\delta_x^2 e_i^k = (R_1)_i^k, & 1 \leqslant i \leqslant m-1, \quad 0 \leqslant k \leqslant n-1, \\ e_i^0 = 0, & 0 \leqslant i \leqslant m, \\ e_0^k = 0, \quad e_m^k = 0, & 1 \leqslant k \leqslant n. \end{cases}$$

Applying Theorem 3.3 and noticing (3.13), when $r \leqslant \frac{1}{2}$, we have

$$\|e^k\|_\infty \leqslant \tau \sum_{s=0}^{k-1} \max_{1 \leqslant i \leqslant m-1} |(R_1)_i^s| \leqslant c_1 \cdot k\tau(\tau + h^2) \leqslant c_1 T(\tau + h^2), \quad 0 \leqslant k \leqslant n,$$

which completes the proof. $\qquad\square$

Stability

If the right-hand side function $f(x_i, t_k)$ contains an error g_i^k, the initial value $\varphi(x_i)$ has an error φ_i, the boundary values $\alpha(t_k)$ and $\beta(t_k)$ involve errors α^k and β^k, respectively, when the difference scheme (3.11) is applied, then the resulting solutions obey the system of difference equations given by

$$\begin{cases} D_t v_i^k - a\delta_x^2 v_i^k = f(x_i, t_k) + g_i^k, & 1 \leqslant i \leqslant m-1, \quad 0 \leqslant k \leqslant n-1, \\ v_i^0 = \varphi(x_i) + \varphi_i, & 0 \leqslant i \leqslant m, \\ v_0^k = \alpha(t_k) + \alpha^k, \quad v_m^k = \beta(t_k) + \beta^k, & 1 \leqslant k \leqslant n. \end{cases}$$

$$\tag{3.16}$$

Let

$$\varepsilon_i^k = v_i^k - u_i^k, \quad 0 \leqslant i \leqslant m, \quad 0 \leqslant k \leqslant n.$$

Subtracting (3.11) from (3.16) gives the **system of perturbation equations**

$$\begin{cases} D_t \varepsilon_i^k - a\delta_x^2 \varepsilon_i^k = g_i^k, & 1 \leqslant i \leqslant m-1, \quad 0 \leqslant k \leqslant n-1, \\ \varepsilon_i^0 = \varphi_i, & 0 \leqslant i \leqslant m, \\ \varepsilon_0^k = \alpha^k, \quad \varepsilon_m^k = \beta^k, & 1 \leqslant k \leqslant n. \end{cases} \tag{3.17}$$

With the application of Theorem 3.2, we have

$$\max_{1\leqslant i\leqslant m-1}|\varepsilon_i^k|\leqslant \max\{\max_{0\leqslant i\leqslant m}|\varphi_i|,\ \max_{1\leqslant s\leqslant k}|\alpha^s|,\ \max_{1\leqslant s\leqslant k}|\beta^s|\}$$

$$+\frac{L^2}{8a}\max_{0\leqslant s\leqslant k-1}\max_{1\leqslant i\leqslant m-1}|g_i^s|,\quad 0\leqslant k\leqslant n$$

when $r\leqslant\frac{1}{2}$.

The above result illustrates that when $\max\limits_{0\leqslant i\leqslant m}|\varphi_i|$, $\max\limits_{1\leqslant k\leqslant n}|\alpha^k|$, $\max\limits_{1\leqslant k\leqslant n}|\beta^k|$, and $\max\limits_{0\leqslant s\leqslant n-1}\max\limits_{1\leqslant i\leqslant m-1}|g_i^s|$ are all small, the error $\max\limits_{1\leqslant k\leqslant n}\|\varepsilon^k\|_\infty$ will also remain small.

The perturbation system (3.17) and the difference scheme (3.11) are in the same form. The results above can be summarized as follows:

Theorem 3.5 *When $r\leqslant\frac{1}{2}$, the solution of the difference scheme* (3.11) *is stable with respect to the initial condition, the right-hand side function, and the boundary values, in the following sense: Let $\{u_i^k\mid 0\leqslant i\leqslant m,\ 0\leqslant k\leqslant n\}$ be the solution of the difference scheme*

$$\begin{cases} D_t u_i^k - a\delta_x^2 u_i^k = f_i^k, & 1\leqslant i\leqslant m-1,\quad 0\leqslant k\leqslant n-1,\\ u_i^0 = \varphi_i, & 0\leqslant i\leqslant m,\\ u_0^k = \alpha^k, & u_m^k = \beta^k,\quad 1\leqslant k\leqslant n. \end{cases}$$

Then we have

$$\max_{1\leqslant i\leqslant m-1}|u_i^k|\leqslant \max\{\max_{0\leqslant i\leqslant m}|\varphi_i|,\ \max_{1\leqslant s\leqslant k}|\alpha^s|,\ \max_{1\leqslant s\leqslant k}|\beta^s|\}$$

$$+\frac{L^2}{8a}\max_{0\leqslant s\leqslant k-1}\max_{1\leqslant i\leqslant m-1}|f_i^s|,\quad 1\leqslant k\leqslant n.$$

We call

$$\begin{cases} D_t\varepsilon_i^k - a\delta_x^2\varepsilon_i^k = 0, & 1\leqslant i\leqslant m-1,\quad 0\leqslant k\leqslant n-1,\\ \varepsilon_i^0 = \varphi_i, & 0\leqslant i\leqslant m,\\ \varepsilon_0^k = 0, & \varepsilon_m^k = 0,\quad 1\leqslant k\leqslant n \end{cases}$$

the system of perturbation equations of the difference scheme (3.11) with respect to the initial value,

$$\begin{cases} D_t\varepsilon_i^k - a\delta_x^2\varepsilon_i^k = 0, & 1\leqslant i\leqslant m-1,\quad 0\leqslant k\leqslant n-1,\\ \varepsilon_i^0 = 0, & 0\leqslant i\leqslant m,\\ \varepsilon_0^k = \alpha^k, & \varepsilon_m^k = \beta^k,\quad 1\leqslant k\leqslant n \end{cases}$$

the system of perturbation equations of the difference scheme (3.11) with respect to
the boundary values, and

$$
\begin{cases}
D_t \varepsilon_i^k - a\delta_x^2 \varepsilon_i^k = g_i^k, & 1 \leqslant i \leqslant m-1, \quad 0 \leqslant k \leqslant n-1, \\
\varepsilon_i^0 = 0, & 0 \leqslant i \leqslant m, \\
\varepsilon_0^k = 0, \quad \varepsilon_m^k = 0, & 1 \leqslant k \leqslant n
\end{cases}
$$

the system of perturbation equations of the difference scheme (3.11) with respect to
the right-hand side term.

For a linear difference scheme, the procedures for analyzing stability and
deriving a priori estimate are essentially the same. Once a priori estimate for the
difference solution is established, the stability of the scheme follows naturally.

In the following, we turn our attention to the case where $r > \frac{1}{2}$ and present the
corresponding conclusion.

Theorem 3.6 When $r > \frac{1}{2}$, the solution of the difference scheme (3.11) is unstable
in the maximum norm.

Proof First note that when $r > \frac{1}{2}$, it yields $4r - 1 > 1$.

Now consider the perturbed system of the difference scheme (3.11) with respect
to a special initial value:

$$
\begin{cases}
D_t \varepsilon_i^k - a\delta_x^2 \varepsilon_i^k = 0, & 1 \leqslant i \leqslant m-1, \quad 0 \leqslant k \leqslant n-1, \quad \text{(3.18a)} \\
\varepsilon_i^0 = (-1)^i \epsilon, & 0 \leqslant i \leqslant m, \quad \text{(3.18b)} \\
\varepsilon_0^k = 0, \quad \varepsilon_m^k = 0, & 1 \leqslant k \leqslant n, \quad \text{(3.18c)}
\end{cases}
$$

where ϵ is a small positive constant.

From the condition $r = a\frac{\tau}{h^2} > \frac{1}{2}$, we deduce that $aTm^2 > \frac{1}{2}L^2 n$. Therefore, as
$n \to +\infty$, it follows that $m \to +\infty$ as well.

Here, $\lfloor A \rfloor$ denotes the greatest integer no more than A.

Case I:

$$
\min\left\{ \left\lfloor \frac{m}{2} \right\rfloor, n \right\} = \left\lfloor \frac{m}{2} \right\rfloor.
$$

Rewrite (3.18a) as

$$
\varepsilon_i^{k+1} = r(\varepsilon_{i-1}^k + \varepsilon_{i+1}^k) + (1 - 2r)\varepsilon_i^k, \quad 1 \leqslant i \leqslant m-1, \quad 0 \leqslant k \leqslant n-1.
$$
$$\text{(3.19)}$$

When $k \leqslant \min\left\{ \lfloor \frac{m}{2} \rfloor, n \right\}$, the analysis shows that the value of ε_i^k ($k \leqslant i \leqslant m-k$)
in the triangle region or trapezoidal region shown in Fig. 3.7 can be totally
determined by (3.18a) and the initial value (3.18b), but independent of the
boundary value (3.18c).

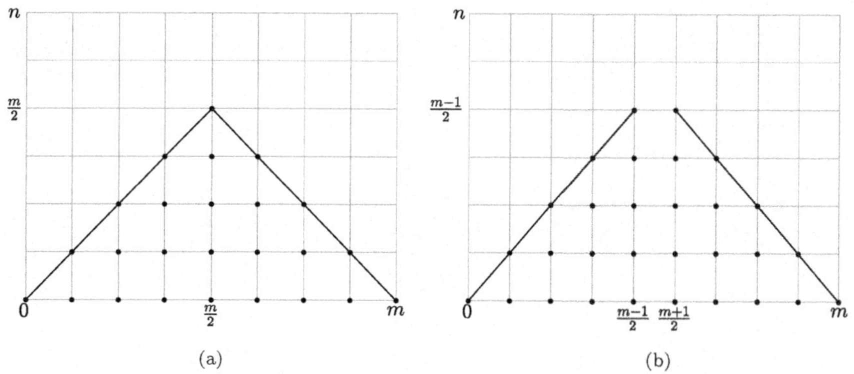

Fig. 3.7 Auxiliary figures of the proof for Theorem 3.6. (**a**) m is even, $n \geqslant \frac{m}{2}$. (**b**) m is odd, $n \geqslant \frac{m-1}{2}$

Suppose that the solutions in the above region have the form of

$$\varepsilon_i^k = (-1)^{i+k} T(k), \quad k \leqslant i \leqslant m - k, \quad k \leqslant \min\left\{\left\lfloor \frac{m}{2} \right\rfloor, n\right\}.$$

Substituting the above equality into (3.19) produces

$$T(k+1) = (4r - 1)T(k), \quad k \leqslant \min\left\{\left\lfloor \frac{m}{2} \right\rfloor, n\right\}.$$

It follows from the recursion that

$$T(k) = (4r - 1)^k T(0) = (4r - 1)^k \epsilon, \quad k \leqslant \min\left\{\left\lfloor \frac{m}{2} \right\rfloor, n\right\}.$$

Thereby, the solution in the triangle region or trapezoidal region shown in Fig. 3.7 is

$$\varepsilon_i^k = (-1)^{i+k}(4r - 1)^k \epsilon, \quad k \leqslant i \leqslant m - k, \quad k \leqslant \min\left\{\left\lfloor \frac{m}{2} \right\rfloor, n\right\}. \tag{3.20}$$

It is easy to know that

$$\left\| \varepsilon^{\min\left\{\left\lfloor \frac{m}{2} \right\rfloor, n\right\}} \right\|_\infty \geqslant (4r - 1)^{\min\left\{\left\lfloor \frac{m}{2} \right\rfloor, n\right\}} \epsilon.$$

Hence,

$$\lim_{n \to +\infty} \left\| \varepsilon^{\min\left\{\left\lfloor \frac{m}{2} \right\rfloor, n\right\}} \right\|_\infty \geqslant \lim_{n \to +\infty} (4r - 1)^{\min\left\{\left\lfloor \frac{m}{2} \right\rfloor, n\right\}} \epsilon = +\infty.$$

It follows from

$$\max_{1\leqslant k\leqslant n}\left\|\varepsilon^k\right\|_\infty \geqslant \left\|\varepsilon^{\min\left\{\left\lfloor\frac{m}{2}\right\rfloor n\right\}}\right\|_\infty$$

that

$$\lim_{n\to+\infty}\max_{1\leqslant k\leqslant n}\left\|\varepsilon^k\right\|_\infty = +\infty. \tag{3.21}$$

Case II:

$$\min\left\{\left\lfloor\frac{m}{2}\right\rfloor, n\right\} = n.$$

The equality (3.20) also holds. Thus, (3.21) is also true.

In summary, when $r > \frac{1}{2}$, the solution of the difference scheme (3.11) is unstable with respect to the initial value in the maximum norm.

$\square$

According to Theorem 3.5 and Theorem 3.6, the difference scheme (3.11) is stable when $r \leqslant \frac{1}{2}$ and unstable when $r > \frac{1}{2}$. The stability in this case is called the **conditional stability**. In the practical calculation, the step ratio must satisfy $r \leqslant \frac{1}{2}$, or equivalently, $a\tau/h^2 \leqslant \frac{1}{2}$.

3.3   The Backward Euler Scheme

The forward Euler scheme requires the step ratio $r \leqslant \frac{1}{2}$, which implies that the temporal step size must be significantly smaller than the spatial step size. To overcome this limitation, an unconditionally stable difference scheme will be introduced in this section.

3.3.1   Derivation of the Difference Scheme

Considering the differential equation (3.1a) at the node point (x_i, t_k), we have

$$\frac{\partial u}{\partial t}(x_i, t_k) - a\frac{\partial^2 u}{\partial x^2}(x_i, t_k) = f(x_i, t_k), \quad 1 \leqslant i \leqslant m-1, \quad 1 \leqslant k \leqslant n. \tag{3.22}$$

Substituting

$$\frac{\partial^2 u}{\partial x^2}(x_i, t_k) = \delta_x^2 U_i^k - \frac{h^2}{12}\frac{\partial^4 u}{\partial x^4}(\xi_{ik}, t_k), \quad x_{i-1} < \xi_{ik} < x_{i+1}$$

and

$$\frac{\partial u}{\partial t}(x_i, t_k) = D_{\bar{t}}U_i^k + \frac{\tau}{2}\frac{\partial^2 u}{\partial t^2}(x_i, \overline{\eta}_{ik}), \quad t_{k-1} < \overline{\eta}_{ik} < t_k$$

into (3.22) arrives at

$$D_{\bar{t}}U_i^k - a\delta_x^2 U_i^k = f(x_i, t_k) - \frac{\tau}{2}\frac{\partial^2 u}{\partial t^2}(x_i, \overline{\eta}_{ik}) - \frac{ah^2}{12}\frac{\partial^4 u}{\partial x^4}(\xi_{ik}, t_k),$$

$$1 \leqslant i \leqslant m - 1, \quad 1 \leqslant k \leqslant n. \qquad (3.23)$$

Noticing the initial-boundary value conditions (3.1b) and (3.1c), we have

$$\begin{cases} U_i^0 = \varphi(x_i), & 0 \leqslant i \leqslant m, \\ U_0^k = \alpha(t_k), \quad U_m^k = \beta(t_k), & 1 \leqslant k \leqslant n. \end{cases} \qquad (3.24)$$

Throwing away the small term

$$(R_2)_i^k = -\frac{\tau}{2}\frac{\partial^2 u}{\partial t^2}(x_i, \overline{\eta}_{ik}) - \frac{ah^2}{12}\frac{\partial^4 u}{\partial x^4}(\xi_{ik}, t_k)$$

in (3.23), and replacing U_i^k with u_i^k, another difference scheme for solving (3.1) reads

$$\begin{cases} D_{\bar{t}}u_i^k - a\delta_x^2 u_i^k = f(x_i, t_k), & 1 \leqslant i \leqslant m - 1, \quad 1 \leqslant k \leqslant n, & (3.25a) \\ u_i^0 = \varphi(x_i), & 0 \leqslant i \leqslant m, & (3.25b) \\ u_0^k = \alpha(t_k), \quad u_m^k = \beta(t_k), & 1 \leqslant k \leqslant n. & (3.25c) \end{cases}$$

The difference scheme (3.25) is known as the **backward Euler scheme**. Here, $(R_2)_i^k$ denotes the local truncation error of the scheme (3.25). Noticing (3.12), we have

$$|(R_2)_i^k| \leqslant c_1(\tau + h^2), \quad 1 \leqslant i \leqslant m - 1, \quad 1 \leqslant k \leqslant n. \qquad (3.26)$$

The computational stencil for the difference scheme (3.25) is illustrated in Fig. 3.8.

Fig. 3.8 The stencil of the backward Euler scheme (3.25)

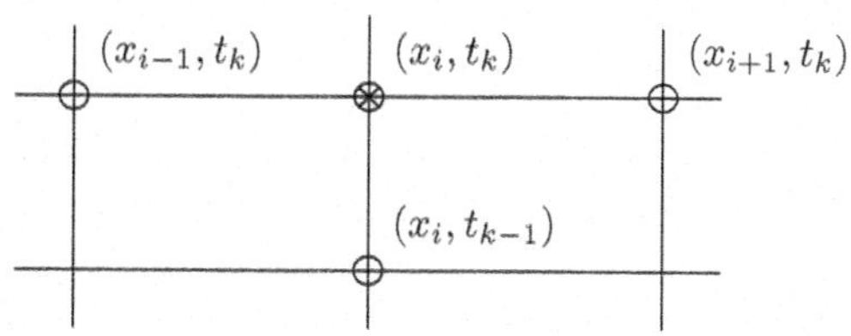

3.3.2   Existence of the Difference Solution

Theorem 3.7 *The difference scheme (3.25) is uniquely solvable.*

Proof Denote

$$u^k = (u_0^k, u_1^k, \ldots, u_{m-1}^k, u_m^k).$$

It follows from (3.25b) that u^0 is already known. Assuming the solution u^{k-1} at the $(k-1)$-th time level has been determined, the difference scheme for computing u^k at the k-th time level is given by

$$\begin{cases} D_{\bar t} u_i^k - a\delta_x^2 u_i^k = f(x_i, t_k), & 1 \leqslant i \leqslant m-1, \\ u_0^k = \alpha(t_k), & u_m^k = \beta(t_k), \end{cases}$$

in other words,

$$\begin{cases} -ru_{i-1}^k + (1+2r)u_i^k - ru_{i+1}^k = u_i^{k-1} + \tau f(x_i, t_k), & 1 \leqslant i \leqslant m-1, \\ u_0^k = \alpha(t_k), & u_m^k = \beta(t_k). \end{cases}$$

The above system can be written in the following matrix form:

$$\begin{pmatrix} 1+2r & -r & & & \\ -r & 1+2r & -r & & \\ & \ddots & \ddots & \ddots & \\ & & -r & 1+2r & -r \\ & & & -r & 1+2r \end{pmatrix} \begin{pmatrix} u_1^k \\ u_2^k \\ \vdots \\ u_{m-2}^k \\ u_{m-1}^k \end{pmatrix}$$

$$= \begin{pmatrix} u_1^{k-1} + \tau f(x_1, t_k) + ru_0^k \\ u_2^{k-1} + \tau f(x_2, t_k) \\ \vdots \\ u_{m-2}^{k-1} + \tau f(x_{m-2}, t_k) \\ u_{m-1}^{k-1} + \tau f(x_{m-1}, t_k) + ru_m^k \end{pmatrix}. \tag{3.27}$$

It is straightforward to verify that the coefficient matrix is strictly diagonally dominant. Consequently, the system admits a unique solution. $\square$

3.3.3   *Implementation of the Difference Scheme and Numerical Examples*

It can be seen from (3.27) that only a tridiagonal system of linear equations needs to be solved at each time level.

Example 3.2   Apply the backward *Euler* scheme (3.25) to compute the problem in Example 3.1:

$$\begin{cases} \dfrac{\partial u}{\partial t} - \dfrac{\partial^2 u}{\partial x^2} = 0, & 0 < x < 1, \quad 0 < t \leqslant 1, \\ u(x,0) = \mathrm{e}^x, & 0 \leqslant x \leqslant 1, \\ u(0,t) = \mathrm{e}^t, \quad u(1,t) = \mathrm{e}^{1+t}, & 0 < t \leqslant 1. \end{cases}$$

The exact solution is $u(x,t) = \mathrm{e}^{x+t}$.

Some of the numerical results are shown in Tables 3.4 and 3.5 for $h = 1/10$, $\tau = 1/200$ (with step ratio $r = 1/2$), and $h = 1/10$, $\tau = 1/100$ (with step ratio $r = 1$), respectively. These results demonstrate that the numerical solutions closely approximate the exact solutions. Table 3.6 presents the maximum errors

$$E_\infty(h,\tau) = \max_{\substack{1 \leqslant i \leqslant m-1 \\ 1 \leqslant k \leqslant n}} |u(x_i, t_k) - u_i^k|$$

for the numerical solutions when $r = 1/2$ with different step sizes.

It can be seen from Table 3.6 that when the spatial step size is halved and the temporal step size is reduced to a quarter of its original value, the maximum error decreases to a quarter of the original error. Similar results can be observed in Table 3.7 when $r = 1$. Figure 3.9 shows the exact solution profile at $t = 1$ along with the numerical solution profile for step sizes $h = 1/10$ and $\tau = 1/200$. Figures 3.10 and 3.11 display the numerical error profiles at $t = 1$, while Fig. 3.12 illustrates the error surfaces of the numerical solutions with different step sizes.

Table 3.4 (Example 3.2) Numerical solutions, exact solutions, and absolute values of errors at some node points ($h = 1/10$, $\tau = 1/200$)

k	(x,t)	NS	ES	$|\mathrm{NS}-\mathrm{ES}|$
20	$(0.5, 0.1)$	1.822566	1.822119	4.471e$-$4
40	$(0.5, 0.2)$	2.014429	2.013753	6.763e$-$4
60	$(0.5, 0.3)$	2.226358	2.225541	8.175e$-$4
80	$(0.5, 0.4)$	2.460534	2.459603	9.304e$-$4
100	$(0.5, 0.5)$	2.719320	2.718282	1.039e$-$3
120	$(0.5, 0.6)$	3.005318	3.004166	1.152e$-$3
140	$(0.5, 0.7)$	3.321391	3.320117	1.275e$-$3
160	$(0.5, 0.8)$	3.670706	3.669297	1.409e$-$3
180	$(0.5, 0.9)$	4.056758	4.055200	1.558e$-$3
200	$(0.5, 1.0)$	4.483411	4.481689	1.722e$-$3

Table 3.5 (Example 3.2)
Numerical solutions, exact
solutions, and absolute values
of errors at some node points
($h = 1/10$, $\tau = 1/100$)

k	(x, t)	NS	ES	|NS−ES|
10	$(0.5, 0.1)$	1.822891	1.822119	7.717e−4
20	$(0.5, 0.2)$	2.014927	2.013753	1.174e−3
30	$(0.5, 0.3)$	2.226965	2.225541	1.424e−3
40	$(0.5, 0.4)$	2.461227	2.459603	1.624e−3
50	$(0.5, 0.5)$	2.720096	2.718282	1.814e−3
60	$(0.5, 0.6)$	3.006178	3.004166	2.012e−3
70	$(0.5, 0.7)$	3.322344	3.320117	2.227e−3
80	$(0.5, 0.8)$	3.671759	3.669297	2.462e−3
90	$(0.5, 0.9)$	4.057922	4.055200	2.722e−3
100	$(0.5, 1.0)$	4.484697	4.481689	3.008e−3

Table 3.6 (Example 3.2)
The maximum error of
numerical solutions with
different step sizes ($r = 1/2$)

h	τ	$E_\infty(h, \tau)$	$E_\infty(2h, 4\tau)/E_\infty(h, \tau)$
1/10	1/200	1.722e−3	
1/20	1/800	4.346e−4	3.962
1/40	1/3200	1.087e−4	3.998
1/80	1/12800	2.718e−5	3.999

Table 3.7 (Example 3.2)
The maximum error of
numerical solutions with
different step sizes ($r = 1$)

h	τ	$E_\infty(h, \tau)$	$E_\infty(2h, 4\tau)/E_\infty(h, \tau)$
1/10	1/100	3.008e−3	
1/20	1/400	7.603e−4	3.956
1/40	1/1600	1.902e−4	3.997
1/80	1/6400	4.757e−5	3.998

Fig. 3.9 (Example 3.2) The
numerical and exact solution
profiles at $t = 1$
($h = 1/10$, $\tau = 1/200$)

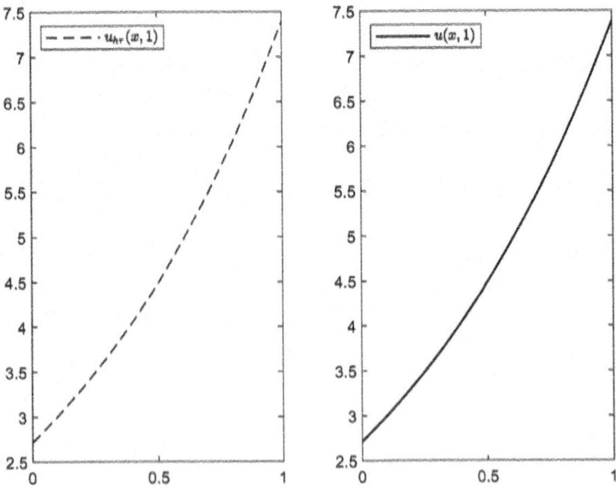

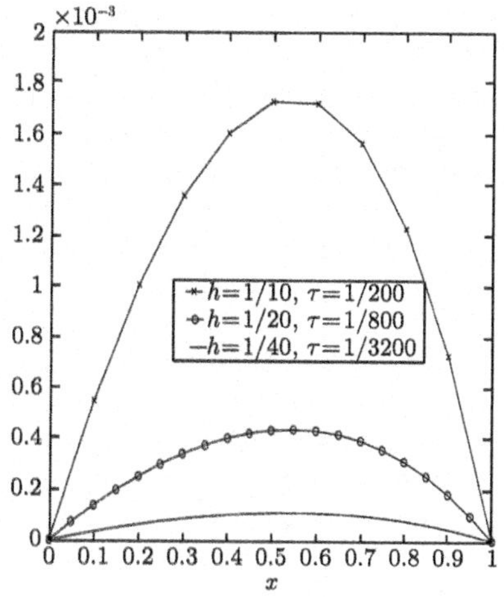

Fig. 3.10 (Example 3.2) The numerical error profiles at $t = 1$ with different step sizes $(r = 1/2)$

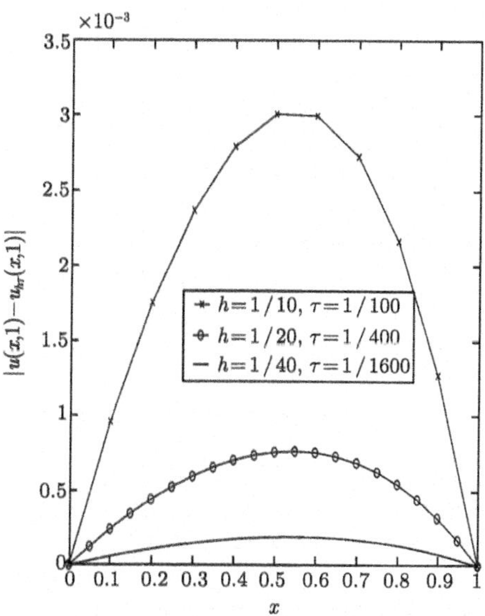

Fig. 3.11 (Example 3.2) The numerical error profiles at $t = 1$ with different step sizes $(r = 1)$

3.3.4 *A Priori Estimate of the Difference Solution*

A priori estimate of the difference solution will be proved by the maximum principle.

Let $v = \{v_i^k \mid 0 \leqslant i \leqslant m, 0 \leqslant k \leqslant n\}$ be a grid function on $\Omega_{h\tau}$. Denote

$$L_{\tau h} v_i^k = D_{\bar{t}} v_i^k - a\delta_x^2 v_i^k, \quad 1 \leqslant i \leqslant m - 1, \quad 1 \leqslant k \leqslant n.$$

Fig. 3.12 (Example 3.2) The numerical error surfaces with different step sizes ($r = 1$)

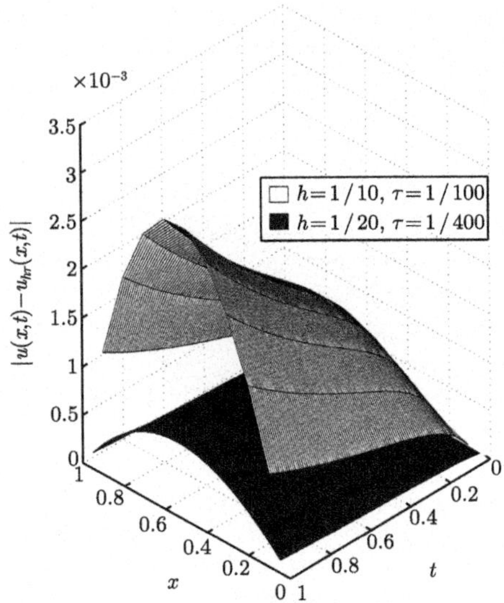

Lemma 3.2 (Maximum Principle) *As shown in Fig. 3.6, for an arbitrary $k \in \{1, 2, \ldots, n\}$, we define*

$$\omega_k = \{(i, s) \mid 1 \leqslant i \leqslant m - 1, 1 \leqslant s \leqslant k\},$$

$$\gamma_k = \{(i, 0) \mid 0 \leqslant i \leqslant m\} \cup \{(0, s) \mid 1 \leqslant s \leqslant k\} \cup \{(m, s) \mid 1 \leqslant s \leqslant k\}.$$

Suppose

$$L_{\tau h} v_i^s = D_{\bar{t}} v_i^s - a\delta_x^2 v_i^s \leqslant 0, \quad 1 \leqslant i \leqslant m - 1, \quad 1 \leqslant s \leqslant k. \tag{3.28}$$

Then for an arbitrary step ratio r, we have

$$\max_{(i,s)\in\omega_k} v_i^s \leqslant \max_{(i,s)\in\gamma_k} v_i^s.$$

Proof The proof is by contradiction. Let

$$\max_{(i,s)\in\omega_k} v_i^s > \max_{(i,s)\in\gamma_k} v_i^s,$$

and denote

$$M = \max_{(i,s)\in\omega_k} v_i^s.$$

Then there are $i_0 \in \{1, 2, \ldots, m-1\}$ and $s_0 \in \{1, 2, \ldots, k\}$ satisfying $v_{i_0}^{s_0} = M$, and at least one of $v_{i_0-1}^{s_0}$, $v_{i_0+1}^{s_0}$, $v_{i_0}^{s_0-1}$ is strictly less than M. Thereby,

$$L_{\tau h} v_{i_0}^{s_0} = \frac{1}{\tau}\left[(1+2r)v_{i_0}^{s_0} - r(v_{i_0-1}^{s_0} + v_{i_0+1}^{s_0}) - v_{i_0}^{s_0-1}\right] > 0,$$

which contradicts the condition (3.28). $\square$

Theorem 3.8 *Let $v = \{v_i^k \mid 0 \leqslant i \leqslant m, \ 0 \leqslant k \leqslant n\}$ be the solution of the system of difference equations*

$$\begin{cases} D_{\bar{t}} v_i^k - a\delta_x^2 v_i^k = g_i^k, & 1 \leqslant i \leqslant m-1, \quad 1 \leqslant k \leqslant n, \\ v_i^0 = \varphi_i, & 0 \leqslant i \leqslant m, \\ v_0^k = \alpha^k, \quad v_m^k = \beta^k, & 1 \leqslant k \leqslant n. \end{cases}$$

Then for an arbitrary step ratio, we have

$$\max_{1 \leqslant i \leqslant m-1} |v_i^k| \leqslant \max\{ \max_{0 \leqslant i \leqslant m} |\varphi_i|, \ \max_{1 \leqslant s \leqslant k} |\alpha^s|, \ \max_{1 \leqslant s \leqslant k} |\beta^s| \}$$

$$+ \frac{L^2}{8a} \max_{1 \leqslant s \leqslant k} \max_{1 \leqslant i \leqslant m-1} |g_i^s|, \quad 1 \leqslant k \leqslant n.$$

Proof As shown in Fig. 3.6, for the arbitrary $k \in \{1, 2, \ldots, n\}$, we define

$$\omega_k = \{(i, s) \mid 1 \leqslant i \leqslant m-1, 1 \leqslant s \leqslant k\},$$

$$\gamma_k = \{(i, 0) \mid 0 \leqslant i \leqslant m\} \cup \{(0, s) \mid 1 \leqslant s \leqslant k\} \cup \{(m, s) \mid 1 \leqslant s \leqslant k\}.$$

Denote

$$C_k = \max_{1 \leqslant s \leqslant k} \max_{1 \leqslant i \leqslant m-1} |g_i^s|, \quad P_k(x) = \frac{C_k}{2a} x(L-x), \quad P_i^s = P_k(x_i), 1 \leqslant s \leqslant k.$$

Calculation yields

$$L_{\tau h}(\pm v - P)_i^s = \pm g_i^s - C_k \leqslant 0, \quad 1 \leqslant i \leqslant m-1, \quad 1 \leqslant s \leqslant k.$$

It follows from Lemma 3.2 that

$$\max_{(i,s)\in\omega_k} (\pm v_i^s - P_i^s) \leqslant \max_{(i,s)\in\gamma_k} (\pm v_i^s - P_i^s)$$

$$\leqslant \max_{(i,s)\in\gamma_k} (\pm v_i^s) + \max_{(i,s)\in\gamma_k} (-P_i^s) \leqslant \max_{(i,s)\in\gamma_k} (\pm v_i^s).$$

Furthermore, we have

$$\max_{(i,s)\in\omega_k}(\pm v_i^s) \leqslant \max_{(i,s)\in\omega_k}(\pm v - P + P)_i^s$$

$$\leqslant \max_{(i,s)\in\omega_k}(\pm v - P)_i^s + \max_{(i,s)\in\omega_k} P_i^s \leqslant \max_{(i,s)\in\gamma_k}(\pm v_i^s) + \max_{(i,s)\in\omega_k} P_i^s$$

$$\leqslant \max\{\max_{0\leqslant i\leqslant m}|\varphi_i|, \max_{1\leqslant s\leqslant k}|\alpha^s|, \max_{1\leqslant s\leqslant k}|\beta^s|\} + \frac{L^2}{8a}\max_{1\leqslant s\leqslant k}\max_{1\leqslant i\leqslant m-1}|g_i^s|.$$

Therefore,

$$\max_{(i,s)\in\omega_k}|v_i^s| \leqslant \max\{\max_{0\leqslant i\leqslant m}|\varphi_i|, \max_{1\leqslant s\leqslant k}|\alpha^s|, \max_{1\leqslant s\leqslant k}|\beta^s|\} + \frac{L^2}{8a}\max_{1\leqslant s\leqslant k}\max_{1\leqslant i\leqslant m-1}|g_i^s|.$$

Particularly,

$$\max_{1\leqslant i\leqslant m-1}|v_i^k| \leqslant \max\{\max_{0\leqslant i\leqslant m}|\varphi_i|, \max_{1\leqslant s\leqslant k}|\alpha^s|, \max_{1\leqslant s\leqslant k}|\beta^s|\} + \frac{L^2}{8a}\max_{1\leqslant s\leqslant k}\max_{1\leqslant i\leqslant m-1}|g_i^s|.$$

$$\square$$

If the boundary values vanish, the following simplified method can be used to derive a priori estimate.

Theorem 3.9 *Let $\{v_i^k \mid 0 \leqslant i \leqslant m, \ 0 \leqslant k \leqslant n\}$ be the solution of the system of difference equations*

$$\begin{cases} D_{\bar{t}}v_i^k - a\delta_x^2 v_i^k = g_i^k, & 1 \leqslant i \leqslant m-1, \quad 1 \leqslant k \leqslant n, & (3.29a) \\[2mm] v_i^0 = \varphi_i, & 0 \leqslant i \leqslant m, & (3.29b) \\[2mm] v_0^k = 0, \quad v_m^k = 0, & 1 \leqslant k \leqslant n. & (3.29c) \end{cases}$$

Then for an arbitrary step ratio r, we have

$$\|v^k\|_\infty \leqslant \|\varphi\|_\infty + \tau \sum_{l=1}^{k} \|g^l\|_\infty, \quad 0 \leqslant k \leqslant n,$$

where $\|g^l\|_\infty = \max\limits_{1\leqslant i\leqslant m-1}|g_i^l|$.

Proof Equation (3.29a) can be rewritten as

$$(1 + 2r)v_i^k = r(v_{i-1}^k + v_{i+1}^k) + v_i^{k-1} + \tau g_i^k, \quad 1 \leqslant i \leqslant m-1, \quad 1 \leqslant k \leqslant n.$$

Then we have

$$(1+2r)|v_i^k| \leqslant r(|v_{i-1}^k| + |v_{i+1}^k|) + |v_i^{k-1}| + \tau|g_i^k|$$
$$\leqslant r(\|v^k\|_\infty + \|v^k\|_\infty) + \|v^{k-1}\|_\infty + \tau\|g^k\|_\infty,$$
$$1 \leqslant i \leqslant m-1, \quad 1 \leqslant k \leqslant n.$$

Hence,

$$(1+2r)\|v^k\|_\infty \leqslant 2r\|v^k\|_\infty + \|v^{k-1}\|_\infty + \tau\|g^k\|_\infty, \quad 1 \leqslant k \leqslant n,$$

or

$$\|v^k\|_\infty \leqslant \|v^{k-1}\|_\infty + \tau\|g^k\|_\infty, \quad 1 \leqslant k \leqslant n.$$

It follows from the recursion that

$$\|v^k\|_\infty \leqslant \|v^0\|_\infty + \tau\sum_{l=1}^{k}\|g^l\|_\infty = \|\varphi\|_\infty + \tau\sum_{l=1}^{k}\|g^l\|_\infty, \quad 1 \leqslant k \leqslant n.$$

$\square$

3.3.5 Convergence and Stability of the Difference Solution

Convergence

Theorem 3.10 *Let $\{u(x,t) \mid (x,t) \in \bar{D}\}$ be the solution of the problem (3.1) and $\{u_i^k \mid 0 \leqslant i \leqslant m,\ 0 \leqslant k \leqslant n\}$ be the solution of the difference scheme (3.25). Then for an arbitrary step ratio r, we have*

$$\max_{0 \leqslant i \leqslant m} \left| u(x_i, t_k) - u_i^k \right| \leqslant c_1 T(\tau + h^2), \quad 0 \leqslant k \leqslant n,$$

where c_1 is defined by (3.12).

Proof Denote

$$e_i^k = u(x_i, t_k) - u_i^k, \quad 0 \leqslant i \leqslant m, \quad 0 \leqslant k \leqslant n.$$

Subtracting (3.25) from (3.23)–(3.24) gives the system of error equations

$$
\begin{cases}
D_{\bar t} e_i^k - a\delta_x^2 e_i^k = (R_2)_i^k, & 1 \leqslant i \leqslant m-1, \quad 1 \leqslant k \leqslant n, \\
e_i^0 = 0, & 0 \leqslant i \leqslant m, \\
e_0^k = 0, \quad e_m^k = 0, & 1 \leqslant k \leqslant n.
\end{cases}
$$

With the application of Theorem 3.9 and noticing (3.26), we have

$$
\|e^k\|_\infty \leqslant \tau \sum_{l=1}^{k} \max_{1 \leqslant i \leqslant m-1} |(R_2)_i^l| \leqslant c_1 \cdot k\tau(\tau+h^2) \leqslant c_1 T(\tau+h^2), \quad 0 \leqslant k \leqslant n.
$$

$\square$

Stability

Theorem 3.11 *The solution of the difference scheme (3.25) is stable for any step ratio r with respect to the initial value, boundary values, and right-hand side term in the following sense: Let $\{u_i^k \mid 0 \leqslant i \leqslant m, 0 \leqslant k \leqslant n\}$ be the solution to the system of difference equations*

$$
\begin{cases}
D_{\bar t} u_i^k - a\delta_x^2 u_i^k = f_i^k, & 1 \leqslant i \leqslant m-1, \quad 1 \leqslant k \leqslant n, \\
u_i^0 = \varphi_i, & 0 \leqslant i \leqslant m, \\
u_0^k = \alpha^k, \quad u_m^k = \beta^k, & 1 \leqslant k \leqslant n.
\end{cases}
$$

Then we have

$$
\max_{1 \leqslant i \leqslant m-1} |u_i^k| \leqslant \max\{ \max_{0 \leqslant i \leqslant m} |\varphi_i|, \max_{1 \leqslant s \leqslant k} |\alpha^s|, \max_{1 \leqslant s \leqslant k} |\beta^s|\}
$$

$$
+ \frac{L^2}{8a} \max_{1 \leqslant s \leqslant k} \max_{1 \leqslant i \leqslant m-1} |f_i^s|, \quad 1 \leqslant k \leqslant n.
$$

Proof A direct application of Theorem 3.8 arrives at the result, which completes the proof. $\square$

3.4   The Richardson Scheme

The convergence orders of both the forward and backward Euler schemes are $O(\tau + h^2)$, indicating that they are first-order accurate in the temporal direction and second-order accurate in the spatial direction. To achieve a convergence order $O(\tau^2 + h^2)$, a natural approach is to approximate the temporal derivative using the central difference quotient.

3.4.1 Derivation of the Difference Scheme

Considering Eq. (3.1a) at the node point (x_i, t_k), we have

$$\frac{\partial u}{\partial t}(x_i, t_k) - a\frac{\partial^2 u}{\partial x^2}(x_i, t_k) = f(x_i, t_k), \quad 1 \leqslant i \leqslant m-1, \quad 1 \leqslant k \leqslant n-1.$$

$$(3.30)$$

Substituting

$$\frac{\partial^2 u}{\partial x^2}(x_i, t_k) = \delta_x^2 U_i^k - \frac{h^2}{12}\frac{\partial^4 u}{\partial x^4}(\xi_{ik}, t_k), \quad x_{i-1} < \xi_{ik} < x_{i+1}$$

and

$$\frac{\partial u}{\partial t}(x_i, t_k) = \Delta_t U_i^k - \frac{\tau^2}{6}\frac{\partial^3 u}{\partial t^3}(x_i, \tilde{\eta}_{ik}), \quad t_{k-1} < \tilde{\eta}_{ik} < t_{k+1}$$

into (3.30), we have

$$\Delta_t U_i^k - a\delta_x^2 U_i^k = f(x_i, t_k) + \frac{\tau^2}{6}\frac{\partial^3 u}{\partial t^3}(x_i, \tilde{\eta}_{ik}) - \frac{ah^2}{12}\frac{\partial^4 u}{\partial x^4}(\xi_{ik}, t_k),$$

$$1 \leqslant i \leqslant m-1, \quad 1 \leqslant k \leqslant n-1. \quad (3.31)$$

Noticing the initial-boundary value conditions (3.1b) and (3.1c), we have

$$\begin{cases} U_i^0 = u(x_i, 0) = \varphi(x_i), & 0 \leqslant i \leqslant m, \\ U_0^k = \alpha(t_k), \quad U_m^k = \beta(t_k), & 1 \leqslant k \leqslant n. \end{cases}$$

Omitting the small term

$$(R_3)_i^k = \frac{\tau^2}{6}\frac{\partial^3 u}{\partial t^3}(x_i, \tilde{\eta}_{ik}) - \frac{ah^2}{12}\frac{\partial^4 u}{\partial x^4}(\xi_{ik}, t_k)$$

in (3.31), and replacing U_i^k with u_i^k, a difference scheme for solving (3.1) reads

$$\begin{cases} \Delta_t u_i^k - a\delta_x^2 u_i^k = f(x_i, t_k), & 1 \leqslant i \leqslant m-1, \quad 1 \leqslant k \leqslant n-1, & (3.32a) \\ u_i^0 = \varphi(x_i), & 0 \leqslant i \leqslant m, & (3.32b) \\ u_0^k = \alpha(t_k), \quad u_m^k = \beta(t_k), & 1 \leqslant k \leqslant n. & (3.32c) \end{cases}$$

The difference scheme (3.32) is referred to as the **Richardson scheme**. Additionally, $(R_3)_i^k$ represents the local truncation error of (3.32a). The Richardson scheme is explicit, and its stencil is illustrated in Fig. 3.13.

Fig. 3.13 The stencil of the Richardson scheme (3.32)

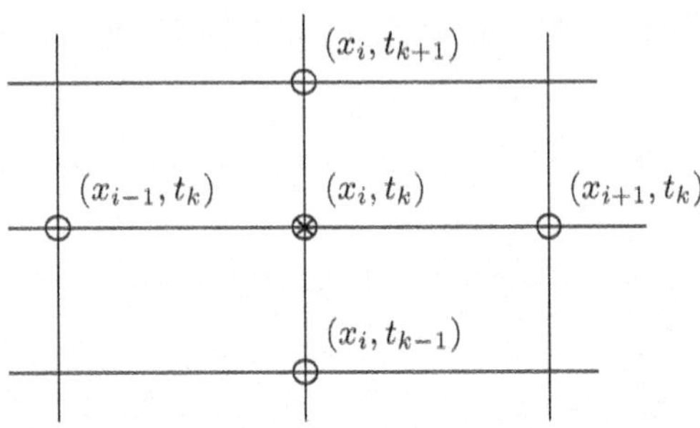

Although Eq. (3.32b) provides the value at the 0-th time level, the value at the first time level still needs to be computed.

From the Taylor expansion, it follows that

$$u(x_i, t_1) = u(x_i, t_0) + \tau u_t(x_i, t_0) + \frac{\tau^2}{2} u_{tt}(x_i, t_0) + \frac{\tau^3}{6} u_{ttt}(x_i, \eta_i), \quad \eta_i \in (t_0, t_1).$$

Equation (3.1a) implies

$$u_t(x, t) = a u_{xx}(x, t) + f(x, t),$$

$$u_{tt}(x, t) = a u_{xxt}(x, t) + f_t(x, t) = a\left[a u_{xxxx}(x, t) + f_{xx}(x, t)\right] + f_t(x, t).$$

According to the initial value condition (3.1b), it yields

$$u_t(x, t_0) = a \varphi''(x) + f(x, t_0),$$

$$u_{tt}(x, t_0) = a^2 \varphi^{(4)}(x) + a f_{xx}(x, t_0) + f_t(x, t_0).$$

One may take

$$u_i^1 = \varphi(x_i) + \tau\left[a \varphi''(x_i) + f(x_i, t_0)\right], \quad 1 \leqslant i \leqslant m - 1 \tag{3.33}$$

or

$$u_i^1 = \varphi(x_i) + \tau\left[a \varphi''(x_i) + f(x_i, t_0)\right]$$

$$+ \frac{\tau^2}{2}\left[a^2 \varphi^{(4)}(x_i) + a f_{xx}(x_i, t_0) + f_t(x_i, t_0)\right], \quad 1 \leqslant i \leqslant m - 1$$

as the approximation of $u(x_i, t_1)$, $1 \leqslant i \leqslant m - 1$.

The equality (3.33) can be replaced by

$$u_i^1 = \varphi(x_i) + \tau\left[a \delta_x^2 \varphi(x_i) + f(x_i, t_0)\right], \quad 1 \leqslant i \leqslant m - 1.$$

Noticing (3.32b), the above equality is equivalent to

$$D_t u_i^0 - a \delta_x^2 u_i^0 = f(x_i, t_0), \quad 1 \leqslant i \leqslant m - 1.$$

3.4.2   *Implementation of the Difference Scheme and Numerical Examples*

The difference equation (3.32a) can be written as

$$u_i^{k+1} = 2r(u_{i-1}^k - 2u_i^k + u_{i+1}^k) + u_i^{k-1} + 2\tau f(x_i, t_k),$$

$$1 \leqslant i \leqslant m - 1, \quad 1 \leqslant k \leqslant n - 1.$$

When applying the three-level explicit Richardson scheme to compute values at the $(k + 1)$-th time level, the values at the k-th time level and the $(k - 1)$-th time level are required. In practice, once the values at the 0-th and 1st time levels are known from Eqs. (3.32b) and (3.33), respectively, the values at the 2nd time level, 3rd time level, and so on, up to the n-th time level, can be computed successively.

Example 3.3   Apply the *Richardson* scheme (3.32)–(3.33) to compute the problem

$$\begin{cases} \dfrac{\partial u}{\partial t} - \dfrac{\partial^2 u}{\partial x^2} = 0, & 0 < x < 1, \quad 0 < t \leqslant 1, \\ u(x, 0) = e^x, & 0 \leqslant x \leqslant 1, \\ u(0, t) = e^t, & u(1, t) = e^{1+t}, \quad 0 < t \leqslant 1. \end{cases}$$

The exact solution to the problem is $u(x, t) = e^{x+t}$.

Tables 3.8 and 3.9 present the numerical solutions at specific node points for step sizes $h = 1/10, \tau = 1/100$ and $h = 1/100, \tau = 1/100$, respectively. As the number of time levels increases, the error grows larger, and the numerical results become impractical. In the following, we will demonstrate that the Richardson scheme is inherently unstable. Specifically, regardless of the step ratio, a small error at the initial time level will inevitably lead to a large error in the numerical solution.

The Richardson scheme is a completely unstable difference scheme and cannot be used to obtain useful numerical results. Its purpose here is to serve as a typical example of an unstable scheme, highlighting the importance of stability.

3.4.3   *Instability of the Difference Scheme*

In this section, we focus exclusively on the impact of the initial error on the solution.

Table 3.8 (Example 3.3) Numerical solutions, exact solutions, and absolute values of errors at some node points ($h = 1/10$, $\tau = 1/100$)

k	(x, t)	NS	ES	|NS−ES|
1	$(0.5, 0.01)$	1.665208	1.665291	8.271e−5
2	$(0.5, 0.02)$	1.682053	1.682028	2.555e−5
3	$(0.5, 0.03)$	1.698878	1.698932	5.472e−5
4	$(0.5, 0.04)$	1.716059	1.716007	5.222e−5
5	$(0.5, 0.05)$	1.733227	1.733253	2.563e−5
6	$(0.5, 0.06)$	1.756722	1.750673	6.049e−3
7	$(0.5, 0.07)$	1.647045	1.768267	1.212e−1
8	$(0.5, 0.08)$	3.408799	1.786038	1.623e+0
9	$(0.5, 0.09)$	−1.619728e+1	1.803988	1.800e+1
10	$(0.5, 0.10)$	1.816445e+2	1.822119	1.798e+2

Table 3.9 (Example 3.3) Numerical solutions, exact solutions, and absolute values of error at some node points ($h = 1/100$, $\tau = 1/100$)

k	(x, t)	NS	ES	|ES−NS|
1	$(0.5, 0.01)$	1.665208	1.665291	8.271e−5
2	$(0.5, 0.02)$	1.682026	1.682028	1.932e−6
3	$(0.5, 0.03)$	1.698849	1.698932	8.303e−5
4	$(0.5, 0.04)$	1.716003	1.716007	3.864e−6
5	$(0.5, 0.05)$	1.733162	1.733253	9.113e−5
6	$(0.5, 0.06)$	1.756219	1.750673	5.546e−3
7	$(0.5, 0.07)$	−2.375314	1.768267	4.144e+0
8	$(0.5, 0.08)$	3.183175e+3	1.786038	3.181e+3
9	$(0.5, 0.09)$	−2.491744e+6	1.803988	2.492e+6
10	$(0.5, 0.10)$	1.978856e+9	1.822119	1.979e+9

When the Richardson scheme (3.32)–(3.33) is applied, suppose that u_i^0 has an error ϕ_i, and u_i^1 has an error ψ_i. Consequently, the solution obtained is actually the solution to the following system of difference equations:

$$
\begin{cases}
\Delta_t v_i^k - a\delta_x^2 v_i^k = f(x_i, t_k), & 1 \leqslant i \leqslant m - 1, \quad 1 \leqslant k \leqslant n - 1, \\
v_i^0 = \varphi(x_i) + \phi_i, \quad 1 \leqslant i \leqslant m - 1, \\
v_i^1 = \varphi(x_i) + \tau\left[a\varphi''(x_i) + f(x_i, t_0)\right] + \psi_i, \quad 1 \leqslant i \leqslant m - 1, \\
v_0^k = \alpha(t_k), \quad v_m^k = \beta(t_k), \quad 0 \leqslant k \leqslant n.
\end{cases}
\tag{3.34}
$$

Let

$$
\varepsilon_i^k = v_i^k - u_i^k, \quad 0 \leqslant i \leqslant m, \quad 0 \leqslant k \leqslant n.
$$

Subtracting (3.32)–(3.33) from (3.34), we have the system of perturbation equations

$$\begin{cases} \Delta_t \varepsilon_i^k - a\delta_x^2 \varepsilon_i^k = 0, & 1 \leqslant i \leqslant m-1, \quad 1 \leqslant k \leqslant n-1, \\ \varepsilon_i^0 = \phi_i, & 1 \leqslant i \leqslant m-1, \\ \varepsilon_i^1 = \psi_i, & 1 \leqslant i \leqslant m-1, \\ \varepsilon_0^k = 0, \quad \varepsilon_m^k = 0, & 0 \leqslant k \leqslant n. \end{cases}$$

Denote

$$\varepsilon^k = \left\{ \varepsilon_i^k \,\middle|\, 0 \leqslant i \leqslant m \right\}.$$

For a three-level difference scheme, if there exist $\varepsilon_0 > 0$, $C > 0$, $\tau_0 > 0$, $h_0 > 0$ such that

$$\|\varepsilon^k\| \leqslant C(\|\varepsilon^0\| + \|\varepsilon^1\|), \quad 2 \leqslant k \leqslant n \tag{3.35}$$

holds for all $\tau \leqslant \tau_0$, $h \leqslant h_0$ and $\|\varepsilon^0\| \leqslant \varepsilon_0$, $\|\varepsilon^1\| \leqslant \varepsilon_0$, then the difference scheme is stable. Otherwise, it is unstable.

Theorem 3.12 *The solution of the Richardson scheme is unstable in the maximum norm for any step ratio r.*

Proof Consider the system of perturbation equations

$$\begin{cases} \Delta_t \varepsilon_i^k - a\delta_x^2 \varepsilon_i^k = 0, & 1 \leqslant i \leqslant m-1, \quad 1 \leqslant k \leqslant n-1, & (3.36a) \\ \varepsilon_i^0 = (-1)^i(-\epsilon), & 1 \leqslant i \leqslant m-1, & (3.36b) \\ \varepsilon_i^1 = (-1)^{i+1}\epsilon, & 1 \leqslant i \leqslant m-1, & (3.36c) \\ \varepsilon_0^k = 0, \quad \varepsilon_m^k = 0, & 0 \leqslant k \leqslant n, & (3.36d) \end{cases}$$

where ϵ is a small positive constant.

It follows from $r = a\frac{\tau}{h^2}$ that $n = \frac{aT}{rL^2}m^2$. When n is suitable large, $n \geqslant \lfloor \frac{m}{2} \rfloor$. Thereby, $\min\left\{ \lfloor \frac{m}{2} \rfloor, n \right\} = \lfloor \frac{m}{2} \rfloor$, and $m \to +\infty$ when $n \to +\infty$.

Equation (3.36a) can be rewritten as

$$\varepsilon_i^{k+1} = 2r(\varepsilon_{i-1}^k + \varepsilon_{i+1}^k) - 4r\varepsilon_i^k + \varepsilon_i^{k-1}, \quad 1 \leqslant i \leqslant m-1, \quad 1 \leqslant k \leqslant n-1. \tag{3.37}$$

It is shown that within the triangular or trapezoidal region illustrated in Fig. 3.14, i.e., when $k \leqslant \min\left\{ \lfloor \frac{m}{2} \rfloor, n \right\}$, the values of ε_i^k ($k \leqslant i \leqslant m-k$) are fully determined by (3.36a), the initial value (3.36b), and (3.36c), but are independent of (3.36d).

Suppose the solution in the above region has the form

$$\varepsilon_i^k = (-1)^{i+k} T(k), \quad k \leqslant i \leqslant m-k, \quad k \leqslant \min\left\{ \lfloor \frac{m}{2} \rfloor, n \right\}.$$

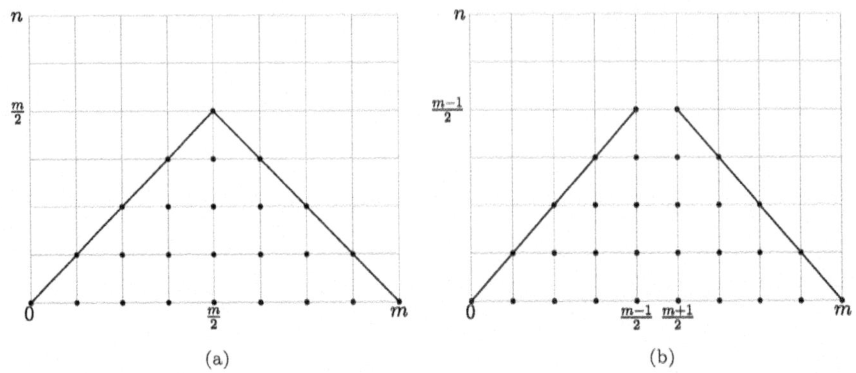

Fig. 3.14 Auxiliary figures of the proof for Theorem 3.12. (a) m is even, $n \geqslant \frac{m}{2}$. (b) m is odd, $n \geqslant \frac{m-1}{2}$

Substituting the above equality into (3.37), we have

$$T(k+1) - 8rT(k) - T(k-1) = 0, \quad k \leqslant \min\left\{\left\lfloor \frac{m}{2} \right\rfloor, n\right\}. \tag{3.38}$$

The quadratic equation

$$\lambda^2 - 8r\lambda - 1 = 0$$

is called the characteristic equation of (3.38), which has two roots

$$\lambda_1 = 4r + \sqrt{16r^2 + 1} \in (1, +\infty), \quad \lambda_2 = 4r - \sqrt{16r^2 + 1} \in (-1, 0).$$

Noticing that $T(0) = -\epsilon$, $T(1) = \epsilon$, we have

$$T(k) = \frac{T(1) - \lambda_2 T(0)}{\lambda_1 - \lambda_2}\lambda_1^k + \frac{\lambda_1 T(0) - T(1)}{\lambda_1 - \lambda_2}\lambda_2^k$$

$$= \left(\frac{1 + \lambda_2}{\lambda_1 - \lambda_2}\lambda_1^k - \frac{1 + \lambda_1}{\lambda_1 - \lambda_2}\lambda_2^k\right)\epsilon, \quad k \leqslant \min\left\{\left\lfloor \frac{m}{2} \right\rfloor, n\right\}.$$

Thus the solution in the triangular or trapezoidal region shown in Fig. 3.14 is

$$\varepsilon_i^k = (-1)^{i+k}\left(\frac{1 + \lambda_2}{\lambda_1 - \lambda_2}\lambda_1^k - \frac{1 + \lambda_1}{\lambda_1 - \lambda_2}\lambda_2^k\right)\epsilon, \quad k \leqslant i \leqslant m - k, \quad k \leqslant \min\left\{\left\lfloor \frac{m}{2} \right\rfloor, n\right\}.$$

It follows from the above equality that

$$\left\|\varepsilon^{\min\left\{\left\lfloor \frac{m}{2} \right\rfloor, n\right\}}\right\|_\infty \geqslant \left(\frac{1 + \lambda_2}{\lambda_1 - \lambda_2}\lambda_1^{\min\left\{\left\lfloor \frac{m}{2} \right\rfloor, n\right\}} - \frac{1 + \lambda_1}{\lambda_1 - \lambda_2}|\lambda_2|^{\min\left\{\left\lfloor \frac{m}{2} \right\rfloor, n\right\}}\right)\epsilon.$$

Therefore,

$$\lim_{n\to+\infty}\left\|\varepsilon^{\min\left\{\left\lfloor\frac{m}{2}\right\rfloor,n\right\}}\right\|_{\infty}=+\infty.$$

Since

$$\max_{1\leqslant k\leqslant n}\left\|\varepsilon^{k}\right\|_{\infty}\geqslant\left\|\varepsilon^{\min\left\{\left\lfloor\frac{m}{2}\right\rfloor,n\right\}}\right\|_{\infty},$$

we have

$$\lim_{n\to+\infty}\max_{1\leqslant k\leqslant n}\left\|\varepsilon^{k}\right\|_{\infty}=+\infty.$$

There is no constant C independent of τ and h such that (3.35) holds. Therefore, the Richardson scheme is unstable for any arbitrary step ratio r. $\qquad\square$

3.5 The Crank-Nicolson Scheme

In this section, an unconditionally stable difference scheme with the accuracy $O(\tau^2+h^2)$ will be established for the problem (3.1).

3.5.1 Derivation of the Difference Scheme

Considering Eq. (3.1a) at the point $(x_i, t_{k+\frac{1}{2}})$, we have

$$\frac{\partial u}{\partial t}(x_i, t_{k+\frac{1}{2}})-a\frac{\partial^2 u}{\partial x^2}(x_i, t_{k+\frac{1}{2}})=f(x_i, t_{k+\frac{1}{2}}),\quad 1\leqslant i\leqslant m-1,\quad 0\leqslant k\leqslant n-1.$$

Employing the formula

$$\frac{\partial^2 u}{\partial x^2}(x_i, t_{k+\frac{1}{2}})=\frac{1}{2}\left[\frac{\partial^2 u}{\partial x^2}(x_i, t_k)+\frac{\partial^2 u}{\partial x^2}(x_i, t_{k+1})\right]$$
$$-\frac{\tau^2}{8}\frac{\partial^4 u}{\partial x^2\partial t^2}(x_i, \zeta_{ik}),\quad \zeta_{ik}\in(t_k, t_{k+1}),$$

we have

$$\frac{\partial u}{\partial t}(x_i, t_{k+\frac{1}{2}}) - \frac{1}{2}a\left[\frac{\partial^2 u}{\partial x^2}(x_i, t_k) + \frac{\partial^2 u}{\partial x^2}(x_i, t_{k+1})\right]$$

$$= f(x_i, t_{k+\frac{1}{2}}) - \frac{a\tau^2}{8}\frac{\partial^4 u}{\partial x^2 \partial t^2}(x_i, \zeta_{ik}), \quad t_k < \zeta_{ik} < t_{k+1}.$$

Combining

$$\frac{\partial^2 u}{\partial x^2}(x_i, t_k) = \delta_x^2 U_i^k - \frac{h^2}{12}\frac{\partial^4 u}{\partial x^4}(\xi_{ik}, t_k), \quad x_{i-1} < \xi_{ik} < x_{i+1}$$

with

$$\frac{\partial u}{\partial t}(x_i, t_{k+\frac{1}{2}}) = \delta_t U_i^{k+\frac{1}{2}} - \frac{\tau^2}{24}\frac{\partial^3 u}{\partial t^3}(x_i, \eta_{ik}), \quad t_k < \eta_{ik} < t_{k+1},$$

we have

$$\delta_t U_i^{k+\frac{1}{2}} - a\delta_x^2 U_i^{k+\frac{1}{2}} = f(x_i, t_{k+\frac{1}{2}}) + (R_4)_i^k, \tag{3.39}$$

where

$$(R_4)_i^k = \left[\frac{1}{24}\frac{\partial^3 u}{\partial t^3}(x_i, \eta_{ik}) - \frac{a}{8}\frac{\partial^4 u}{\partial x^2 \partial t^2}(x_i, \zeta_{ik})\right]\tau^2$$

$$- \frac{a}{24}\left[\frac{\partial^4 u}{\partial x^4}(\xi_{ik}, t_k) + \frac{\partial^4 u}{\partial x^4}(\xi_{i,k+1}, t_{k+1})\right]h^2.$$

Noticing the initial-boundary value conditions (3.1b) and (3.1c), we have

$$\begin{cases} U_i^0 = \varphi(x_i), & 0 \leqslant i \leqslant m, \\ U_0^k = \alpha(t_k), & U_m^k = \beta(t_k), \quad 1 \leqslant k \leqslant n. \end{cases} \tag{3.40}$$

Omitting the small term $(R_4)_i^k$ in (3.39) and replacing U_i^k with u_i^k, a difference scheme reads

$$\begin{cases} \delta_t u_i^{k+\frac{1}{2}} - a\delta_x^2 u_i^{k+\frac{1}{2}} = f(x_i, t_{k+\frac{1}{2}}), & 1 \leqslant i \leqslant m-1, \quad 0 \leqslant k \leqslant n-1, & \text{(3.41a)} \\ u_i^0 = \varphi(x_i), \quad 0 \leqslant i \leqslant m, & & \text{(3.41b)} \\ u_0^k = \alpha(t_k), \quad u_m^k = \beta(t_k), \quad 1 \leqslant k \leqslant n. & & \text{(3.41c)} \end{cases}$$

Fig. 3.15 The stencil of the
Crank-Nicolson scheme
(3.41)

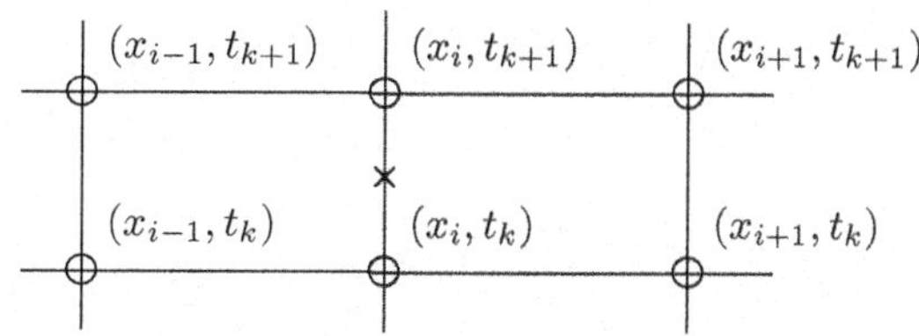

The difference scheme (3.41) is referred to as the **Crank-Nicolson scheme**, and $(R_4)_i^k$ represents the local truncation error of the difference scheme (3.41).

Denote

$$c_2 = \max \left\{ \frac{1}{24} \max_{(x,y) \in \bar{D}} \left| \frac{\partial^3 u(x,t)}{\partial t^3} \right| + \frac{a}{8} \max_{(x,y) \in \bar{D}} \left| \frac{\partial^4 u(x,t)}{\partial x^2 \partial t^2} \right|, \ \frac{a}{12} \max_{(x,y) \in \bar{D}} \left| \frac{\partial^4 u(x,t)}{\partial x^4} \right| \right\},$$

$$(3.42)$$

and then we have

$$\left| (R_4)_i^k \right| \leqslant c_2(\tau^2 + h^2), \quad 1 \leqslant i \leqslant m - 1, \quad 0 \leqslant k \leqslant n - 1. \tag{3.43}$$

The computational stencil of the two-level Crank-Nicolson scheme (3.41) is shown in Fig. 3.15.

3.5.2   Existence of the Difference Solution

Theorem 3.13 *The difference scheme (3.41) is uniquely solvable.*

Proof Denote

$$u^k = (u_0^k, u_1^k, \ldots, u_{m-1}^k, u_m^k).$$

The value at the 0-th time level has been determined by (3.41b). Suppose the value of u^k has been determined. Then, the difference scheme for u^{k+1} is given by

$$\begin{cases} \delta_t u_i^{k+\frac{1}{2}} - a\delta_x^2 u_i^{k+\frac{1}{2}} = f(x_i, t_{k+\frac{1}{2}}), & 1 \leqslant i \leqslant m - 1, \\ u_0^{k+1} = \alpha(t_{k+1}), & u_m^{k+1} = \beta(t_{k+1}). \end{cases}$$

It can be written in the following matrix form

$$
\begin{pmatrix}
1+r & -\dfrac{r}{2} \\
-\dfrac{r}{2} & 1+r & -\dfrac{r}{2} \\
 & \ddots & \ddots & \ddots \\
 & & -\dfrac{r}{2} & 1+r & -\dfrac{r}{2} \\
 & & & -\dfrac{r}{2} & 1+r
\end{pmatrix}
\begin{pmatrix}
u_1^{k+1} \\
u_2^{k+1} \\
\vdots \\
u_{m-2}^{k+1} \\
u_{m-1}^{k+1}
\end{pmatrix}
$$

$$
=
\begin{pmatrix}
1-r & \dfrac{r}{2} \\
\dfrac{r}{2} & 1-r & \dfrac{r}{2} \\
 & \ddots & \ddots & \ddots \\
 & & \dfrac{r}{2} & 1-r & \dfrac{r}{2} \\
 & & & \dfrac{r}{2} & 1-r
\end{pmatrix}
\begin{pmatrix}
u_1^{k} \\
u_2^{k} \\
\vdots \\
u_{m-2}^{k} \\
u_{m-1}^{k}
\end{pmatrix}
$$

$$
+
\begin{pmatrix}
\dfrac{r}{2}\left(u_0^k + u_0^{k+1}\right) + \tau f(x_1, t_{k+\frac{1}{2}}) \\
\tau f(x_2, t_{k+\frac{1}{2}}) \\
\vdots \\
\tau f(x_{m-2}, t_{k+\frac{1}{2}}) \\
\dfrac{r}{2}\left(u_m^k + u_m^{k+1}\right) + \tau f(x_{m-1}, t_{k+\frac{1}{2}})
\end{pmatrix}. \tag{3.44}
$$

It is easy to see that the coefficient matrix is strictly diagonally dominant; therefore, a unique solution exists.

By induction, the result follows. $\qquad\square$

3.5.3   Implementation of the Difference Scheme and Numerical Examples

It follows from (3.44) that (3.41) is a tridiagonal system of linear equations, which can be solved by the double sweep method.

Example 3.4   Apply the *Crank-Nicolson* scheme (3.41) to compute the problem

$$
\begin{cases}
\dfrac{\partial u}{\partial t} - \dfrac{\partial^2 u}{\partial x^2} = 0, & 0 < x < 1, \quad 0 < t \leqslant 1, \\
u(x,0) = e^x, & 0 \leqslant x \leqslant 1, \\
u(0,t) = e^t, & u(1,t) = e^{1+t}, \quad 0 < t \leqslant 1.
\end{cases}
$$

Table 3.10 (Example 3.4) Numerical solutions (NS), exact solutions (ES), and absolute values of errors at part of node points ($h = 1/10, \tau = 1/10$)

k	(x, t)	NS	ES	|ES−NS|
1	(0.5, 0.1)	1.822349	1.822119	2.305e−4
2	(0.5, 0.2)	2.014105	2.013753	3.522e−4
3	(0.5, 0.3)	2.225953	2.225541	4.124e−4
4	(0.5, 0.4)	2.460072	2.459603	4.692e−4
5	(0.5, 0.5)	2.718802	2.718282	5.204e−4
6	(0.5, 0.6)	3.004743	3.004166	5.770e−4
7	(0.5, 0.7)	3.320755	3.320117	6.379e−4
8	(0.5, 0.8)	3.670002	3.669297	7.051e−4
9	(0.5, 0.9)	4.055979	4.055200	7.795e−4
10	(0.5, 1.0)	4.482550	4.481689	8.612e−4

Table 3.11 (Example 3.4) Numerical solutions (NS), exact solutions (ES), and the absolute values of errors at part of node points ($h = 1/100, \tau = 1/100$)

k	(x, t)	NS	ES	|ES−NS|
10	(0.5, 0.1)	1.822121	1.822119	2.281e−6
20	(0.5, 0.2)	2.013756	2.013753	3.420e−6
30	(0.5, 0.3)	2.225545	2.225541	4.115e−6
40	(0.5, 0.4)	2.459608	2.459603	4.672e−6
50	(0.5, 0.5)	2.718287	2.718282	5.210e−6
60	(0.5, 0.6)	3.004172	3.004166	5.776e−6
70	(0.5, 0.7)	3.320123	3.320117	6.389e−6
80	(0.5, 0.8)	3.669304	3.669297	7.064e−6
90	(0.5, 0.9)	4.055208	4.055200	7.808e−6
100	(0.5, 1.0)	4.481698	4.481689	8.629e−6

The exact solution of the above problem is $u(x, t) = e^{x+t}$.

Table 3.10 presents part of the numerical results calculated with the step sizes $h = 1/10$ and $\tau = 1/10$. Table 3.11 gives some numerical results calculated with the step sizes $h = 1/100$ and $\tau = 1/100$. The numerical solution approximates the exact solution very well. Table 3.12 shows the maximum error

$$E_\infty(h, \tau) = \max_{1 \leqslant i \leqslant m-1, 1 \leqslant k \leqslant n} |u(x_i, t_k) - u_i^k|,$$

of the numerical solutions with different step sizes.

It can be seen from Table 3.12 that when both the spatial and temporal step sizes are halved, the maximum error is reduced to one-fourth of the original. Figure 3.16 illustrates the exact solution profile and the numerical solution profile at $t = 1$ with step sizes $h = 1/10, \tau = 1/10$. Figure 3.17 displays the numerical error profiles at $t = 1$ for different step sizes. Figure 3.18 shows the error surfaces of the numerical solutions with various step sizes.

Table 3.12 (Example 3.4) The maximum errors of numerical solutions with different step sizes

h	τ	$E_\infty(h, \tau)$	$E_\infty(2h, 2\tau)/E_\infty(h, \tau)$
1/10	1/10	8.612e−4	
1/20	1/20	2.174e−4	3.961
1/40	1/40	5.436e−5	3.999
1/80	1/80	1.359e−5	4.000
1/160	1/160	3.398e−6	3.999
1/320	1/320	8.495e−7	4.000
1/640	1/640	2.123e−7	4.001

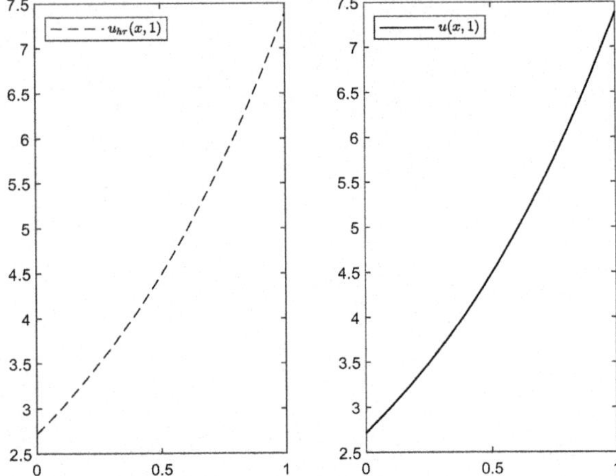

Fig. 3.16 (Example 3.4) The numerical solution profile and the exact solution profile at $t = 1$ with $h = 1/10$, $\tau = 1/10$

3.5.4 A Priori Estimate of the Difference Solution

Theorem 3.14 *Let $\{v_i^k \mid 0 \leqslant i \leqslant m,\ 0 \leqslant k \leqslant n\}$ be the solution of the system of difference equations*

$$
\begin{cases}
\delta_t v_i^{k+\frac{1}{2}} - a\delta_x^2 v_i^{k+\frac{1}{2}} = g_i^k, & 1 \leqslant i \leqslant m-1, \quad 0 \leqslant k \leqslant n-1, & (3.45a) \\[2mm]
v_i^0 = \varphi_i, & 1 \leqslant i \leqslant m-1, & (3.45b) \\[2mm]
v_0^k = 0, \quad v_m^k = 0, & 0 \leqslant k \leqslant n. & (3.45c)
\end{cases}
$$

Then for any step ratio r, we have

$$
\|v^k\|^2 \leqslant \|v^0\|^2 + \frac{L^2}{12a}\tau \sum_{l=0}^{k-1} \|g^l\|^2, \quad 0 \leqslant k \leqslant n, \tag{3.46}
$$

$$
|v^k|_1^2 \leqslant |v^0|_1^2 + \frac{1}{2a}\tau \sum_{l=0}^{k-1} \|g^l\|^2, \quad 0 \leqslant k \leqslant n, \tag{3.47}
$$

Fig. 3.17 (Example 3.4) The numerical error profiles at $t = 1$ with different step sizes

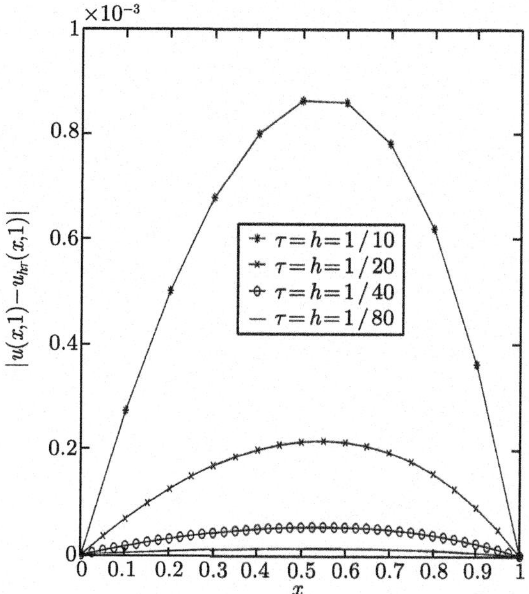

Fig. 3.18 (Example 3.4) The numerical error surfaces with different step sizes

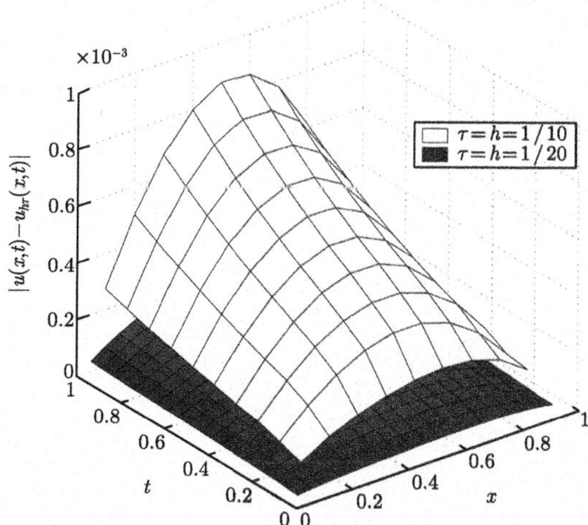

$$\|v^k\|_\infty^2 \leqslant \frac{L}{4}\left(|v^0|_1^2 + \frac{1}{2a}\tau\sum_{l=0}^{k-1}\|g^l\|^2\right), \quad 0 \leqslant k \leqslant n, \tag{3.48}$$

where $\|g^l\|^2 = h\sum_{i=1}^{m-1}(g_i^l)^2$.

Proof

(a) Taking an inner product of (3.45a) with $2v^{k+\frac{1}{2}}$, we have

$$2(\delta_t v^{k+\frac{1}{2}}, v^{k+\frac{1}{2}}) - 2a(\delta_x^2 v^{k+\frac{1}{2}}, v^{k+\frac{1}{2}}) = 2(g^k, v^{k+\frac{1}{2}}), \quad 0 \leqslant k \leqslant n-1.$$

It follows by noticing

$$2(\delta_t v^{k+\frac{1}{2}}, v^{k+\frac{1}{2}}) = \frac{1}{\tau}(\|v^{k+1}\|^2 - \|v^k\|^2), \quad (\delta_x^2 v^{k+\frac{1}{2}}, v^{k+\frac{1}{2}}) = -|v^{k+\frac{1}{2}}|_1^2,$$

that

$$\frac{1}{\tau}(\|v^{k+1}\|^2 - \|v^k\|^2) = -2a|v^{k+\frac{1}{2}}|_1^2 + 2(g^k, v^{k+\frac{1}{2}})$$

$$\leqslant -2a|v^{k+\frac{1}{2}}|_1^2 + \frac{12a}{L^2}\|v^{k+\frac{1}{2}}\|^2 + \frac{L^2}{12a}\|g^k\|^2.$$

With the application of Lemma 1.4, we have

$$\frac{1}{\tau}\left(\|v^{k+1}\|^2 - \|v^k\|^2\right) \leqslant -2a|v^{k+\frac{1}{2}}|_1^2 + 2a|v^{k+\frac{1}{2}}|_1^2 + \frac{L^2}{12a}\|g^k\|^2$$

$$= \frac{L^2}{12a}\|g^k\|^2, \quad 0 \leqslant k \leqslant n-1.$$

Multiplying both sides of the above inequality by τ and rearranging the result produce

$$\|v^{k+1}\|^2 \leqslant \|v^k\|^2 + \frac{L^2}{12a}\tau\|g^k\|^2, \quad 0 \leqslant k \leqslant n-1. \tag{3.49}$$

It follows from the recursion of (3.49) that

$$\|v^k\|^2 \leqslant \|v^0\|^2 + \frac{L^2}{12a}\tau\sum_{l=0}^{k-1}\|g^l\|^2, \quad 0 \leqslant k \leqslant n,$$

which completes the proof of (3.46).

(b) Taking an inner product of (3.45a) with $\delta_t v^{k+\frac{1}{2}}$, we have

$$\|\delta_t v^{k+\frac{1}{2}}\|^2 - a(\delta_x^2 v^{k+\frac{1}{2}}, \delta_t v^{k+\frac{1}{2}}) = (g^k, \delta_t v^{k+\frac{1}{2}}). \tag{3.50}$$

Using Lemma 1.4, we have

$$-(\delta_x^2 v^{k+\frac{1}{2}}, \delta_t v^{k+\frac{1}{2}}) = \frac{1}{2\tau}(|v^{k+1}|_1^2 - |v^k|_1^2).$$

Substituting the above equality into (3.50) gives

$$\|\delta_t v^{k+\frac{1}{2}}\|^2 + \frac{a}{2\tau}(|v^{k+1}|_1^2 - |v^k|_1^2) = h \sum_{i=1}^{m-1} g_i^k \left(\delta_t v_i^{k+\frac{1}{2}}\right)$$

$$\leqslant \frac{1}{4}\|g^k\|^2 + \|\delta_t v^{k+\frac{1}{2}}\|^2, \quad 0 \leqslant k \leqslant n-1,$$

which implies

$$|v^{k+1}|_1^2 \leqslant |v^k|_1^2 + \frac{1}{2a}\tau\|g^k\|^2, \quad 0 \leqslant k \leqslant n-1.$$

Consequently,

$$|v^k|_1^2 \leqslant |v^0|_1^2 + \frac{1}{2a}\tau \sum_{l=0}^{k-1} \|g^l\|^2, \quad 0 \leqslant k \leqslant n.$$

That is, (3.47) holds.

(c) Using Lemma 1.4 and noticing (3.45c), we have $\|v^k\|_\infty \leqslant \frac{\sqrt{L}}{2}|v^k|_1$. Applying (3.47), it leads to (3.48).

$\square$

3.5.5   Convergence and Stability of the Difference Solution

Convergence

Theorem 3.15 *Let $\{u(x,t) \mid (x,t) \in \bar{D}\}$ be the solution of the problem (3.1) and $\{u_i^k \mid 0 \leqslant i \leqslant m, 0 \leqslant k \leqslant n\}$ be the solution of the difference scheme (3.41). Then for an arbitrary step ratio r, we have*

$$\max_{1 \leqslant k \leqslant n, 1 \leqslant i \leqslant m-1} \left|u(x_i, t_k) - u_i^k\right| \leqslant \frac{c_2 L}{2} \cdot \sqrt{\frac{T}{2a}} (\tau^2 + h^2),$$

where c_2 is defined by (3.42).

Proof Denote

$$e_i^k = u(x_i, t_k) - u_i^k, \quad 0 \leqslant i \leqslant m, \quad 0 \leqslant k \leqslant n.$$

Subtracting (3.41) from (3.39) and (3.40), we have the system of error equations

$$\begin{cases} \delta_t e_i^{k+\frac{1}{2}} - a\delta_x^2 e_i^{k+\frac{1}{2}} = (R_4)_i^k, & 1 \leqslant i \leqslant m-1, \quad 0 \leqslant k \leqslant n-1, \\ e_i^0 = 0, & 0 \leqslant i \leqslant m, \\ e_0^k = 0, \quad e_m^k = 0, & 1 \leqslant k \leqslant n. \end{cases} \tag{3.51}$$

Applying Theorem 3.14 and noticing (3.43) yield

$$\|e^k\|_\infty^2 \leqslant \frac{L}{4} \cdot \frac{1}{2a}\tau \sum_{l=0}^{k-1} \|(R_4)^l\|^2 \leqslant \frac{L}{4} \cdot \frac{T}{2a} Lc_2^2(\tau^2 + h^2)^2.$$

Taking the square root on both sides of the above inequality, we obtain

$$\|e^k\|_\infty \leqslant \frac{c_2 L}{2} \cdot \sqrt{\frac{T}{2a}}\,(\tau^2 + h^2), \quad 1 \leqslant k \leqslant n.$$

$\square$

Stability

Theorem 3.16 *The solution of the difference scheme (3.41) is stable for any step ratio r with respect to the initial value and the right-hand side term, in the following sense: Let $\{u_i^k \,|\, 0 \leqslant i \leqslant m,\ 0 \leqslant k \leqslant n\}$ be the solution of the difference scheme*

$$\begin{cases} \delta_t u_i^{k+\frac{1}{2}} - a\delta_x^2 u_i^{k+\frac{1}{2}} = f_i^{k+\frac{1}{2}}, & 1 \leqslant i \leqslant m-1, \quad 0 \leqslant k \leqslant n-1, \\ u_i^0 = \varphi_i, & 1 \leqslant i \leqslant m-1, \\ u_0^k = 0, \quad u_m^k = 0, & 0 \leqslant k \leqslant n. \end{cases}$$

Then we have

$$\|u^k\|^2 \leqslant \|u^0\|^2 + \frac{L^2}{12a}\tau \sum_{l=0}^{k-1} \|f^{l+\frac{1}{2}}\|^2, \quad 1 \leqslant k \leqslant n,$$

$$\|u^k\|_\infty^2 \leqslant \frac{L}{4}\left(|u^0|_1^2 + \frac{1}{2a}\tau \sum_{l=0}^{k-1} \|f^{l+\frac{1}{2}}\|^2\right), \quad 1 \leqslant k \leqslant n.$$

Proof A direct application of Theorem 3.14 yields the results. $\square$

3.5.6 *The Richardson Extrapolation*

Denote the solution of the difference scheme (3.41) by $u_i^k(h, \tau)$.

Theorem 3.17 *Suppose the problems*

$$
\begin{cases}
\dfrac{\partial v}{\partial t} - a\dfrac{\partial^2 v}{\partial x^2} = -p(x, t), & 0 < x < L, \quad 0 < t \leqslant T, \\[2mm]
v(x, 0) = 0, & 0 \leqslant x \leqslant L, \\[2mm]
v(0, t) = 0, \quad v(L, t) = 0, & 0 < t \leqslant T
\end{cases}
\tag{3.52}
$$

and

$$
\begin{cases}
\dfrac{\partial w}{\partial t} - a\dfrac{\partial^2 w}{\partial x^2} = -q(x, t), & 0 < x < L, \quad 0 < t \leqslant T, \\[2mm]
w(x, 0) = 0, & 0 \leqslant x \leqslant L, \\[2mm]
w(0, t) = 0, \quad w(L, t) = 0, & 0 < t \leqslant T
\end{cases}
\tag{3.53}
$$

admit smooth solutions, where

$$
p(x, t) = \frac{1}{24}\frac{\partial^3 u}{\partial t^3}(x, t) - \frac{a}{8}\frac{\partial^4 u}{\partial x^2 \partial t^2}(x, t), \quad q(x, t) = -\frac{a}{12}\frac{\partial^4 u}{\partial x^4}(x, t).
$$

Then we have

$$
u_i^k(h, \tau) = u(x_i, t_k) + \tau^2 v(x_i, t_k) + h^2 w(x_i, t_k) + O(\tau^4 + h^4),
$$
$$
1 \leqslant i \leqslant m - 1, \quad 1 \leqslant k \leqslant n;
$$

$$
\max_{1 \leqslant i \leqslant m-1, 1 \leqslant k \leqslant n} \left| u(x_i, t_k) - \left[\frac{4}{3}u_{2i}^{2k}\left(\frac{h}{2}, \frac{\tau}{2}\right) - \frac{1}{3}u_i^k(h, \tau) \right] \right| = O(\tau^4 + h^4).
$$

Proof Detailed analysis shows that

$$
(R_4)_i^k = p(x_i, t_{k+\frac{1}{2}})\tau^2 + q(x_i, t_{k+\frac{1}{2}})h^2 + O(\tau^4 + h^4),
$$
$$
1 \leqslant i \leqslant m - 1, \quad 0 \leqslant k \leqslant n - 1.
$$

Therefore, the system of error equations (3.51) can be written as

$$
\begin{cases}
\delta_t e_i^{k+\frac{1}{2}} - a\delta_x^2 e_i^{k+\frac{1}{2}} = p(x_i, t_{k+\frac{1}{2}})\tau^2 + q(x_i, t_{k+\frac{1}{2}})h^2 + O(\tau^4 + h^4), \\[2mm]
\qquad\qquad\qquad\qquad 1 \leqslant i \leqslant m - 1, \quad 0 \leqslant k \leqslant n - 1, \\[2mm]
e_i^0 = 0, \quad 0 \leqslant i \leqslant m, \\[2mm]
e_0^k = 0, \quad e_m^k = 0, \quad 1 \leqslant k \leqslant n.
\end{cases}
\tag{3.54}
$$

Denote

$$V_i^k = v(x_i, t_k), \quad W_i^k = w(x_i, t_k), \quad 0 \leqslant i \leqslant m, \quad 0 \leqslant k \leqslant n.$$

By discretizing (3.52), we have

$$\begin{cases} \delta_t V_i^{k+\frac{1}{2}} - a\delta_x^2 V_i^{k+\frac{1}{2}} = -p(x_i, t_{k+\frac{1}{2}}) + O(\tau^2 + h^2), \\ \qquad\qquad\qquad 1 \leqslant i \leqslant m-1, \quad 0 \leqslant k \leqslant n-1, \\ V_i^0 = 0, \quad 0 \leqslant i \leqslant m, \\ V_0^k = 0, \quad V_m^k = 0, \quad 1 \leqslant k \leqslant n. \end{cases} \tag{3.55}$$

By discretizing (3.53), we have

$$\begin{cases} \delta_t W_i^{k+\frac{1}{2}} - a\delta_x^2 W_i^{k+\frac{1}{2}} = -q(x_i, t_{k+\frac{1}{2}}) + O(\tau^2 + h^2), \\ \qquad\qquad\qquad 1 \leqslant i \leqslant m-1, \quad 0 \leqslant k \leqslant n-1, \\ W_i^0 = 0, \quad 0 \leqslant i \leqslant m, \\ W_0^k = 0, \quad W_m^k = 0, \quad 1 \leqslant k \leqslant n. \end{cases} \tag{3.56}$$

Denote

$$r_i^k = e_i^k + \tau^2 V_i^k + h^2 W_i^k, \quad 0 \leqslant i \leqslant m, \quad 0 \leqslant k \leqslant n.$$

Multiplying (3.55) by τ^2 and (3.56) by h^2, respectively, then summing up the results with (3.54), we have

$$\begin{cases} \delta_t r_i^{k+\frac{1}{2}} - a\delta_x^2 r_i^{k+\frac{1}{2}} = O(\tau^4 + h^4), \quad 1 \leqslant i \leqslant m-1, \quad 0 \leqslant k \leqslant n-1, \\ r_i^0 = 0, \quad 0 \leqslant i \leqslant m, \\ r_0^k = 0, \quad r_m^k = 0, \quad 1 \leqslant k \leqslant n. \end{cases}$$

According to Theorem 3.14, we have

$$\|r^k\|_\infty = O(\tau^4 + h^4), \quad 1 \leqslant k \leqslant n,$$

namely,

$$u(x_i, t_k) - u_i^k(h, \tau) + \tau^2 V_i^k + h^2 W_i^k = O(\tau^4 + h^4), \quad 1 \leqslant i \leqslant m-1, \quad 1 \leqslant k \leqslant n.$$

Rearranging the above equality yields

$$u_i^k(h, \tau) = u(x_i, t_k) + \tau^2 v(x_i, t_k) + h^2 w(x_i, t_k) + O(\tau^4 + h^4),$$
$$1 \leqslant i \leqslant m-1, \quad 1 \leqslant k \leqslant n. \tag{3.57}$$

Table 3.13 (Example 3.5)
The maximum errors of
numerical solutions
calculated by using the
extrapolation once

h	τ	$\widetilde{E}_\infty(h, \tau)$
1/10	1/10	8.160e−6
1/20	1/20	1.076e−6
1/40	1/40	1.383e−7
1/80	1/80	1.765e−8

In a similar way, we have

$$
u_{2i}^{2k}\left(\frac{h}{2}, \frac{\tau}{2}\right) = u(x_i, t_k) + \left(\frac{\tau}{2}\right)^2 v(x_i, t_k) + \left(\frac{h}{2}\right)^2 w(x_i, t_k)
$$

$$
+ O\left(\left(\frac{\tau}{2}\right)^4 + \left(\frac{h}{2}\right)^4\right), \quad 1 \leqslant i \leqslant m - 1, \ 1 \leqslant k \leqslant n. \quad (3.58)
$$

Multiplying both sides of (3.58) by 4/3 and (3.57) by 1/3, respectively, and then subtracting the results, we have

$$
\frac{4}{3}u_{2i}^{2k}\left(\frac{h}{2}, \frac{\tau}{2}\right) - \frac{1}{3}u_i^k(h, \tau) = u(x_i, t_k) + O(\tau^4 + h^4), \quad 1 \leqslant i \leqslant m-1, \quad 1 \leqslant k \leqslant n.
$$

It completes the proof. $\qquad\qquad\qquad\qquad\qquad\qquad\qquad\qquad\qquad\qquad$ □

Example 3.5 Apply the Richardson extrapolation to compute the problem in Example 3.4.

Table 3.13 presents the maximum error

$$
\widetilde{E}_\infty(h, \tau) = \max_{\substack{1 \leqslant i \leqslant m-1 \\ 1 \leqslant k \leqslant n}} \left| u(x_i, t_k) - \left[\frac{4}{3}u_{2i}^{2k}\left(\frac{h}{2}, \frac{\tau}{2}\right) - \frac{1}{3}u_i^k(h, \tau)\right] \right|
$$

of the numerical results computed using the extrapolation once with different step sizes. By comparing the results in Table 3.13 with those in Table 3.12, it can be concluded that the extrapolation significantly improves the accuracy of the numerical solution.

3.6   The Compact Difference Scheme

In this section, we establish an unconditionally stable difference scheme with accuracy $O(\tau^2 + h^4)$ for the problem (3.1).

3.6.1 Derivation of the Difference Scheme

For $w = \{w_i \mid 0 \leqslant i \leqslant m\} \in \mathcal{U}_h$, define an operator

$$(\mathcal{A}w)_i = \begin{cases} \dfrac{1}{12}(w_{i-1} + 10w_i + w_{i+1}), & 1 \leqslant i \leqslant m-1, \\ w_i, & i = 0,\, m. \end{cases}$$

Considering Eq. (3.1a) at the point $(x_i, t_{k+\frac{1}{2}})$, we have

$$\frac{\partial u}{\partial t}(x_i, t_{k+\frac{1}{2}}) - a\frac{\partial^2 u}{\partial x^2}(x_i, t_{k+\frac{1}{2}}) = f(x_i, t_{k+\frac{1}{2}}), \quad 0 \leqslant i \leqslant m, \quad 0 \leqslant k \leqslant n-1.$$

Using Lemma 1.2, we have

$$\left[\delta_t U_i^{k+\frac{1}{2}} - \frac{\tau^2}{24}\frac{\partial^3 u}{\partial t^3}(x_i, \theta_{ik})\right] - a \cdot \left[\frac{1}{2}\left(\frac{\partial^2 u}{\partial x^2}(x_i, t_k) + \frac{\partial^2 u}{\partial x^2}(x_i, t_{k+1})\right)\right.$$

$$\left. - \frac{\tau^2}{8}\cdot\frac{\partial^4 u}{\partial x^2 \partial t^2}(x_i, \widetilde{\theta}_{ik})\right] = f(x_i, t_{k+\frac{1}{2}}), \quad 0 \leqslant i \leqslant m, \quad 0 \leqslant k \leqslant n-1,$$

where $\theta_{ik}, \widetilde{\theta}_{ik} \in (t_k, t_{k+1})$. In other words,

$$\delta_t U_i^{k+\frac{1}{2}} - a \cdot \frac{1}{2}\left(\frac{\partial^2 u}{\partial x^2}(x_i, t_k) + \frac{\partial^2 u}{\partial x^2}(x_i, t_{k+1})\right)$$

$$= f(x_i, t_{k+\frac{1}{2}}) + \left[\frac{1}{24}\frac{\partial^3 u}{\partial t^3}(x_i, \theta_{ik}) - \frac{a}{8}\frac{\partial^4 u}{\partial x^2 \partial t^2}(x_i, \widetilde{\theta}_{ik})\right]\tau^2,$$

$$0 \leqslant i \leqslant m, \quad 0 \leqslant k \leqslant n-1.$$

Performing the operator $\mathcal{A}$ on both sides of the above equality, we have

$$\mathcal{A}\delta_t U_i^{k+\frac{1}{2}} - a \cdot \frac{1}{2}\left(\mathcal{A}\frac{\partial^2 u}{\partial x^2}(x_i, t_k) + \mathcal{A}\frac{\partial^2 u}{\partial x^2}(x_i, t_{k+1})\right)$$

$$= \mathcal{A}f(x_i, t_{k+\frac{1}{2}}) + \mathcal{A}\left[\frac{1}{24}\frac{\partial^3 u}{\partial t^3}(x_i, \theta_{ik}) - \frac{a}{8}\frac{\partial^4 u}{\partial x^2 \partial t^2}(x_i, \widetilde{\theta}_{ik})\right]\tau^2,$$

$$1 \leqslant i \leqslant m-1, \quad 0 \leqslant k \leqslant n-1. \tag{3.59}$$

With the help of Lemma 1.2, it yields

$$\mathcal{A}\frac{\partial^2 u}{\partial x^2}(x_i, t_k) = \delta_x^2 U_i^k + \frac{h^4}{240}\frac{\partial^6 u}{\partial x^6}(\xi_{ik}, t_k),$$

where $\xi_{ik} \in (x_{i-1}, x_{i+1})$. Averaging the above equalities with $t = t_k$ and $t = t_{k+1}$ produces

$$\mathcal{A}\left[\frac{1}{2}\left(\frac{\partial^2 u}{\partial x^2}(x_i, t_k) + \frac{\partial^2 u}{\partial x^2}(x_i, t_{k+1})\right)\right] = \frac{1}{2}(\delta_x^2 U_i^k + \delta_x^2 U_i^{k+1})$$

$$+ \frac{h^4}{240}\frac{\partial^6 u}{\partial x^6}(\bar{\bar{\xi}}_{ik}, \bar{t}_{ik}), \quad 1 \leqslant i \leqslant m-1, 0 \leqslant k \leqslant n-1, \tag{3.60}$$

where $\bar{\bar{\xi}}_{ik} \in (x_{i-1}, x_{i+1})$ and $\bar{t}_{ik} \in (t_k, t_{k+1})$.

Substituting (3.60) into (3.59), we have

$$\mathcal{A}\delta_t U_i^{k+\frac{1}{2}} - a\delta_x^2 U_i^{k+\frac{1}{2}} = \mathcal{A}f(x_i, t_{k+\frac{1}{2}}) + (R_5)_i^k,$$

$$1 \leqslant i \leqslant m-1, \quad 0 \leqslant k \leqslant n-1, \tag{3.61}$$

where

$$(R_5)_i^k = \mathcal{A}\left[\frac{1}{24}\frac{\partial^3 u}{\partial t^3}(x_i, \theta_{ik}) - \frac{a}{8}\frac{\partial^4 u}{\partial x^2 \partial t^2}(x_i, \widetilde{\theta}_{ik})\right]\tau^2 + \frac{a}{240}\frac{\partial^6 u}{\partial x^6}(\bar{\xi}_{ik}, \bar{t}_{ik})h^4.$$

Noticing the initial-boundary value conditions (3.1b) and (3.1c), we have

$$\begin{cases} U_i^0 = \varphi(x_i), & 0 \leqslant i \leqslant m, \\ U_0^k = \alpha(t_k), & U_m^k = \beta(t_k), \quad 1 \leqslant k \leqslant n. \end{cases} \tag{3.62}$$

Omitting the small term $(R_5)_i^k$ in (3.61) and replacing U_i^k with u_i^k, a difference scheme reads

$$\begin{cases} \mathcal{A}\delta_t u_i^{k+\frac{1}{2}} - a\delta_x^2 u_i^{k+\frac{1}{2}} = \mathcal{A}f(x_i, t_{k+\frac{1}{2}}), \\ \qquad 1 \leqslant i \leqslant m-1, \quad 0 \leqslant k \leqslant n-1, \tag{3.63a} \\ u_i^0 = \varphi(x_i), \quad 0 \leqslant i \leqslant m, \tag{3.63b} \\ u_0^k = \alpha(t_k), \quad u_m^k = \beta(t_k), \quad 1 \leqslant k \leqslant n. \tag{3.63c} \end{cases}$$

The stencil of the two-level difference scheme (3.63) is shown in Fig. 3.19.

Fig. 3.19 The stencil of the compact difference scheme (3.63)

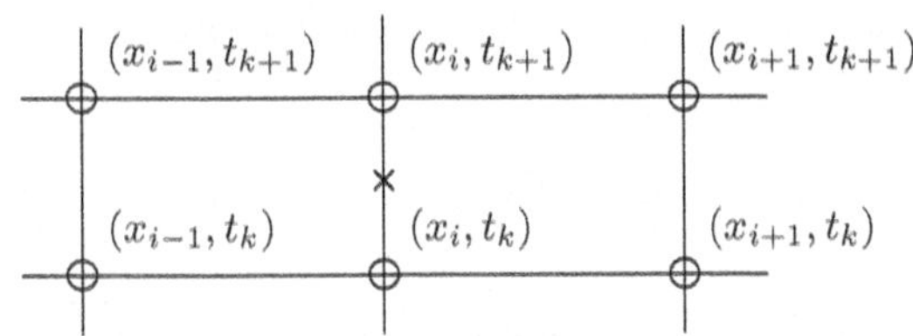

Let

$$
c_3 = \max \left\{ \frac{a}{240} \max_{(x,t)\in\bar{D}} \left| \frac{\partial^6 u(x,t)}{\partial x^6} \right|, \; \frac{1}{24} \max_{(x,t)\in\bar{D}} \left| \frac{\partial^3 u(x,t)}{\partial t^3} \right| \right.
$$

$$
\left. + \frac{a}{8} \max_{(x,t)\in\bar{D}} \left| \frac{\partial^4 u(x,t)}{\partial x^2 \partial t^2} \right| \right\}.
\tag{3.64}
$$

Then we have

$$
\left| (R_5)_i^k \right| \leqslant c_3(\tau^2 + h^4), \quad 1 \leqslant i \leqslant m-1, \quad 0 \leqslant k \leqslant n-1.
\tag{3.65}
$$

Among all possible difference schemes constructed using the six grid points shown in Fig. 3.19, the local truncation error of the difference scheme (3.63) attains the highest order. Therefore, the difference scheme (3.63) is referred to as the **compact difference scheme**.

3.6.2   Existence of the Difference Solution

Theorem 3.18 *The difference scheme (3.63) is uniquely solvable.*

Proof Denote

$$
u^k = (u_0^k, u_1^k, \ldots, u_{m-1}^k, u_m^k).
$$

The value of u^0 at the 0-th time level is known from (3.63b). Suppose the value of u^k has been determined. Then, with the help of (3.63a) and (3.63c), the difference scheme for u^{k+1} is

$$
\begin{cases}
\mathcal{A}\delta_t u_i^{k+\frac{1}{2}} - a\delta_x^2 u_i^{k+\frac{1}{2}} = \mathcal{A}f(x_i, t_{k+\frac{1}{2}}), & 1 \leqslant i \leqslant m-1, \\
u_0^{k+1} = \alpha(t_{k+1}), \quad u_m^{k+1} = \beta(t_{k+1}).
\end{cases}
$$

It can be written in the following matrix form

$$
\begin{pmatrix}
\frac{5}{6}+r & \frac{1}{12}-\frac{1}{2}r \\
\frac{1}{12}-\frac{1}{2}r & \frac{5}{6}+r & \frac{1}{12}-\frac{1}{2}r \\
 & \ddots & \ddots & \ddots \\
 & & \frac{1}{12}-\frac{1}{2}r & \frac{5}{6}+r & \frac{1}{12}-\frac{1}{2}r \\
 & & & \frac{1}{12}-\frac{1}{2}r & \frac{5}{6}+r
\end{pmatrix}
\begin{pmatrix}
u_1^{k+1} \\ u_2^{k+1} \\ \vdots \\ u_{m-2}^{k+1} \\ u_{m-1}^{k+1}
\end{pmatrix}
$$

$$
=
\begin{pmatrix}
\frac{5}{6}-r & \frac{1}{12}+\frac{1}{2}r \\
\frac{1}{12}+\frac{1}{2}r & \frac{5}{6}-r & \frac{1}{12}+\frac{1}{2}r \\
 & \ddots & \ddots & \ddots \\
 & & \frac{1}{12}+\frac{1}{2}r & \frac{5}{6}-r & \frac{1}{12}+\frac{1}{2}r \\
 & & & \frac{1}{12}+\frac{1}{2}r & \frac{5}{6}-r
\end{pmatrix}
\begin{pmatrix}
u_1^{k} \\ u_2^{k} \\ \vdots \\ u_{m-2}^{k} \\ u_{m-1}^{k}
\end{pmatrix}
$$

$$
+
\begin{pmatrix}
\left(\frac{1}{12}+\frac{1}{2}r\right)u_0^k - \left(\frac{1}{12}-\frac{1}{2}r\right)u_0^{k+1} + \tau\mathcal{A}f(x_1,t_{k+\frac{1}{2}}) \\
\tau\mathcal{A}f(x_2,t_{k+\frac{1}{2}}) \\
\vdots \\
\tau\mathcal{A}f(x_{m-2},t_{k+\frac{1}{2}}) \\
\left(\frac{1}{12}+\frac{1}{2}r\right)u_m^k - \left(\frac{1}{12}-\frac{1}{2}r\right)u_m^{k+1} + \tau\mathcal{A}f(x_{m-1},t_{k+\frac{1}{2}})
\end{pmatrix}. \tag{3.66}
$$

It is easy to see that the coefficient matrix is strictly diagonally dominant, and therefore, there exists a unique solution. $\qquad\square$

3.6.3 Implementation of the Difference Scheme and Numerical Examples

By means of (3.66), we know that (3.63) forms a tridiagonal system of linear equations at each time level. Therefore, it can be efficiently solved using the double sweep method.

Table 3.14 (Example 3.6) Numerical solutions (NS), exact solutions (ES), and absolute values of errors at part of node points ($h = 1/10$, $\tau = 1/100$)

k	(x, t)	NS	ES	$|ES-NS|$
10	$(0.5, 0.1)$	1.822120	1.822119	1.084e−6
20	$(0.5, 0.2)$	2.013754	2.013753	1.625e−6
30	$(0.5, 0.3)$	2.225543	2.225541	1.955e−6
40	$(0.5, 0.4)$	2.459605	2.459603	2.219e−6
50	$(0.5, 0.5)$	2.718284	2.718282	2.475e−6
60	$(0.5, 0.6)$	3.004169	3.004166	2.743e−6
70	$(0.5, 0.7)$	3.320120	3.320117	3.035e−6
80	$(0.5, 0.8)$	3.669300	3.669297	3.355e−6
90	$(0.5, 0.9)$	4.055204	4.055200	3.709e−6
100	$(0.5, 1.0)$	4.481693	4.481689	4.099e−6

Table 3.15 (Example 3.6) The maximum errors of numerical solutions with different step sizes

h	τ	$E_\infty(h, \tau)$	$E_\infty(2h, 4\tau)/E_\infty(h, \tau)$
1/10	1/100	4.099e−06	
1/20	1/400	2.582e−07	15.88
1/40	1/1600	1.614e−08	16.00
1/80	1/6400	1.009e−09	16.00
1/160	1/25600	6.571e−11	15.36

Example 3.6 Apply the compact difference scheme (3.63) to compute the problem

$$
\begin{cases}
\dfrac{\partial u}{\partial t} - \dfrac{\partial^2 u}{\partial x^2} = 0, & 0 < x < 1, \quad 0 < t \leqslant 1, \\[2mm]
u(x, 0) = e^x, & 0 \leqslant x \leqslant 1, \\[2mm]
u(0, t) = e^t, \quad u(1, t) = e^{1+t}, & 0 < t \leqslant 1.
\end{cases}
$$

The exact solution of the above problem is $u(x, t) = e^{x+t}$.

Table 3.14 lists a portion of the numerical results with $h = 1/10$ and $\tau = 1/100$, which demonstrates that the numerical solutions closely approximate the exact solutions. Table 3.15 presents the maximum errors

$$
E_\infty(h, \tau) = \max_{\substack{1 \leqslant i \leqslant m-1 \\ 1 \leqslant k \leqslant n}} |u(x_i, t_k) - u_i^k|
$$

of the numerical solutions for different step sizes. As observed from this table, when the spatial step size is halved and the temporal step size is reduced to one-fourth of the original, the maximum error decreases to approximately $1/16$ of the original.

The readers can compare the numerical results in Table 3.15 with those in Table 3.12 and examine the difference in the computational effort between the compact difference scheme (3.63) and the Crank-Nicolson scheme (3.41) in order to achieve an approximate solution with the same accuracy.

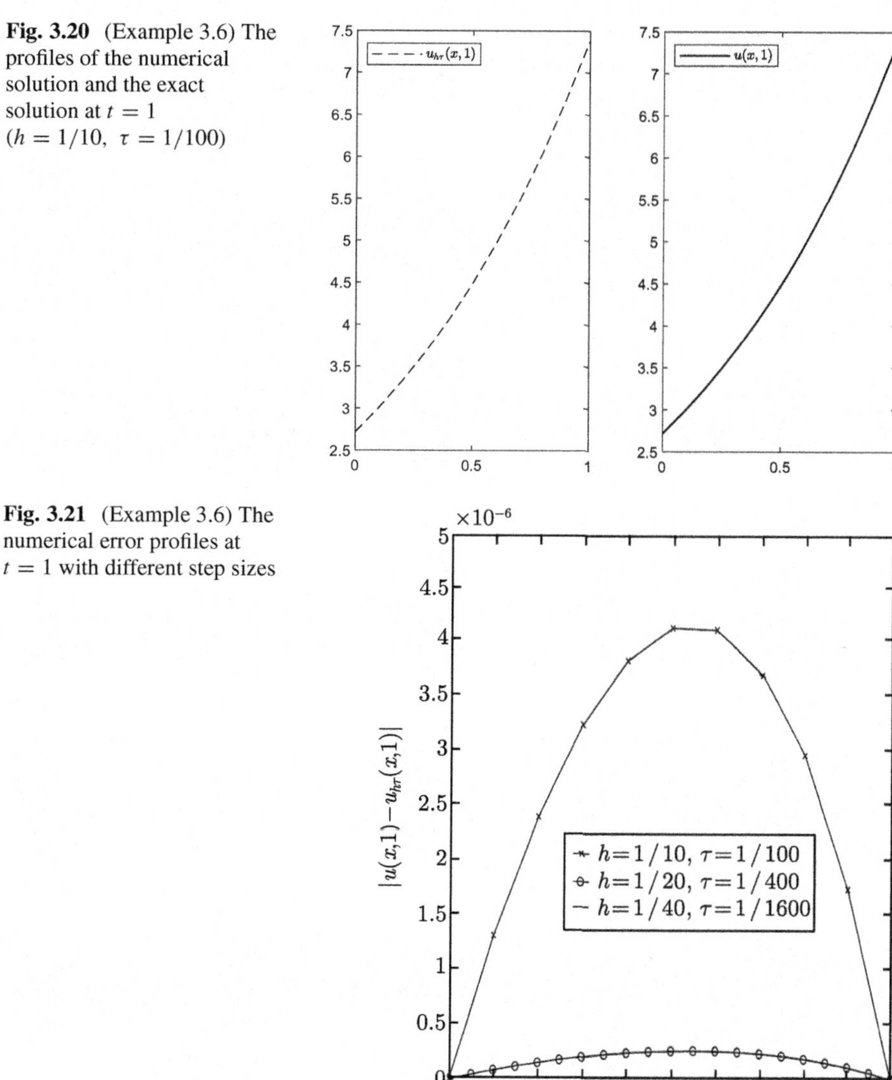

Fig. 3.20 (Example 3.6) The profiles of the numerical solution and the exact solution at $t = 1$ ($h = 1/10, \ \tau = 1/100$)

Fig. 3.21 (Example 3.6) The numerical error profiles at $t = 1$ with different step sizes

Figure 3.20 shows the numerical solution profiles at $t = 1$ with step sizes $h = 1/10$, $\tau = 1/100$, along with the exact solution profile. Figure 3.21 displays the numerical error profiles at $t = 1$ for different step sizes. Figure 3.22 illustrates the error surfaces of the numerical solutions with varying step sizes.

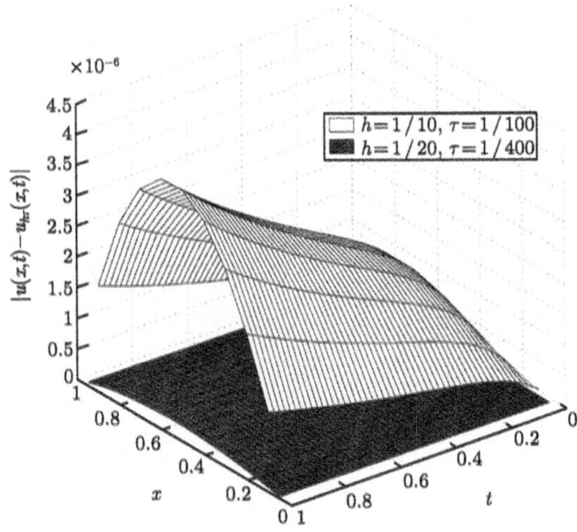

Fig. 3.22 (Example 3.6) The numerical error surfaces with different step sizes

3.6.4   A Priori Estimate of the Difference Solution

Theorem 3.19 *Let $\{v_i^k \mid 0 \leqslant i \leqslant m, \ 0 \leqslant k \leqslant n\}$ be the solution of the difference scheme*

$$
\begin{cases}
\mathcal{A}\delta_t v_i^{k+\frac{1}{2}} - a\delta_x^2 v_i^{k+\frac{1}{2}} = g_i^k, & 1 \leqslant i \leqslant m-1, \quad 0 \leqslant k \leqslant n-1, & (3.67a) \\[2mm]
v_i^0 = \varphi(x_i), & 1 \leqslant i \leqslant m-1, & (3.67b) \\[2mm]
v_0^k = 0, \quad v_m^k = 0, & 0 \leqslant k \leqslant n. & (3.67c)
\end{cases}
$$

Then we have

$$
|v^k|_1^2 \leqslant |v^0|_1^2 + \frac{3}{4a}\tau \sum_{l=0}^{k-1} \|g^l\|^2, \quad 0 \leqslant k \leqslant n, \tag{3.68}
$$

$$
\|v^k\|_\infty^2 \leqslant \frac{L}{4}\left(|v^0|_1^2 + \frac{3}{4a}\tau \sum_{l=0}^{k-1} \|g^l\|^2 \right), \quad 0 \leqslant k \leqslant n. \tag{3.69}
$$

Proof

(a) Taking an inner product of (3.67a) with $\delta_t v^{k+\frac{1}{2}}$, we have

$$
(\mathcal{A}\delta_t v^{k+\frac{1}{2}}, \delta_t v^{k+\frac{1}{2}}) - a(\delta_x^2 v^{k+\frac{1}{2}}, \delta_t v^{k+\frac{1}{2}}) = (g^k, \delta_t v^{k+\frac{1}{2}}), \quad 0 \leqslant k \leqslant n-1. \tag{3.70}
$$

Now, we estimate each term in the above equality. Noticing that $\delta_t v_0^{k+\frac{1}{2}} = 0$ and $\delta_t v_m^{k+\frac{1}{2}} = 0$, and using summation by parts, both terms on the left-hand side are estimated as

$$\left(\mathcal{A}\delta_t v^{k+\frac{1}{2}}, \delta_t v^{k+\frac{1}{2}}\right) = \left(\left(\mathcal{I} + \frac{h^2}{12}\delta_x^2\right)\delta_t v^{k+\frac{1}{2}}, \delta_t v^{k+\frac{1}{2}}\right)$$

$$= \|\delta_t v^{k+\frac{1}{2}}\|^2 + \frac{h^2}{12}(\delta_x^2\delta_t v^{k+\frac{1}{2}}, \delta_t v^{k+\frac{1}{2}}) = \|\delta_t v^{k+\frac{1}{2}}\|^2 - \frac{h^2}{12}|\delta_t v^{k+\frac{1}{2}}|_1^2$$

$$\geqslant \|\delta_t v^{k+\frac{1}{2}}\|^2 - \frac{h^2}{12}\cdot\frac{4}{h^2}\|\delta_t v^{k+\frac{1}{2}}\|^2 = \frac{2}{3}\|\delta_t v^{k+\frac{1}{2}}\|^2$$

and

$$-a\left(\delta_x^2 v^{k+\frac{1}{2}}, \delta_t v^{k+\frac{1}{2}}\right) = a\left(\delta_x v^{k+\frac{1}{2}}, \delta_x \delta_t v^{k+\frac{1}{2}}\right) = \frac{a}{2\tau}\left(|v^{k+1}|_1^2 - |v^k|_1^2\right).$$

With the help of the Cauchy-Schwarz inequality, the right-hand side term is estimated as

$$(g^k, \delta_t v^{k+\frac{1}{2}}) \leqslant \frac{2}{3}\|\delta_t v^{k+\frac{1}{2}}\|^2 + \frac{3}{8}\|g^k\|^2.$$

Substituting the above three estimates into (3.70), we have

$$\frac{a}{2\tau}(|v^{k+1}|_1^2 - |v^k|_1^2) \leqslant \frac{3}{8}\|g^k\|^2,$$

i.e.,

$$|v^{k+1}|_1^2 \leqslant |v^k|_1^2 + \frac{3}{4a}\tau\|g^k\|^2, \quad 0 \leqslant k \leqslant n-1.$$

By recursion, it yields

$$|v^k|_1^2 \leqslant |v^0|_1^2 + \frac{3}{4a}\tau\sum_{l=0}^{k-1}\|g^l\|^2, \quad 1 \leqslant k \leqslant n.$$

(b) With the application of Lemma 1.4 and (3.68), the inequality (3.69) holds.

$\square$

3.6.5   Convergence and Stability of the Difference Solution

Convergence

Theorem 3.20 *Let $\{u(x,t) \mid (x,t) \in \bar{D}\}$ be the solution of the problem (3.1) and $\{u_i^k \mid 0 \leqslant i \leqslant m, \ 0 \leqslant k \leqslant n\}$ be the solution of the difference scheme (3.63). Then* *for an arbitrary step ratio r, we have*

$$\max_{1\leqslant i\leqslant m-1,\, 1\leqslant k\leqslant n}\left|u(x_i,t_k)-u_i^k\right|\leqslant \frac{Lc_3}{4}\cdot\sqrt{\frac{3T}{a}}(\tau^2+h^4),$$

where c_3 is defined by (3.64).

Proof Denote

$$e_i^k=u(x_i,t_k)-u_i^k,\quad 0\leqslant i\leqslant m,\quad 0\leqslant k\leqslant n.$$

Subtracting (3.63) from (3.61) and (3.62), we have the system of error equations

$$\begin{cases}\mathcal{A}\delta_t e_i^{k+\frac{1}{2}}-a\delta_x^2 e_i^{k+\frac{1}{2}}=(R_5)_i^k,\quad 1\leqslant i\leqslant m-1,\quad 0\leqslant k\leqslant n-1,\\ e_i^0=0,\quad 0\leqslant i\leqslant m,\\ e_0^k=0,\quad e_m^k=0,\quad 1\leqslant k\leqslant n.\end{cases}$$

Applying Theorem 3.19 and noticing (3.65), we have

$$\|e^k\|_\infty^2\leqslant \frac{L}{4}\cdot\frac{3}{4a}\tau\sum_{l=0}^{k-1}\|(R_5)^l\|^2\leqslant \frac{L}{4}\cdot\frac{3T}{4a}Lc_3^2(\tau^2+h^4)^2,\quad 1\leqslant k\leqslant n,$$

which implies

$$\|e^k\|_\infty\leqslant \frac{Lc_3}{4}\cdot\sqrt{\frac{3T}{a}}(\tau^2+h^4),\quad 1\leqslant k\leqslant n.$$

$$\square$$

Stability

Theorem 3.21 *The solution of the difference scheme (3.63) is stable with respect to the initial value and the right-hand side term in the following sense: Let $\{u_i^k\mid 0\leqslant i\leqslant m,\ 0\leqslant k\leqslant n\}$ be the solution of the difference scheme*

$$\begin{cases}\mathcal{A}\delta_t u_i^{k+\frac{1}{2}}-a\delta_x^2 u_i^{k+\frac{1}{2}}=f_i^{k+\frac{1}{2}},\quad 1\leqslant i\leqslant m-1,\quad 0\leqslant k\leqslant n-1,\\ u_i^0=\varphi_i,\quad 1\leqslant i\leqslant m-1,\\ u_0^k=0,\quad u_m^k=0,\quad 0\leqslant k\leqslant n.\end{cases}$$

Then we have

$$\|u^k\|_\infty^2\leqslant \frac{L}{4}\left(|u^0|_1^2+\frac{3}{4a}\tau\sum_{l=0}^{k-1}\|f^{l+\frac{1}{2}}\|^2\right),\quad 0\leqslant k\leqslant n.$$

Proof A direct application of Theorem 3.19 gives the result.   □

3.7 Nonlinear Parabolic Equations

In this section, we consider the initial-boundary value problem of the nonlinear parabolic equation

$$
\begin{cases}
\dfrac{\partial u}{\partial t} - a(u)\dfrac{\partial^2 u}{\partial x^2} = f(x,t), & 0 < x < L, \quad 0 < t \leqslant T, & (3.71\text{a}) \\[2mm]
u(x,0) = \varphi(x), & 0 \leqslant x \leqslant L, & (3.71\text{b}) \\[2mm]
u(0,t) = \alpha(t), \quad u(L,t) = \beta(t), & 0 < t \leqslant T, & (3.71\text{c})
\end{cases}
$$

where $a(u)$, $f(x,t)$, $\varphi(x)$, $\alpha(t)$, and $\beta(t)$ are known functions. Suppose that the problem (3.71) has a smooth solution $u(x,t)$ and that there are positive constants ϵ, a_0, a_1, and L_p such that when $(x,t) \in \bar{D}$, and $|\epsilon_1| \leqslant \epsilon$, $|\epsilon_2| \leqslant \epsilon$, it holds

$$
a_0 \leqslant a(u(x,t) + \epsilon_1) \leqslant a_1, \tag{3.72}
$$

$$
|a(u(x,t) + \epsilon_1) - a(u(x,t) + \epsilon_2)| \leqslant L_p|\epsilon_1 - \epsilon_2|, \tag{3.73}
$$

i.e., $a(u)$ has positive upper and lower bounds in an ϵ-neighborhood of the solution and satisfies the Lipschitz condition with a Lipschitz constant L_p.

Before introducing the difference scheme, we first present some important Gronwall inequalities.

Lemma 3.3 (Gronwall Inequalities)

(a) *Suppose $\{F^k\}_{k=0}^{\infty}$ is a nonnegative sequence, c and g are two nonnegative constants, satisfying*

$$
F^{k+1} \leqslant (1 + c\tau)F^k + \tau g, \quad k = 0, 1, 2, \ldots.
$$

Then we have

$$
F^k \leqslant e^{ck\tau}\left(F^0 + \frac{g}{c}\right), \quad k = 1, 2, 3, \ldots.
$$

(b) *Suppose $\{F^k\}_{k=0}^{\infty}$ and $\{g^k\}_{k=0}^{\infty}$ are two nonnegative sequences, and c is a nonnegative constant, satisfying*

$$
F^{k+1} \leqslant (1 + c\tau)F^k + \tau g^k, \quad k = 0, 1, 2, \ldots.
$$

Then we have

$$F^k \leqslant e^{ck\tau} \left(F^0 + \tau \sum_{l=0}^{k-1} g^l \right), \quad k = 0, 1, 2, \ldots.$$

Proof

(a)

$$F^{k+1}$$

$$\leqslant (1 + c\tau) F^k + \tau g$$

$$\leqslant (1 + c\tau)[(1 + c\tau) F^{k-1} + \tau g] + \tau g$$

$$= (1 + c\tau)^2 F^{k-1} + [(1 + c\tau) + 1]\tau g$$

$$\leqslant (1 + c\tau)^2 [(1 + c\tau) F^{k-2} + \tau g] + [(1 + c\tau) + 1]\tau g$$

$$= (1 + c\tau)^3 F^{k-2} + [(1 + c\tau)^2 + (1 + c\tau) + 1]\tau g$$

$$\leqslant \cdots$$

$$\leqslant (1 + c\tau)^k F^1 + [(1 + c\tau)^{k-1} + (1 + c\tau)^{k-2} + \cdots + 1]\tau g$$

$$\leqslant (1 + c\tau)^k [(1 + c\tau) F^0 + \tau g] + [(1 + c\tau)^{k-1} + (1 + c\tau)^{k-2} + \cdots + 1]\tau g$$

$$= (1 + c\tau)^{k+1} F^0 + [(1 + c\tau)^k + (1 + c\tau)^{k-1} + \cdots + 1]\tau g$$

$$= (1 + c\tau)^{k+1} F^0 + \frac{(1 + c\tau)^{k+1} - 1}{c\tau} \cdot \tau g$$

$$\leqslant e^{c(k+1)\tau} \left(F^0 + \frac{g}{c} \right), \quad k = 0, 1, \ldots.$$

(b)

$$F^{k+1}$$

$$\leqslant (1 + c\tau) F^k + \tau g^k$$

$$\leqslant (1 + c\tau)[(1 + c\tau) F^{k-1} + \tau g^{k-1}] + \tau g^k$$

$$= (1 + c\tau)^2 F^{k-1} + (1 + c\tau)\tau g^{k-1} + \tau g^k$$

$$\leqslant (1 + c\tau)^2 [(1 + c\tau) F^{k-2} + \tau g^{k-2}] + (1 + c\tau)\tau g^{k-1} + \tau g^k$$

$$= (1 + c\tau)^3 F^{k-2} + (1 + c\tau)^2 \tau g^{k-2} + (1 + c\tau)\tau g^{k-1} + \tau g^k$$

$$\leqslant (1 + c\tau)^3 [(1 + c\tau) F^{k-3} + \tau g^{k-3}] + (1 + c\tau)^2 \tau g^{k-2}$$

$$+ (1 + c\tau)\tau g^{k-1} + \tau g^k$$

$$= (1 + c\tau)^4 F^{k-3} + (1 + c\tau)^3 \tau g^{k-3} + (1 + c\tau)^2 \tau g^{k-2}$$

$$+ (1 + c\tau)\tau g^{k-1} + \tau g^k$$

$$\leqslant \cdots$$

$$\leqslant (1 + c\tau)^{k+1} F^0 + \tau \sum_{l=0}^{k} (1 + c\tau)^{k-l} g^l$$

$$\leqslant (1 + c\tau)^{k+1} \left(F^0 + \tau \sum_{l=0}^{k} g^l \right), \quad k = 0, 1, 2, \ldots .$$

Thereby,

$$F^k \leqslant (1 + c\tau)^k \left(F^0 + \tau \sum_{l=0}^{k-1} g^l \right) \leqslant \mathrm{e}^{ck\tau} \left(F^0 + \tau \sum_{l=0}^{k-1} g^l \right), \quad k = 0, 1, 2, \ldots .$$

$\square$

3.7.1 The Forward Euler Scheme

Considering Eq. (3.71a) at the node point (x_i, t_k), we have

$$\frac{\partial u}{\partial t}(x_i, t_k) - a\big(u(x_i, t_k)\big)\frac{\partial^2 u}{\partial x^2}(x_i, t_k) = f(x_i, t_k),$$

$$1 \leqslant i \leqslant m - 1, \quad 0 \leqslant k \leqslant n - 1. \tag{3.74}$$

Substituting

$$\frac{\partial^2 u}{\partial x^2}(x_i, t_k) = \delta_x^2 U_i^k - \frac{h^2}{12}\frac{\partial^4 u}{\partial x^4}(\xi_{ik}, t_k), \quad \xi_{ik} \in (x_{i-1}, x_{i+1})$$

and

$$\frac{\partial u}{\partial t}(x_i, t_k) = D_t U_i^k - \frac{\tau}{2}\frac{\partial^2 u}{\partial t^2}(x_i, \eta_{ik}), \quad \eta_{ik} \in (t_k, t_{k+1})$$

into (3.74) produces

$$D_t U_i^k - a(U_i^k)\delta_x^2 U_i^k = f(x_i, t_k) + \frac{\tau}{2}\frac{\partial^2 u}{\partial t^2}(x_i, \eta_{ik}) - \frac{h^2}{12}a(U_i^k)\frac{\partial^4 u}{\partial x^4}(\xi_{ik}, t_k),$$

$$1 \leqslant i \leqslant m - 1, \quad 0 \leqslant k \leqslant n - 1. \tag{3.75}$$

Denote

$$(R_6)_i^k = \frac{\tau}{2}\frac{\partial^2 u}{\partial t^2}(x_i, \eta_{ik}) - \frac{h^2}{12}a(U_i^k)\frac{\partial^4 u}{\partial x^4}(\xi_{ik}, t_k).$$

Let

$$\max_{(x,t)\in\bar{D}}\left|\frac{\partial u(x,t)}{\partial t}\right| = M_t, \qquad \max_{(x,t)\in\bar{D}}\left|\frac{\partial^2 u(x,t)}{\partial t^2}\right| = M_{tt},$$

$$\max_{(x,t)\in\bar{D}}\left|\frac{\partial^2 u(x,t)}{\partial x^2}\right| = M_{xx}, \qquad \max_{(x,t)\in\bar{D}}\left|\frac{\partial^4 u(x,t)}{\partial x^4}\right| = M_{xxxx},$$

$$c_4 = \max\left\{\frac{1}{2}M_{tt}, \frac{a_1}{12}M_{xxxx}\right\}.$$

Then we have

$$|(R_6)_i^k| \leqslant c_4(\tau + h^2), \quad 1 \leqslant i \leqslant m - 1, \quad 0 \leqslant k \leqslant n - 1. \tag{3.76}$$

Noticing the initial-boundary value conditions (3.71b) and (3.71c), we have

$$\begin{cases} U_i^0 = \varphi(x_i), & 0 \leqslant i \leqslant m, \\ U_0^k = \alpha(t_k), \quad U_m^k = \beta(t_k), & 1 \leqslant k \leqslant n. \end{cases} \tag{3.77}$$

Omitting the small term $(R_6)_i^k$ in (3.75) and replacing U_i^k with u_i^k, a finite difference scheme for the problem (3.71) reads

$$\begin{cases} D_t u_i^k - a(u_i^k)\delta_x^2 u_i^k = f(x_i, t_k), & 1 \leqslant i \leqslant m - 1, \quad 0 \leqslant k \leqslant n - 1, & \text{(3.78a)} \\ u_i^0 = \varphi(x_i), & 0 \leqslant i \leqslant m, & \text{(3.78b)} \\ u_0^k = \alpha(t_k), \quad u_m^k = \beta(t_k), & 1 \leqslant k \leqslant n. & \text{(3.78c)} \end{cases}$$

Denote

$$\mu = \frac{\tau}{h^2}.$$

The difference equation (3.78a) can then be rewritten as

$$u_i^{k+1} = \left[1 - 2\mu a(u_i^k)\right]u_i^k + \mu a(u_i^k)\left(u_{i-1}^k + u_{i+1}^k\right) + \tau f(x_i, t_k),$$

$$1 \leqslant i \leqslant m - 1, \quad 0 \leqslant k \leqslant n - 1.$$

Therefore, the difference scheme (3.78) is an explicit one.

As for the convergence of the difference scheme (3.78), the following result holds.

Theorem 3.22 *Let $\{u(x,t)\mid 0 \leqslant x \leqslant L, 0 \leqslant t \leqslant T\}$ be the solution of the problem (3.71) and $\{u_i^k \mid 0 \leqslant i \leqslant m, 0 \leqslant k \leqslant n\}$ be the solution of the difference scheme (3.78), and μ satisfies the condition*

$$2\mu \cdot a(u_i^k) \leqslant 1, \quad 1 \leqslant i \leqslant m-1, \quad 0 \leqslant k \leqslant n-1. \tag{3.79}$$

Denote

$$e_i^k = U_i^k - u_i^k, \quad 0 \leqslant i \leqslant m, \quad 0 \leqslant k \leqslant n,$$

$$c_5 = \frac{c_4}{L_p M_{xx}} e^{L_p M_{xx} T}.$$

Then when τ and h satisfy

$$\tau \leqslant \frac{\epsilon}{2c_5}, \quad h \leqslant \sqrt{\frac{\epsilon}{2c_5}}, \tag{3.80}$$

we have

$$\|e^k\|_\infty \leqslant c_5 \left(\tau + h^2\right), \quad 0 \leqslant k \leqslant n. \tag{3.81}$$

Proof Subtracting (3.78) from (3.75) and (3.77), we have the system of error equations

$$
\begin{cases}
D_t e_i^k - \left[a(U_i^k)\delta_x^2 U_i^k - a(u_i^k)\delta_x^2 u_i^k\right] = (R_6)_i^k, \\
\qquad\qquad 1 \leqslant i \leqslant m-1, \quad 0 \leqslant k \leqslant n-1, & (3.82a) \\
e_i^0 = 0, \quad 0 \leqslant i \leqslant m, & (3.82b) \\
e_0^k = 0, \quad e_m^k = 0, \quad 1 \leqslant k \leqslant n. & (3.82c)
\end{cases}
$$

Using (3.82b), we have $\|e^0\|_\infty = 0$, and thus (3.81) holds for $k = 0$.

Now suppose (3.81) holds for $0 \leqslant k \leqslant l$, and then when h and τ satisfy (3.80), we have

$$\|e^k\|_\infty \leqslant \epsilon, \quad 0 \leqslant k \leqslant l,$$

With the application of the conditions (3.72)–(3.73), it yields

$$a_0 \leqslant a(u_i^k) \leqslant a_1, \quad |a(U_i^k) - a(u_i^k)| \leqslant L_p|e_i^k|, \quad 1 \leqslant i \leqslant m-1, \quad 0 \leqslant k \leqslant l. \tag{3.83}$$

Rewriting (3.82a) as

$$
\begin{aligned}
e_i^{k+1} &= e_i^k + \tau \left[a(U_i^k)\delta_x^2 U_i^k - a(u_i^k)\delta_x^2 u_i^k \right] + \tau (R_6)_i^k \\
&= e_i^k + \tau \left[a(u_i^k)\delta_x^2(U_i^k - u_i^k) + (a(U_i^k) - a(u_i^k))\delta_x^2 U_i^k \right] + \tau (R_6)_i^k \\
&= \left[1 - 2\mu a(u_i^k) \right] e_i^k + \mu a(u_i^k)(e_{i-1}^k + e_{i+1}^k) \\
&\quad + \tau \left[a(U_i^k) - a(u_i^k) \right] \delta_x^2 U_i^k + \tau (R_6)_i^k, \quad 1 \leqslant i \leqslant m - 1,
\end{aligned}
$$

and combining (3.79), (3.83), and (3.76), we have

$$
\begin{aligned}
|e_i^{k+1}| &\leqslant \left[1 - 2\mu a(u_i^k) \right] |e_i^k| + \mu a(u_i^k)(|e_{i-1}^k| + |e_{i+1}^k|) \\
&\quad + \tau |a(U_i^k) - a(u_i^k)| \cdot |\delta_x^2 U_i^k| + \tau |(R_6)_i^k| \\
&\leqslant \left[1 - 2\mu a(u_i^k) \right] \|e^k\|_\infty + \mu a(u_i^k)(\|e^k\|_\infty + \|e^k\|_\infty) \\
&\quad + \tau L_p M_{xx} \|e^k\|_\infty + c_4 \tau (\tau + h^2) \\
&= (1 + L_p M_{xx} \tau) \|e^k\|_\infty + c_4 \tau (\tau + h^2), \quad 1 \leqslant i \leqslant m - 1, \quad 0 \leqslant k \leqslant l,
\end{aligned}
$$

namely,

$$
\|e^{k+1}\|_\infty \leqslant (1 + L_p M_{xx} \tau) \|e^k\|_\infty + c_4 \tau (\tau + h^2), \quad 0 \leqslant k \leqslant l.
$$

Applying the Gronwall inequality (Lemma 3.3) arrives at

$$
\begin{aligned}
\|e^{l+1}\|_\infty &\leqslant e^{L_p M_{xx}(l+1)\tau} \left[\|e^0\|_\infty + \frac{c_4}{L_p M_{xx}}(\tau + h^2) \right] \\
&\leqslant \frac{c_4}{L_p M_{xx}} e^{L_p M_{xx} T} \left(\tau + h^2 \right) = c_5 \left(\tau + h^2 \right).
\end{aligned}
$$

Therefore, (3.81) holds for $k = l + 1$.

By induction, (3.81) holds for $0 \leqslant k \leqslant n$, which completes the proof. $\qquad\square$

Under the condition (3.79), we can prove that the difference scheme (3.78) is stable with respect to the initial value. Thus, (3.79) is referred to as the stability condition. It can be seen that the stability condition is not only dependent on $\mu = \dfrac{\tau}{h^2}$, but also on the numerical solution $\{u_i^k\}$.

A difference scheme with a time-varying step size can be derived as follows:

$$\begin{cases} \dfrac{1}{\Delta t_k}(u_i^{k+1} - u_i^k) - a(u_i^k)\delta_x^2 u_i^k = f(x_i, t_k), & 1 \leqslant i \leqslant m-1, \quad 0 \leqslant k \leqslant n-1, \\ u_i^0 = \varphi(x_i), & 0 \leqslant i \leqslant m, \\ u_0^k = \alpha(t_k), \quad u_m^k = \beta(t_k), & 1 \leqslant k \leqslant n, \end{cases}$$

where $t_{k+1} = t_k + \Delta t_k$. When $\{u_i^k \mid 0 \leqslant i \leqslant m\}$ has been solved, Δt_k can be determined by

$$\frac{\Delta t_k}{h^2} \max_{1 \leqslant i \leqslant m-1} a(u_i^k) \leqslant \frac{1}{2}.$$

Example 3.7 Apply the forward *Euler* scheme (3.78) to compute the problem

$$\begin{cases} \dfrac{\partial u}{\partial t} - u^4 \cdot \dfrac{\partial^2 u}{\partial x^2} = \mathrm{e}^{-t} \sin\left(\dfrac{\pi}{4} + \dfrac{\pi}{2}x\right)\left[\dfrac{\pi^2}{4}\mathrm{e}^{-4t}\sin^4\left(\dfrac{\pi}{4} + \dfrac{\pi}{2}x\right) - 1\right], \\ \qquad\qquad\qquad 0 < x < 1, \quad 0 < t \leqslant 1, \\ u(x, 0) = \sin\left(\dfrac{\pi}{4} + \dfrac{\pi}{2}x\right), \quad 0 \leqslant x \leqslant 1, \\ u(0, t) = \dfrac{\sqrt{2}}{2}\mathrm{e}^{-t}, \quad u(1, t) = \dfrac{\sqrt{2}}{2}\mathrm{e}^{-t}, \quad 0 < t \leqslant 1. \end{cases}$$

$$(3.84)$$

The exact solution of the above problem is $u(x, t) = \mathrm{e}^{-t} \sin\left(\dfrac{\pi}{4} + \dfrac{\pi}{2}x\right)$.

Table 3.16 lists part of the numerical results with step sizes $h = 1/10$ and $\tau = 1/200$ (with the step ratio $\mu = 1/2$), showing that the numerical solution approximates the exact solution very well. Table 3.17 presents part of the numerical results with step sizes $h = 1/10$ and $\tau = 1/100$ (with the step ratio $\mu = 1$). As the number of computed time levels increases, the error grows larger, and the numerical results become less meaningful in practice. Table 3.18 shows the maximum error

$$E_\infty(h, \tau) = \max_{1 \leqslant i \leqslant m-1, 1 \leqslant k \leqslant n} |u(x_i, t_k) - u_i^k|$$

with $\mu = 1/2$ and different step sizes.

As can be seen from Table 3.18, when the spatial step size is reduced by half and the temporal step size is reduced to a quarter of the original, the maximum error is reduced to a quarter of the original. Figure 3.23 shows the exact solution profile and the numerical solution profile with step sizes $h = 1/10$ and $\tau = 1/200$. In this figure, $u_{h\tau}(x, t)$ represents the numerical solution at the point (x, t) with step sizes h and τ. Since the relative error is about two in ten thousand, the difference between the numerical solution profile and the exact solution profile is indistinguishable to the naked eye. Figure 3.24 displays the error profiles of the numerical solution at $t = 1$. Figure 3.25 illustrates the error surfaces for different step sizes.

Table 3.16 (Example 3.7) Numerical solutions (NS), exact solutions (ES), and absolute values of errors at part of node points ($h = 1/10$, $\tau = 1/200$)

k	(x, t)	NS	ES	$|ES - NS|$
20	$(0.5, 0.1)$	0.9048948	0.9048374	5.737e−5
40	$(0.5, 0.2)$	0.8187399	0.8187308	9.192e−6
60	$(0.5, 0.3)$	0.7407599	0.7408182	5.831e−5
80	$(0.5, 0.4)$	0.6701864	0.6703200	1.337e−4
100	$(0.5, 0.5)$	0.6063173	0.6065307	2.134e−4
120	$(0.5, 0.6)$	0.5485161	0.5488116	2.955e−4
140	$(0.5, 0.7)$	0.4962070	0.4965853	3.783e−4
160	$(0.5, 0.8)$	0.4488686	0.4493290	4.604e−4
180	$(0.5, 0.9)$	0.4060293	0.4065697	5.404e−4
200	$(0.5, 1.0)$	0.3672621	0.3678794	6.173e−4

Table 3.17 (Example 3.7) Numerical solutions (NS), exact solutions (ES), and the absolute values of errors at part of node points ($h = 1/10$, $\tau = 1/100$)

k	(x, t)	NS	ES	$|ES - NS|$
1	$(0.5, 0.01)$	0.9900127	0.9900498	3.715e−5
2	$(0.5, 0.02)$	0.9801275	0.9801987	7.122e−5
3	$(0.5, 0.03)$	0.9703430	0.9704455	1.025e−4
4	$(0.5, 0.04)$	0.9606578	0.9607894	1.317e−4
5	$(0.5, 0.05)$	0.9510706	0.9512294	1.589e−4
6	$(0.5, 0.06)$	0.9415796	0.9417645	1.849e−4
7	$(0.5, 0.07)$	0.9321851	0.9323938	2.087e−4
8	$(0.5, 0.08)$	0.9228816	0.9231163	2.347e−4
9	$(0.5, 0.09)$	0.9136845	0.9139312	2.467e−4
10	$(0.5, 0.10)$	0.9045210	0.9048374	3.164e−4
11	$(0.5, 0.11)$	1.077546	0.8958341	1.817e−1
12	$(0.5, 0.12)$	− 12.35406	0.8869204	1.324e+1
13	$(0.5, 0.13)$	2.273332e+6	0.8780954	2.273e+6

Table 3.18 (Example 3.7) The maximum errors of numerical solutions with different step sizes ($\mu = 1/2$)

h	τ	$E_\infty(h, \tau)$	$E_\infty(2h, 4\tau)/E_\infty(h, \tau)$
1/10	1/200	6.173e−4	
1/20	1/800	1.557e−4	3.965
1/40	1/3200	3.902e−5	3.991
1/80	1/12800	9.760e−6	3.998

3.7.2   The Backward Euler Scheme

Considering Eq. (3.71a) at the node point (x_i, t_k), we have

$$\frac{\partial u}{\partial t}(x_i, t_k) - a\big(u(x_i, t_k)\big)\frac{\partial^2 u}{\partial x^2}(x_i, t_k) = f(x_i, t_k), \ 1 \leqslant i \leqslant m - 1, \quad 1 \leqslant k \leqslant n,$$

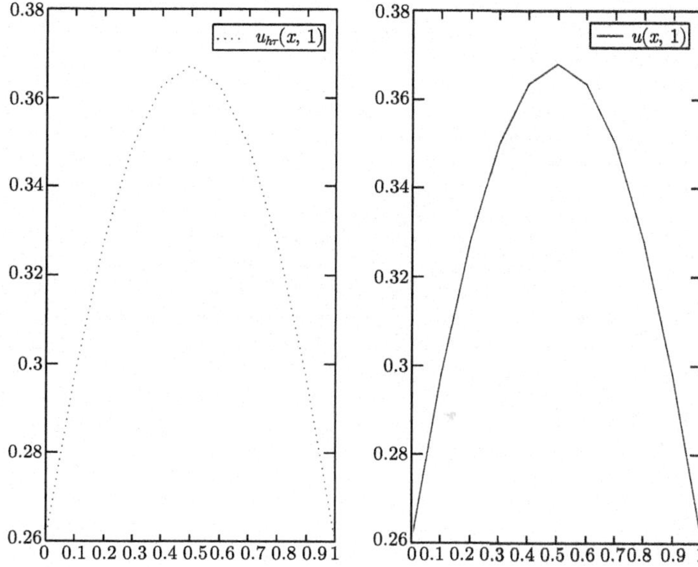

Fig. 3.23 (Example 3.7) The numerical solution profile ($h = 1/10$, $\tau = 1/200$) and the exact solution profile at $t = 1$

Fig. 3.24 (Example 3.7) The error profiles of numerical solutions with different step sizes at $t = 1$

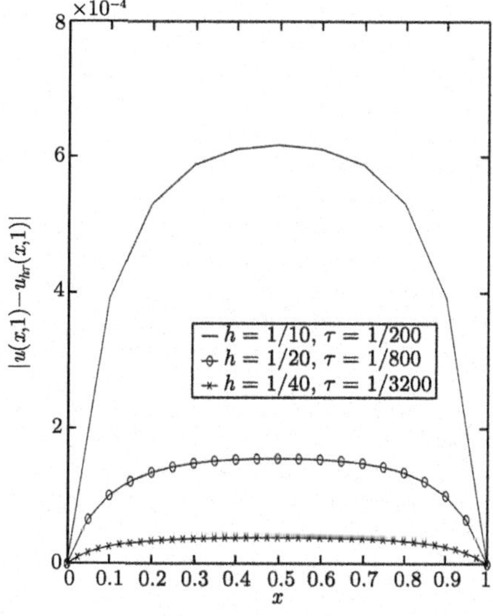

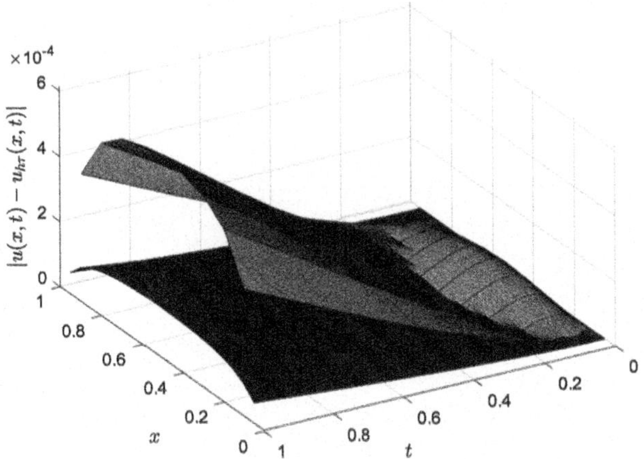

Fig. 3.25 (Example 3.7) The error surfaces with different step sizes ($\mu = 1/2$)

or

$$\frac{\partial u}{\partial t}(x_i, t_k) - a\big(u(x_i, t_{k-1})\big)\frac{\partial^2 u}{\partial x^2}(x_i, t_k)$$

$$= f(x_i, t_k) + \Big[a\big(u(x_i, t_k)\big) - a\big(u(x_i, t_{k-1})\big)\Big]\frac{\partial^2 u}{\partial x^2}(x_i, t_k),$$

$$1 \leqslant i \leqslant m-1, \quad 1 \leqslant k \leqslant n. \qquad (3.85)$$

Substituting

$$\frac{\partial^2 u}{\partial x^2}(x_i, t_k) = \delta_x^2 U_i^k - \frac{h^2}{12}\frac{\partial^4 u}{\partial x^4}(\xi_{ik}, t_k), \quad \xi_{ik} \in (x_{i-1}, x_{i+1})$$

and

$$\frac{\partial u}{\partial t}(x_i, t_k) = D_{\bar t}U_i^k + \frac{\tau}{2}\frac{\partial^2 u}{\partial t^2}(x_i, \bar\eta_{ik}), \quad \bar\eta_{ik} \in (t_{k-1}, t_k)$$

into (3.85) produces

$$D_{\bar t}U_i^k - a(U_i^{k-1})\delta_x^2 U_i^k = f(x_i, t_k) + \Big[a(U_i^k) - a(U_i^{k-1})\Big]\frac{\partial^2 u}{\partial x^2}(x_i, t_k)$$

$$- \frac{\tau}{2}\frac{\partial^2 u}{\partial t^2}(x_i, \bar\eta_{ik}) - \frac{h^2}{12}a(U_i^{k-1})\frac{\partial^4 u}{\partial x^4}(\xi_{ik}, t_k),$$

$$1 \leqslant i \leqslant m-1, \quad 1 \leqslant k \leqslant n.$$

Denote

$$(R_7)_i^k = \left[a(U_i^k) - a(U_i^{k-1})\right]\frac{\partial^2 u}{\partial x^2}(x_i, t_k) - \frac{\tau}{2}\frac{\partial^2 u}{\partial t^2}(x_i, \bar{\eta}_{ik})$$
$$- \frac{h^2}{12}a(U_i^{k-1})\frac{\partial^4 u}{\partial x^4}(\xi_{ik}, t_k). \tag{3.86}$$

Then we have

$$D_{\bar t}U_i^k - a(U_i^{k-1})\delta_x^2 U_i^k = f(x_i, t_k) + (R_7)_i^k, \quad 1 \leqslant i \leqslant m - 1, \quad 1 \leqslant k \leqslant n. \tag{3.87}$$

Denote

$$c_6 = \max\left\{L_p M_t M_{xx} + \frac{1}{2}M_{tt}, \frac{1}{12}a_1 M_{xxxx}\right\}.$$

It follows from (3.86) that

$$|(R_7)_i^k| \leqslant c_6(\tau + h^2), \quad 1 \leqslant i \leqslant m - 1, \quad 1 \leqslant k \leqslant n. \tag{3.88}$$

Noticing the initial-boundary value conditions (3.71b) and (3.71c), we have

$$\begin{cases} U_i^0 = \varphi(x_i), & 0 \leqslant i \leqslant m, \\ U_0^k = \alpha(t_k), & U_m^k = \beta(t_k), \quad 1 \leqslant k \leqslant n. \end{cases} \tag{3.89}$$

Omitting the small term $(R_7)_i^k$ in (3.87) and replacing U_i^k with u_i^k, another difference scheme reads

$$\begin{cases} D_{\bar t}u_i^k - a(u_i^{k-1})\delta_x^2 u_i^k = f(x_i, t_k), & 1 \leqslant i \leqslant m - 1, \quad 1 \leqslant k \leqslant n, \\ u_i^0 = \varphi(x_i), & 0 \leqslant i \leqslant m, \\ u_0^k = \alpha(t_k), & u_m^k = \beta(t_k), \quad 1 \leqslant k \leqslant n. \end{cases} \tag{3.90}$$

There is the following convergence theorem.

Theorem 3.23 *Let* $\{u(x, t) \mid (x, t) \in \bar{D}\}$ *be the solution of the problem* (3.71) *and* $\{u_i^k \mid 0 \leqslant i \leqslant m, 0 \leqslant k \leqslant n\}$ *be the solution of the difference scheme* (3.90). *Denote*

$$e_i^k = u(x_i, t_k) - u_i^k, \quad 0 \leqslant i \leqslant m, \quad 0 \leqslant k \leqslant n,$$

$$c_7 = \frac{c_6}{L_p M_{xx}}e^{L_p M_{xx}T}.$$

Then when the step sizes τ and h satisfy

$$\tau \leqslant \frac{\epsilon}{2c_7}, \quad h \leqslant \sqrt{\frac{\epsilon}{2c_7}}, \tag{3.91}$$

we have

$$\|e^k\|_\infty \leqslant c_7\left(\tau + h^2\right), \quad 0 \leqslant k \leqslant n. \tag{3.92}$$

Proof Subtracting (3.90) from (3.87) and (3.89), we have the system of error equations

$$\begin{cases} D_{\bar{t}}e_i^k - a(u_i^{k-1})\delta_x^2 e_i^k = \left[a(U_i^{k-1}) - a(u_i^{k-1})\right]\delta_x^2 U_i^k + (R_7)_i^k, \\ \qquad\qquad 1 \leqslant i \leqslant m-1, \quad 1 \leqslant k \leqslant n, & (3.93a) \\ e_i^0 = 0, \quad 0 \leqslant i \leqslant m, & (3.93b) \\ e_0^k = 0, \quad e_m^k = 0, \quad 1 \leqslant k \leqslant n. & (3.93c) \end{cases}$$

It follows from (3.93b) that (3.92) holds for $k = 0$.

Now suppose (3.92) holds for $0 \leqslant k \leqslant l$, and then when h and τ satisfy (3.91), we have

$$\|e^k\|_\infty \leqslant \epsilon, \quad 0 \leqslant k \leqslant l.$$

Therefore,

$$a_0 \leqslant a(u_i^k) \leqslant a_1, \quad |a(U_i^k) - a(u_i^k)| \leqslant L_p|e_i^k|, \quad 1 \leqslant i \leqslant m-1, \quad 0 \leqslant k \leqslant l. \tag{3.94}$$

Reformulating (3.93a), we have

$$\left[1 + 2\mu a(u_i^{k-1})\right]e_i^k$$

$$= \mu a(u_i^{k-1})\left(e_{i-1}^k + e_{i+1}^k\right) + e_i^{k-1}$$

$$+ \tau\left[a(U_i^{k-1}) - a(u_i^{k-1})\right]\delta_x^2 U_i^k + \tau(R_7)_i^k, \quad 1 \leqslant i \leqslant m-1, \ 1 \leqslant k \leqslant n. \tag{3.95}$$

Taking absolute values on both sides of (3.95), with the application of the triangle inequality, and noticing (3.94), (3.88), we have

$$\left[1 + 2\mu a(u_i^{k-1})\right]|e_i^k|$$

$$\leqslant \mu a(u_i^{k-1})(|e_{i-1}^k| + |e_{i+1}^k|) + |e_i^{k-1}| + \tau|a(U_i^{k-1}) - a(u_i^{k-1})| \cdot |\delta_x^2 U_i^k| + \tau|(R_7)_i^k|$$

$$\leqslant 2\mu a(u_i^{k-1})\|e^k\|_\infty + \|e^{k-1}\|_\infty + \tau L_p M_{xx}\|e^{k-1}\|_\infty + c_6\tau(\tau + h^2),$$

$$1 \leqslant i \leqslant m-1, \quad 1 \leqslant k \leqslant l+1.$$

Assume $\|e^k\|_\infty = |e_{i_k}^k|$, $i_k \in \{1, 2, \ldots, m-1\}$. Taking $i = i_k$ in the above inequality, we have

$$\|e^k\|_\infty \leqslant (1 + L_p M_{xx}\tau)\|e^{k-1}\|_\infty + c_6\tau(\tau + h^2), \quad 1 \leqslant k \leqslant l+1.$$

By the Gronwall inequality (Lemma 3.3), we have

$$\|e^{l+1}\|_\infty \leqslant e^{L_p M_{xx}(l+1)\tau} \cdot \left[\|e^0\|_\infty + \frac{c_6(\tau + h^2)}{L_p M_{xx}}\right]$$

$$\leqslant \frac{c_6}{L_p M_{xx}} e^{L_p M_{xx}T}\left(\tau + h^2\right) = c_7\left(\tau + h^2\right).$$

Thereby, (3.92) holds for $k = l + 1$.

By induction, (3.92) holds. $\qquad\square$

Example 3.8 Apply the backward *Euler* scheme (3.90) to compute the problem (3.84) in Example 3.7.

Tables 3.19 and 3.20 list part of the numerical results with step sizes $h = 1/10$, $\tau = 1/200$ (with the step ratio $\mu = 1/2$) and $h = 1/10$, $\tau = 1/100$ (with the step ratio $\mu = 1$), respectively, showing that the numerical solution approximates the exact solution very well. Table 3.21 provides the maximum error

$$E_\infty(h, \tau) = \max_{1\leqslant i\leqslant m-1, 1\leqslant k\leqslant n} |u(x_i, t_k) - u_i^k|$$

with $\mu = 1/2$ and different step sizes.

As can be seen from Table 3.21, when the spatial step size is reduced by half and the temporal step size is reduced to a quarter of the original, the maximum error

Table 3.19 (Example 3.8) Numerical solutions (NS), exact solutions (ES), and absolute values of errors at part of node points ($h = 1/10$, $\tau = 1/200$)

k	(x, t)	NS	ES	$|\text{ES} - \text{NS}|$
20	$(0.5, 0.1)$	0.9032027	0.9048374	1.6347e−3
40	$(0.5, 0.2)$	0.8169320	0.8187308	1.7988e−3
60	$(0.5, 0.3)$	0.7391011	0.7408182	1.7171e−3
80	$(0.5, 0.4)$	0.6687385	0.6703200	1.5815e−3
100	$(0.5, 0.5)$	0.6050933	0.6065307	1.4374e−3
120	$(0.5, 0.6)$	0.5475135	0.5488116	1.2981e−3
140	$(0.5, 0.7)$	0.4954172	0.4965853	1.1681e−3
160	$(0.5, 0.8)$	0.4482804	0.4493290	1.0485e−3
180	$(0.5, 0.9)$	0.4056301	0.4065697	9.3957e−4
200	$(0.5, 1.0)$	0.3670387	0.3678794	8.4072e−4

Table 3.20 (Example 3.8) Numerical solutions (NS), exact solutions (ES), and absolute values of errors at part of node points ($h = 1/10$, $\tau = 1/100$)

k	(x, t)	NS	ES	$|\text{ES} - \text{NS}|$
10	$(0.5, 0.1)$	0.9013752	0.9048374	3.4622e$-$3
20	$(0.5, 0.2)$	0.8149169	0.8187308	3.8139e$-$3
30	$(0.5, 0.3)$	0.7371693	0.7408182	3.6489e$-$3
40	$(0.5, 0.4)$	0.6669493	0.6703200	3.3708e$-$3
50	$(0.5, 0.5)$	0.6034557	0.6065307	3.0750e$-$3
60	$(0.5, 0.6)$	0.5460219	0.5488116	2.7897e$-$3
70	$(0.5, 0.7)$	0.4940612	0.4965853	2.5241e$-$3
80	$(0.5, 0.8)$	0.4470484	0.4493290	2.2806e$-$3
90	$(0.5, 0.9)$	0.4045103	0.4065697	2.0593e$-$3
100	$(0.5, 1.0)$	0.3660203	0.3678794	1.8591e$-$3

Table 3.21 (Example 3.8) The maximum errors of numerical solutions with different step sizes ($\mu = 1/2$)

h	τ	$E_\infty(h, \tau)$	$E_\infty(2h, 4\tau)/E_\infty(h, \tau)$
1/10	1/200	1.800e$-$3	
1/20	1/800	4.520e$-$4	3.9812
1/40	1/3200	1.131e$-$4	3.9950
1/80	1/12800	2.829e$-$5	3.9989

Table 3.22 (Example 3.8) The maximum errors of numerical solutions with different step sizes with ($\mu = 1$)

h	τ	$E_\infty(h, \tau)$	$E_\infty(2h, 4\tau)/E_\infty(h, \tau)$
1/10	1/100	3.815e$-$3	
1/20	1/400	9.611e$-$4	3.9538
1/40	1/1600	2.408e$-$4	3.9918
1/80	1/6400	6.022e$-$5	3.9980

is reduced to 1/4 of the original. Similar results are presented in Table 3.22 with $\mu = 1$.

Figure 3.26 shows the exact solution profile and the numerical solution profile with step sizes $h = 1/10$, $\tau = 1/200$ at $t = 1$. Figures 3.27 and 3.28 display the error profiles between the exact solution and the numerical solution at $t = 1$ with $\mu = \frac{1}{2}$ and $\mu = 1$, respectively. Figure 3.29 illustrates error surfaces of numerical solutions with different step sizes and $\mu = 1$.

3.7.3 The Crank-Nicolson Scheme

Considering Eq. (3.71a) at the point $(x_i, t_{\frac{1}{2}})$, we have

$$\frac{\partial u}{\partial t}(x_i, t_{\frac{1}{2}}) - a\big(u(x_i, t_{\frac{1}{2}})\big)\frac{\partial^2 u}{\partial x^2}(x_i, t_{\frac{1}{2}}) = f(x_i, t_{\frac{1}{2}}), \quad 1 \leqslant i \leqslant m - 1,$$

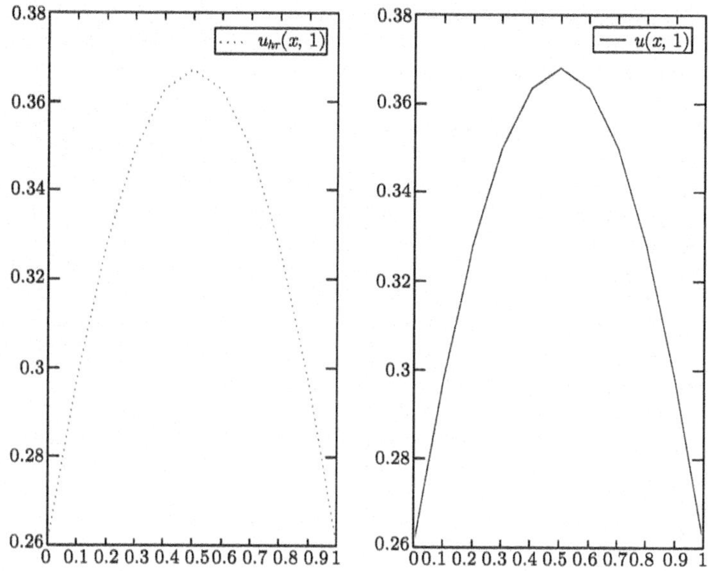

Fig. 3.26 (Example 3.8) The numerical solution profile ($h = 1/10$, $\tau = 1/200$) and the exact solution profile at $t = 1$

namely,

$$\frac{\partial u}{\partial t}(x_i, t_{\frac{1}{2}}) - a\big(u(x_i, 0) + \frac{\tau}{2} u_t(x_i, 0)\big) \frac{\partial^2 u}{\partial x^2}(x_i, t_{\frac{1}{2}})$$

$$= f(x_i, t_{\frac{1}{2}}) + \left[a\big(u(x_i, t_{\frac{1}{2}})\big) - a\big(u(x_i, 0) + \frac{\tau}{2} u_t(x_i, 0)\big) \right] \frac{\partial^2 u}{\partial x^2}(x_i, t_{\frac{1}{2}}),$$

$$1 \leqslant i \leqslant m - 1. \tag{3.96}$$

From the Taylor expansion, we have

$$\frac{\partial u}{\partial t}(x_i, t_{\frac{1}{2}}) = \delta_t U_i^{\frac{1}{2}} - \frac{\tau^2}{24} \frac{\partial^3 u}{\partial t^3}(x_i, \eta_{i0}), \quad \eta_{i0} \in (t_0, t_1),$$

$$\frac{\partial^2 u}{\partial x^2}(x_i, t_{\frac{1}{2}}) = \frac{1}{2}\left[\frac{\partial^2 u}{\partial x^2}(x_i, t_1) + \frac{\partial^2 u}{\partial x^2}(x_i, t_0) \right] - \frac{\tau^2}{8} \frac{\partial^4 u}{\partial x^2 \partial t^2}(x_i, \zeta_{i0})$$

$$= \frac{1}{2}\left[\delta_x^2 U_i^1 - \frac{h^2}{12} \frac{\partial^4 u}{\partial x^4}(\xi_{i1}, t_1) + \delta_x^2 U_i^0 - \frac{h^2}{12} \frac{\partial^4 u}{\partial x^4}(\xi_{i0}, t_0) \right]$$

$$\quad - \frac{\tau^2}{8} \frac{\partial^4 u}{\partial x^2 \partial t^2}(x_i, \zeta_{i0})$$

$$= \delta_x^2 U_i^{\frac{1}{2}} - \frac{h^2}{24}\left[\frac{\partial^4 u}{\partial x^4}(\xi_{i1}, t_1) + \frac{\partial^4 u}{\partial x^4}(\xi_{i0}, t_0) \right] - \frac{\tau^2}{8} \frac{\partial^4 u}{\partial x^2 \partial t^2}(x_i, \zeta_{i0}),$$

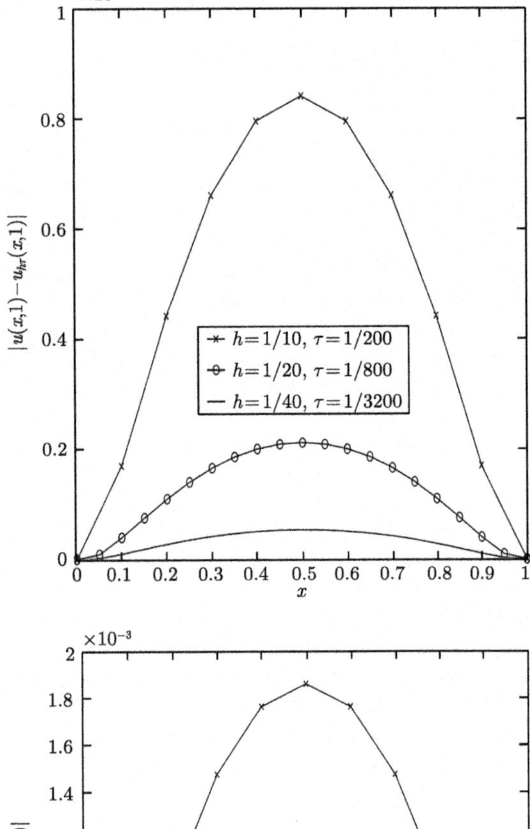

Fig. 3.27 (Example 3.8) The numerical error profiles at $t = 1$ with different step sizes $(\mu = 1/2)$

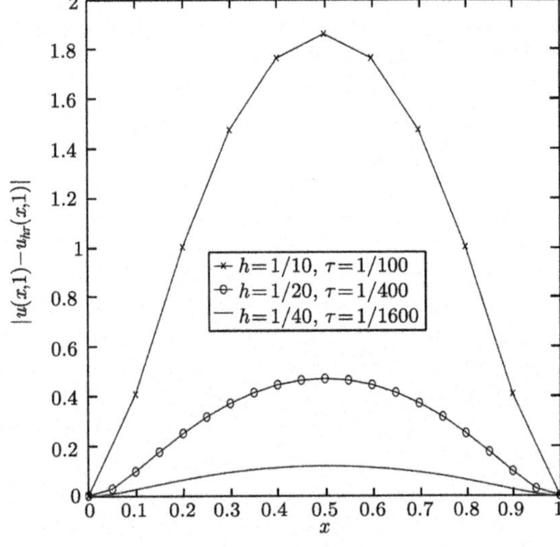

Fig. 3.28 (Example 3.8) The numerical error profiles at $t = 1$ with different step sizes $(\mu = 1)$

where ξ_{i1}, $\xi_{i0} \in (x_{i-1}, x_{i+1})$ and $\zeta_{i0} \in (t_0, t_1)$. Substituting the above two equalities into (3.96), we have

$$\delta_t U_i^{\frac{1}{2}} - a\big(u(x_i, 0) + \frac{\tau}{2} u_t(x_i, 0)\big)\delta_x^2 U_i^{\frac{1}{2}} = f(x_i, t_{\frac{1}{2}}) + (R_8)_i^0, \quad 1 \leqslant i \leqslant m - 1,$$

$$(3.97)$$

Fig. 3.29 (Example 3.8) The numerical error surfaces with different step sizes at $t = 1$ ($\mu = 1$)

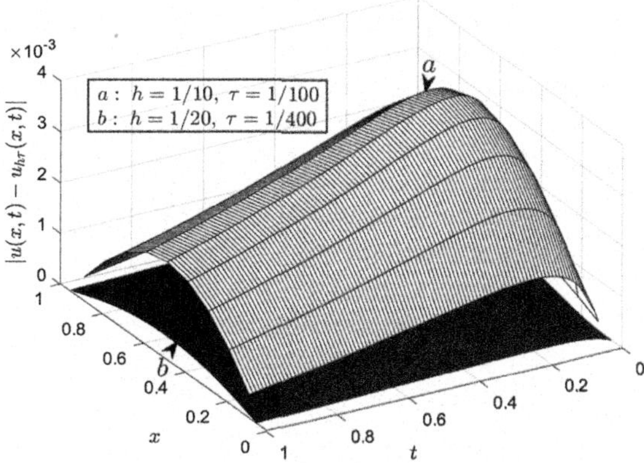

where

$$(R_8)_i^0$$

$$= \left[a\big(u(x_i, t_{\frac{1}{2}})\big) - a\big(u(x_i, 0) + \frac{\tau}{2} u_t(x_i, 0)\big) \right] \frac{\partial^2 u}{\partial x^2}(x_i, t_{\frac{1}{2}}) + \frac{\tau^2}{24} \frac{\partial^3 u}{\partial t^3}(x_i, \eta_{i0})$$

$$- \frac{\tau^2}{8} a\big(u(x_i, 0) + \frac{\tau}{2} u_t(x_i, 0)\big) \frac{\partial^4 u}{\partial x^2 \partial t^2}(x_i, \zeta_{i0})$$

$$- \frac{h^2}{24} a\big(u(x_i, 0) + \frac{\tau}{2} u_t(x_i, 0)\big) \left[\frac{\partial^4 u}{\partial x^4}(\xi_{i,1}, t_1) + \frac{\partial^4 u}{\partial x^4}(\xi_{i,0}, t_0) \right], \quad 1 \leqslant i \leqslant m - 1.$$

Considering Eq. (3.71a) at the node point (x_i, t_k), we have

$$\frac{\partial u}{\partial t}(x_i, t_k) - a\big(u(x_i, t_k)\big) \frac{\partial^2 u}{\partial x^2}(x_i, t_k) = f(x_i, t_k),$$

$$1 \leqslant i \leqslant m - 1, \quad 1 \leqslant k \leqslant n - 1. \tag{3.98}$$

It follows from the Taylor expansion that

$$\frac{\partial u}{\partial t}(x_i, t_k) = \Delta_t U_i^k - \frac{\tau^2}{6} \frac{\partial^3 u}{\partial t^3}(x_i, \eta_{ik}),$$

$$\frac{\partial^2 u}{\partial x^2}(x_i, t_k) = \frac{1}{2} \left[\frac{\partial^2 u}{\partial x^2}(x_i, t_{k+1}) + \frac{\partial^2 u}{\partial x^2}(x_i, t_{k-1}) \right] - \frac{\tau^2}{2} \frac{\partial^4 u}{\partial x^2 \partial t^2}(x_i, \zeta_{ik})$$

$$= \frac{1}{2} \left[\delta_x^2 U_i^{k+1} - \frac{h^2}{12} \frac{\partial^4 u}{\partial x^4}(\xi_{i,k+1}, t_{k+1}) + \delta_x^2 U_i^{k-1} - \frac{h^2}{12} \frac{\partial^4 u}{\partial x^4}(\xi_{i,k-1}, t_{k-1}) \right]$$

$$- \frac{\tau^2}{2} \frac{\partial^4 u}{\partial x^2 \partial t^2}(x_i, \zeta_{ik})$$

$$
= \delta_x^2 U_i^{\bar{k}} - \frac{h^2}{24}\left[\frac{\partial^4 u}{\partial x^4}(\xi_{i,k+1}, t_{k+1}) + \frac{\partial^4 u}{\partial x^4}(\xi_{i,k-1}, t_{k-1})\right]
$$

$$
- \frac{\tau^2}{2}\frac{\partial^4 u}{\partial x^2 \partial t^2}(x_i, \zeta_{ik}),
$$

where $\xi_{i,k+1}, \xi_{i,k-1} \in (x_{i-1}, x_{i+1})$ and $\eta_{ik}, \zeta_{ik} \in (t_{k-1}, t_{k+1})$. Substituting the above two equalities into (3.98), we have

$$
\Delta_t U_i^k - a(U_i^k)\delta_x^2 U_i^{\bar{k}} = f(x_i, t_k) + (R_8)_i^k, \quad 1 \leqslant i \leqslant m-1, \quad 1 \leqslant k \leqslant n-1,
\tag{3.99}
$$

where

$$
(R_8)_i^k = \tau^2\left[\frac{1}{6}\frac{\partial^3 u}{\partial t^3}(x_i, \eta_{ik}) - \frac{1}{2}a(U_i^k)\frac{\partial^4 u}{\partial x^2 \partial t^2}(x_i, \zeta_{ik})\right]
$$

$$
- \frac{h^2}{24}a(U_i^k)\left[\frac{\partial^4 u}{\partial x^4}(\xi_{i,k+1}, t_{k+1}) + \frac{\partial^4 u}{\partial x^4}(\xi_{i,k-1}, t_{k-1})\right],
$$

$$
1 \leqslant i \leqslant m-1, \quad 1 \leqslant k \leqslant n-1.
$$

Denote

$$
\max_{(x,t)\in\bar{D}}\left|\frac{\partial^2 u(x,t)}{\partial x^2}\right| = M_{xx}, \quad \max_{(x,t)\in\bar{D}}\left|\frac{\partial^2 u(x,t)}{\partial t^2}\right| = M_{tt}, \quad \max_{(x,t)\in\bar{D}}\left|\frac{\partial^3 u(x,t)}{\partial t^3}\right| = M_{ttt},
$$

$$
\max_{(x,t)\in\bar{D}}\left|\frac{\partial^3 u(x,t)}{\partial x \partial t^2}\right| = M_{xtt}, \quad \max_{(x,t)\in\bar{D}}\left|\frac{\partial^4 u(x,t)}{\partial x^2 \partial t^2}\right| = M_{xxtt},
$$

$$
c_8 = \max\left\{\frac{1}{8}L_p M_{xx}M_{tt} + \frac{1}{8}a_1 M_{xxtt} + \frac{1}{24}M_{ttt}, \frac{a_1}{12}M_{xxxx}, \frac{1}{2}a_1 M_{xxtt} + \frac{1}{6}M_{ttt}\right\},
$$

and then we have

$$
|(R_8)_i^k| \leqslant c_8(\tau^2 + h^2), \quad 1 \leqslant i \leqslant m-1, \quad 0 \leqslant k \leqslant n-1.
\tag{3.100}
$$

Noticing the initial-boundary value conditions (3.71b) and (3.71c), we have

$$
\begin{cases}
U_i^0 = \varphi(x_i), & 0 \leqslant i \leqslant m, \\
U_0^k = \alpha(t_k), \quad U_m^k = \beta(t_k), & 1 \leqslant k \leqslant n.
\end{cases}
\tag{3.101}
$$

Omitting the small terms $(R_8)_i^0$ and $(R_8)_i^k$ in (3.97) and (3.99), respectively, and replacing U_i^k with u_i^k, a Crank-Nicolson type scheme reads

$$
\left\{
\begin{array}{l}
\delta_t u_i^{\frac{1}{2}} - a\big(u(x_i,0) + \dfrac{\tau}{2} u_t(x_i,0)\big)\delta_x^2 u_i^{\frac{1}{2}} = f(x_i, t_{\frac{1}{2}}), \quad 1 \leqslant i \leqslant m-1, \text{(3.102a)} \\[2mm]
\Delta_t u_i^k - a(u_i^k)\delta_x^2 u_i^{\bar{k}} = f(x_i, t_k), \quad 1 \leqslant i \leqslant m-1, \quad 1 \leqslant k \leqslant n-1, \text{ (3.102b)} \\[2mm]
u_i^0 = \varphi(x_i), \quad 0 \leqslant i \leqslant m, \hfill \text{(3.102c)} \\[2mm]
u_0^k = \alpha(t_k), \quad u_m^k = \beta(t_k), \quad 1 \leqslant k \leqslant n. \hfill \text{(3.102d)}
\end{array}
\right.
$$

Theorem 3.24 *Let $\{u(x,t)\,|\,0 \leqslant x \leqslant L, 0 \leqslant t \leqslant T\}$ be the solution of the problem (3.71) and $\{u_i^k\,|\,0 \leqslant i \leqslant m, 0 \leqslant k \leqslant n\}$ be the solution of the difference scheme (3.102). Denote*

$$
e_i^k = u(x_i, t_k) - u_i^k, \quad 0 \leqslant i \leqslant m, \quad 0 \leqslant k \leqslant n,
$$

$$
c_9 = \frac{L c_8}{2} \sqrt{\frac{1}{2a_0} + \frac{6}{L^2 L_p^2 M_{xx}^2}} \, e^{\frac{L^2 L_p^2 M_{xx}^2 T}{6a_0}}.
$$

Then when the step sizes τ and h satisfy

$$
\tau \leqslant \sqrt{\frac{\epsilon}{2c_9}}, \quad h \leqslant \sqrt{\frac{\epsilon}{2c_9}}, \tag{3.103}
$$

we have

$$
\|e^k\|_\infty \leqslant c_9(\tau^2 + h^2), \quad 1 \leqslant k \leqslant n. \tag{3.104}
$$

Proof Subtracting (3.102) from (3.97), (3.99), and (3.101), respectively, we have the system of error equations

$$
\left\{
\begin{array}{l}
\delta_t e_i^{\frac{1}{2}} - a\big(u(x_i,0) + \dfrac{\tau}{2} u_t(x_i,0)\big)\delta_x^2 e_i^{\frac{1}{2}} = (R_8)_i^0, \quad 1 \leqslant i \leqslant m-1, \text{ (3.105a)} \\[2mm]
\Delta_t e_i^k - \Big[a(U_i^k)\delta_x^2 U_i^{\bar{k}} - a(u_i^k)\delta_x^2 u_i^{\bar{k}}\Big] = (R_8)_i^k, \\[2mm]
\qquad\qquad\qquad\qquad 1 \leqslant i \leqslant m-1, \quad 1 \leqslant k \leqslant n-1, \hfill \text{(3.105b)} \\[2mm]
e_i^0 = 0, \quad 0 \leqslant i \leqslant m, \hfill \text{(3.105c)} \\[2mm]
e_0^k = 0, \quad e_m^k = 0, \quad 1 \leqslant k \leqslant n. \hfill \text{(3.105d)}
\end{array}
\right.
$$

It follows from (3.105c)–(3.105d) that

$$
\left\{
\begin{array}{l}
\delta_t e_0^{\frac{1}{2}} = 0, \quad \delta_t e_m^{\frac{1}{2}} = 0, \hfill \text{(3.106a)} \\[2mm]
\Delta_t e_0^k = 0, \quad \Delta_t e_m^k = 0, \quad 1 \leqslant k \leqslant n-1. \hfill \text{(3.106b)}
\end{array}
\right.
$$

From (3.105c), we have

$$\|e^0\| = 0, \quad |e^0|_1 = 0. \tag{3.107}$$

Taking an inner product of (3.105a) with $-\delta_x^2 e^{\frac{1}{2}}$, we have

$$-\left(\delta_t e^{\frac{1}{2}}, \delta_x^2 e^{\frac{1}{2}}\right) + h \sum_{i=1}^{m-1} a\left(u(x_i, 0) + \frac{\tau}{2} u_t(x_i, 0)\right)(\delta_x^2 e_i^{\frac{1}{2}})^2 = -\left((R_8)^0, \delta_x^2 e^{\frac{1}{2}}\right).$$

With the application of the summation by parts and noticing (3.106a) and $a\left(u(x_i, 0)+\frac{\tau}{2} u_t(x_i, 0)\right) \geqslant a_0$, we have

$$\frac{1}{2\tau}(|e^1|_1^2 - |e^0|_1^2) + a_0\|\delta_x^2 e^{\frac{1}{2}}\|^2 \leqslant a_0\|\delta_x^2 e^{\frac{1}{2}}\|^2 + \frac{1}{4a_0}\|(R_8)^0\|^2.$$

Combining (3.107) with (3.100), we have

$$\frac{1}{2\tau}|e^1|_1^2 \leqslant \frac{L}{4a_0}\left[c_8(\tau^2 + h^2)\right]^2,$$

namely,

$$|e^1|_1^2 \leqslant \frac{L}{2a_0}c_8^2\tau(\tau^2 + h^2)^2 \leqslant \frac{Lc_8^2}{2a_0}(\tau^2 + h^2)^2. \tag{3.108}$$

By means of Lemma 1.4, it yields

$$\|e^1\|_\infty^2 \leqslant \frac{L}{4}|e^1|_1^2 \leqslant \frac{L^2 c_8^2}{8a_0}(\tau^2 + h^2)^2.$$

Therefore, (3.104) holds for $k = 1$.

Now suppose (3.104) holds for $1 \leqslant k \leqslant l$, where $1 \leqslant l \leqslant n - 1$. When τ and h satisfy (3.103), it holds that

$$\|e^k\|_\infty \leqslant \epsilon, \quad 0 \leqslant k \leqslant l.$$

Applying the conditions (3.72) and (3.73) yields

$$a_0 \leqslant a(u_i^k) \leqslant a_1, \quad |a(U_i^k) - a(u_i^k)| \leqslant L_p|e_i^k|, \quad 1 \leqslant i \leqslant m - 1, \quad 0 \leqslant k \leqslant l.$$

$$\tag{3.109}$$

Rewriting the error Eq. (3.105b) as follows

$$\Delta_t e_i^k - a(u_i^k)\delta_x^2 e_i^{\bar{k}} = \left[a(U_i^k) - a(u_i^k)\right]\delta_x^2 U_i^{\bar{k}} + (R_8)_i^k, \quad 1 \leqslant i \leqslant m-1,\ 1 \leqslant k \leqslant n-1.$$

Taking an inner product of the above equality with $-\delta_x^2 e^{\bar{k}}$, we have

$$-\left(\Delta_t e^k, \delta_x^2 e^{\bar{k}}\right) + h \sum_{i=1}^{m-1} a(u_i^k)(\delta_x^2 e_i^{\bar{k}})^2$$

$$= -h \sum_{i=1}^{m-1}\left[a(U_i^k) - a(u_i^k)\right](\delta_x^2 U_i^{\bar{k}})(\delta_x^2 e_i^{\bar{k}}) - \left((R_8)^k, \delta_x^2 e^{\bar{k}}\right), \quad 1 \leqslant k \leqslant l.$$

$$(3.110)$$

Now, we estimate each term in the above equality.

Using summation by parts and noticing (3.106b), the first term on the left-hand side becomes

$$-\left(\Delta_t e^k, \delta_x^2 e^{\bar{k}}\right) = \left(\Delta_t \delta_x e^k, \delta_x e^{\bar{k}}\right) = \frac{1}{4\tau}(|e^{k+1}|_1^2 - |e^{k-1}|_1^2), \quad 1 \leqslant k \leqslant l.$$

$$(3.111)$$

Using the first inequality in (3.109), the second term on the left-hand side has the following lower bound estimate

$$h \sum_{i=1}^{m-1} a(u_i^k)(\delta_x^2 e_i^{\bar{k}})^2 \geqslant a_0 \|\delta_x^2 e^{\bar{k}}\|^2, \quad 1 \leqslant k \leqslant l.$$

With the help of the second inequality in (3.109), the first term on the right-hand side has the following upper bound estimate:

$$-h \sum_{i=1}^{m-1}[a(U_i^k) - a(u_i^k)](\delta_x^2 U_i^{\bar{k}})(\delta_x^2 e_i^{\bar{k}})$$

$$\leqslant L_p M_{xx} h \sum_{i=1}^{m-1} |e_i^k| \cdot |\delta_x^2 e_i^{\bar{k}}|$$

$$\leqslant L_p M_{xx}\left(\frac{L_p M_{xx}}{2a_0}\|e^k\|^2 + \frac{a_0}{2L_p M_{xx}}\|\delta_x^2 e^{\bar{k}}\|^2\right)$$

$$= \frac{L_p^2 M_{xx}^2}{2a_0}\|e^k\|^2 + \frac{a_0}{2}\|\delta_x^2 e^{\bar{k}}\|^2.$$

$$(3.112)$$

Combining the Cauchy-Schwarz inequality with (3.100), the second term on the right-hand side becomes

$$- \left((R_8)^k, \delta_x^2 e^{\bar{k}} \right) \leqslant \frac{1}{2a_0} \| (R_8)^k \|^2 + \frac{a_0}{2} \| \delta_x^2 e^{\bar{k}} \|^2 \leqslant \frac{Lc_8^2}{2a_0} \left(\tau^2 + h^2 \right)^2 + \frac{a_0}{2} \| \delta_x^2 e^{\bar{k}} \|^2.$$

$$(3.113)$$

Inserting (3.111)–(3.113) into (3.110), we have

$$\frac{1}{4\tau} \left(|e^{k+1}|_1^2 - |e^{k-1}|_1^2 \right) \leqslant \frac{L_p^2 M_{xx}^2}{2a_0} \| e^k \|^2 + \frac{Lc_8^2}{2a_0} \left(\tau^2 + h^2 \right)^2, \quad 1 \leqslant k \leqslant l.$$

With the help of Lemma 1.4, we have

$$\frac{1}{4\tau} \left(|e^{k+1}|_1^2 - |e^{k-1}|_1^2 \right) \leqslant \frac{L^2 L_p^2 M_{xx}^2}{12a_0} |e^k|_1^2 + \frac{Lc_8^2}{2a_0} \left(\tau^2 + h^2 \right)^2, \quad 1 \leqslant k \leqslant l,$$

or

$$|e^{k+1}|_1^2 \leqslant |e^{k-1}|_1^2 + \frac{L^2 L_p^2 M_{xx}^2}{3a_0} \tau |e^k|_1^2 + \frac{2Lc_8^2}{a_0} \tau \left(\tau^2 + h^2 \right)^2, \quad 1 \leqslant k \leqslant l.$$

Let $F^k = \max\{ |e^{k+1}|_1^2, |e^k|_1^2 \}$, and then we have

$$F^k \leqslant \left(1 + \frac{L^2 L_p^2 M_{xx}^2}{3a_0} \tau \right) F^{k-1} + \frac{2Lc_8^2}{a_0} \tau \left(\tau^2 + h^2 \right)^2, \quad 1 \leqslant k \leqslant l.$$

Employing the Gronwall inequality (Lemma 3.3) and noticing (3.108), we have

$$F^l \leqslant e^{\frac{L^2 L_p^2 M_{xx}^2 \, l\tau}{3a_0}} \cdot \left[F^0 + \frac{6Lc_8^2}{L_p^2 L^2 M_{xx}^2} \left(\tau^2 + h^2 \right)^2 \right]$$

$$\leqslant e^{\frac{L^2 L_p^2 M_{xx}^2 T}{3a_0}} \cdot \left(\frac{Lc_8^2}{2a_0} + \frac{6c_8^2}{LL_p^2 M_{xx}^2} \right) \left(\tau^2 + h^2 \right)^2, \quad 1 \leqslant k \leqslant l.$$

Therefore,

$$|e^{l+1}|_1^2 \leqslant e^{\frac{L^2 L_p^2 M_{xx}^2 T}{3a_0}} \cdot \left(\frac{Lc_8^2}{2a_0} + \frac{6c_8^2}{L_p^2 L M_{xx}^2} \right) \left(\tau^2 + h^2 \right)^2.$$

Applying Lemma 1.4 again, we have

$$\| e^{l+1} \|_\infty^2 \leqslant \frac{L}{4} |e^{l+1}|_1^2 \leqslant \frac{L}{4} e^{\frac{L^2 L_p^2 M_{xx}^2 T}{3a_0}} \cdot \left(\frac{Lc_8^2}{2a_0} + \frac{6c_8^2}{LL_p^2 M_{xx}^2} \right) \left(\tau^2 + h^2 \right)^2,$$

or

$$\|e^{l+1}\|_\infty \leqslant \frac{Lc_8}{2}\sqrt{\frac{1}{2a_0} + \frac{6}{L^2 L_p^2 M_{xx}^2}}\, e^{\frac{L^2 L_p^2 M_{xx}^2 T}{6a_0}}\left(\tau^2 + h^2\right),$$

which implies that (3.104) holds for $k = l + 1$.

By induction, it completes the proof. $\qquad\square$

Theorem 3.25 *The difference scheme* (3.102) *is uniquely solvable.*

Proof Denote the value of solutions at the k-th time level by $u^k = (u_0^k, u_1^k, \ldots, u_{m-1}^k, u_m^k)$.

The value at the 0-th time level has been determined by (3.102c). From (3.102a) and (3.102d), it follows that the system of linear equations in u^1 is

$$\begin{cases} \delta_t u_i^{\frac{1}{2}} - a\big(u(x_i, 0) + \frac{\tau}{2}u_t(x_i, 0)\big)\delta_x^2 u_i^{\frac{1}{2}} = f(x_i, t_{\frac{1}{2}}), & 1 \leqslant i \leqslant m - 1, \\ u_0^1 = \alpha(t_1), \quad u_m^1 = \beta(t_1). \end{cases}$$

Consider its homogeneous one

$$\begin{cases} \dfrac{1}{\tau}u_i^1 - \dfrac{1}{2}a\big(u(x_i, 0) + \dfrac{\tau}{2}u_t(x_i, 0)\big)\delta_x^2 u_i^1 = 0, & 1 \leqslant i \leqslant m - 1, & (3.114a) \\ u_0^1 = 0, \quad u_m^1 = 0. & (3.114b) \end{cases}$$

Taking an inner product of (3.114a) with $\delta_x^2 u^1$, we have

$$-\frac{1}{\tau}(u^1, \delta_x^2 u^1) + \frac{1}{2}h\sum_{i=1}^{m-1} a\big(u(x_i, 0) + \frac{\tau}{2}u_t(x_i, 0)\big)(\delta_x^2 u_i^1)^2 = 0.$$

Applying the summation by parts with (3.114b) and noticing $a\big(u(x_i, 0) + \frac{\tau}{2}u_t(x_i, 0)\big) \geqslant a_0$, we have

$$|u^1|_1^2 = 0.$$

Thus,

$$u_i^1 = 0, \quad 0 \leqslant i \leqslant m.$$

Therefore, u^1 is uniquely determined by the difference scheme (3.102).

Suppose the values of u^{k-1} and u^k have been determined. From (3.102b) and (3.102d), it follows that the system of difference equations in u^{k+1} is

$$\begin{cases} \Delta_t u_i^k - a(u_i^k)\delta_x^2 u_i^{\bar{k}} = f(x_i, t_k), & 1 \leqslant i \leqslant m-1, \\ u_0^{k+1} = \alpha(t_{k+1}), & u_m^{k+1} = \beta(t_{k+1}). \end{cases}$$

Consider the homogeneous one

$$\begin{cases} \dfrac{1}{2\tau} u_i^{k+1} - \dfrac{1}{2} a(u_i^k)\delta_x^2 u_i^{k+1} = 0, & 1 \leqslant i \leqslant m-1, & (3.115a) \\ u_0^{k+1} = 0, & u_m^{k+1} = 0. & (3.115b) \end{cases}$$

Taking an inner product of (3.115a) with $-\delta_x^2 u^{k+1}$, we have

$$-\frac{1}{2\tau}(u^{k+1}, \delta_x^2 u^{k+1}) + \frac{1}{2}h \sum_{i=1}^{m-1} a(u_i^k)(\delta_x^2 u_i^{k+1})^2 = 0.$$

Combining the summation by parts with (3.115b) and noticing $a(u_i^k) \geqslant a_0$, we get

$$|u^{k+1}|_1^2 = 0.$$

Thus,

$$u_i^{k+1} = 0, \quad 0 \leqslant i \leqslant m.$$

Therefore, u^{k+1} is uniquely determined by the difference scheme (3.102).

By induction, the result is obtained. $\qquad\qquad\qquad\qquad\qquad\qquad\qquad\qquad\quad\square$

Example 3.9 Apply the *Crank-Nicolson* scheme (3.102) to compute the problem (3.84) given in Example 3.7.

Define

$$E_\infty(h, \tau) = \max_{0 \leqslant i \leqslant m, 1 \leqslant k \leqslant n} |u(x_i, t_k) - u_i^k|.$$

Table 3.23 lists the numerical results calculated with different step sizes. According to the results in Table 3.23, the convergence orders with respect to both the spatial and temporal step sizes are two. Figure 3.30 shows the error profiles at $t = 1$ with different step sizes.

Table 3.23 (Example 3.9) The maximum errors of numerical solutions with different step sizes

h	τ	$E_\infty(h, \tau)$	$\dfrac{E_\infty(h, \tau)}{E_\infty(h/2, \tau/2)}$	$\log_2 \dfrac{E_\infty(h, \tau)}{E_\infty(h/2, \tau/2)}$
1/10	1/10	1.105e−3	6.0131	2.5881
1/20	1/20	1.838e−4	5.7700	2.5286
1/40	1/40	3.185e−5	5.2153	2.3828
1/80	1/80	6.108e−6	4.7247	2.2402
1/160	1/160	1.293e−6	4.3984	2.1370
1/320	1/320	2.939e−7	4.1963	2.0691
1/640	1/640	7.004e−8	4.0934	2.0333
1/1280	1/1280	1.711e−8		

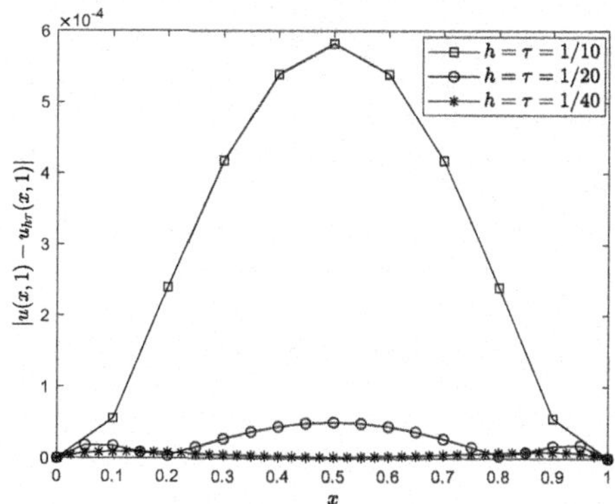

Fig. 3.30 (Example 3.9) The numerical error profiles calculated with different step sizes at $t = 1$

3.8   The Derivative Boundary Value Problem

Consider the following problem with the derivative boundary value conditions

$$
\begin{cases}
\dfrac{\partial u}{\partial t} - a\dfrac{\partial^2 u}{\partial x^2} = f(x, t), & 0 < x < L, \quad 0 < t \leqslant T, & (3.116a) \\[2mm]
u(x, 0) = \varphi(x), & 0 \leqslant x \leqslant L, & (3.116b) \\[2mm]
\left[-a\dfrac{\partial u}{\partial x} + \lambda_1(t)u \right]\Bigg|_{x=0} = \alpha(t), & 0 < t \leqslant T, & (3.116c) \\[2mm]
\left[a\dfrac{\partial u}{\partial x} + \lambda_2(t)u \right]\Bigg|_{x=L} = \beta(t), & 0 < t \leqslant T, & (3.116d)
\end{cases}
$$

where $\lambda_1(t), \lambda_2(t), \alpha(t), \beta(t), \varphi(x)$ and $f(x, t)$ are given functions. Generally, $\lambda_1(t) \geqslant 0, \lambda_2(t) \geqslant 0$.

Applying the Taylor expansion to the differential equation (3.116a), we have

$$\delta_t U_i^{k+\frac{1}{2}} - a\delta_x^2 U_i^{k+\frac{1}{2}} = f(x_i, t_{k+\frac{1}{2}}) + O(\tau^2 + h^2), \quad 1 \leqslant i \leqslant m-1, \quad 0 \leqslant k \leqslant n-1.$$

$$(3.117)$$

Using the initial value condition (3.116b), we have

$$u(x_i, 0) = \varphi(x_i), \quad 0 \leqslant i \leqslant m.$$

Now we consider the discretization of the boundary value condition (3.116c). By means of (3.116c), we have

$$a\frac{\partial u}{\partial x}(x_0, t) = \lambda_1(t)u(x_0, t) - \alpha(t). \tag{3.118}$$

Using (3.116a), we have

$$a\frac{\partial^2 u}{\partial x^2}(x_0, t) = \frac{\partial u}{\partial t}(x_0, t) - f(x_0, t). \tag{3.119}$$

With the application of the Taylor expansion and (3.118), we obtain

$$a\frac{\partial u(x_0, t_k)}{\partial x} = a\frac{U_1^k - U_0^k}{h} - \frac{ah}{2}\frac{\partial^2 u(x_0, t_k)}{\partial x^2} + O(h^2) = \lambda_1(t_k)U_0^k - \alpha(t_k).$$

Taking the weighted average of $t = t_k$ and $t = t_{k+1}$, respectively, in the above equality, and applying (3.119), we obtain

$$\frac{a}{2}\left(\frac{U_1^k - U_0^k}{h} + \frac{U_1^{k+1} - U_0^{k+1}}{h}\right)$$

$$= \frac{1}{2}\left[\lambda_1(t_k)U_0^k + \lambda_1(t_{k+1})U_0^{k+1}\right] - \frac{1}{2}\left[\alpha(t_k) + \alpha(t_{k+1})\right]$$

$$+ \frac{ah}{4}\left[\frac{\partial^2 u(x_0, t_k)}{\partial x^2} + \frac{\partial^2 u(x_0, t_{k+1})}{\partial x^2}\right] + O(h^2)$$

$$= \frac{1}{2}\left[\lambda_1(t_k)U_0^k + \lambda_1(t_{k+1})U_0^{k+1}\right] - \frac{1}{2}\left[\alpha(t_k) + \alpha(t_{k+1})\right]$$

$$+ \frac{h}{2}\left[a\frac{\partial^2 u(x_0, t_{k+\frac{1}{2}})}{\partial x^2} + O(\tau^2)\right] + O(h^2)$$

$$= \frac{1}{2}\left[\lambda_1(t_k)U_0^k + \lambda_1(t_{k+1})U_0^{k+1}\right] - \frac{1}{2}[\alpha(t_k) + \alpha(t_{k+1})]$$

$$+ \frac{h}{2}\left[\delta_t U_0^{k+\frac{1}{2}} - f(x_0, t_{k+\frac{1}{2}}) + O(\tau^2)\right] + O(h^2), \quad 0 \leqslant k \leqslant n-1. \qquad (3.120)$$

Similarly, it follows from (3.116d) that

$$a\frac{\partial u}{\partial x}(x_m, t) = -\lambda_2(t)u(x_m, t) + \beta(t)$$

and from (3.116a) that

$$a\frac{\partial^2 u}{\partial x^2}(x_m, t) = \frac{\partial u}{\partial t}(x_m, t) - f(x_m, t).$$

It can be derived from the above two equalities that

$$\frac{a}{2}\left(\frac{U_m^k - U_{m-1}^k}{h} + \frac{U_m^{k+1} - U_{m-1}^{k+1}}{h}\right)$$

$$= -\frac{1}{2}\left[\lambda_2(t_k)U_m^k + \lambda_2(t_{k+1})U_m^{k+1}\right] + \frac{1}{2}[\beta(t_k) + \beta(t_{k+1})]$$

$$- \frac{h}{2}\left[\delta_t U_m^{k+\frac{1}{2}} - f(x_m, t_{k+\frac{1}{2}}) + O(\tau^2)\right] + O(h^2), \quad 0 \leqslant k \leqslant n-1. \quad (3.121)$$

Omitting the small terms in (3.117), (3.120), and (3.121), and replacing U_i^k with u_i^k, a difference scheme reads

$$\left\{
\begin{aligned}
&\delta_t u_i^{k+\frac{1}{2}} - a\delta_x^2 u_i^{k+\frac{1}{2}} = f(x_i, t_{k+\frac{1}{2}}), \quad 1 \leqslant i \leqslant m-1, \quad 0 \leqslant k \leqslant n-1, && (3.122a) \\[2mm]
&u_i^0 = \varphi(x_i), \quad 0 \leqslant i \leqslant m, && (3.122b) \\[2mm]
&-aD_x u_0^{k+\frac{1}{2}} + \frac{1}{2}\left[\lambda_1(t_k)u_0^k + \lambda_1(t_{k+1})u_0^{k+1}\right] + \frac{h}{2}\delta_t u_0^{k+\frac{1}{2}} \\[1mm]
&\quad = \frac{1}{2}[\alpha(t_k) + \alpha(t_{k+1})] + \frac{h}{2}f(x_0, t_{k+\frac{1}{2}}), \quad 0 \leqslant k \leqslant n-1, && (3.122c) \\[2mm]
&aD_{\bar{x}} u_m^{k+\frac{1}{2}} + \frac{1}{2}\left[\lambda_2(t_k)u_m^k + \lambda_2(t_{k+1})u_m^{k+1}\right] + \frac{h}{2}\delta_t u_m^{k+\frac{1}{2}} \\[1mm]
&\quad = \frac{1}{2}[\beta(t_k) + \beta(t_{k+1})] + \frac{h}{2}f(x_m, t_{k+\frac{1}{2}}), \quad 0 \leqslant k \leqslant n-1. && (3.122d)
\end{aligned}
\right.$$

The difference equations (3.122c)–(3.122d) can also be rewritten as

$$
\begin{cases}
\delta_t u_0^{k+\frac{1}{2}} - \frac{2}{h}\left\{ aD_x u_0^{k+\frac{1}{2}} - \frac{1}{2}\left[\lambda_1(t_k)u_0^k + \lambda_1(t_{k+1})u_0^{k+1} \right] \right. \\
\left. + \frac{1}{2}\left[\alpha(t_k) + \alpha(t_{k+1}) \right] \right\} = f(x_0, t_{k+\frac{1}{2}}), \quad 0 \leqslant k \leqslant n-1, \\[2mm]
\delta_t u_m^{k+\frac{1}{2}} - \frac{2}{h}\left\{ \frac{1}{2}\left[\beta(t_k) + \beta(t_{k+1}) \right] - \frac{1}{2}\left[\lambda_2(t_k)u_m^k + \lambda_2(t_{k+1})u_m^{k+1} \right] \right. \\
\left. - aD_{\bar{x}}u_m^{k+\frac{1}{2}} \right\} = f(x_m, t_{k+\frac{1}{2}}), \quad 0 \leqslant k \leqslant n-1.
\end{cases}
$$

3.9 Summary and Extension

In this chapter, we introduce several difference schemes for the initial-boundary value problem of one-dimensional parabolic equation. First, a priori estimate for the Dirichlet initial-boundary value problem is analyzed by the energy method, and then five difference schemes are introduced in detail, which are the forward Euler scheme, backward Euler scheme, Richardson scheme, Crank-Nicolson scheme, and compact difference scheme, respectively. The forward Euler scheme is a two-level explicit one, which is conditionally stable with the stability condition $r \leqslant 1/2$. The Richardson scheme is a three-level explicit one, which is unstable for any step ratio r. The backward Euler scheme, the Crank-Nicolson scheme, and the compact difference scheme are unconditionally stable two-level schemes, and a tridiagonal system of linear equations needs to be solved at each time level. However, their convergence rates are different. In theoretical analysis, the maximum principle is utilized for the analysis of the forward Euler scheme and backward Euler scheme, and the energy method is used for the analysis of the Crank-Nicolson scheme and the compact difference scheme. The proof for the instability of the forward Euler scheme and the Richardson scheme is based on a counter-example. The Crank-Nicolson extrapolation method is introduced, and the difference scheme for the derivative boundary value problem is also given. In addition, the forward Euler scheme, backward Euler scheme, and Crank-Nicolson scheme for solving nonlinear parabolic equations are also covered. The convergence for the first two difference schemes is analyzed by the maximum principle. The convergence of the Crank-Nicolson scheme is analyzed by the energy method.

It is suggested that readers can compare the numerical results of different difference schemes in terms of the computational cost and accuracy. Readers can also try to analyze the forward and backward Euler schemes with the energy method and analyze the Crank-Nicolson scheme and the compact difference scheme using the maximum principle. The optimal convergence rate of the corrected forward Euler method for solving one-dimensional parabolic equations with a special step ratio is referred to [5].

Although the Richardson scheme is unconditionally unstable, the following improved scheme

$$
\begin{cases}
\Delta_t u_i^k - a\delta_x^2 u_i^k = f(x_i, t_k), \quad 1 \leqslant i \leqslant m-1,\ 1 \leqslant k \leqslant n-1, \\[2mm]
\delta_t u_i^{n-\frac{1}{2}} - a\delta_x^2 u_i^{n-\frac{1}{2}} = f(x_i, t_{n-\frac{1}{2}}), \quad 1 \leqslant i \leqslant m-1, \\[2mm]
u_i^0 = \varphi(x_i), \quad 0 \leqslant i \leqslant m, \\[2mm]
u_0^k = \alpha(t_k), \quad u_m^k = \beta(t_k), \quad 1 \leqslant k \leqslant n
\end{cases}
$$

is unconditionally stable and second-order convergent [4].

For the general linear parabolic equation with variable coefficients

$$
\frac{\partial u}{\partial t} - a(x, t)\frac{\partial^2 u}{\partial x^2} - b(x, t)\frac{\partial u}{\partial x} - c(x, t)u = f(x, t),
$$

the forward Euler scheme

$$
D_t u_i^k - a(x_i, t_k)\delta_x^2 u_i^k - \frac{1}{2}b(x_i, t_k)(D_x u_i^k + D_{\bar{x}} u_i^k) - c(x_i, t_k)u_i^k = f(x_i, t_k),
$$

the backward Euler scheme

$$
D_{\bar{t}} u_i^k - a(x_i, t_k)\delta_x^2 u_i^k - \frac{1}{2}b(x_i, t_k)(D_x u_i^k + D_{\bar{x}} u_i^k) - c(x_i, t_k)u_i^k = f(x_i, t_k),
$$

and the Crank-Nicolson scheme

$$
\delta_t u_i^{k+\frac{1}{2}} - a(x_i, t_{k+\frac{1}{2}})\delta_x^2 u_i^{k+\frac{1}{2}} - \frac{1}{2}b(x_i, t_{k+\frac{1}{2}})\left(D_x u_i^{k+\frac{1}{2}} + D_{\bar{x}} u_i^{k+\frac{1}{2}} \right)
$$

$$
- c(x_i, t_{k+\frac{1}{2}})u_i^{k+\frac{1}{2}} = f(x_i, t_{k+\frac{1}{2}})
$$

can be established. The maximum principle can be used for the numerical analysis
of the forward Euler scheme and the backward Euler scheme; meanwhile, the energy
method can be used for the analysis of the Crank-Nicolson scheme.

For the parabolic equation with variable coefficients

$$
r(x, t)\frac{\partial u}{\partial t} - \frac{\partial^2 u}{\partial x^2} = f(x, t),
$$

a compact difference scheme

$$\frac{1}{12}\Big[r(x_{i-1}, t_{k+\frac{1}{2}})\delta_t u_{i-1}^{k+\frac{1}{2}} + 10 r(x_i, t_{k+\frac{1}{2}})\delta_t u_i^{k+\frac{1}{2}}$$

$$+ r(x_{i+1}, t_{k+\frac{1}{2}})\delta_t u_{i+1}^{k+\frac{1}{2}} \Big] - \delta_x^2 u_i^{k+\frac{1}{2}}$$

$$= \frac{1}{12}\Big[f(x_{i-1}, t_{k+\frac{1}{2}}) + 10 f(x_i, t_{k+\frac{1}{2}}) + f(x_{i+1}, t_{k+\frac{1}{2}}) \Big]$$

can be established. The energy method can be used for the analysis of the difference scheme with the convergence order $O(\tau^2 + h^4)$, see [2] for the details. The compact difference scheme for the derivative boundary value problems is referred to [1, 3].

3.10   Exercise

3.1 Consider the following initial-boundary value problem

$$\begin{cases} \dfrac{\partial u}{\partial t} - \dfrac{\partial^2 u}{\partial x^2} + \dfrac{\partial u}{\partial x} + u = f(x, t), & 0 < x < L, \quad 0 < t \leqslant T, \\ u(x, 0) = \varphi(x), & 0 \leqslant x \leqslant L, \\ u(0, t) = 0, \quad u(L, t) = 0, & 0 < t \leqslant T, \end{cases}$$

where $\varphi(0) = \varphi(L) = 0$. Derive the following forward Euler scheme

$$\begin{cases} D_t u_i^k - \delta_x^2 u_i^k + \Delta_x u_i^k + u_i^k = f(x_i, t_k), & 1 \leqslant i \leqslant m - 1, \quad 0 \leqslant k \leqslant n - 1, \\ u_i^0 = \varphi(x_i), & 0 \leqslant i \leqslant m, \\ u_0^k = 0, \quad u_m^k = 0, & 1 \leqslant k \leqslant n, \end{cases}$$

where $\Delta_x u_i^k = \dfrac{1}{2h}(u_{i+1}^k - u_{i-1}^k)$. Denote $r = \dfrac{\tau}{h^2}$.

(1) Show the expression of the local truncation error.
(2) Suppose $r \leqslant 1/2$. Show a priori estimate of the difference solution.
(3) Suppose $r \leqslant 1/2$. Show the convergence of the difference solution.

3.2 Consider the following problem:

$$\begin{cases} \dfrac{\partial u}{\partial t} - a(x, t)\dfrac{\partial^2 u}{\partial x^2} = f(x, t), & 0 < x < L, \quad 0 < t \leqslant T, \\ u(x, 0) = \varphi(x), & 0 \leqslant x \leqslant L, \\ u(0, t) = \alpha(t), \quad u(L, t) = \beta(t), & 0 < t \leqslant T, \end{cases}$$

where $a(x, t) \geqslant a_0 > 0$. Derive the backward Euler scheme

$$\begin{cases} D_{\bar{t}} u_i^k - a(x_i, t_k)\delta_x^2 u_i^k = f(x_i, t_k), & 1 \leqslant i \leqslant m - 1, \quad 1 \leqslant k \leqslant n, \\ u_i^0 = \varphi(x_i), & 0 \leqslant i \leqslant m, \\ u_0^k = \alpha(t_k), \quad u_m^k = \beta(t_k), & 1 \leqslant k \leqslant n. \end{cases}$$

(1) Show the expression of the local truncation error.
(2) Let $\alpha(t) \equiv 0$, $\beta(t) \equiv 0$ and give a priori estimate of the solution.
(3) Show that the difference scheme is uniquely solvable and unconditionally stable.

3.3 For the parabolic equation

$$\frac{\partial u}{\partial t} - a\frac{\partial^2 u}{\partial x^2} = f(x, t),$$

try to derive the local truncation error of the following Du Fort-Frankel scheme

$$\frac{u_i^{k+1} - u_i^{k-1}}{2\tau} - a\frac{u_{i-1}^k - (u_i^{k+1} + u_i^{k-1}) + u_{i+1}^k}{h^2} = f(x_i, t_k).$$

3.4 For the problem

$$\begin{cases} \dfrac{\partial u}{\partial t} - \dfrac{\partial^2 u}{\partial x^2} = f(x, t), & 0 < x < L, \quad 0 < t \leqslant T, \\ u(x, 0) = \varphi(x), & 0 \leqslant x \leqslant L, \\ u(0, t) = \alpha(t), \quad u(L, t) = \beta(t), & 0 < t \leqslant T, \end{cases}$$

where $\varphi(0) = \alpha(0)$, $\varphi(L) = \beta(0)$. Derive the following difference scheme

$$\begin{cases} \dfrac{u_i^1 - u_i^0}{\tau} - \dfrac{u_{i-1}^1 - 2u_i^1 + u_{i+1}^1}{h^2} = f(x_i, t_1), & 1 \leqslant i \leqslant m - 1, \\ \dfrac{3u_i^k - 4u_i^{k-1} + u_i^{k-2}}{2\tau} - \dfrac{u_{i-1}^k - 2u_i^k + u_{i+1}^k}{h^2} = f(x_i, t_k), \\ \qquad 1 \leqslant i \leqslant m - 1, \quad 2 \leqslant k \leqslant n, \\ u_i^0 = \varphi(x_i), & 0 \leqslant i \leqslant m, \\ u_0^k = \alpha(t_k), \quad u_m^k = \beta(t_k), & 1 \leqslant k \leqslant n. \end{cases}$$

(1) Analyze the local truncation error of the difference scheme.

(2) Let $\alpha(t) \equiv 0, \beta(t) \equiv 0$. Show the difference solution has the following priori estimate

$$|u^k|_1^2 \leqslant |u^0|_1^2 + \tau \sum_{l=1}^{k} \|f^l\|^2, \quad 1 \leqslant k \leqslant n.$$

(Hint: Noticing

$$\frac{3u_i^k - 4u_i^{k-1} + u_i^{k-2}}{2\tau} = \frac{3}{2} D_{\bar{t}} u_i^k - \frac{1}{2} D_{\bar{t}} u_i^{k-1},$$

and taking an inner product with $D_{\bar{t}} u^k$ on both sides of the difference scheme.)

(3) Show the convergence of the difference solution.

3.5 Consider the following problem:

$$\begin{cases} \dfrac{\partial u}{\partial t} - \dfrac{\partial^2 u}{\partial x^2} = f(x, t), & 0 < x < L, \quad 0 < t \leqslant T, \\ u(x, 0) = \varphi(x), & 0 \leqslant x \leqslant L, \\ \dfrac{\partial u}{\partial x}(0, t) = 0, \quad \dfrac{\partial u}{\partial x}(L, t) = 0, & 0 < t \leqslant T, \end{cases}$$

where $\varphi'(0) = \varphi'(L) = 0$. Derive the following Crank-Nicolson scheme

$$\begin{cases} \delta_t u_i^{k+\frac{1}{2}} - \delta_x^2 u_i^{k+\frac{1}{2}} = f(x_i, t_{k+\frac{1}{2}}), & 1 \leqslant i \leqslant m-1, \quad 0 \leqslant k \leqslant n-1, \\ u_i^0 = \varphi(x_i), & 1 \leqslant i \leqslant m-1, \\ D_x u_0^k = 0, \quad D_{\bar{x}} u_m^k = 0, & 0 \leqslant k \leqslant n. \end{cases}$$

(1) Show a priori estimate of the difference solution.
(2) Show that the difference scheme is uniquely solvable.

3.6 For the parabolic equation

$$\frac{\partial u}{\partial t} - \frac{\partial^2 u}{\partial x^2} = 0,$$

derive the following difference scheme

$$\delta_t u_i^{k+\frac{1}{2}} - \left[\theta \delta_x^2 u_i^{k+1} + (1-\theta) \delta_x^2 u_i^k \right] = 0,$$

where θ is a constant and $\theta \in [0, 1]$. Try to select θ such that the local truncation error is $O(\tau^2 + h^4)$, and compare the obtained difference scheme with the compact difference scheme (3.63a).

3.7 For the nonlinear parabolic differential equation (3.71a), derive the following difference scheme:

$$\frac{u_i^{k+1} - u_i^{k-1}}{2\tau} - a(u_i^k)\left(\frac{1}{4}\delta_x^2 u_i^{k+1} + \frac{1}{2}\delta_x^2 u_i^k + \frac{1}{4}\delta_x^2 u_i^{k-1}\right) = f(x_i, t_k).$$

Show the local truncation error.

3.8 Apply the forward Euler scheme (3.11) to compute the following problem:

$$\begin{cases} \dfrac{\partial u}{\partial t} - 2\dfrac{\partial^2 u}{\partial x^2} = -e^x\left[\cos\left(\frac{1}{2} - t\right) + 2\sin\left(\frac{1}{2} - t\right)\right], \\[2mm] \quad 0 < x < 1, \quad 0 < t \leqslant 1, \\[2mm] u(x, 0) = e^x \sin\dfrac{1}{2}, \quad 0 \leqslant x \leqslant 1, \\[2mm] u(0, t) = \sin\left(\frac{1}{2} - t\right), \quad u(1, t) = e\sin\left(\frac{1}{2} - t\right), \quad 0 < t \leqslant 1. \end{cases}$$

The exact solution of the problem is $u(x, t) = e^x \sin\left(\dfrac{1}{2} - t\right)$.

(1) Fill in Tables 3.24, 3.25, 3.26, 3.27, and 3.28 and analyze the results.
(2) Take the step sizes $(h, \tau) = (1/10, 1/400)$, $(1/20, 1/1600)$, $(1/40, 1/6400)$, and draw the numerical error profiles at $t = 1$ in the same coordinate system.

3.9 Apply the backward Euler scheme (3.25) to compute the problem in Exercise 3.8:

(1) Fill in Tables 3.29 and 3.30.

Table 3.24 (Exercise 3.8) Numerical results at part of node points ($h = 1/100$, $\tau = 1/100$)

k	(x, t)	NS	ES	|ES−NS|
1	$(0.5, 0.01)$			
2	$(0.5, 0.02)$			
3	$(0.5, 0.03)$			
4	$(0.5, 0.04)$			
5	$(0.5, 0.05)$			
6	$(0.5, 0.06)$			
7	$(0.5, 0.07)$			
8	$(0.5, 0.08)$			
9	$(0.5, 0.09)$			
10	$(0.5, 0.10)$			

Table 3.25 (Exercise 3.8) Numerical results at part of node points ($h = 1/100$, $\tau = 1/1000$)

k	(x, t)	NS	ES	$|$ES$-$NS$|$
1	(0.5, 0.001)			
2	(0.5, 0.002)			
3	(0.5, 0.003)			
4	(0.5, 0.004)			
5	(0.5, 0.005)			
6	(0.5, 0.006)			
7	(0.5, 0.007)			
8	(0.5, 0.008)			
9	(0.5, 0.009)			
10	(0.5, 0.010)			

Table 3.26 (Exercise 3.8) Numerical results at part of node points ($h = 1/100$, $\tau = 1/10000$)

k	(x, t)	NS	ES	$|$ES$-$NS$|$
1	(0.5, 0.0001)			
2	(0.5, 0.0002)			
3	(0.5, 0.0003)			
4	(0.5, 0.0004)			
5	(0.5, 0.0005)			
6	(0.5, 0.0006)			
7	(0.5, 0.0007)			
8	(0.5, 0.0008)			
9	(0.5, 0.0009)			
10	(0.5, 0.0010)			

(2) Take the step sizes $(h, \tau) = (1/10, 1/100)$, $(1/20, 1/400)$, $(1/40, 1/1600)$, and draw the error profiles of numerical solutions at $t = 1$ in the same coordinate system.

3.10 Apply the Richardson scheme (3.32) and (3.33) to compute the problem in Exercise 3.8. Fill in Tables 3.31 and 3.32 and analyze the calculated results.

3.11 Apply the Crank-Nicolson scheme (3.41) to compute the problem in Exercise 3.8:

(1) Fill in Tables 3.33, 3.34, 3.35, and 3.36 and analyze the calculated results.

(2) Take the step sizes $(h, \tau) = (1/10, 1/10)$, $(1/20, 1/20)$, $(1/40, 1/40)$, and draw the error profiles of numerical solutions at $t = 1$ in the same coordinate system.

(3) Using the extrapolation to numerical solutions in Tables 3.33 and 3.34 to compute new approximate solutions and observe the practical errors of these approximations.

3.12 Apply the compact difference scheme (3.63) to compute the problem in Exercise 3.8:

(1) Fill in Tables 3.37, 3.38, 3.39, and 3.40 and analyze the computed numerical results.

Table 3.27 (Exercise 3.8)
Numerical results at part of
node points
($h = 1/100$, $\tau = 1/40000$)

k	(x, t)	NS	ES	$|ES-NS|$
1	$(0.5, 0.000025)$			
2	$(0.5, 0.000050)$			
3	$(0.5, 0.000075)$			
4	$(0.5, 0.000100)$			
5	$(0.5, 0.000125)$			
6	$(0.5, 0.000150)$			
7	$(0.5, 0.000175)$			
8	$(0.5, 0.000200)$			
9	$(0.5, 0.000225)$			
10	$(0.5, 0.000250)$			
1000	$(0.5, 0.025000)$			
5000	$(0.5, 0.125000)$			
10000	$(0.5, 0.250000)$			
15000	$(0.5, 0.375000)$			
20000	$(0.5, 0.500000)$			
25000	$(0.5, 0.625000)$			
30000	$(0.5, 0.750000)$			
35000	$(0.5, 0.875000)$			
40000	$(0.5, 1.000000)$			

Table 3.28 (Exercise 3.8)
The maximum errors of
numerical solutions

h	τ	$E_\infty(h, \tau)$	$E_\infty(2h, 4\tau)/E_\infty(h, \tau)$
1/10	1/400		
1/20	1/1600		
1/40	1/6400		

Table 3.29 (Exercise 3.9)
Numerical results at part of
node points
($h = 1/100$, $\tau = 1/10000$)

(x, t)	NS	ES	$|ES-NS|$
$(0.5, 0.1)$			
$(0.5, 0.2)$			
$(0.5, 0.3)$			
$(0.5, 0.4)$			
$(0.5, 0.5)$			
$(0.5, 0.6)$			
$(0.5, 0.7)$			
$(0.5, 0.8)$			
$(0.5, 0.9)$			
$(0.5, 1.0)$			

Table 3.30 (Exercise 3.9) Numerical results at part of node points ($h = 1/100$, $\tau = 1/40000$)

(x, t)	NS	ES	$|ES-NS|$
$(0.5, 0.1)$			
$(0.5, 0.2)$			
$(0.5, 0.3)$			
$(0.5, 0.4)$			
$(0.5, 0.5)$			
$(0.5, 0.6)$			
$(0.5, 0.7)$			
$(0.5, 0.8)$			
$(0.5, 0.9)$			
$(0.5, 1.0)$			

Table 3.31 (Exercise 3.10) Numerical results at part of node points ($h = 1/100$, $\tau = 1/100$)

k	(x, t)	NS	ES	$|ES-NS|$
1	$(0.5, 0.01)$			
2	$(0.5, 0.02)$			
3	$(0.5, 0.03)$			
4	$(0.5, 0.04)$			
5	$(0.5, 0.05)$			
6	$(0.5, 0.06)$			
7	$(0.5, 0.07)$			
8	$(0.5, 0.08)$			
9	$(0.5, 0.09)$			
10	$(0.5, 0.10)$			

Table 3.32 (Exercise 3.10) Numerical results at part of node points ($h = 1/100$, $\tau = 1/1000$)

k	(x, t)	NS	ES	$|ES-NS|$
1	$(0.5, 0.001)$			
2	$(0.5, 0.002)$			
3	$(0.5, 0.003)$			
4	$(0.5, 0.004)$			
5	$(0.5, 0.005)$			
6	$(0.5, 0.006)$			
7	$(0.5, 0.007)$			
8	$(0.5, 0.008)$			
9	$(0.5, 0.009)$			
10	$(0.5, 0.010)$			

Table 3.33 (Exercise 3.11) Numerical results at part of node points ($h = 1/100, \tau = 1/100$)

(x, t)	NS	ES	$|ES-NS|$
(0.5, 0.1)			
(0.5, 0.2)			
(0.5, 0.3)			
(0.5, 0.4)			
(0.5, 0.5)			
(0.5, 0.6)			
(0.5, 0.7)			
(0.5, 0.8)			
(0.5, 0.9)			
(0.5, 1.0)			

Table 3.34 (Exercise 3.11) Numerical results at part of node points ($h = 1/200, \tau = 1/200$)

(x, t)	NS	ES	$|ES-NS|$
(0.5, 0.1)			
(0.5, 0.2)			
(0.5, 0.3)			
(0.5, 0.4)			
(0.5, 0.5)			
(0.5, 0.6)			
(0.5, 0.7)			
(0.5, 0.8)			
(0.5, 0.9)			
(0.5, 1.0)			

Table 3.35 (Exercise 3.11) Numerical results at part of node points ($h = 1/200, \tau = 1/2000$)

(x, t)	NS	ES	$|ES-NS|$
(0.5, 0.1)			
(0.5, 0.2)			
(0.5, 0.3)			
(0.5, 0.4)			
(0.5, 0.5)			
(0.5, 0.6)			
(0.5, 0.7)			
(0.5, 0.8)			
(0.5, 0.9)			
(0.5, 1.0)			

Table 3.36 (Exercise 3.11) The maximum errors of the numerical solutions

h	τ	$E_\infty(h, \tau)$	$E_\infty(2h, 2\tau)/E_\infty(h, \tau)$
1/10	1/10		
1/20	1/20		
1/40	1/40		
1/80	1/80		
1/160	1/160		

Table 3.37 (Exercise 3.12) Numerical results at part of node points ($h = 1/10$, $\tau = 1/100$)

(x, t)	NS	ES	$|ES-NS|$
(0.5, 0.1)			
(0.5, 0.2)			
(0.5, 0.3)			
(0.5, 0.4)			
(0.5, 0.5)			
(0.5, 0.6)			
(0.5, 0.7)			
(0.5, 0.8)			
(0.5, 0.9)			
(0.5, 1.0)			

Table 3.38 (Exercise 3.12) Numerical results at part of node points ($h = 1/20$, $\tau = 1/400$)

(x, t)	NS	ES	$|ES-NS|$
(0.5, 0.1)			
(0.5, 0.2)			
(0.5, 0.3)			
(0.5, 0.4)			
(0.5, 0.5)			
(0.5, 0.6)			
(0.5, 0.7)			
(0.5, 0.8)			
(0.5, 0.9)			
(0.5, 1.0)			

Table 3.39 (Exercise 3.12)
Numerical results at part of
node points
($h = 1/200$, $\tau = 1/400$)

(x, t)	NS	ES	$|ES-NS|$
$(0.5, 0.1)$			
$(0.5, 0.2)$			
$(0.5, 0.3)$			
$(0.5, 0.4)$			
$(0.5, 0.5)$			
$(0.5, 0.6)$			
$(0.5, 0.7)$			
$(0.5, 0.8)$			
$(0.5, 0.9)$			
$(0.5, 1.0)$			

Table 3.40 (Exercise 3.12)
The maximum errors of the
numerical solutions

h	τ	$E_\infty(h, \tau)$	$E_\infty(2h, 4\tau)/E_\infty(h, \tau)$
1/10	1/10		
1/20	1/400		
1/40	1/1600		
1/80	1/6400		
1/160	1/25600		

(2) Take the step sizes $(h, \tau) = (1/10, 1/100)$, $(1/20, 1/400)$, $(1/40, 1/1600)$, and draw the error profiles of numerical solutions at $t = 1$ in the same coordinate system.

References

1. Gao, G.H., Sun, Z.Z.: Compact difference schemes for heat equation with Neumann boundary conditions (II). Numer. Methods Partial Differ. Equ. **29**, 1459–1486 (2013)
2. Sun, Z.Z.: An unconditionally stable and $O(\tau^2+h^4)$ order L_∞ convergent difference scheme for parabolic equations with variable coefficients. Numer. Methods Partial Differ. Equ. **17**, 619–631 (2001)
3. Sun, Z.Z.: Compact difference schemes for heat equation with the Neumann boundary conditions. Numer. Methods Partial Differ. Equ. **25**, 1320–1341 (2009)
4. Zhang, Q.F.: Error estimates of compact and hybrid Richardson schemes for the parabolic equation. Appl. Math. Lett. **153**, 109078 (2024)
5. Zhang, Q.F., Zhang, J.Y., Sun, Z.Z.: Optimal convergence rate of the explicit Euler method for convection-diffusion equations. Appl. Math. Lett. **131**, 108048 (2022)

Chapter 4
Finite Difference Methods for Hyperbolic Equations

Many problems in fluid mechanics, such as those in aviation, meteorology, oceanography, and water conservancy, come down to hyperbolic equations and hyperbolic systems. These problems are often nonstationary and nonlinear, coupled with viscosity, turbulence, and shock waves (discontinuous interfaces). The resulting complex phenomena make these problems particularly challenging to solve. Given that the complexity of the differential equations may obscure the essence of the numerical solution, this chapter focuses on finite difference schemes specifically designed to solve the wave equation. The convergence and stability of these schemes are also discussed.

4.1 The Dirichlet Initial-Boundary Value Problem

Taking the wave equation as a model, the difference method for the Dirichlet initial-boundary value problem (the first boundary value problem) is discussed. Consider

$$
\begin{cases}
\dfrac{\partial^2 u}{\partial t^2} - c^2 \dfrac{\partial^2 u}{\partial x^2} = f(x,t), & 0 < x < L, \quad 0 < t \leqslant T, & \text{(4.1a)} \\[3mm]
u(x,0) = \varphi(x), \quad \dfrac{\partial u}{\partial t}(x,0) = \psi(x), & 0 \leqslant x \leqslant L, & \text{(4.1b)} \\[3mm]
u(0,t) = \alpha(t), \quad u(L,t) = \beta(t), & 0 < t \leqslant T, & \text{(4.1c)}
\end{cases}
$$

where c is a positive constant (generally called wave speed), and $f(x,t)$, $\varphi(x)$, $\psi(x)$, $\alpha(t)$, and $\beta(t)$ are all given functions. These functions are assumed to satisfy the compatibility conditions: $\varphi(0) = \alpha(0)$, $\varphi(L) = \beta(0)$, $\psi(0) = \alpha'(0)$, and $\psi(L) = \beta'(0)$. Equation (4.1b) specifies the initial value conditions, while Eq. (4.1c) defines the boundary value conditions.

© Science Press 2026

Z.-Z. Sun et al., *Numerical Solutions to Partial Differential Equations with Finite Difference Methods*, Springer Asia Pacific Mathematics Series 9, https://doi.org/10.1007/978-981-95-5563-5_4

With respect to the solution of the homogeneous boundary value problem, the following priori estimate can be obtained.

Theorem 4.1 *Let $v(x, t)$ be the solution of the first boundary value problem of the hyperbolic equation*

$$
\begin{cases}
\dfrac{\partial^2 v}{\partial t^2} - c^2 \dfrac{\partial^2 v}{\partial x^2} = g(x, t), & 0 < x < L, \quad 0 < t \leqslant T, & (4.2a) \\[3mm]
v(x, 0) = \varphi(x), \quad \dfrac{\partial v}{\partial t}(x, 0) = \psi(x), & 0 \leqslant x \leqslant L, & (4.2b) \\[3mm]
v(0, t) = 0, \quad v(L, t) = 0, & 0 < t \leqslant T, & (4.2c)
\end{cases}
$$

where $\varphi(0) = \varphi(L) = \psi(0) = \psi(L) = 0$. Then we have

$$
\int_0^L \left[\frac{\partial v(x, t)}{\partial t} \right]^2 dx + c^2 \int_0^L \left[\frac{\partial v(x, t)}{\partial x} \right]^2 dx
$$

$$
\leqslant e^t \left\{ \int_0^L \psi^2(x)dx + c^2 \int_0^L [\varphi'(x)]^2 dx \right.
$$

$$
\left. + \int_0^t e^{-s} \left[\int_0^L g^2(x, s)dx \right] ds \right\}, \quad 0 < t \leqslant T.
$$

Proof Multiplying both sides of (4.2a) by $2\frac{\partial v}{\partial t}$ and integrating the result with respect to x, we obtain

$$
2 \int_0^L \frac{\partial^2 v(x, t)}{\partial t^2} \cdot \frac{\partial v(x, t)}{\partial t} dx - 2c^2 \int_0^L \frac{\partial^2 v(x, t)}{\partial x^2} \cdot \frac{\partial v(x, t)}{\partial t} dx
$$

$$
= 2 \int_0^L g(x, t) \frac{\partial v(x, t)}{\partial t} dx. \tag{4.3}
$$

Substituting

$$
2 \int_0^L \frac{\partial^2 v(x, t)}{\partial t^2} \cdot \frac{\partial v(x, t)}{\partial t} dx = \frac{d}{dt} \int_0^L \left[\frac{\partial v(x, t)}{\partial t} \right]^2 dx
$$

and

$$
-2 \int_0^L \frac{\partial^2 v(x, t)}{\partial x^2} \cdot \frac{\partial v(x, t)}{\partial t} dx
$$

$$
= -2 \frac{\partial v(x, t)}{\partial x} \cdot \frac{\partial v(x, t)}{\partial t} \bigg|_{x=0}^{L} + 2 \int_0^L \frac{\partial v(x, t)}{\partial x} \cdot \frac{\partial^2 v(x, t)}{\partial x \partial t} dx
$$

$$
= \frac{d}{dt} \int_0^L \left[\frac{\partial v(x, t)}{\partial x} \right]^2 dx
$$

into (4.3) derives to

$$\frac{d}{dt}\left\{\int_0^L \left[\frac{\partial v(x,t)}{\partial t}\right]^2 dx + c^2 \int_0^L \left[\frac{\partial v(x,t)}{\partial x}\right]^2 dx\right\} = 2\int_0^L g(x,t)\frac{\partial v(x,t)}{\partial t}dx.$$

Applying the Cauchy-Schwarz inequality to the right-hand side term of the above equality, we get

$$\frac{d}{dt}\left\{\int_0^L \left[\frac{\partial v(x,t)}{\partial t}\right]^2 dx + c^2 \int_0^L \left[\frac{\partial v(x,t)}{\partial x}\right]^2 dx\right\}$$

$$\leqslant \int_0^L \left[\frac{\partial v(x,t)}{\partial t}\right]^2 dx + \int_0^L g^2(x,t)dx.$$

Denote

$$E(t) = \int_0^L \left[\frac{\partial v(x,t)}{\partial t}\right]^2 dx + c^2 \int_0^L \left[\frac{\partial v(x,t)}{\partial x}\right]^2 dx, \quad G(t) = \int_0^L g^2(x,t)dx.$$

Then we have

$$\frac{dE(t)}{dt} \leqslant E(t) + G(t), \quad 0 < t \leqslant T.$$

Multiplying both sides of the above inequality by e^{-t} and then rearranging the terms, we have

$$\frac{d}{dt}\left[E(t)e^{-t}\right] \leqslant e^{-t}G(t), \quad 0 < t \leqslant T. \tag{4.4}$$

Integrating on both sides of (4.4) with respect to t produces

$$E(t)e^{-t} \leqslant E(0) + \int_0^t e^{-s}G(s)ds, \quad 0 < t \leqslant T.$$

Multiplying both sides of the above inequality by e^t derives to

$$E(t) \leqslant e^t\left[E(0) + \int_0^t e^{-s}G(s)ds\right], \quad 0 < t \leqslant T,$$

namely,

$$\int_0^L \left[\frac{\partial v(x,t)}{\partial t}\right]^2 \mathrm{d}x + c^2 \int_0^L \left[\frac{\partial v(x,t)}{\partial x}\right]^2 \mathrm{d}x$$

$$\leqslant \mathrm{e}^t \left\{\int_0^L \psi^2(x)\mathrm{d}x + c^2 \int_0^L [\varphi'(x)]^2 \mathrm{d}x\right.$$

$$\left. + \int_0^t \mathrm{e}^{-s}\left[\int_0^L g^2(x,s)\mathrm{d}x\right]\mathrm{d}s\right\}, \quad 0 < t \leqslant T.$$

$$\square$$

4.2   The Explicit Difference Scheme

4.2.1   Derivation of the Difference Scheme

In order to solve (4.1) by the difference method, firstly, the calculated region

$$D = \{(x,t) \mid 0 < x < L, \ 0 < t \leqslant T\}$$

is divided. Take two positive integers m and n and denote $x_i = ih$, $0 \leqslant i \leqslant m$, $t_k = k\tau$, $0 \leqslant k \leqslant n$, where $h = L/m$ and $\tau = T/n$ are called the spatial step size and the temporal step size, respectively. Denote $s = c\tau/h$, which is called the **step ratio**.

Two clusters of parallel lines

$$x = x_i, \quad 0 \leqslant i \leqslant m;$$

$$t = t_k, \quad 0 \leqslant k \leqslant n$$

are utilized to divide $\bar{D}$ into the rectangular grids, which is shown in Fig. 4.1. Denote $\Omega_h = \{x_i \mid 0 \leqslant i \leqslant m\}$, $\Omega_\tau = \{t_k \mid 0 \leqslant k \leqslant n\}$ and $\Omega_{h\tau} = \Omega_h \times \Omega_\tau$. We call (x_i, t_k) a node point. The node points located at $t = t_k$ are called node points at the k-th time level. For any grid function

$$v = \{v_i^k \mid 0 \leqslant i \leqslant m, \ 0 \leqslant k \leqslant n\}$$

defined on $\Omega_{h\tau}$, we follow the notation in Sect. 3.2. In addition, denote

$$\delta_t^2 v_i^k = \frac{1}{\tau^2}(v_i^{k+1} - 2v_i^k + v_i^{k-1}).$$

Fig. 4.1 Grid subdivision

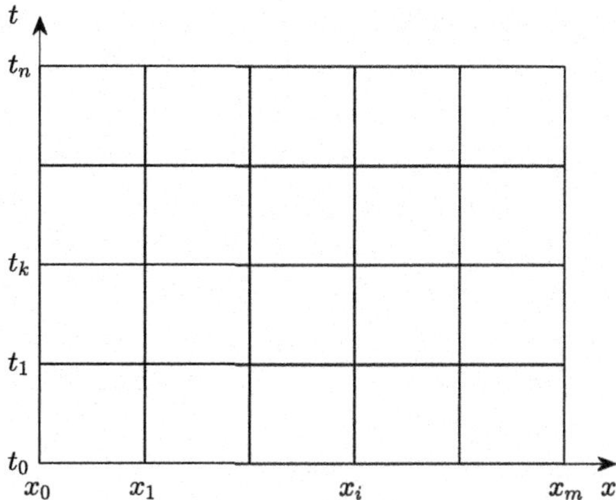

Define the grid function on $\Omega_{h\tau}$ by

$$U = \{U_i^k \mid 0 \leqslant i \leqslant m,\ 0 \leqslant k \leqslant n\},$$

where

$$U_i^k = u(x_i, t_k), \quad 0 \leqslant i \leqslant m, \quad 0 \leqslant k \leqslant n.$$

(I) It follows from (4.1a) that

$$\frac{\partial^2 u(x_i, t_0)}{\partial t^2} = c^2 \frac{\partial^2 u(x_i, t_0)}{\partial x^2} + f(x_i, t_0), \quad 1 \leqslant i \leqslant m - 1. \tag{4.5}$$

Using Lemma 1.2, we have

$$\frac{\partial^2 u(x_i, t_0)}{\partial t^2} = \frac{2}{\tau}\left[\delta_t U_i^{\frac{1}{2}} - \frac{\partial u(x_i, t_0)}{\partial t}\right] - \frac{\tau}{3}\frac{\partial^3 u(x_i, \eta_{i0})}{\partial t^3}, \quad \eta_{i0} \in (t_0, t_1),$$

$$\frac{\partial^2 u(x_i, t_0)}{\partial x^2} = \delta_x^2 U_i^0 - \frac{h^2}{12}\frac{\partial^4 u(\xi_{i0}, t_0)}{\partial x^4}, \quad x_{i-1} < \xi_{i0} < x_{i+1}.$$

Substituting the above two equalities into (4.5) generates

$$\frac{2}{\tau}\left[\delta_t U_i^{\frac{1}{2}} - \frac{\partial u(x_i, t_0)}{\partial t}\right] - c^2\delta_x^2 U_i^0 = f(x_i, t_0) + (R_1)_i^0, \quad 1 \leqslant i \leqslant m - 1, \tag{4.6}$$

where

$$(R_1)_i^0 = \frac{\tau}{3}\frac{\partial^3 u(x_i, \eta_{i0})}{\partial t^3} - \frac{c^2 h^2}{12}\frac{\partial^4 u(\xi_{i0}, t_0)}{\partial x^4}, \quad 1 \leqslant i \leqslant m - 1. \tag{4.7}$$

(II) Considering Eq. (4.1a) at the node point (x_i, t_k), we have

$$\frac{\partial^2 u(x_i, t_k)}{\partial t^2} - c^2 \frac{\partial^2 u(x_i, t_k)}{\partial x^2} = f(x_i, t_k), \quad 1 \leqslant i \leqslant m-1, \quad 1 \leqslant k \leqslant n-1. \tag{4.8}$$

Substituting

$$\frac{\partial^2 u(x_i, t_k)}{\partial x^2} = \frac{u(x_{i-1}, t_k) - 2u(x_i, t_k) + u(x_{i+1}, t_k)}{h^2} - \frac{h^2}{12} \frac{\partial^4 u(\xi_{ik}, t_k)}{\partial x^4}$$

$$= \delta_x^2 U_i^k - \frac{h^2}{12} \frac{\partial^4 u(\xi_{ik}, t_k)}{\partial x^4}, \quad x_{i-1} < \xi_{ik} < x_{i+1}$$

and

$$\frac{\partial^2 u(x_i, t_k)}{\partial t^2} = \frac{u(x_i, t_{k-1}) - 2u(x_i, t_k) + u(x_i, t_{k+1})}{\tau^2} - \frac{\tau^2}{12} \frac{\partial^4 u(x_i, \eta_{ik})}{\partial t^4}$$

$$= \delta_t^2 U_i^k - \frac{\tau^2}{12} \frac{\partial^4 u(x_i, \eta_{ik})}{\partial t^4}, \quad t_{k-1} < \eta_{ik} < t_{k+1}$$

into (4.8), one has

$$\delta_t^2 U_i^k - c^2 \delta_x^2 U_i^k = f(x_i, t_k) + (R_1)_i^k, \quad 1 \leqslant i \leqslant m-1, \quad 1 \leqslant k \leqslant n-1, \tag{4.9}$$

where

$$(R_1)_i^k = \frac{\tau^2}{12} \frac{\partial^4 u(x_i, \eta_{ik})}{\partial t^4} - \frac{c^2 h^2}{12} \frac{\partial^4 u(\xi_{ik}, t_k)}{\partial x^4},$$

$$1 \leqslant i \leqslant m-1, \quad 1 \leqslant k \leqslant n-1. \tag{4.10}$$

By observing (4.7) and (4.10), there is a constant c_1 such that

$$\begin{cases} |(R_1)_i^0| \leqslant c_1(\tau + h^2), & 1 \leqslant i \leqslant m-1, \\ |(R_1)_i^k| \leqslant c_1(\tau^2 + h^2), & 1 \leqslant i \leqslant m-1, \quad 1 \leqslant k \leqslant n-1. \end{cases} \tag{4.11}$$

According to the initial value condition (4.1b), it yields

$$U_i^0 = \varphi(x_i), \quad 1 \leqslant i \leqslant m-1. \tag{4.12}$$

Noticing the boundary value condition (4.1c), we have

$$U_0^k = \alpha(t_k), \quad U_m^k = \beta(t_k), \quad 0 \leqslant k \leqslant n. \tag{4.13}$$

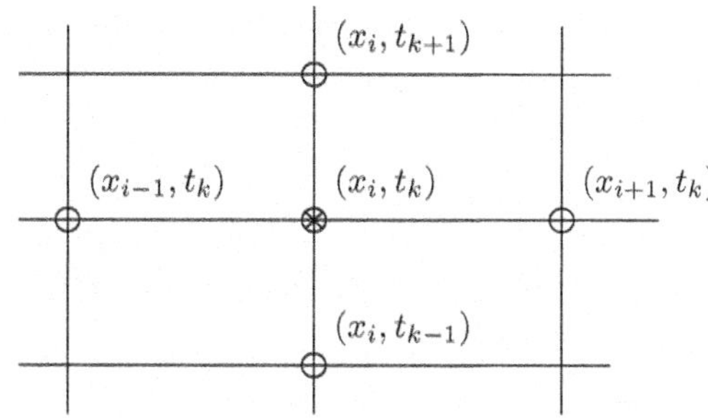

Fig. 4.2 The stencil of the explicit difference scheme (4.14)

Omitting the small terms in (4.6) and (4.9), noticing (4.12)–(4.13), and replacing U_i^k with u_i^k, a difference scheme for solving the problem (4.1) reads

$$
\begin{cases}
\dfrac{2}{\tau}\left[\delta_t u_i^{\frac{1}{2}} - \psi(x_i)\right] - c^2 \delta_x^2 u_i^0 = f(x_i, t_0), & 1 \leqslant i \leqslant m-1, & (4.14a) \\[2mm]
\delta_t^2 u_i^k - c^2 \delta_x^2 u_i^k = f(x_i, t_k), & 1 \leqslant i \leqslant m-1, \quad 1 \leqslant k \leqslant n-1, & (4.14b) \\[2mm]
u_i^0 = \varphi(x_i), & 1 \leqslant i \leqslant m-1, & (4.14c) \\[2mm]
u_0^k = \alpha(t_k), \quad u_m^k = \beta(t_k), & 0 \leqslant k \leqslant n. & (4.14d)
\end{cases}
$$

The stencil of the difference scheme (4.14) is shown in Fig. 4.2, which illustrates that the explicit difference scheme (4.14) is a three-time-level and five-point one.

4.2.2   Existence of the Difference Solution

Denote

$$
u^k = (u_0^k, u_1^k, \ldots, u_{m-1}^k, u_m^k).
$$

It follows from (4.14c) and (4.14d) that u^0 is determined.
Rewriting (4.14a) gives

$$
u_i^1 = u_i^0 + \tau \psi(x_i) + \frac{\tau^2}{2}\left[c^2 \delta_x^2 u_i^0 + f(x_i, t_0)\right], \quad 1 \leqslant i \leqslant m-1.
$$

In combination of (4.14d), one knows that u^1 is determined.
   Now suppose the values of u^{k-1} and u^k have been determined. From (4.14b), we have

$$
u_i^{k+1} = s^2 u_{i-1}^k + 2(1-s^2)u_i^k + s^2 u_{i+1}^k - u_i^{k-1} + \tau^2 f(x_i, t_k), \quad 1 \leqslant i \leqslant m-1,
$$

which reveals that the difference scheme (4.14) is an explicit one and the value of $\{u_i^{k+1} \mid 1 \leqslant i \leqslant m-1\}$ can be obtained directly. Consequently, for an arbitrary step ratio s, the difference scheme (4.14) is uniquely solvable.

4.2.3   Implementation of the Difference Scheme and Numerical Examples

Equation (4.14b) can be rewritten as the following matrix form:

$$
\begin{pmatrix} u_1^{k+1} \\ u_2^{k+1} \\ \vdots \\ u_{m-2}^{k+1} \\ u_{m-1}^{k+1} \end{pmatrix} = \begin{pmatrix} 2(1-s^2) & s^2 & & & \\ s^2 & 2(1-s^2) & s^2 & & \\ & \ddots & \ddots & \ddots & \\ & & s^2 & 2(1-s^2) & s^2 \\ & & & s^2 & 2(1-s^2) \end{pmatrix} \begin{pmatrix} u_1^k \\ u_2^k \\ \vdots \\ u_{m-2}^k \\ u_{m-1}^k \end{pmatrix}
$$

$$
- \begin{pmatrix} u_1^{k-1} \\ u_2^{k-1} \\ \vdots \\ u_{m-2}^{k-1} \\ u_{m-1}^{k-1} \end{pmatrix} + \begin{pmatrix} \tau^2 f(x_1, t_k) + s^2 u_0^k \\ \tau^2 f(x_2, t_k) \\ \vdots \\ \tau^2 f(x_{m-2}, t_k) \\ \tau^2 f(x_{m-1}, t_k) + s^2 u_m^k \end{pmatrix}, \quad 1 \leqslant k \leqslant n-1.
$$

Example 4.1 Apply the explicit difference scheme (4.14) to compute the problem

$$
\begin{cases} \dfrac{\partial^2 u}{\partial t^2} - \dfrac{\partial^2 u}{\partial x^2} = 0, & 0 < x < 1, \quad 0 < t \leqslant 1, \\[2mm] u(x,0) = e^x, \quad \dfrac{\partial u}{\partial t}(x,0) = e^x, & 0 \leqslant x \leqslant 1, \\[2mm] u(0,t) = e^t, \quad u(1,t) = e^{1+t}, & 0 < t \leqslant 1. \end{cases} \tag{4.15}
$$

The exact solution of the problem is $u(x,t) = e^{x+t}$.

Tables 4.1 and 4.2 list part of the numerical results calculated with the step sizes $h = 1/100$, $\tau = 1/200$ (the step ratio $s = 1/2$) and $h = 1/100$, $\tau = 1/100$ (the step ratio $s = 1$), which show that the numerical solution approximates the exact solution very well. Table 4.3 gives part of the numerical results calculated with $h = 1/100$, $\tau = 1/50$ (the step ratio $s = 2$), which demonstrates that numerical errors become larger and larger with the increase of the calculated time level such

Table 4.1 (Example 4.1)
The numerical solutions,
exact solutions, and absolute
values of the errors at part of
the node points
($h = 1/100$, $\tau = 1/200$)

k	(x, t)	NS	ES	$|ES - NS|$
20	(0.5,0.1)	1.8221182	1.8221188	6.348e−7
40	(0.5,0.2)	2.0137515	2.0137527	1.162e−6
60	(0.5,0.3)	2.2255394	2.2255409	1.574e−6
80	(0.5,0.4)	2.4596012	2.4596031	1.864e−6
100	(0.5,0.5)	2.7182799	2.7182818	1.951e−6
120	(0.5,0.6)	3.0041654	3.0041660	5.895e−7
140	(0.5,0.7)	3.3201177	3.3201169	7.712e−7
160	(0.5,0.8)	3.6692987	3.6692967	2.043e−6
180	(0.5,0.9)	4.0552032	4.0552000	3.265e−6
200	(0.5,1.0)	4.4816935	4.4816891	4.426e−6

Table 4.2 (Example 4.1)
The numerical solutions,
exact solutions, and absolute
values of the errors at part of
the node points
($h = 1/100$, $\tau = 1/100$)

k	(x, t)	NS	ES	$|ES - NS|$
10	(0.5,0.1)	1.8221160	1.8221188	2.752e−6
20	(0.5,0.2)	2.0137472	2.0137527	5.532e−6
30	(0.5,0.3)	2.2255326	2.2255409	8.368e−6
40	(0.5,0.4)	2.4595918	2.4596031	1.129e−5
50	(0.5,0.5)	2.7182675	2.7182818	1.432e−5
60	(0.5,0.6)	3.0041547	3.0041660	1.129e−5
70	(0.5,0.7)	3.3201086	3.3201169	8.368e−6
80	(0.5,0.8)	3.6692911	3.6692967	5.532e−6
90	(0.5,0.9)	4.0551972	4.0552000	2.752e−6
100	(0.5,1.0)	4.4816891	4.4816891	1.210e−14

Table 4.3 (Example 4.1)
The numerical solutions,
exact solutions, and absolute
values of the errors at part of
the node points
($h = 1/100$, $\tau = 1/50$)

k	(x, t)	NS	ES	$|ES - NS|$
5	(0.5,0.1)	1.8221076e+00	1.8221188	1.122e−05
10	(0.5,0.2)	2.0137310e+00	2.0137527	2.175e−05
15	(0.5,0.3)	1.4923732e+00	2.2255409	7.332e−01
20	(0.5,0.4)	4.3303094e+05	2.4596031	4.330e+05
25	(0.5,0.5)	−2.5136946e+11	2.7182818	2.514e+11
30	(0.5,0.6)	1.4323236e+17	3.0041660	1.432e+17
35	(0.5,0.7)	−8.0432242e+22	3.3201169	8.043e+22
40	(0.5,0.8)	4.4673157e+28	3.6692967	4.467e+28
45	(0.5,0.9)	−2.4604668e+34	4.0552000	2.461e+34
50	(0.5,1.0)	1.3462410e+40	4.4816891	1.346e+40

Table 4.4 (Example 4.1) The maximum errors of numerical solutions with difference step sizes ($s = 1/2$)

h	τ	$E_\infty(h, \tau)$	$E_\infty(2h, 2\tau)/E_\infty(h, \tau)$
1/10	1/20	4.370e−4	*
1/20	1/40	1.106e−4	3.9493
1/40	1/80	2.771e−5	3.9927
1/80	1/160	6.926e−6	4.0010
1/160	1/320	1.732e−6	3.9982
1/320	1/640	4.331e−7	4.0002
1/640	1/1280	1.083e−7	4.0003
1/1280	1/2560	2.706e−8	4.0000

Fig. 4.3 (Example 4.1) The curves of the numerical and exact solutions ($h = 1/10, \tau = 1/20$)

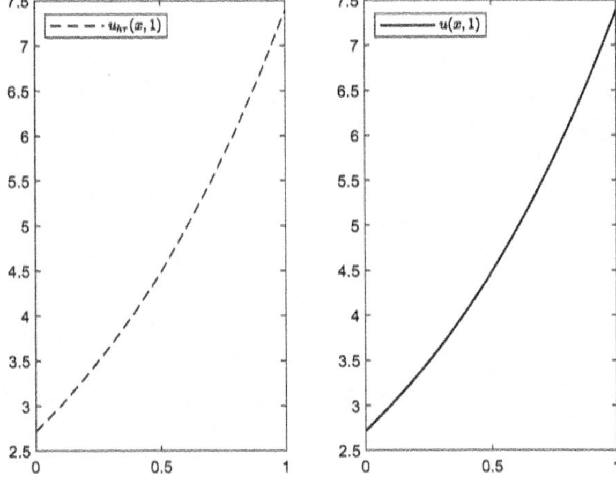

that the numerical results are meaningless in practice. Table 4.4 gives the maximum errors

$$E_\infty(h, \tau) = \max_{\substack{1 \leqslant i \leqslant m-1 \\ 1 \leqslant k \leqslant n}} \left| u(x_i, t_k) - u_i^k \right|$$

of numerical solutions with different step sizes ($s = 1/2$).

As we see from Table 4.4, when the spatial step size and the temporal step size are both reduced by half, the maximum errors are reduced to about a quarter of the original. Figure 4.3 shows the curve of the exact solution at $t = 1$ and the curve of the numerical solution with $h = 1/10$, $\tau = 1/20$. Figure 4.4 displays the numerical error curves at $t = 1$ with different step sizes. Figure 4.5 illustrates the error surfaces with different step sizes.

Fig. 4.4 (Example 4.1) The error curves at $t = 1$ with different step sizes ($s = 1/2$)

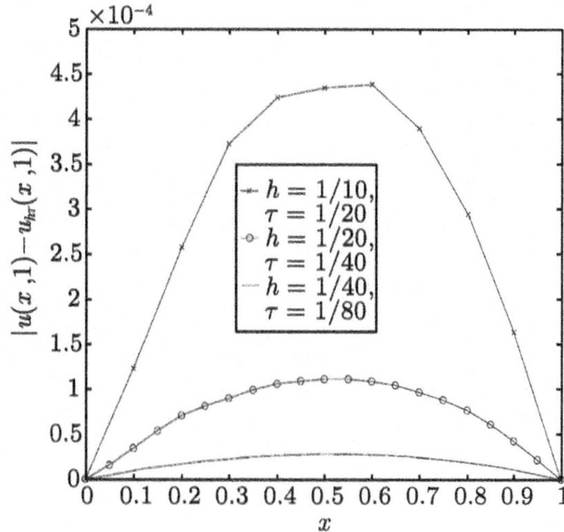

Fig. 4.5 (Example 4.1) The error surfaces with different step sizes

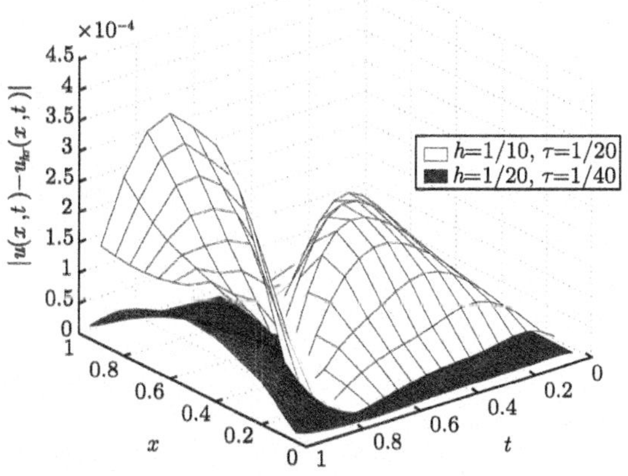

4.2.4   A Priori Estimate of the Difference Solution

Theorem 4.2 *Let $\{v_i^k \mid 0 \leqslant i \leqslant m,\ 0 \leqslant k \leqslant n\}$ be the solution of the difference scheme*

$$\begin{cases} \dfrac{2}{\tau}\left(\delta_t v_i^{\frac{1}{2}} - \psi_i\right) - c^2 \delta_x^2 v_i^0 = g_i^0, & 1 \leqslant i \leqslant m-1, & (4.16\text{a}) \\[2mm] \delta_t^2 v_i^k - c^2 \delta_x^2 v_i^k = g_i^k, & 1 \leqslant i \leqslant m-1, \quad 1 \leqslant k \leqslant n-1, & (4.16\text{b}) \\[2mm] v_i^0 = \varphi_i, & 1 \leqslant i \leqslant m-1, & (4.16\text{c}) \\[2mm] v_0^k = 0, \quad v_m^k = 0, & 0 \leqslant k \leqslant n. & (4.16\text{d}) \end{cases}$$

Then when $s < 1$, we have

$$(1 - s^2)\|\delta_t v^{k+\frac{1}{2}}\|^2 + c^2 |v^{k+\frac{1}{2}}|_1^2$$

$$\leqslant e^{\frac{3}{2}k\tau}\left[c^2\|\delta_x v^0\|^2 + 2\|\psi\|^2 + \frac{1}{2}\tau^2\|g^0\|^2 + \frac{3\tau}{2(1-s^2)}\sum_{l=1}^{k}\|g^l\|^2\right], \quad 0 \leqslant k \leqslant n-1,$$

where

$$\|\psi\|^2 = h\sum_{i=1}^{m-1}(\psi_i)^2, \quad \|g^k\|^2 = h\sum_{i=1}^{m-1}(g_i^k)^2.$$

Proof

(I) Denote

$$E^k = \|\delta_t v^{k+\frac{1}{2}}\|^2 + c^2(\delta_x v^k, \delta_x v^{k+1}).$$

Then we have

$$E^k = \|\delta_t v^{k+\frac{1}{2}}\|^2 + c^2\|\delta_x v^{k+\frac{1}{2}}\|^2 - c^2\left[\|\delta_x v^{k+\frac{1}{2}}\|^2 - (\delta_x v^k, \delta_x v^{k+1})\right]$$

$$= \|\delta_t v^{k+\frac{1}{2}}\|^2 + c^2\|\delta_x v^{k+\frac{1}{2}}\|^2 - \frac{1}{4}c^2\|\delta_x v^{k+1} - \delta_x v^k\|^2$$

$$= \|\delta_t v^{k+\frac{1}{2}}\|^2 + c^2\|\delta_x v^{k+\frac{1}{2}}\|^2 - \frac{1}{4}c^2\tau^2\|\delta_x \delta_t v^{k+\frac{1}{2}}\|^2$$

$$\geqslant \|\delta_t v^{k+\frac{1}{2}}\|^2 + c^2\|\delta_x v^{k+\frac{1}{2}}\|^2 - \frac{1}{4}c^2\tau^2 \cdot \frac{4}{h^2}\|\delta_t v^{k+\frac{1}{2}}\|^2$$

$$= (1 - s^2)\|\delta_t v^{k+\frac{1}{2}}\|^2 + c^2\|\delta_x v^{k+\frac{1}{2}}\|^2.$$

Therefore, when $s < 1$,

$$\|\delta_t v^{k+\frac{1}{2}}\|^2 \leqslant \frac{1}{1-s^2}E^k. \tag{4.17}$$

(II) Taking an inner product on both sides of (4.16a) with $\delta_t v^{\frac{1}{2}}$ produces

$$\frac{2}{\tau}\left(\delta_t v^{\frac{1}{2}} - \psi, \delta_t v^{\frac{1}{2}}\right) - c^2\left(\delta_x^2 v^0, \delta_t v^{\frac{1}{2}}\right) = \left(g^0, \delta_t v^{\frac{1}{2}}\right),$$

which can be reformulated as

$$
2\|\delta_t v^{\frac{1}{2}}\|^2 + c^2(\delta_x v^0, \delta_x v^1)
$$

$$
= 2\big(\psi, \delta_t v^{\frac{1}{2}}\big) + c^2\|\delta_x v^0\|^2 + \tau\big(g^0, \delta_t v^{\frac{1}{2}}\big)
$$

$$
\leqslant \frac{1}{2}\|\delta_t v^{\frac{1}{2}}\|^2 + 2\|\psi\|^2 + c^2\|\delta_x v^0\|^2 + \frac{1}{2}\|\delta_t v^{\frac{1}{2}}\|^2 + \frac{1}{2}\tau^2\|g^0\|^2.
$$

Therefore,

$$
E^0 \leqslant c^2\|\delta_x v^0\|^2 + 2\|\psi\|^2 + \frac{1}{2}\tau^2\|g^0\|^2. \tag{4.18}
$$

(III) Taking an inner product on both sides of (4.16b) with $2\Delta_t v^k$, we obtain

$$
2(\delta_t^2 v^k, \Delta_t v^k) - 2c^2(\delta_x^2 v^k, \Delta_t v^k) = 2(g^k, \Delta_t v^k), \quad 1 \leqslant k \leqslant n-1. \tag{4.19}
$$

Now we estimate each term in the above equality. The first term on the left-hand side is

$$
2\big(\delta_t^2 v^k, \Delta_t v^k\big) = \frac{1}{\tau}\big(\delta_t v^{k+\frac{1}{2}} - \delta_t v^{k-\frac{1}{2}}, \delta_t v^{k+\frac{1}{2}} + \delta_t v^{k-\frac{1}{2}}\big)
$$

$$
= \frac{1}{\tau}\big(\|\delta_t v^{k+\frac{1}{2}}\|^2 - \|\delta_t v^{k-\frac{1}{2}}\|^2\big).
$$

Applying Lemma 1.4 to the second term on the left-hand side, and noticing $\Delta_t v_0^k = 0$, $\Delta_t v_m^k = 0$, we have

$$
-2(\delta_x^2 v^k, \Delta_t v^k) = 2\big(\delta_x v^k, \Delta_t \delta_x v^k\big)
$$

$$
= \frac{1}{\tau}(\delta_x v^k, \delta_x v^{k+1} - \delta_x v^{k-1})
$$

$$
= \frac{1}{\tau}\Big[\big(\delta_x v^k, \delta_x v^{k+1}\big) - \big(\delta_x v^{k-1}, \delta_x v^k\big)\Big].
$$

Substituting the above two equalities into (4.19) and applying the Cauchy-Schwarz inequality to the right-hand side term, we arrive at

$$
\frac{1}{\tau}\Big\{\Big[\|\delta_t v^{k+\frac{1}{2}}\|^2 + c^2(\delta_x v^k, \delta_x v^{k+1})\Big] - \Big[\|\delta_t v^{k-\frac{1}{2}}\|^2 + c^2(\delta_x v^{k-1}, \delta_x v^k)\Big]\Big\}
$$

$$
= 2(g^k, \Delta_t v^k)
$$

$$\leqslant (1 - s^2)\|\Delta_t v^k\|^2 + \frac{1}{1 - s^2}\|g^k\|^2$$

$$\leqslant \frac{1 - s^2}{2}\left(\|\delta_t v^{k+\frac{1}{2}}\|^2 + \|\delta_t v^{k-\frac{1}{2}}\|^2\right) + \frac{1}{1 - s^2}\|g^k\|^2,$$

which can be rearranged as

$$\|\delta_t v^{k+\frac{1}{2}}\|^2 + c^2(\delta_x v^k, \delta_x v^{k+1})$$

$$\leqslant \|\delta_t v^{k-\frac{1}{2}}\|^2 + c^2(\delta_x v^{k-1}, \delta_x v^k) + \frac{1 - s^2}{2}\tau\left(\|\delta_t v^{k+\frac{1}{2}}\|^2 + \|\delta_t v^{k-\frac{1}{2}}\|^2\right)$$

$$+ \frac{\tau}{1 - s^2}\|g^k\|^2, \quad 1 \leqslant k \leqslant n - 1.$$

Combining the above inequality with (4.17), one has

$$E^k \leqslant E^{k-1} + \frac{\tau}{2}(E^k + E^{k-1}) + \frac{\tau}{1 - s^2}\|g^k\|^2, \quad 1 \leqslant k \leqslant n - 1,$$

or

$$\left(1 - \frac{\tau}{2}\right)E^k \leqslant \left(1 + \frac{\tau}{2}\right)E^{k-1} + \frac{\tau}{1 - s^2}\|g^k\|^2, \quad 1 \leqslant k \leqslant n - 1.$$

When $\tau \leqslant 2/3$, we have

$$E^k \leqslant \left(1 + \frac{3}{2}\tau\right)E^{k-1} + \frac{3\tau}{2(1 - s^2)}\|g^k\|^2, \quad 1 \leqslant k \leqslant n - 1.$$

According to the Gronwall inequality (Lemma 3.3), we have

$$E^k \leqslant e^{\frac{3}{2}k\tau}\left[E^0 + \frac{3\tau}{2(1 - s^2)}\sum_{l=1}^{k}\|g^l\|^2\right], \quad 0 \leqslant k \leqslant n - 1.$$

Combining (4.18) with the definition of E^k, we easily get

$$(1 - s^2)\|\delta_t v^{k+\frac{1}{2}}\|^2 + c^2|v^{k+\frac{1}{2}}|_1^2$$

$$\leqslant e^{\frac{3}{2}k\tau}\left[c^2\|\delta_x v^0\|^2 + 2\|\psi\|^2 + \frac{1}{2}\tau^2\|g^0\|^2 + \frac{3\tau}{2(1 - s^2)}\sum_{l=1}^{k}\|g^l\|^2\right], \quad 0 \leqslant k \leqslant n - 1.$$

$$\square$$

Colollary 4.1 *Let $\{v_i^k \mid 0 \leqslant i \leqslant m,\ 0 \leqslant k \leqslant n\}$ be the solution of the difference scheme (4.16). When $s < 1$, we have*

$$|v^{k+1}|_1^2 \leqslant \frac{2}{c^2(1-s^2)} e^{\frac{3}{2}T}\left[c^2\|\delta_x v^0\|^2 + 2\|\psi\|^2 + \frac{1}{2}\tau^2\|g^0\|^2 + \frac{3\tau}{2(1-s^2)}\sum_{l=1}^{k}\|g^l\|^2 \right],$$

$$0 \leqslant k \leqslant n-1. \tag{4.20}$$

Proof Denote

$$G^k = e^{\frac{3}{2}T}\left[c^2\|\delta_x v^0\|^2 + 2\|\psi\|^2 + \frac{1}{2}\tau^2\|g^0\|^2 + \frac{3\tau}{2(1-s^2)}\sum_{l=1}^{k}\|g^l\|^2 \right].$$

It follows from Theorem 4.2 that

$$\|\delta_t v^{k+\frac{1}{2}}\|^2 \leqslant \frac{1}{1-s^2}G^k, \quad |v^{k+\frac{1}{2}}|_1^2 \leqslant \frac{1}{c^2}G^k, \quad 0 \leqslant k \leqslant n-1. \tag{4.21}$$

On the other hand, it follows from

$$v_i^{k+1} = \frac{1}{2}(v_i^{k+1} + v_i^k + v_i^{k+1} - v_i^k) = v_i^{k+\frac{1}{2}} + \frac{1}{2}\tau\delta_t v_i^{k+\frac{1}{2}}, \quad 0 \leqslant i \leqslant m$$

that

$$|v^{k+1}|_1^2 \leqslant 2|v^{k+\frac{1}{2}}|_1^2 + 2\cdot\frac{1}{4}\tau^2|\delta_t v^{k+\frac{1}{2}}|_1^2$$

$$\leqslant 2|v^{k+\frac{1}{2}}|_1^2 + \frac{\tau^2}{2}\cdot\frac{4}{h^2}\|\delta_t v^{k+\frac{1}{2}}\|^2$$

$$= 2|v^{k+\frac{1}{2}}|_1^2 + \frac{2s^2}{c^2}\|\delta_t v^{k+\frac{1}{2}}\|^2.$$

With the application of (4.21), it yields

$$|v^{k+1}|_1^2 \leqslant 2\cdot\frac{1}{c^2}G^k + \frac{2s^2}{c^2}\cdot\frac{1}{1-s^2}G^k = \frac{2}{c^2(1-s^2)}G^k, \quad 0 \leqslant k \leqslant n-1.$$

Noticing the definition of G^k in the above inequality, it yields (4.20). $\square$

4.2.5   Convergence and Stability of the Difference Solution

Convergence

Theorem 4.3 *Let $\{u(x,t) \mid (x,t) \in \bar{D}\}$ be the solution of the problem (4.1) and $\{u_i^k \mid 0 \leqslant i \leqslant m,\ 0 \leqslant k \leqslant n\}$ be the solution of the difference scheme (4.14). Denote*

$$e_i^k = u(x_i, t_k) - u_i^k, \quad 0 \leqslant i \leqslant m, \quad 0 \leqslant k \leqslant n.$$

Then when $s < 1$, we have

$$\|e^k\|_\infty \leqslant \frac{Lc_1}{2c(1-s^2)}\, e^{\frac{3}{4}T}\, \sqrt{1 - s^2 + 3T}\, (\tau^2 + h^2), \quad 1 \leqslant k \leqslant n.$$

Proof Subtracting (4.14) from (4.6), (4.9), (4.12), and (4.13), we have the system of error equations

$$\begin{cases}
\dfrac{2}{\tau}\left(\delta_t e_i^{\frac{1}{2}} - 0\right) - c^2 \delta_x^2 e_i^0 = (R_1)_i^0, & 1 \leqslant i \leqslant m-1, \\[2mm]
\delta_t^2 e_i^k - c^2 \delta_x^2 e_i^k = (R_1)_i^k, & 1 \leqslant i \leqslant m-1, \quad 1 \leqslant k \leqslant n-1, \\[2mm]
e_i^0 = 0, & 1 \leqslant i \leqslant m-1, \\[2mm]
e_0^k = 0, \quad e_m^k = 0, & 0 \leqslant k \leqslant n.
\end{cases} \qquad (4.22)$$

Noticing (4.11) and applying Corollary 4.1 to (4.22), we have

$$|e^{k+1}|_1^2$$

$$\leqslant \frac{2}{c^2(1-s^2)}\, e^{\frac{3}{2}T}\left[c^2\|\delta_x e^0\|^2 + \frac{1}{2}\tau^2\|(R_1)^0\|^2 + \frac{3}{2(1-s^2)}\tau \sum_{l=1}^{k} \|(R_1)^l\|^2\right]$$

$$\leqslant \frac{2}{c^2(1-s^2)}\, e^{\frac{3}{2}T}\left[\frac{1}{2}\tau^2 L c_1^2(\tau + h^2)^2 + \frac{3}{2(1-s^2)}k\tau L c_1^2(\tau^2 + h^2)^2\right]$$

$$\leqslant \frac{1}{c^2(1-s^2)^2}\, e^{\frac{3}{2}T} L\left(1 - s^2 + 3T\right) c_1^2(\tau^2 + h^2)^2, \quad 0 \leqslant k \leqslant n-1.$$

It easily follows from Lemma 1.4 when $s < 1$ that

$$\|e^k\|_\infty \leqslant \frac{\sqrt{L}}{2}|e^k|_1 \leqslant \frac{Lc_1}{2c(1-s^2)}\, e^{\frac{3}{4}T}\, \sqrt{1 - s^2 + 3T}\, (\tau^2 + h^2), \quad 1 \leqslant k \leqslant n,$$

which completes the proof. $\square$

Stability

Theorem 4.4 *Let $s < 1$. The solution of the difference scheme (4.14) is stable with respect to the initial value and right-hand side function in the following sense: Let $\{u_i^k \mid 0 \leqslant i \leqslant m,\ 0 \leqslant k \leqslant n\}$ be the solution of the system of difference equations*

$$
\begin{cases}
\dfrac{2}{\tau}\left(\delta_t v_i^{\frac{1}{2}} - \psi_i\right) - c^2 \delta_x^2 v_i^0 = f_i^0, & 1 \leqslant i \leqslant m-1, \\[2mm]
\delta_t^2 u_i^k - c^2 \delta_x^2 u_i^k = f_i^k, & 1 \leqslant i \leqslant m-1, \quad 1 \leqslant k \leqslant n-1, \\[2mm]
u_i^0 = \varphi_i, & 1 \leqslant i \leqslant m-1, \\[2mm]
u_0^k = 0, \quad u_m^k = 0, & 0 \leqslant k \leqslant n.
\end{cases}
$$

Then we have

$$
|u^{k+1}|_1^2 \leqslant \frac{2}{c^2(1-s^2)} e^{\frac{3}{2}T}\left[c^2 \|\delta_x u^0\|^2 + 2\|\psi\|^2 + \frac{1}{2}\tau^2 \|f^0\|^2 + \frac{3\tau}{2(1-s^2)} \sum_{l=1}^{k} \|f^l\|^2\right],
$$

$$
0 \leqslant k \leqslant n-1.
$$

Proof It can be proved by a direct application of Corollary 4.1. □

Theorem 4.5 *Let $s \geqslant 1$. Then the difference scheme (4.14) is unstable with respect to the initial value in the maximum norm.*

Proof Consider the system of perturbed equations

$$
\begin{cases}
\delta_t^2 \varepsilon_i^k - c^2 \delta_x^2 \varepsilon_i^k = 0, & 1 \leqslant i \leqslant m-1, \quad 1 \leqslant k \leqslant n-1, & (4.23\text{a}) \\[2mm]
\varepsilon_i^0 = (-1)^i(-\epsilon), & 1 \leqslant i \leqslant m-1, & (4.23\text{b}) \\[2mm]
\varepsilon_i^1 = (-1)^{i+1}\epsilon, & 1 \leqslant i \leqslant m-1, & (4.23\text{c}) \\[2mm]
\varepsilon_0^k = 0, \quad \varepsilon_m^k = 0, & 0 \leqslant k \leqslant n, & (4.23\text{d})
\end{cases}
$$

where ϵ is a small positive constant.

Note that $s = c\frac{\tau}{h} \geqslant 1$ implies $cTm \geqslant Ln$. Therefore, when $n \to +\infty$, one gets $m \to +\infty$.

Equation (4.23a) can be rewritten as

$$
\varepsilon_i^{k+1} = s^2(\varepsilon_{i-1}^k + \varepsilon_{i+1}^k) + 2(1-s^2)\varepsilon_i^k - \varepsilon_i^{k-1}, \quad 1 \leqslant i \leqslant m-1, \quad 1 \leqslant k \leqslant n-1.
$$

$$
(4.24)
$$

In a triangular region or trapezoidal region shown in Fig. 4.6, when $k \leqslant \min\left\{\left\lfloor \frac{m}{2} \right\rfloor, n\right\}$, the analysis shows that the value of ε_i^k ($k \leqslant i \leqslant m-k$) is independent of the boundary value (4.23d), but totally determined by (4.23a), the initial values (4.23b) and (4.23c).

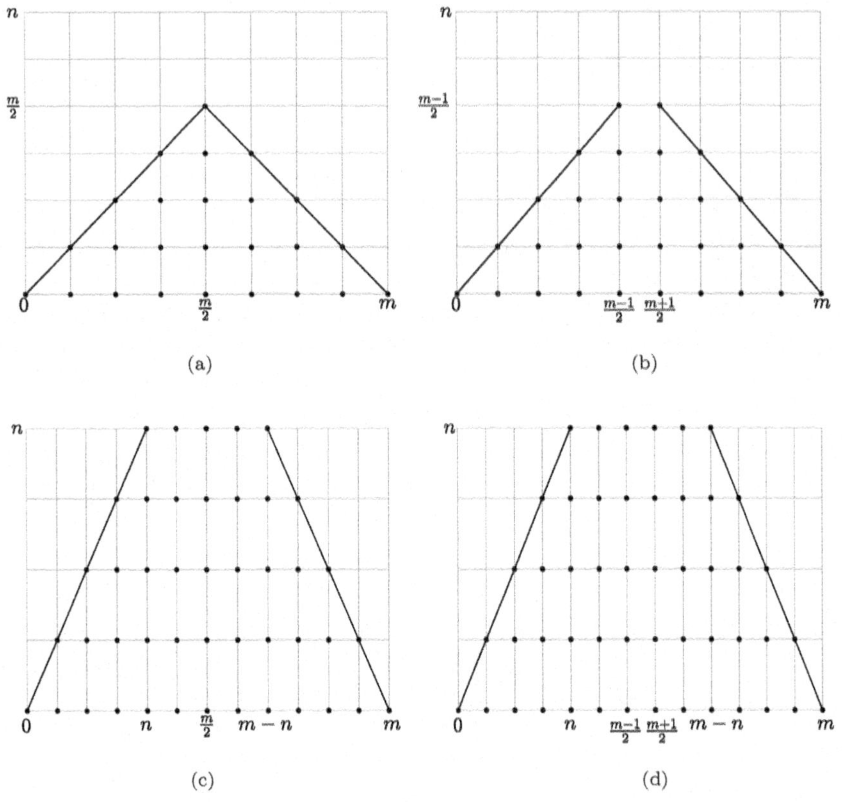

Fig. 4.6 Auxiliary figures for the proof of Theorem 4.5. **(a)** m is even, $n \geqslant \frac{m}{2}$. **(b)** m is odd, $n \geqslant \frac{m-1}{2}$. **(c)** m is even, $n < \frac{m}{2}$. **(d)** m is odd, $n < \frac{m-1}{2}$

(a) When $s = 1$, it follows from (4.24) that

$$\varepsilon_i^{k+1} = \varepsilon_{i-1}^{k} + \varepsilon_{i+1}^{k} - \varepsilon_i^{k-1}, \quad k \leqslant i \leqslant m-k, \quad k \leqslant \min\left\{\left\lfloor \frac{m}{2} \right\rfloor, n\right\}. \tag{4.25}$$

Suppose the solution of (4.25) has the form

$$\varepsilon_i^k = (-1)^{i+k} T(k).$$

Substituting it into (4.25), we arrive at

$$T(k+1) - 2T(k) + T(k-1) = 0, \quad k \leqslant \min\left\{\left\lfloor \frac{m}{2} \right\rfloor, n\right\}.$$

It follows from the recursion that

$$T(k) = T(1)k + T(0)(1 - k), \quad k \leqslant \min\left\{\left\lfloor \frac{m}{2} \right\rfloor, n\right\}.$$

Noticing $T(0) = -\epsilon$ and $T(1) = \epsilon$, we have

$$T(k) = (2k - 1)\epsilon, \quad k \leqslant \min\left\{\left\lfloor \frac{m}{2} \right\rfloor, n\right\}.$$

The solution in the triangular region or trapezoidal region shown in Fig. 4.6 can be achieved in the form of

$$\varepsilon_i^k = (-1)^{i+k}(2k - 1)\epsilon, \quad k \leqslant i \leqslant m - k, \quad k \leqslant \min\left\{\left\lfloor \frac{m}{2} \right\rfloor, n\right\}.$$

It follows from the above equality that

$$\left\| \varepsilon^{\min\left\{\left\lfloor \frac{m}{2} \right\rfloor, n\right\}} \right\|_\infty \geqslant \left(2 \min\left\{\left\lfloor \frac{m}{2} \right\rfloor, n\right\} - 1\right) \epsilon.$$

Hence,

$$\lim_{n \to +\infty} \left\| \varepsilon^{\min\left\{\left\lfloor \frac{m}{2} \right\rfloor, n\right\}} \right\|_\infty = +\infty.$$

Because of

$$\max_{1 \leqslant k \leqslant n} \left\| \varepsilon^k \right\|_\infty \geqslant \left\| \varepsilon^{\min\left\{\left\lfloor \frac{m}{2} \right\rfloor, n\right\}} \right\|_\infty,$$

it yields

$$\lim_{n \to +\infty} \max_{1 \leqslant k \leqslant n} \left\| \varepsilon^k \right\|_\infty = +\infty.$$

(b) When $s > 1$, it follows from (4.24) that

$$\varepsilon_i^{k+1} = s^2(\varepsilon_{i-1}^k + \varepsilon_{i+1}^k) + 2(1 - s^2)\varepsilon_i^k - \varepsilon_i^{k-1},$$

$$k \leqslant i \leqslant m - k, \quad k \leqslant \min\left\{\left\lfloor \frac{m}{2} \right\rfloor, n\right\}. \tag{4.26}$$

Suppose the solution of (4.26) has the form of

$$\varepsilon_i^k = (-1)^{i+k} T(k).$$

Substituting it into (4.26), it yields

$$T(k+1) + 2(1 - 2s^2)T(k) + T(k-1) = 0, \quad k \leqslant \min\left\{\left\lfloor \frac{m}{2}\right\rfloor, n\right\}.$$

Two roots of its characteristic equation

$$\lambda^2 + 2(1 - 2s^2)\lambda + 1 = 0$$

are, respectively,

$$\lambda_1 = 2s^2 - 1 + 2s\sqrt{s^2 - 1} \in (1, +\infty), \quad \lambda_2 = 2s^2 - 1 - 2s\sqrt{s^2 - 1} \in (0, 1).$$

Noticing $T(0) = -\epsilon$ and $T(1) = \epsilon$, we have

$$T(k) = \frac{T(1) - \lambda_2 T(0)}{\lambda_1 - \lambda_2}\lambda_1^k + \frac{\lambda_1 T(0) - T(1)}{\lambda_1 - \lambda_2}\lambda_2^k$$

$$= \left(\frac{1 + \lambda_2}{\lambda_1 - \lambda_2}\lambda_1^k - \frac{1 + \lambda_1}{\lambda_1 - \lambda_2}\lambda_2^k\right)\epsilon, \quad k \leqslant \min\left\{\left\lfloor \frac{m}{2}\right\rfloor, n\right\}.$$

Therefore, the solution in the triangular or trapezoidal region shown in Fig. 4.6 can be expressed by

$$\varepsilon_i^k = (-1)^{i+k}\left(\frac{1 + \lambda_2}{\lambda_1 - \lambda_2}\lambda_1^k - \frac{1 + \lambda_1}{\lambda_1 - \lambda_2}\lambda_2^k\right)\epsilon, \quad k \leqslant i \leqslant m - k, \quad k \leqslant \min\left\{\left\lfloor \frac{m}{2}\right\rfloor, n\right\}.$$

It follows from the above equality that

$$\left\|\varepsilon^{\min\left\{\left\lfloor \frac{m}{2}\right\rfloor, n\right\}}\right\|_\infty \geqslant \left(\frac{1 + \lambda_2}{\lambda_1 - \lambda_2}\lambda_1^{\min\left\{\left\lfloor \frac{m}{2}\right\rfloor n\right\}} - \frac{1 + \lambda_1}{\lambda_1 - \lambda_2}\lambda_2^{\min\left\{\left\lfloor \frac{m}{2}\right\rfloor, n\right\}}\right)\epsilon.$$

Therefore,

$$\lim_{n \to +\infty}\left\|\varepsilon^{\min\left\{\left\lfloor \frac{m}{2}\right\rfloor, n\right\}}\right\|_\infty = +\infty.$$

It follows from

$$\max_{1 \leqslant k \leqslant n}\left\|\varepsilon^k\right\|_\infty \geqslant \left\|\varepsilon^{\min\left\{\left\lfloor \frac{m}{2}\right\rfloor, n\right\}}\right\|_\infty$$

that

$$\lim_{n \to +\infty}\max_{1 \leqslant k \leqslant n}\left\|\varepsilon^k\right\|_\infty = +\infty.$$

In summary, when $s \geqslant 1$, the difference scheme (4.14) is unstable with respect to the initial value in the maximum norm. $\square$

4.3   The Implicit Difference Scheme

The explicit difference scheme introduced in the previous section requires a step ratio $s < 1$. In this section we introduce an unconditionally stable difference scheme.

4.3.1   Derivation of the Difference Scheme

(I) Considering the differential equation (4.1a) at the node point (x_i, t_0), we have

$$\frac{\partial^2 u(x_i, t_0)}{\partial t^2} - \frac{1}{2} c^2 \left[\frac{\partial^2 u(x_i, t_1)}{\partial x^2} + \frac{\partial^2 u(x_i, t_0)}{\partial x^2} \right]$$

$$= f(x_i, t_0) - \frac{1}{2} c^2 \left[\frac{\partial^2 u(x_i, t_1)}{\partial x^2} - \frac{\partial^2 u(x_i, t_0)}{\partial x^2} \right], \quad 1 \leqslant i \leqslant m - 1. \quad (4.27)$$

By means of Lemma 1.2, we have

$$\frac{\partial^2 u(x_i, t_0)}{\partial t^2} = \frac{2}{\tau} \left[\delta_t U_i^{\frac{1}{2}} - u_t(x_i, t_0) \right] - \frac{\tau}{3} \frac{\partial^3 u(x_i, \eta_{i0})}{\partial t^3}, \quad \eta_{i0} \in (t_0, t_1),$$

$$\frac{\partial^2 u(x_i, t_k)}{\partial x^2} = \delta_x^2 U_i^k - \frac{h^2}{12} \frac{\partial^4 u(\xi_{ik}, t_k)}{\partial x^4}, \quad \xi_{ik} \in (x_{i-1}, x_{i+1}).$$

Substituting the above two equalities into (4.27), we have

$$\frac{2}{\tau} \left[\delta_t U_i^{\frac{1}{2}} - u_t(x_i, t_0) \right] - c^2 \delta_x^2 U_i^{\frac{1}{2}} = f(x_i, t_0) + (R_2)_i^0, \quad 1 \leqslant i \leqslant m - 1,$$

$$(4.28)$$

where

$$(R_2)_i^0 = \frac{\tau}{3} \frac{\partial^3 u(x_i, \eta_{i0})}{\partial t^3} - \frac{1}{2} c^2 \left[\frac{\partial^2 u(x_i, t_1)}{\partial x^2} - \frac{\partial^2 u(x_i, t_0)}{\partial x^2} \right]$$

$$- \frac{c^2 h^2}{24} \left[\frac{\partial^4 u(\xi_{i0}, t_0)}{\partial x^4} + \frac{\partial^4 u(\xi_{i1}, t_1)}{\partial x^4} \right], \quad 1 \leqslant i \leqslant m - 1. \quad (4.29)$$

(II) Considering Eq. (4.1a) at the node point (x_i, t_k), we have

$$\frac{\partial^2 u(x_i, t_k)}{\partial t^2} - c^2 \frac{\partial^2 u(x_i, t_k)}{\partial x^2} = f(x_i, t_k), \quad 1 \leqslant i \leqslant m-1, \quad 1 \leqslant k \leqslant n-1. \tag{4.30}$$

According to Lemma 1.2, we have

$$\begin{aligned}
\frac{\partial^2 u(x_i, t_k)}{\partial x^2} &= \frac{1}{2}\left[\frac{\partial^2 u(x_i, t_{k-1})}{\partial x^2} + \frac{\partial^2 u(x_i, t_{k+1})}{\partial x^2}\right] - \frac{\tau^2}{2}\frac{\partial^4 u(x_i, \zeta_{ik})}{\partial x^2 \partial t^2} \\
&= \frac{1}{2}\left[\delta_x^2 U_i^{k-1} - \frac{h^2}{12}\frac{\partial^2 u(\xi_{i,k-1}, t_{k-1})}{\partial x^4} + \delta_x^2 U_i^{k+1} - \frac{h^2}{12}\frac{\partial^2 u(\xi_{i,k+1}, t_{k+1})}{\partial x^4}\right] \\
&\quad - \frac{\tau^2}{2}\frac{\partial^4 u(x_i, \zeta_{ik})}{\partial x^2 \partial t^2}, \quad \zeta_{ik} \in (t_{k-1}, t_{k+1}), \quad \xi_{i,k-1}, \xi_{i,k+1} \in (x_{i-1}, x_{i+1}),
\end{aligned}$$

$$\frac{\partial^2 u(x_i, t_k)}{\partial t^2} = \delta_t^2 U_i^k - \frac{\tau^2}{12}\frac{\partial^4 u(x_i, \eta_{ik})}{\partial t^4}, \quad \eta_{ik} \in (t_{k-1}, t_{k+1}).$$

Substituting the above two equalities into (4.30), we arrive at

$$\delta_t^2 U_i^k - \frac{c^2}{2}(\delta_x^2 U_i^{k-1} + \delta_x^2 U_i^{k+1}) = f(x_i, t_k) + (R_2)_i^k,$$

$$1 \leqslant i \leqslant m-1, \quad 1 \leqslant k \leqslant n-1, \tag{4.31}$$

where

$$\begin{aligned}
(R_2)_i^k = \tau^2 &\left[\frac{1}{12}\frac{\partial^4 u(x_i, \eta_{ik})}{\partial t^4} - \frac{c^2}{2}\frac{\partial^4 u(x_i, \zeta_{ik})}{\partial x^2 \partial t^2}\right] \\
&- \frac{c^2 h^2}{24}\left[\frac{\partial^4 u(\xi_{i,k-1}, t_{k-1})}{\partial x^4} + \frac{\partial^4 u(\xi_{i,k+1}, t_{k+1})}{\partial x^4}\right]. \tag{4.32}
\end{aligned}$$

Noticing (4.29) and (4.32), there is a positive constant c_2 such that

$$\begin{cases} |(R_2)_i^0| \leqslant c_2(\tau + h^2), & 1 \leqslant i \leqslant m-1, \\ |(R_2)_i^k| \leqslant c_2(\tau^2 + h^2), & 1 \leqslant i \leqslant m-1, \quad 1 \leqslant k \leqslant n-1. \end{cases} \tag{4.33}$$

Using the initial value condition (4.1b), we have

$$U_i^0 = \varphi(x_i), \quad 1 \leqslant i \leqslant m-1. \tag{4.34}$$

Again using the boundary value condition (4.1c), we have

$$U_0^k = \alpha(t_k), \quad U_m^k = \beta(t_k), \quad 0 \leqslant k \leqslant n. \tag{4.35}$$

Fig. 4.7 The stencil of the implicit difference scheme (4.36)

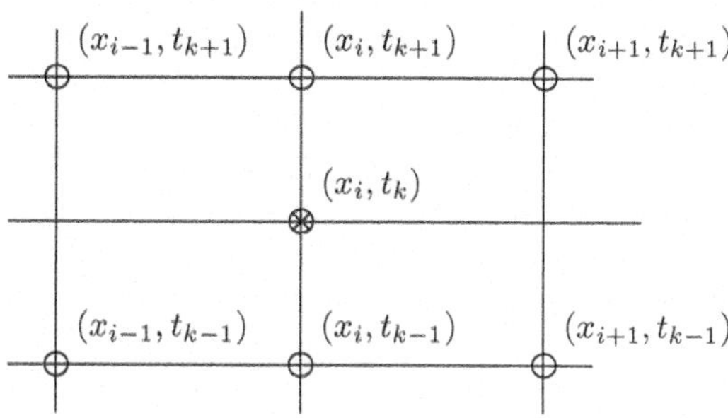

Omitting the small terms $(R_2)_i^0$ in (4.28) and $(R_2)_i^k$ in (4.31), respectively, noticing (4.34) and (4.35), and replacing U_i^k with u_i^k, the following difference scheme for (4.1) reads

$$
\begin{cases}
\dfrac{2}{\tau}\left[\delta_t u_i^{\frac{1}{2}} - \psi(x_i)\right] - c^2 \delta_x^2 u_i^{\frac{1}{2}} = f(x_i, t_0), \quad 1 \leqslant i \leqslant m-1, & (4.36a) \\[2mm]
\delta_t^2 u_i^k - \dfrac{c^2}{2}(\delta_x^2 u_i^{k-1} + \delta_x^2 u_i^{k+1}) = f(x_i, t_k), & \\[2mm]
\qquad\qquad\qquad 1 \leqslant i \leqslant m-1, \quad 1 \leqslant k \leqslant n-1, & (4.36b) \\[2mm]
u_i^0 = \varphi(x_i), \quad 1 \leqslant i \leqslant m-1, & (4.36c) \\[2mm]
u_0^k = \alpha(t_k), \quad u_m^k = \beta(t_k), \quad 0 \leqslant k \leqslant n. & (4.36d)
\end{cases}
$$

The difference equation (4.36a) is a two-time-level and six-node-point implicit one, and the difference equation (4.36b) is a three-time-level and seven-node-point implicit one. The stencil is shown in Fig. 4.7.

Remark 4.1 *The difference equation (4.36a) can be replaced with*

$$
\frac{2}{\tau}\left[\delta_t u_i^{\frac{1}{2}} - \psi(x_i)\right] - c^2 \delta_x^2 u_i^1 = f(x_i, t_0), \quad 1 \leqslant i \leqslant m-1. \qquad (4.37)
$$

The difference equations (4.37) and (4.36a) have the same approximate accuracy.

4.3.2   Existence of the Difference Solution

Theorem 4.6 *The difference scheme (4.36) is uniquely solvable.*

Proof According to (4.36c) and (4.36d), u^0 has been determined.

One can rewrite (4.36a) as

$$u_i^1 - u_i^0 - \frac{1}{4}s^2(u_{i-1}^0 - 2u_i^0 + u_{i+1}^0 + u_{i-1}^1 - 2u_i^1 + u_{i+1}^1)$$

$$= \tau\psi(x_i) + \frac{\tau^2}{2}f(x_i, t_0), \quad 1 \leqslant i \leqslant m - 1,$$

or

$$-\frac{1}{2}s^2 u_{i-1}^1 + (2 + s^2)u_i^1 - \frac{1}{2}s^2 u_{i+1}^1$$

$$= \frac{1}{2}s^2 u_{i-1}^0 + (2 - s^2)u_i^0 + \frac{1}{2}s^2 u_{i+1}^0 + 2\tau\psi(x_i) + \tau^2 f(x_i, t_0), \quad 1 \leqslant i \leqslant m - 1.$$

The above equality can be further rewritten in the following matrix form:

$$
\begin{pmatrix}
2+s^2 & -\frac{1}{2}s^2 & & & \\
-\frac{1}{2}s^2 & 2+s^2 & -\frac{1}{2}s^2 & & \\
& \ddots & \ddots & \ddots & \\
& & -\frac{1}{2}s^2 & 2+s^2 & -\frac{1}{2}s^2 \\
& & & -\frac{1}{2}s^2 & 2+s^2
\end{pmatrix}
\begin{pmatrix}
u_1^1 \\ u_2^1 \\ \vdots \\ u_{m-2}^1 \\ u_{m-1}^1
\end{pmatrix}
$$

$$
=
\begin{pmatrix}
2-s^2 & \frac{1}{2}s^2 & & & \\
\frac{1}{2}s^2 & 2-s^2 & \frac{1}{2}s^2 & & \\
& \ddots & \ddots & \ddots & \\
& & \frac{1}{2}s^2 & 2-s^2 & \frac{1}{2}s^2 \\
& & & \frac{1}{2}s^2 & 2-s^2
\end{pmatrix}
\begin{pmatrix}
u_1^0 \\ u_2^0 \\ \vdots \\ u_{m-2}^0 \\ u_{m-1}^0
\end{pmatrix}
$$

$$
+
\begin{pmatrix}
\frac{1}{2}s^2(u_0^1 + u_0^0) + 2\tau\psi(x_1) + \tau^2 f(x_1, t_0) \\
2\tau\psi(x_2) + \tau^2 f(x_2, t_0) \\
\vdots \\
2\tau\psi(x_{m-2}) + \tau^2 f(x_{m-2}, t_0) \\
\frac{1}{2}s^2(u_m^1 + u_m^0) + 2\tau\psi(x_{m-1}) + \tau^2 f(x_{m-1}, t_0)
\end{pmatrix}.
$$

The coefficient matrix is strictly diagonally dominant at a glance. Consequently, (4.36a) and (4.36d) have a unique solution u^1.

Suppose we have obtained u^{k-1} and u^k ($k \geq 1$). The difference equation (4.36b) can be rewritten as

$$u_i^{k+1} - 2u_i^k + u_i^{k-1} - \frac{1}{2}s^2(u_{i-1}^{k-1} - 2u_i^{k-1} + u_{i+1}^{k-1} + u_{i-1}^{k+1} - 2u_i^{k+1} + u_{i+1}^{k+1})$$

$$= \tau^2 f(x_i, t_k), \quad 1 \leq i \leq m - 1,$$

or

$$-\frac{1}{2}s^2 u_{i-1}^{k+1} + (1 + s^2)u_i^{k+1} - \frac{1}{2}s^2 u_{i+1}^{k+1}$$

$$= \frac{1}{2}s^2 u_{i-1}^{k-1} - (1 + s^2)u_i^{k-1} + \frac{1}{2}s^2 u_{i+1}^{k-1} + 2u_i^k + \tau^2 f(x_i, t_k), \quad 1 \leq i \leq m - 1.$$

Thus, (4.36b) can be further rewritten in the matrix form

$$
\begin{pmatrix}
1+s^2 & -\frac{1}{2}s^2 & & & \\
-\frac{1}{2}s^2 & 1+s^2 & -\frac{1}{2}s^2 & & \\
& \ddots & \ddots & \ddots & \\
& & -\frac{1}{2}s^2 & 1+s^2 & -\frac{1}{2}s^2 \\
& & & -\frac{1}{2}s^2 & 1+s^2
\end{pmatrix}
\begin{pmatrix}
u_1^{k+1} \\
u_2^{k+1} \\
\vdots \\
u_{m-2}^{k+1} \\
u_{m-1}^{k+1}
\end{pmatrix}
$$

$$
=
\begin{pmatrix}
-(1+s^2) & \frac{1}{2}s^2 & & & \\
\frac{1}{2}s^2 & -(1+s^2) & \frac{1}{2}s^2 & & \\
& \ddots & \ddots & \ddots & \\
& & \frac{1}{2}s^2 & -(1+s^2) & \frac{1}{2}s^2 \\
& & & \frac{1}{2}s^2 & -(1+s^2)
\end{pmatrix}
\begin{pmatrix}
u_1^{k-1} \\
u_2^{k-1} \\
\vdots \\
u_{m-2}^{k-1} \\
u_{m-1}^{k-1}
\end{pmatrix}
$$

$$
+
\begin{pmatrix}
\frac{1}{2}s^2(u_0^{k+1} + u_0^{k-1}) + 2u_1^k + \tau^2 f(x_1, t_k) \\
2u_2^k + \tau^2 f(x_2, t_k) \\
\vdots \\
2u_{m-2}^k + \tau^2 f(x_{m-2}, t_k) \\
\frac{1}{2}s^2(u_m^{k+1} + u_m^{k-1}) + 2u_{m-1}^k + \tau^2 f(x_{m-1}, t_k)
\end{pmatrix}.
$$

By observing its coefficient matrix, one can show that it is strictly diagonally dominant. As a result, there exists a unique solution u^{k+1}.

By induction, (4.36b) and (4.36d) are uniquely solvable.   □

4.3.3 Implementation of the Difference Scheme and Numerical Examples

From the process of proving Theorem 4.6, it is easy to know that only a tridiagonal system of linear equations needs to be solved at each time level.

Example 4.2 Apply the implicit difference scheme (4.36) to solve the problem (4.15) in Example 4.1.

Table 4.5 lists some numerical results with $h = 1/100$ and $\tau = 1/100$. Table 4.6 gives some numerical results with $h = 1/200$ and $\tau = 1/200$. The numerical solutions approximate the exact solutions very well. Table 4.7 provides the maximum errors

$$E_\infty(h, \tau) = \max_{\substack{1 \leqslant i \leqslant m-1 \\ 1 \leqslant k \leqslant n}} \left| u(x_i, t_k) - u_i^k \right|$$

of numerical solutions with different step sizes.

As we see from Table 4.7, when the step sizes h and τ are both reduced by half, the maximum errors are reduced to about a quarter of the original.

Figure 4.8 shows the exact solution curve at $t = 1$ and the numerical solution curve with step sizes $h = \tau = 1/10$. Figure 4.9 draws the error curves of numerical solutions at $t = 1$ with different step sizes. Figure 4.10 illustrates the numerical error surfaces with different step sizes.

Table 4.5 (Example 4.2) Part of the numerical solutions, exact solutions, and absolute values of the errors ($h = 1/100, \tau = 1/100$)

k	(x, t)	NS	ES	$|ES-NS|$
10	$(0.5, 0.1)$	1.8221206	1.8221188	1.782e−6
20	$(0.5, 0.2)$	2.0137572	2.0137527	4.495e−6
30	$(0.5, 0.3)$	2.2255492	2.2255409	8.261e−6
40	$(0.5, 0.4)$	2.4596163	2.4596031	1.322e−5
50	$(0.5, 0.5)$	2.7183011	2.7182818	1.926e−5
60	$(0.5, 0.6)$	3.0041894	3.0041660	2.335e−5
70	$(0.5, 0.7)$	3.3201439	3.3201169	2.696e−5
80	$(0.5, 0.8)$	3.6693268	3.6692967	3.013e−5
90	$(0.5, 0.9)$	4.0552329	4.0552000	3.296e−5
100	$(0.5, 1.0)$	4.4817245	4.4816891	3.543e−5

Table 4.6 (Example 4.2)
Part of the numerical
solutions, exact solutions, and
absolute values of the errors
($h = 1/200$, $\tau = 1/200$)

k	(x, t)	NS	ES	$|ES-NS|$
20	(0.5, 0.1)	1.8221192	1.8221188	4.481e−7
40	(0.5, 0.2)	2.0137538	2.0137527	1.129e−6
60	(0.5, 0.3)	2.2255430	2.2255409	2.073e−6
80	(0.5, 0.4)	2.4596064	2.4596031	3.317e−6
100	(0.5, 0.5)	2.7182867	2.7182818	4.856e−6
120	(0.5, 0.6)	3.0041719	3.0041660	5.875e−6
140	(0.5, 0.7)	3.3201237	3.3201169	6.746e−6
160	(0.5, 0.8)	3.6693042	3.6692967	7.550e−6
180	(0.5, 0.9)	4.0552082	4.0552000	8.250e−6
200	(0.5, 1.0)	4.4816979	4.4816891	8.855e−6

Table 4.7 (Example 4.2)
The maximum errors of
numerical solutions with
different step sizes ($s = 1$)

h	τ	$E_\infty(h, \tau)$	$E_\infty(2h, 2\tau)/E_\infty(h, \tau)$
1/10	1/10	3.5077e−3	
1/20	1/20	8.8597e−4	3.959
1/40	1/40	2.2195e−4	3.992
1/80	1/80	5.5441e−5	4.003
1/160	1/160	1.3854e−5	4.002
1/320	1/320	3.4646e−6	3.999
1/640	1/640	8.6607e−7	4.000

Fig. 4.8 (Example 4.2) The
curves of the numerical and
exact solutions at $t = 1$
($h = \tau = 1/10$)

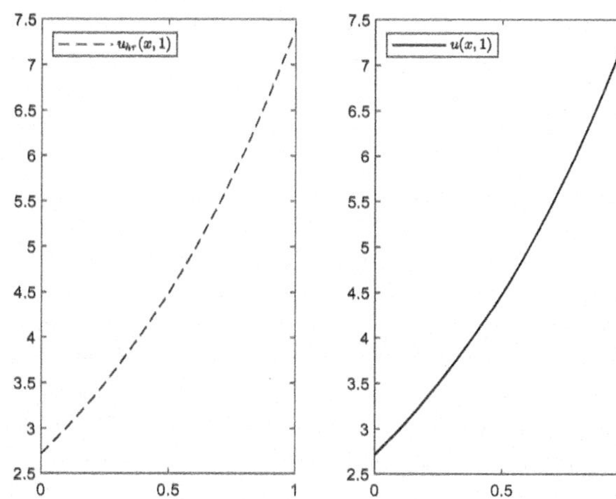

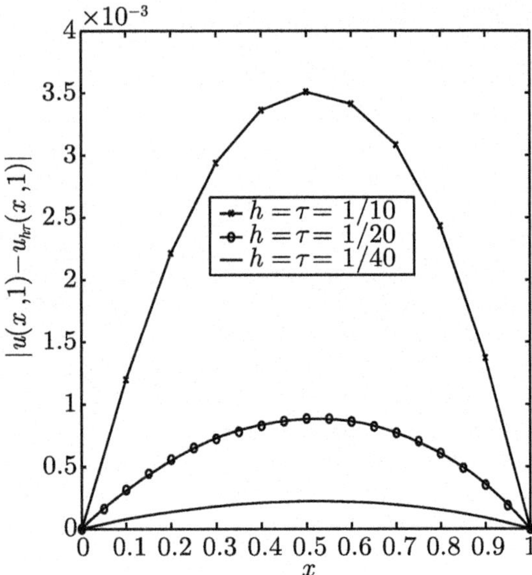

Fig. 4.9 (Example 4.2) The error curves of the numerical solution at $t = 1$ with different step sizes ($s = 1$)

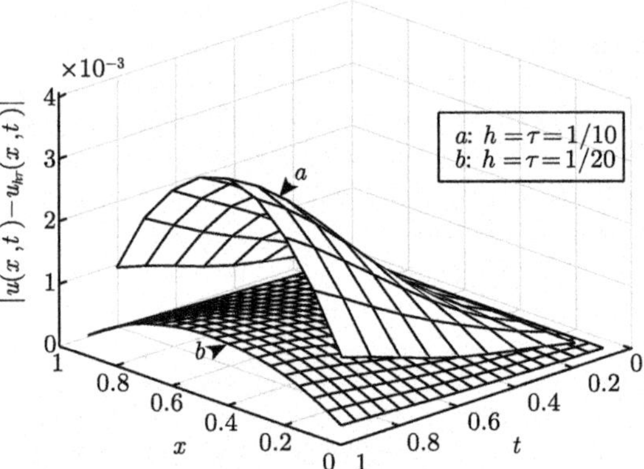

Fig. 4.10 (Example 4.2) The numerical error surfaces with different step sizes

4.3.4   A Priori Estimate of the Difference Solution

Theorem 4.7 *Let* $\{v_i^k \mid 0 \leqslant i \leqslant m, 0 \leqslant k \leqslant n\}$ *be the solution of the difference scheme*

$$
\begin{cases}
\dfrac{2}{\tau}\left(\delta_t v_i^{\frac{1}{2}} - \psi_i\right) - \dfrac{c^2}{2}(\delta_x^2 v_i^0 + \delta_x^2 v_i^1) = g_i^0, & 1 \leqslant i \leqslant m-1, & (4.38\text{a}) \\[2ex]
\delta_t^2 v_i^k - \dfrac{c^2}{2}(\delta_x^2 v_i^{k-1} + \delta_x^2 v_i^{k+1}) = g_i^k, & & \\[1ex]
\qquad 1 \leqslant i \leqslant m-1, \quad 1 \leqslant k \leqslant n-1, & & (4.38\text{b}) \\[2ex]
v_i^0 = \varphi_i, \quad 1 \leqslant i \leqslant m-1, & & (4.38\text{c}) \\[2ex]
v_0^k = 0, \quad v_m^k = 0, \quad 0 \leqslant k \leqslant n. & & (4.38\text{d})
\end{cases}
$$

Denote

$$E^k = \|\delta_t v^{k+\frac{1}{2}}\|^2 + \frac{c^2}{2}(|v^{k+1}|_1^2 + |v^k|_1^2).$$

Then for an arbitrary step ratio s, we have

$$E^k \leqslant e^{\frac{3}{2}k\tau}\left[c^2|v^0|_1^2 + 2\|\psi\|^2 + \frac{\tau^2}{2}\|g^0\|^2 + \frac{3}{2}\tau\sum_{l=1}^{k}\|g^l\|^2\right], \quad 0 \leqslant k \leqslant n-1.$$

Proof Using (4.38d), one gets

$$\delta_t v_0^{\frac{1}{2}} = 0, \quad \delta_t v_m^{\frac{1}{2}} = 0, \tag{4.39}$$

$$\Delta_t v_0^k = 0, \quad \Delta_t v_m^k = 0, \quad 1 \leqslant k \leqslant n-1. \tag{4.40}$$

Taking an inner product on both sides of (4.38a) with $\delta_t v^{\frac{1}{2}}$, we have

$$\frac{2}{\tau}\left[\|\delta_t v^{\frac{1}{2}}\|^2 - \left(\psi, \delta_t v^{\frac{1}{2}}\right)\right] - \frac{c^2}{2}\left(\delta_x^2 v^0 + \delta_x^2 v^1, \delta_t v^{\frac{1}{2}}\right) = \left(g^0, \delta_t v^{\frac{1}{2}}\right).$$

Using Lemma 1.4 and noticing (4.39), one arrives at

$$\frac{2}{\tau}\|\delta_t v^{\frac{1}{2}}\|^2 + \frac{c^2}{2\tau}\left(|v^1|_1^2 - |v^0|_1^2\right)$$

$$= \frac{2}{\tau}\left(\psi, \delta_t v^{\frac{1}{2}}\right) + \left(g^0, \delta_t v^{\frac{1}{2}}\right)$$

$$\leqslant \left(\frac{1}{2\tau}\|\delta_t v^{\frac{1}{2}}\|^2 + \frac{2}{\tau}\|\psi\|^2\right) + \left(\frac{1}{2\tau}\|\delta_t v^{\frac{1}{2}}\|^2 + \frac{\tau}{2}\|g^0\|^2\right).$$

It easily follows from the above inequality that

$$E^0 = \|\delta_t v^{\frac{1}{2}}\|^2 + \frac{c^2}{2}\left(|v^1|_1^2 + |v^0|_1^2\right) \leqslant c^2|v^0|_1^2 + 2\|\psi\|^2 + \frac{\tau^2}{2}\|g^0\|^2. \tag{4.41}$$

Taking an inner product on both sides of (4.38b) with $2\Delta_t v^k$, we have

$$2(\delta_t^2 v^k, \Delta_t v^k) - c^2(\delta_x^2 v^{k-1} + \delta_x^2 v^{k+1}, \Delta_t v^k) = 2(g^k, \Delta_t v^k), \quad 1 \leqslant k \leqslant n-1. \tag{4.42}$$

Now we estimate each term in the above equality. The first term on the left-hand side is

$$2(\delta_t^2 v^k, \Delta_t v^k) = \frac{1}{\tau}\left(\|\delta_t v^{k+\frac{1}{2}}\|^2 - \|\delta_t v^{k-\frac{1}{2}}\|^2\right). \tag{4.43}$$

For the second term on the left-hand side, applying Lemma 1.4 and noticing (4.40), we get

$$\begin{aligned}
&-(\delta_x^2 v^{k-1} + \delta_x^2 v^{k+1}, \Delta_t v^k) \\
&= (\delta_x v^{k-1} + \delta_x v^{k+1}, \Delta_t \delta_x v^k) \\
&= \frac{1}{2\tau}(\delta_x v^{k-1} + \delta_x v^{k+1}, \delta_x v^{k+1} - \delta_x v^{k-1}) \\
&= \frac{1}{2\tau}(\|\delta_x v^{k+1}\|^2 - \|\delta_x v^{k-1}\|^2) \\
&= \frac{1}{\tau}\left(\frac{\|\delta_x v^{k+1}\|^2 + \|\delta_x v^k\|^2}{2} - \frac{\|\delta_x v^k\|^2 + \|\delta_x v^{k-1}\|^2}{2}\right). \tag{4.44}
\end{aligned}$$

The right-hand side term becomes

$$2(g^k, \Delta_t v^k) \leqslant \|\Delta_t v^k\|^2 + \|g^k\|^2 \leqslant \frac{1}{2}\left(\|\delta_t v^{k+\frac{1}{2}}\|^2 + \|\delta_t v^{k-\frac{1}{2}}\|^2\right) + \|g^k\|^2. \tag{4.45}$$

Substituting (4.43), (4.44), and (4.45) into (4.42), we arrive at

$$\frac{1}{\tau}(E^k - E^{k-1}) \leqslant \frac{1}{2}(E^k + E^{k-1}) + \|g^k\|^2, \quad 1 \leqslant k \leqslant n-1,$$

or

$$\left(1 - \frac{\tau}{2}\right) E^k \leqslant \left(1 + \frac{\tau}{2}\right) E^{k-1} + \tau\|g^k\|^2, \quad 1 \leqslant k \leqslant n-1.$$

When $\tau \leqslant 2/3$, we have

$$E^k \leqslant \left(1 + \frac{3\tau}{2}\right) E^{k-1} + \frac{3}{2}\tau\|g^k\|^2, \quad 1 \leqslant k \leqslant n-1.$$

A direct application of the Gronwall inequality (Lemma 3.3) gives

$$E^k \leqslant e^{\frac{3}{2}k\tau}\left[E^0 + \frac{3}{2}\tau \sum_{l=1}^{k} \|g^l\|^2\right], \quad 1 \leqslant k \leqslant n-1.$$

Noticing (4.41), we have the desired inequality. $\square$

4.3.5 *Convergence and Stability of the Difference Solution*

Convergence

Theorem 4.8 *Let $\{u(x,t) \mid (x,t) \in \bar{D}\}$ be the solution of the problem (4.1) and $\{u_i^k \mid 0 \leqslant i \leqslant m,\ 0 \leqslant k \leqslant n\}$ be the solution of the difference scheme (4.36). Denote*

$$e_i^k = u(x_i, t_k) - u_i^k, \quad 0 \leqslant i \leqslant m, \quad 0 \leqslant k \leqslant n.$$

Then for an arbitrary step ratio s, we have

$$\|e^k\|_\infty \leqslant \frac{\sqrt{2}}{c}\frac{L}{4}e^{\frac{3}{4}T}\sqrt{1+6T}\,c_2(\tau^2 + h^2), \quad 1 \leqslant k \leqslant n.$$

Proof Subtracting (4.36) from (4.28), (4.31), (4.34), and (4.35), we get the system of error equations

$$\begin{cases} \dfrac{2}{\tau}\delta_t e_i^{\frac{1}{2}} - \dfrac{c^2}{2}(\delta_x^2 e_i^0 + \delta_x^2 e_i^1) = (R_2)_i^0, \quad 1 \leqslant i \leqslant m-1, \\[2mm] \delta_t^2 e_i^k - \dfrac{c^2}{2}(\delta_x^2 e_i^{k-1} + \delta_x^2 e_i^{k+1}) = (R_2)_i^k, \\[2mm] \qquad\qquad 1 \leqslant i \leqslant m-1, \quad 1 \leqslant k \leqslant n-1, \\[2mm] e_i^0 = 0, \quad 1 \leqslant i \leqslant m-1, \\[2mm] e_0^k = 0, \quad e_m^k = 0, \quad 0 \leqslant k \leqslant n. \end{cases}$$

By Theorem 4.7, one has

$$\|\delta_t e^{k+\frac{1}{2}}\|^2 + \frac{c^2}{2}(|e^{k+1}|_1^2 + |e^k|_1^2)$$

$$\leqslant e^{\frac{3}{2}k\tau}\left[c^2|e^0|_1^2 + \frac{\tau^2}{4}\|(R_2)^0\|^2 + \frac{3}{2}\tau\sum_{l=1}^{k}\|(R_2)^l\|^2 \right], \quad 0 \leqslant k \leqslant n-1.$$

In combination of (4.33), the above inequality becomes

$$\|\delta_t e^{k+\frac{1}{2}}\|^2 + \frac{c^2}{2}(|e^{k+1}|_1^2 + |e^k|_1^2)$$

$$\leqslant e^{\frac{3}{2}k\tau}\left[\frac{\tau^2}{4}Lc_2^2(\tau + h^2)^2 + \frac{3}{2}k\tau Lc_2^2(\tau^2 + h^2)^2 \right]$$

$$\leqslant e^{\frac{3}{2}T}\left(\frac{1}{4} + \frac{3}{2}T\right)Lc_2^2(\tau^2 + h^2)^2, \quad 0 \leqslant k \leqslant n-1.$$

It is easy to get

$$|e^k|_1 \leqslant \frac{\sqrt{2}}{c} \cdot e^{\frac{3}{4}T} \sqrt{\left(\frac{1}{4} + \frac{3}{2}T\right)L}\, c_2(\tau^2 + h^2), \quad 1 \leqslant k \leqslant n.$$

Therefore,

$$\|e^k\|_\infty \leqslant \frac{\sqrt{L}}{2}|e^k|_1 \leqslant \frac{\sqrt{2}}{c} \cdot \frac{L}{4} e^{\frac{3}{4}T} \sqrt{1+6T}\, c_2(\tau^2 + h^2), \quad 1 \leqslant k \leqslant n.$$

$\square$

Stability

Theorem 4.9 *The solution of the difference scheme (4.36) is stable with respect to the initial value and right-hand side term for any step ratio s in the following sense: Suppose $\{u_i^k \mid 0 \leqslant i \leqslant m, 0 \leqslant k \leqslant n\}$ is the solution of the difference scheme*

$$\begin{cases} \dfrac{2}{\tau}\left(\delta_t u_i^{\frac{1}{2}} - \psi_i\right) - \dfrac{c^2}{2}\left(\delta_x^2 u_i^0 + \delta_x^2 u_i^1\right) = f_i^0, & 1 \leqslant i \leqslant m-1, \\[2mm] \delta_t^2 u_i^k - \dfrac{c^2}{2}\left(\delta_x^2 u_i^{k-1} + \delta_x^2 u_i^{k+1}\right) = f_i^k, & 1 \leqslant i \leqslant m-1, \quad 1 \leqslant k \leqslant n-1, \\[2mm] u_i^0 = \varphi_i, & 1 \leqslant i \leqslant m-1, \\[2mm] u_0^k = 0, \quad u_m^k = 0, & 0 \leqslant k \leqslant n. \end{cases}$$

Then for an arbitrary step ratio s, we have

$$\|\delta_t u^{k+\frac{1}{2}}\|^2 + \frac{c^2}{2}\left(|u^{k+1}|_1^2 + |u^k|_1^2\right)$$

$$\leqslant e^{\frac{3}{2}k\tau}\left[c^2|u^0|_1^2 + 2\|\psi\|^2 + \frac{\tau^2}{2}\|f^0\|^2 + \frac{3}{2}\tau \sum_{l=1}^{k}\left\|f^l\right\|^2\right], \quad 0 \leqslant k \leqslant n-1.$$

Proof A direct application of Theorem 4.7 arrives at the conclusion. $\square$

Remark 4.2 *We can derive the following weighted difference scheme for (4.1)*

$$\begin{cases} \dfrac{2}{\tau}\left[\delta_t u_i^{\frac{1}{2}} - \psi(x_i)\right] - c^2 \delta_x^2 u_i^{\frac{1}{2}} = f(x_i, t_0), & 1 \leqslant i \leqslant m-1, \\[2mm] \delta_t^2 u_i^k - c^2\left[\theta \delta_x^2 u_i^{k-1} + (1-2\theta)\delta_x^2 u_i^k + \theta \delta_x^2 u_i^{k+1}\right] = f(x_i, t_k), & \\[1mm] \qquad\qquad\qquad 1 \leqslant i \leqslant m-1, \quad 1 \leqslant k \leqslant n-1, \\[2mm] u_i^0 = \varphi(x_i), & 1 \leqslant i \leqslant m-1, \\[2mm] u_0^k = \alpha(t_k), \quad u_m^k = \beta(t_k), & 0 \leqslant k \leqslant n, \end{cases} \tag{4.46}$$

Fig. 4.11 The stencil of the compact difference scheme (4.47)

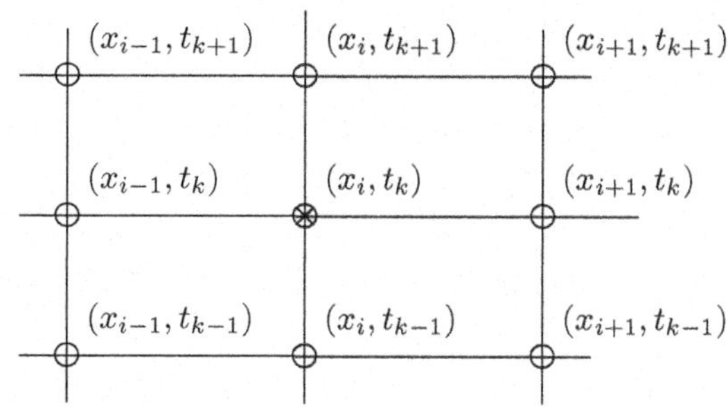

where $\theta \in [0, \frac{1}{2}]$. When $\theta = 0$, (4.46) simplifies to the explicit difference scheme (4.14); when $\theta = \frac{1}{2}$, (4.46) becomes the implicit difference scheme (4.36). It can be proved that the difference scheme (4.46) is stable and convergent when $(1 - 4\theta)s^2 < 1$.

4.4   The Compact Difference Scheme

The following compact difference scheme can be derived for the problem (4.1) in the form of

$$
\begin{cases}
\dfrac{2}{\tau}\mathcal{A}\left(\delta_t u_i^{\frac{1}{2}} - \psi(x_i)\right) - \dfrac{c^2}{2}\left(\delta_x^2 u_i^0 + \delta_x^2 u_i^1\right) = \mathcal{A}f(x_i, t_0), \quad 1 \leqslant i \leqslant m-1, \\[2mm]
\mathcal{A}\delta_t^2 u_i^k - \dfrac{c^2}{2}\left(\delta_x^2 u_i^{k-1} + \delta_x^2 u_i^{k+1}\right) = \mathcal{A}f(x_i, t_k), \\[2mm]
\qquad\qquad\qquad\qquad 1 \leqslant i \leqslant m-1, \quad 1 \leqslant k \leqslant n-1, \\[2mm]
u_i^0 = \varphi(x_i), \quad 1 \leqslant i \leqslant m-1, \\[2mm]
u_0^k = \alpha(t_k), \quad u_m^k = \beta(t_k), \quad 0 \leqslant k \leqslant n.
\end{cases}
$$

$$\tag{4.47}$$

The stencil of the compact difference scheme (4.47) is illustrated in Fig. 4.11.

It can be proved that the difference scheme (4.47) is uniquely solvable, convergent, and unconditionally stable for any step ratio s, and has the error estimate

$$
\max_{0 \leqslant i \leqslant m, 0 \leqslant k \leqslant n} |u(x_i, t_k) - u_i^k| = O(\tau^2 + h^4).
$$

Example 4.3 Apply the compact difference scheme (4.47) to solve the problem (4.15) in Example 4.1.

Table 4.8 lists some numerical results calculated with $h = 1/10$ and $\tau = 1/100$. Table 4.9 gives some numerical results calculated with $h = 1/20$ and $\tau = 1/400$. They indicate that the numerical solutions approximate the exact solutions well.

Table 4.8 (Example 4.3) Numerical solutions, exact solutions, and absolute values of the errors at part of node points ($h = 1/10$, $\tau = 1/100$)

(x, t)	NS	ES	$|ES - NS|$
$(0.5, 0.1)$	1.8221205	1.8221188	1.708e$-$6
$(0.5, 0.2)$	2.0137569	2.0137527	4.184e$-$6
$(0.5, 0.3)$	2.2255485	2.2255409	7.540e$-$6
$(0.5, 0.4)$	2.4596151	2.4596031	1.195e$-$5
$(0.5, 0.5)$	2.7182987	2.7182818	1.686e$-$5
$(0.5, 0.6)$	3.0041866	3.0041660	2.057e$-$5
$(0.5, 0.7)$	3.3201398	3.3201169	2.288e$-$5
$(0.5, 0.8)$	3.6693220	3.6692967	2.530e$-$5
$(0.5, 0.9)$	4.0552275	4.0552000	2.757e$-$5
$(0.5, 1.0)$	4.4817182	4.4816891	2.914e$-$5

Table 4.9 (Example 4.3) Numerical solutions, exact solutions, and absolute values of the errors at part of node points ($h = 1/20$, $\tau = 1/400$)

(x, t)	NS	ES	$|ES-NS|$
$(0.5, 0.1)$	1.8221189	1.8221188	1.077e$-$7
$(0.5, 0.2)$	2.0137530	2.0137527	2.635e$-$7
$(0.5, 0.3)$	2.2255414	2.2255409	4.740e$-$7
$(0.5, 0.4)$	2.4596039	2.4596031	7.473e$-$7
$(0.5, 0.5)$	2.7182829	2.7182818	1.073e$-$6
$(0.5, 0.6)$	3.0041673	3.0041660	1.271e$-$6
$(0.5, 0.7)$	3.3201184	3.3201169	1.443e$-$6
$(0.5, 0.8)$	3.6692983	3.6692967	1.585e$-$6
$(0.5, 0.9)$	4.0552017	4.0552000	1.716e$-$6
$(0.5, 1.0)$	4.4816909	4.4816891	1.828e$-$6

Table 4.10 (Example 4.3) The maximum errors of numerical solutions with different step sizes

h	τ	$E_\infty(h, \tau)$	$E_\infty(2h, 4\tau)/E_\infty(h, \tau)$
1/10	1/100	2.9143e$-$5	
1/20	1/400	1.8277e$-$6	15.945
1/40	1/1600	1.1427e$-$7	15.994
1/80	1/6400	7.4753e$-$9	15.287

Table 4.10 provides the maximum errors

$$E_\infty(h, \tau) = \max_{\substack{1 \leqslant i \leqslant m-1 \\ 0 \leqslant k \leqslant n}} |u(x_i, t_k) - u_i^k|$$

with different step sizes. We see from Table 4.10 that when h is reduced by half and τ is reduced to a quarter of the original, the maximum errors are reduced to 1/16 of the original.

Figure 4.12 shows the curves of the exact solutions and numerical solutions at $t = 1$ with the step sizes $h = 1/10$, $\tau = 1/100$. Figure 4.13 draws the error curves of the numerical solutions at $t = 1$. Figure 4.14 illustrates the error surfaces of numerical solutions with different step sizes.

Fig. 4.12 (Example 4.3) The curves of the numerical solutions and exact solutions at $t = 1$ ($h = 1/10$, $\tau = 1/100$)

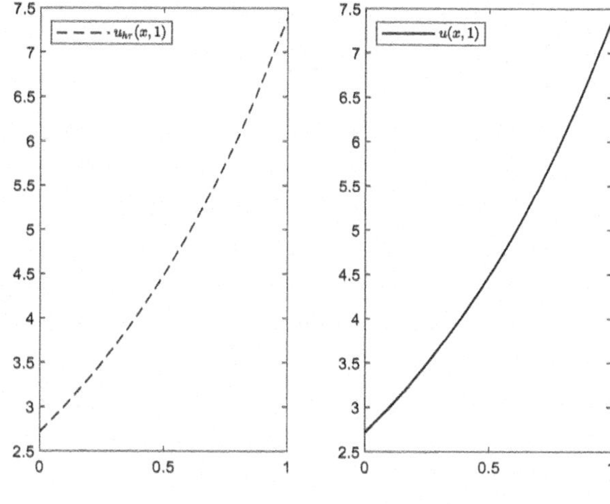

Fig. 4.13 (Example 4.3) The numerical error curves with different step sizes at $t = 1$

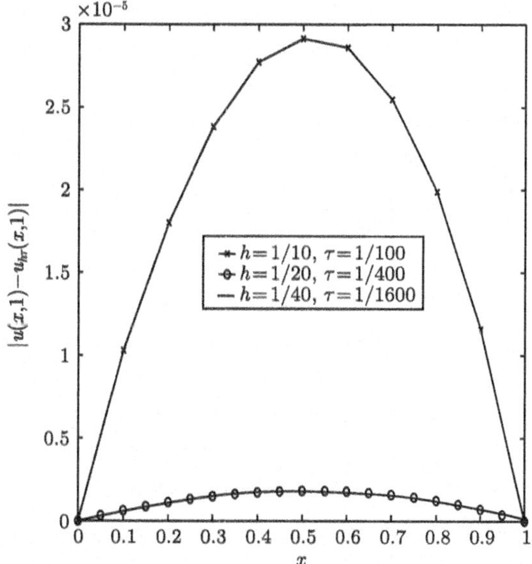

4.5   Finite Fourier Series with Applications

4.5.1   Finite Fourier Series

For the function $f(x)$ defined on $[0, L]$, if $f(0) = 0$ and $f(L) = 0$, then under certain conditions, it can be expanded into the following Fourier series (sine series)

$$f(x) = \sum_{l=1}^{\infty} a_l \sin \frac{l\pi x}{L}, \quad x \in [0, L], \tag{4.48}$$

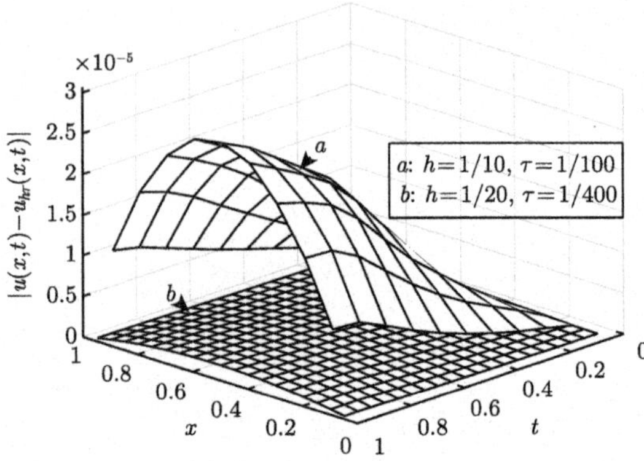

Fig. 4.14 (Example 4.3) The numerical error surfaces with different step sizes

where the coefficient $\{a_l\}$ is determined by

$$a_l = \frac{2}{L} \int_0^L f(x) \sin \frac{l\pi x}{L} \mathrm{d}x, \quad l = 1, 2, \ldots,$$

and we further have the Parseval identity

$$\int_0^L f^2(x)\mathrm{d}x = \frac{L}{2} \sum_{l=1}^{\infty} a_l^2.$$

Similarly, we can divide the interval $[0, L]$ into m equal subintervals. Denote $h = L/m, x_i = ih, 0 \leqslant i \leqslant m$. Consider the function $f(x)$ only at the node point in $\Omega_h \equiv \{x_i \mid 0 \leqslant i \leqslant m\}$. We can take the sum of the finite terms in (4.48)

$$f(x_i) = \sum_{l=1}^{m-1} A_l \sin \frac{l\pi x_i}{L}, \quad 1 \leqslant i \leqslant m - 1. \tag{4.49}$$

The expression (4.49) is called the **finite Fourier series expansion** of the function $f(x)$ on the interval $[0, L]$ with respect to the node point in Ω_h, and A_l is called the **finite Fourier coefficient**.

Lemma 4.1 *In regard to* $\theta \in (-2, 2)$, *we have*

$$\sum_{l=1}^{m-1} \cos(l\pi\theta) = \begin{cases} \dfrac{1}{2}\left[\dfrac{\sin\left(\left(m - \frac{1}{2}\right)\pi\theta\right)}{\sin\left(\frac{1}{2}\pi\theta\right)} - 1\right], & \theta \neq 0, \\[4mm] m - 1, & \theta = 0. \end{cases}$$

Proof When $\theta = 0$, the result is true obviously.

When $\theta \neq 0$, with the application of the product to sum formula, we have

$$2\sin\left(\frac{1}{2}\pi\theta\right)\sum_{l=1}^{m-1}\cos(l\pi\theta) = \sum_{l=1}^{m-1}\left[\sin\left(\left(l+\frac{1}{2}\right)\pi\theta\right) - \sin\left(\left(l-\frac{1}{2}\right)\pi\theta\right)\right]$$

$$= \sin\left(\left(m-\frac{1}{2}\right)\pi\theta\right) - \sin\left(\frac{1}{2}\pi\theta\right).$$

Dividing by $2\sin\left(\frac{1}{2}\pi\theta\right)$ on both sides, it yields the result. $\square$

Lemma 4.2

$$h\sum_{l=1}^{m-1}\left(\sin\frac{l\pi x_i}{L}\right)\left(\sin\frac{l\pi x_j}{L}\right) = \begin{cases} 0, & 1 \leqslant i, j \leqslant m-1, \quad i \neq j, \\ \dfrac{L}{2}, & 1 \leqslant i, j \leqslant m-1, \quad i = j. \end{cases}$$

Proof

(I) When $1 \leqslant i, j \leqslant m-1$, $i \neq j$, using Lemma 4.1, we have

$$h\sum_{l=1}^{m-1}\left(\sin\frac{l\pi x_i}{L}\right)\left(\sin\frac{l\pi x_j}{L}\right)$$

$$= \frac{1}{2}h\sum_{l-1}^{m-1}\left[\cos\frac{l\pi(x_i-x_j)}{L} - \cos\frac{l\pi(x_i+x_j)}{L}\right]$$

$$= \frac{1}{2}h\left[\sum_{l=1}^{m-1}\cos\frac{l\pi(x_i-x_j)}{L} - \sum_{l=1}^{m-1}\cos\frac{l\pi(x_i+x_j)}{L}\right]$$

$$= \frac{1}{4}h\left[\frac{\sin\left(\left(m-\frac{1}{2}\right)\pi\frac{x_i-x_j}{L}\right)}{\sin\frac{\pi(x_i-x_j)}{2L}} - \frac{\sin\left(\left(m-\frac{1}{2}\right)\pi\frac{x_i+x_j}{L}\right)}{\sin\frac{\pi(x_i+x_j)}{2L}}\right]$$

$$= \frac{1}{4}h\left[\frac{\sin\left((i-j)\pi - \frac{\pi(x_i-x_j)}{2L}\right)}{\sin\frac{\pi(x_i-x_j)}{2L}} - \frac{\sin\left((i+j)\pi - \frac{\pi(x_i+x_j)}{2L}\right)}{\sin\frac{\pi(x_i+x_j)}{2L}}\right]$$

$$= \frac{1}{4}h\left[\frac{(-1)^{i-j-1}\sin\frac{\pi(x_i-x_j)}{2L}}{\sin\frac{\pi(x_i-x_j)}{2L}} - \frac{(-1)^{i+j-1}\sin\frac{\pi(x_i+x_j)}{2L}}{\sin\frac{\pi(x_i+x_j)}{2L}}\right]$$

$$= \frac{1}{4}h\left[(-1)^{i-j-1} - (-1)^{i+j-1}\right] = 0.$$

(II) When $1 \leqslant i, j \leqslant m - 1$, $i = j$, applying Lemma 4.1, we have

$$h \sum_{l=1}^{m-1} \sin^2 \frac{l\pi x_i}{L}$$

$$= \frac{1}{2} h \sum_{l=1}^{m-1} \left(1 - \cos \frac{2l\pi x_i}{L} \right)$$

$$= \frac{1}{2} h \left[(m-1) - \sum_{l=1}^{m-1} \cos \frac{2l\pi x_i}{L} \right]$$

$$= \frac{1}{2} h \left\{ (m-1) - \frac{1}{2} \left[\frac{\sin\left(2\left(m-\frac{1}{2}\right)\frac{\pi x_i}{L}\right)}{\sin \frac{\pi x_i}{L}} - 1 \right] \right\}$$

$$= \frac{1}{2} h \left\{ m - 1 - \frac{1}{2} \left[\frac{\sin\left(2i\pi - \frac{\pi x_i}{L}\right)}{\sin \frac{\pi x_i}{L}} - 1 \right] \right\}$$

$$= \frac{1}{2} h \left[m - 1 - \frac{1}{2} \left(\frac{-\sin \frac{\pi x_i}{L}}{\sin \frac{\pi x_i}{L}} - 1 \right) \right]$$

$$= \frac{1}{2} mh = \frac{L}{2},$$

which completes the proof.

$\square$

Theorem 4.10 *For the finite* Fourier *series (4.49), we have*

$$A_l = \frac{2}{L} \cdot h \sum_{i=1}^{m-1} f(x_i) \sin \frac{l\pi x_i}{L}, \quad 1 \leqslant l \leqslant m - 1,$$

$$h \sum_{i=1}^{m-1} f^2(x_i) = \frac{L}{2} \sum_{l=1}^{m-1} A_l^2. \tag{4.50}$$

Proof

(a) The equality (4.49) can be written as

$$f(x_i) = \sum_{j=1}^{m-1} A_j \sin \frac{j\pi x_i}{L}, \quad 1 \leqslant i \leqslant m - 1.$$

Multiplying by $\sin \frac{l\pi x_i}{L}$ on both sides of the above equality, summing over i from 1 to $m-1$ and applying Lemma 4.2, we have

$$
h \sum_{i=1}^{m-1} f(x_i) \sin \frac{l\pi x_i}{L}
$$

$$
= h \sum_{i=1}^{m-1} \left[\sum_{j=1}^{m-1} A_j \left(\sin \frac{j\pi x_i}{L} \right) \left(\sin \frac{l\pi x_i}{L} \right) \right]
$$

$$
= \sum_{j=1}^{m-1} A_j \left[h \sum_{i=1}^{m-1} \left(\sin \frac{i\pi x_j}{L} \right) \left(\sin \frac{i\pi x_l}{L} \right) \right]
$$

$$
= A_l \left(h \sum_{i=1}^{m-1} \sin^2 \frac{i\pi x_l}{L} \right)
$$

$$
= \frac{L}{2} A_l, \quad 1 \leqslant l \leqslant m-1.
$$

Therefore,

$$
A_l = \frac{2}{L} h \sum_{i=1}^{m-1} f(x_i) \sin \frac{l\pi x_i}{L}, \quad 1 \leqslant l \leqslant m-1.
$$

(b)

$$
h \sum_{i=1}^{m-1} f^2(x_i) = h \sum_{i=1}^{m-1} \left(\sum_{j=1}^{m-1} A_j \sin \frac{j\pi x_i}{L} \right)^2
$$

$$
= h \sum_{i=1}^{m-1} \left(\sum_{j=1}^{m-1} A_j \sin \frac{j\pi x_i}{L} \right) \left(\sum_{l=1}^{m-1} A_l \sin \frac{l\pi x_i}{L} \right)
$$

$$
= \sum_{j=1}^{m-1} A_j \left(h \sum_{i=1}^{m-1} \sin \frac{j\pi x_i}{L} \right) \left(\sum_{l=1}^{m-1} A_l \sin \frac{l\pi x_i}{L} \right)
$$

$$
= \sum_{j=1}^{m-1} \sum_{l=1}^{m-1} A_j A_l \left[h \sum_{i=1}^{m-1} \left(\sin \frac{j\pi x_i}{L} \right) \left(\sin \frac{l\pi x_i}{L} \right) \right]
$$

$$
= \sum_{j=1}^{m-1} \sum_{l=1}^{m-1} A_j A_l \left[h \sum_{i=1}^{m-1} \left(\sin \frac{i\pi x_j}{L} \right) \left(\sin \frac{i\pi x_l}{L} \right) \right]
$$

$$= \sum_{l=1}^{m-1} A_l^2 \left(h \sum_{i=1}^{m-1} \sin^2 \frac{i\pi x_l}{L} \right)$$

$$= \frac{L}{2} \sum_{l=1}^{m-1} A_l^2.$$

This completes the proof.

$\square$

From (4.50), it follows that the evaluation of the 2-norm of a grid function

$$f = \{ f_i \mid 0 \leqslant i \leqslant m, \text{ and } f_0 = 0, \ f_m = 0 \}$$

can be transformed into computing the sum of squares of the coefficients of the finite Fourier series. Thus, the finite Fourier series can be employed to obtain a priori estimates for the solution of the constant coefficient difference scheme, as well as to analyze the stability and convergence of the difference scheme in the 2-norm.

4.5.2 A Priori Estimate of the Difference Solution for the Two-Point Boundary Value Problem

Consider the two-point boundary value problem

$$\begin{cases} -v'' + qv = f(x), & 0 < x < L, \\ v(0) = 0, & v(L) = 0, \end{cases}$$

where q is a nonnegative constant.

Denote $h = L/m$, $x_i = ih$, $0 \leqslant i \leqslant m$. A difference scheme for the above problem reads

$$\begin{cases} -\delta_x^2 v_i + q v_i = f_i, & 1 \leqslant i \leqslant m-1, & \text{(4.51a)} \\ v_0 = 0, & v_m = 0. & \text{(4.51b)} \end{cases}$$

Theorem 4.11 *Let* $v = \{ v_i \mid 0 \leqslant i \leqslant m \}$ *be the solution of the difference scheme (4.51). Then we have*

$$\|v\| \leqslant \frac{L^2}{4} \|f\|,$$

where $\|f\| = \sqrt{h \sum_{i=1}^{m-1} (f_i)^2}.$

Proof Let $f_0 = 0$ and $f_m = 0$. Then we have

$$v_i = \sum_{l=1}^{m-1} A_l \sin \frac{l\pi x_i}{L}, \quad f_i = \sum_{l=1}^{m-1} \alpha_l \sin \frac{l\pi x_i}{L}, \quad 1 \leqslant i \leqslant m-1. \tag{4.52}$$

Substituting the above equalities into (4.51a), we arrive at

$$-\frac{1}{h^2}\left(\sum_{l=1}^{m-1} A_l \sin \frac{l\pi x_{i+1}}{L} - 2\sum_{l=1}^{m-1} A_l \sin \frac{l\pi x_i}{L} + \sum_{l=1}^{m-1} A_l \sin \frac{l\pi x_{i-1}}{L}\right)$$

$$+q\sum_{l=1}^{m-1} A_l \sin \frac{l\pi x_i}{L}$$

$$= \sum_{l=1}^{m-1} \alpha_l \sin \frac{l\pi x_i}{L}, \quad 1 \leqslant i \leqslant m-1,$$

or

$$\sum_{l=1}^{m-1} A_l \left(-\frac{1}{h^2}\right)\left(\sin \frac{l\pi x_{i+1}}{L} - 2\sin \frac{l\pi x_i}{L} + \sin \frac{l\pi x_{i-1}}{L}\right) + q\sum_{l=1}^{m-1} A_l \sin \frac{l\pi x_i}{L}$$

$$= \sum_{l-1}^{m-1} \alpha_l \sin \frac{l\pi x_i}{L}, \quad 1 \leqslant i \leqslant m-1.$$

By simplification of the above equality, we get

$$\sum_{l=1}^{m-1}\left(\frac{4}{h^2}\sin^2 \frac{l\pi h}{2L} + q\right) A_l \sin \frac{l\pi x_i}{L} = \sum_{l=1}^{m-1} \alpha_l \sin \frac{l\pi x_i}{L}, \quad 1 \leqslant i \leqslant m-1.$$

Comparing the coefficients on both sides of the above equality, we have

$$\left(\frac{4}{h^2}\sin^2 \frac{l\pi h}{2L} + q\right) A_l = \alpha_l, \quad 1 \leqslant l \leqslant m-1, \tag{4.53}$$

which implies

$$A_l = \frac{\alpha_l}{\dfrac{4}{h^2}\sin^2 \dfrac{l\pi h}{2L} + q}, \quad 1 \leqslant l \leqslant m-1.$$

When $0 < x < \pi/2$,

$$\frac{x}{\sin x} < \frac{\pi}{2}.$$

It further deduces that

$$|A_l| \leqslant \frac{|\alpha_l|}{\dfrac{4}{h^2}\sin^2\dfrac{l\pi h}{2L}} \leqslant \frac{L^2}{4}|\alpha_l|, \quad 1 \leqslant l \leqslant m-1.$$

Hence,

$$\|v\|^2 = \frac{L}{2}\sum_{l=1}^{m-1} A_l^2 \leqslant \frac{L}{2}\sum_{l=1}^{m-1}\left(\frac{L^2}{4}|\alpha_l|\right)^2 = \left(\frac{L^2}{4}\right)^2 \|f\|^2.$$

Taking the square root on both sides of the above inequality, we have

$$\|v\| \leqslant \frac{L^2}{4}\|f\|.$$

$\square$

It can be seen from the above procedure of analysis that it suffices to substitute the general terms

$$v_i = A_l \sin\frac{l\pi x_i}{L}, \quad f_i = \alpha_l \sin\frac{l\pi x_i}{L}$$

of (4.52) into (4.51a). Then we immediately get

$$A_l\left(-\frac{1}{h^2}\right)\left(\sin\frac{l\pi x_{i+1}}{L} - 2\sin\frac{l\pi x_i}{L} + \sin\frac{l\pi x_{i-1}}{L}\right) + qA_l\sin\frac{l\pi x_i}{L} = \alpha_l\sin\frac{l\pi x_i}{L},$$

$$1 \leqslant i \leqslant m-1.$$

Dividing by $\sin(\frac{l\pi x_i}{L})$ on both sides of the above equality directly gives (4.53).

4.5.3 A Priori Estimate of the Difference Solution for the First Boundary Value Problem of the Parabolic Equation

Consider the first boundary value problem of the parabolic equation

$$
\begin{cases}
\dfrac{\partial v}{\partial t} - a \dfrac{\partial^2 v}{\partial x^2} = 0, & 0 < x < L, \quad 0 < t \leqslant T, \\[2mm]
v(x,0) = \varphi(x), & 0 \leqslant x \leqslant L, \\[2mm]
v(0,t) = 0, \quad v(L,t) = 0, & 0 < t \leqslant T,
\end{cases}
$$

where a is a positive constant and $\varphi(0) = \varphi(L) = 0$. Denote $h = \frac{L}{m}$, $\tau = \frac{T}{n}$, $r = a\frac{\tau}{h^2}$, $x_i = ih$, $0 \leqslant i \leqslant m$, $t_k = k\tau$, $0 \leqslant k \leqslant n$. A difference scheme reads

$$
\begin{cases}
\dfrac{1}{\tau}(v_i^{k+1} - v_i^{k}) - a\left[\theta \delta_x^2 v_i^{k+1} + (1-\theta)\delta_x^2 v_i^{k}\right] = 0, & \\[2mm]
\qquad\qquad\qquad\qquad 1 \leqslant i \leqslant m-1, \quad 0 \leqslant k \leqslant n-1, & (4.54\text{a}) \\[2mm]
v_i^0 = \varphi(x_i), \quad 1 \leqslant i \leqslant m-1, & (4.54\text{b}) \\[2mm]
v_0^k = 0, \quad v_m^k = 0, \quad 0 \leqslant k \leqslant n, & (4.54\text{c})
\end{cases}
$$

where $\theta \in [0, 1]$. When $\theta = 0$, (4.54) corresponds to the forward Euler scheme; when $\theta = 1$, (4.54) corresponds to the backward Euler scheme; when $\theta = 1/2$, (4.54) corresponds to the Crank-Nicolson scheme; and when $\theta = \frac{1}{2}(1 - \frac{1}{6r})$, (4.54) represents the compact difference scheme.

Theorem 4.12 *Let $\{v_i^k \mid 0 \leqslant i \leqslant m, 0 \leqslant k \leqslant n\}$ be the solution of the difference scheme (4.54). Denote $r = a\tau/h^2$. Then when $2r(1 - 2\theta) \leqslant 1$, we have*

$$
\|v^k\| \leqslant \|\varphi\|, \quad 1 \leqslant k \leqslant n.
$$

Proof It follows from (4.54c) that the solution of (4.54) can be rewritten as

$$
v_i^k = \sum_{l=1}^{m-1} T_l(k) \sin \frac{l\pi x_i}{L}, \quad 1 \leqslant i \leqslant m-1, \quad 0 \leqslant k \leqslant n. \tag{4.55}
$$

Substituting (4.55) into (4.54a), we have

$$
\frac{1}{\tau} \sum_{l=1}^{m-1} \left[T_l(k+1) - T_l(k)\right] \sin \frac{l\pi x_i}{L} - a \sum_{l=1}^{m-1} \left[\theta T_l(k+1) + (1-\theta)T_l(k)\right]
$$

$$
\cdot \frac{1}{h^2}\left(\sin \frac{l\pi x_{i+1}}{L} - 2\sin \frac{l\pi x_i}{L} + \sin \frac{l\pi x_{i-1}}{L}\right) = 0, \quad 1 \leqslant i \leqslant m-1.
$$

The simplification of the above equality gives

$$\sum_{l=1}^{m-1}\left\{\left(1+4r\theta\sin^2\frac{l\pi h}{2}L\right)T_l(k+1)\right.$$
$$\left.-\left[1-4r(1-\theta)\sin^2\frac{l\pi h}{2}L\right]T_l(k)\right\}\sin\frac{l\pi x_i}{2}=0,\quad 1\leqslant i\leqslant m-1.$$

Consequently,

$$\left(1+4r\theta\sin^2\frac{l\pi h}{2L}\right)T_l(k+1)-\left[1-4r(1-\theta)\sin^2\frac{l\pi h}{2L}\right]T_l(k)=0,$$
$$1\leqslant l\leqslant m-1.$$

Denote

$$G(l)=\frac{1-4r(1-\theta)\sin^2\dfrac{l\pi h}{2L}}{1+4r\theta\sin^2\dfrac{l\pi h}{2L}},$$

and then we have

$$T_l(k+1)=G(l)T_l(k),\quad 1\leqslant l\leqslant m-1,\quad 0\leqslant k\leqslant n-1. \tag{4.56}$$

When $2r(1-2\theta)\leqslant 1$,

$$2r(1-2\theta)\sin^2\frac{l\pi h}{2L}\leqslant 1,\quad 1\leqslant l\leqslant m-1.$$

It is easy to know

$$-1\leqslant G(l)\leqslant 1,\quad 1\leqslant l\leqslant m-1.$$

Consequently, with the help of (4.56), we have

$$|T_l(k+1)|\leqslant|T_l(k)|,\quad 1\leqslant l\leqslant m-1,\quad 0\leqslant k\leqslant n-1. \tag{4.57}$$

Combining (4.55) with (4.57), we have

$$\|v^{k+1}\|^2=\frac{L}{2}\sum_{l=1}^{m-1}\left(T_l(k+1)\right)^2\leqslant\frac{L}{2}\sum_{l=1}^{m-1}\left(T_l(k)\right)^2=\|v^k\|^2,\quad 0\leqslant k\leqslant n-1.$$

It follows from the recursion that

$$\|v^k\| \leqslant \|v^0\| = \|\varphi\|, \quad 1 \leqslant k \leqslant n.$$

$\square$

4.5.4 A Priori Estimate of the Difference Solution for the First Boundary Value Problem of the Hyperbolic Equation

Consider the first boundary value problem of the hyperbolic equation

$$\begin{cases} \dfrac{\partial^2 v}{\partial t^2} - c^2 \dfrac{\partial^2 v}{\partial x^2} = 0, \quad 0 < x < L, \quad 0 < t \leqslant T, \\[2mm] v(x,0) = \varphi(x), \quad \dfrac{\partial v}{\partial t}(x,0) = \psi(x), \quad 0 \leqslant x \leqslant L, \\[2mm] v(0,t) = 0, \quad v(L,t) = 0, \quad 0 < t \leqslant T, \end{cases}$$

where c is a positive constant and $\varphi(0) = \varphi(L) = \psi(0) = \psi(L) = 0$. A difference scheme for the above problem reads

$$\begin{cases} \delta_t^2 v_i^k - c^2 \left[\theta \delta_x^2 v_i^{k+1} + (1 - 2\theta)\delta_x^2 v_i^k + \theta \delta_x^2 v_i^{k-1} \right] = 0, \\[2mm] \qquad\qquad\qquad 1 \leqslant i \leqslant m-1, \quad 1 \leqslant k \leqslant n-1, \qquad (4.58\text{a}) \\[2mm] v_i^0 = \varphi(x_i), \quad v_i^1 = \varphi(x_i) + \tau \psi(x_i), \quad 1 \leqslant i \leqslant m-1, \qquad (4.58\text{b}) \\[2mm] v_0^k = 0, \quad v_m^k = 0, \quad 0 \leqslant k \leqslant n. \qquad\qquad\qquad\qquad (4.58\text{c}) \end{cases}$$

It can be proved that the local truncation error of (4.58) is

$$R_i^k = \begin{cases} O(\tau^2 + h^2), \ \theta \neq \dfrac{1}{12}\left(1 - \dfrac{1}{s^2}\right), \\[3mm] O(\tau^4 + h^4), \ \theta = \dfrac{1}{12}\left(1 - \dfrac{1}{s^2}\right), \end{cases}$$

where $s = \frac{c\tau}{h}$. When $\theta = 0$, (4.58) is an explicit scheme, and when $\theta \neq 0$, (4.58) is an implicit one.

Theorem 4.13 *Let $\{v_i^k \mid 0 \leqslant i \leqslant m, 0 \leqslant k \leqslant n\}$ be the solution of the difference scheme (4.58). Then when*

$$(1 - 4\theta)s^2 \leqslant 1,$$

we have

$$\|v^k\|^2 \leqslant 4\|\varphi\|^2 + 2T^2\|\psi\|^2, \quad 0 \leqslant k \leqslant n.$$

Proof It follows from (4.58c) that the solution of (4.58) can be written as

$$v_i^k = \sum_{l=1}^{m-1} T_l(k) \sin \frac{l\pi x_i}{L}, \quad 1 \leqslant i \leqslant m-1, \quad 0 \leqslant k \leqslant n. \tag{4.59}$$

Substituting (4.59) into (4.58a), we get

$$\sum_{l=1}^{m-1} \frac{1}{\tau^2} \Big[T_l(k+1) - 2T_l(k) + T_l(k-1) \Big] \sin \frac{l\pi x_i}{L}$$

$$- c^2 \sum_{l=1}^{m-1} \Big[\theta T_l(k+1) + (1-2\theta)T_l(k) + \theta T_l(k-1) \Big]$$

$$\cdot \frac{1}{h^2} \Big(\sin \frac{l\pi x_{i+1}}{L} - 2\sin \frac{l\pi x_i}{L} + \sin \frac{l\pi x_{i-1}}{L} \Big) = 0.$$

By simplification, we get

$$\sum_{l=1}^{m-1} \Big\{ \Big[T_l(k+1) - 2T_l(k) + T_l(k-1) \Big]$$

$$+ 4s^2 \sin^2 \frac{l\pi h}{2L} \Big[\theta T_l(k+1) + (1-2\theta)T_l(k) + \theta T_l(k-1) \Big] \Big\} \sin \frac{l\pi x_i}{L} = 0,$$

namely,

$$\sum_{l=1}^{m-1} \Big[\Big(1 + 4\theta s^2 \sin^2 \frac{l\pi h}{2L} \Big) T_l(k+1) - 2 \Big(1 - 2(1-2\theta)s^2 \sin^2 \frac{l\pi h}{2L} \Big) T_l(k)$$

$$+ \Big(1 + 4\theta s^2 \sin^2 \frac{l\pi h}{2L} \Big) T_l(k-1) \Big] \sin \frac{l\pi x_i}{2} = 0,$$

$$1 \leqslant i \leqslant m-1, \quad 1 \leqslant k \leqslant n-1.$$

Consequently,

$$\Big(1 + 4\theta s^2 \sin^2 \frac{l\pi h}{2L} \Big) T_l(k+1) - 2 \Big(1 - 2(1-2\theta)s^2 \sin^2 \frac{l\pi h}{2L} \Big) T_l(k)$$

$$+ \Big(1 + 4\theta s^2 \sin^2 \frac{l\pi h}{2L} \Big) T_l(k-1) = 0, \quad 1 \leqslant l \leqslant m-1, \quad 1 \leqslant k \leqslant n-1.$$

Denote

$$C_l = \frac{1 - 2(1 - 2\theta)s^2 \sin^2 \dfrac{l\pi h}{2L}}{1 + 4\theta s^2 \sin^2 \dfrac{l\pi h}{2L}}.$$

Then we have

$$T_l(k+1) - 2C_l T_l(k) + T_l(k-1) = 0, \quad 1 \leqslant l \leqslant m-1, \quad 1 \leqslant k \leqslant n-1. \tag{4.60}$$

Note that (4.60) is a second-order constant coefficient difference equation. Suppose the solution of (4.60) has the form $T_l(k) = \lambda^k$. Substituting this into (4.60), we obtain the quadratic equation

$$\lambda^2 - 2C_l\lambda + 1 = 0, \tag{4.61}$$

which is called the **characteristic equation** of the difference equation (4.60).
   When $(1 - 4\theta)s^2 \leqslant 1$,

$$(1 - 4\theta)s^2 \sin^2 \frac{l\pi h}{2L} < 1, \quad 1 \leqslant l \leqslant m-1.$$

It is easy to know

$$-1 < C_l < 1.$$

Denote

$$\theta_l = \arctan \frac{\sqrt{1 - C_l^2}}{C_l}.$$

The roots of the characteristic equation (4.61) (generally called the characteristic root) are

$$\lambda = \cos\theta_l \pm i\sin\theta_l.$$

The general solution of (4.60) is thus obtained as

$$T_l(k) = \mu_l \cos(k\theta_l) + \nu_l \sin(k\theta_l), \tag{4.62}$$

where μ_l and ν_l are undetermined constants.

It follows from $T_l(0)$ and $T_l(1)$ that

$$\mu_l = T_l(0), \quad \nu_l = \frac{T_l(1) - T_l(0)\cos\theta_l}{\sin\theta_l}. \tag{4.63}$$

Substituting (4.63) into (4.62), one knows that the solution of (4.60) is

$$T_l(k) = -\frac{\sin((k-1)\theta_l)}{\sin\theta_l} T_l(0) + \frac{\sin(k\theta_l)}{\sin\theta_l} T_l(1), \quad 1 \leqslant l \leqslant m-1, \quad 0 \leqslant k \leqslant n. \tag{4.64}$$

Let

$$\varphi_i = \sum_{l=1}^{m-1} a_l \sin\frac{l\pi x_i}{L}, \quad \psi_i = \sum_{l=1}^{m-1} b_l \sin\frac{l\pi x_i}{L}, \quad 1 \leqslant i \leqslant m-1.$$

Then we have

$$T_l(0) = a_l, \quad T_l(1) = a_l + \tau b_l, \quad 1 \leqslant l \leqslant m-1. \tag{4.65}$$

Substituting (4.65) into (4.64), we have

$$T_l(k) = \frac{\sin(k\theta_l) - \sin\big((k-1)\theta_l\big)}{\sin\theta_l} a_l + \tau \frac{\sin(k\theta_l)}{\sin\theta_l} b_l, \quad 1 \leqslant l \leqslant m-1, \quad 0 \leqslant k \leqslant n. \tag{4.66}$$

Noticing

$$\frac{\sin(k\theta_l) - \sin\big((k-1)\theta_l\big)}{\sin\theta_l} = \frac{2\cos\big(\big(k-\frac{1}{2}\big)\theta_l\big)\sin\frac{\theta_l}{2}}{\sin\theta_l} = \frac{\cos\big(\big(k-\frac{1}{2}\big)\theta_l\big)}{\cos\frac{\theta_l}{2}}$$

and

$$\frac{1}{\sqrt{2}} \leqslant \cos\frac{x}{2} \leqslant 1, \quad x \in \left(-\frac{\pi}{2}, \frac{\pi}{2}\right),$$

we have

$$\left| \frac{\sin(k\theta_l) - \sin\big((k-1)\theta_l\big)}{\sin\theta_l} \right| \leqslant \frac{1}{\cos\frac{\theta_l}{2}} \leqslant \sqrt{2}. \tag{4.67}$$

On the other hand, we can prove by induction that

$$|\sin k\theta_l| \leqslant k |\sin\theta_l|, \quad 0 \leqslant k \leqslant n.$$

Therefore,

$$\left| \frac{\tau \sin(k\theta_l)}{\sin \theta_l} \right| \leqslant k\tau \leqslant T, \quad 0 \leqslant k \leqslant n. \tag{4.68}$$

Substituting (4.67) and (4.68) into (4.66), we can get

$$|T_l(k)| \leqslant \sqrt{2}\, |a_l| + T|b_l|, \quad 1 \leqslant l \leqslant m-1, \quad 0 \leqslant k \leqslant n.$$

Thus,

$$\left(T_l(k)\right)^2 \leqslant 2\left[(\sqrt{2}\,a_l)^2 + (Tb_l)^2\right] = 4a_l^2 + 2T^2 b_l^2, \quad 1 \leqslant l \leqslant m-1, \quad 0 \leqslant k \leqslant n.$$

Summing over l on both sides of the above inequality produces

$$\|v^k\|^2 = \frac{L}{2}\sum_{l=1}^{m-1}\left(T_l(k)\right)^2 \leqslant \frac{L}{2}\sum_{l=1}^{m-1}\left(4a_l^2 + 2T^2 b_l^2\right)$$

$$= 4\|\varphi\|^2 + 2T^2\|\psi\|^2, \quad 0 \leqslant k \leqslant n.$$

$$\square$$

Remark 4.3 *The stability derived through finite Fourier series analysis is understood in the sense of the 2-norm. For the explicit difference scheme (4.14) applied to the hyperbolic equation, the step ratio $s = 1$ represents a critical point that separates stability from instability. According to Theorem 4.13, the scheme is stable in the 2-norm when the step ratio $s = 1$. However, as shown in Theorem 4.5, it is unstable in the maximum norm for the same step ratio. Both results are not contradictory, as the 2-norm is weaker than the maximum norm.*

4.6 Summary and Extension

In this chapter, we have examined finite difference methods for solving the initial-boundary value problem of the one-dimensional hyperbolic equation. First, a priori estimate for the homogeneous problem is derived using the energy method. Then, we introduce in detail an explicit three-time-level difference scheme utilizing five node points, along with a three-time-level implicit difference scheme using seven node points. A priori estimate for the solutions of both schemes is provided using the energy method. The uniqueness, convergence, and stability of the difference schemes are proved. Finally, we derive a three-time-level implicit difference scheme (compact difference scheme) employing nine node points. Readers can perform the theoretical analysis of the compact difference scheme using the method outlined in Sect. 4.3. The difference method for solving the initial-boundary value problem of the first-order hyperbolic equation is left as an exercise for readers. Additionally, the Richardson extrapolation method is also applicable to the hyperbolic equation.

In Sect. 4.5, we begin by discussing the finite Fourier series and its properties. The finite Fourier series is then applied to derive a priori estimate of the difference solution for the two-point boundary value problem of the ordinary differential equation, the first initial-boundary value problem of the parabolic equation, and the first initial-boundary value problem of the hyperbolic equation. Finite Fourier series analysis, Von Neumann analysis, and Fourier series analysis are essentially equivalent [1–5]. Readers are encouraged to extend the concept of finite Fourier series to two-dimensional rectangular grid functions and apply it to study a priori estimate of the difference solution for the first boundary value problem of a two-dimensional elliptic equation, the first boundary value problem of a two-dimensional parabolic equation, and the first boundary value problem of a two-dimensional hyperbolic equation. Additionally, for the difference solutions of parabolic and hyperbolic equations, a priori estimate with respect to the source function can also be considered.

This book emphasizes the maximum principle, the energy method, and the finite Fourier series method for the analysis of difference schemes, with the energy method being one of the most powerful tools. For other analytical methods, readers are referred to the references [1–5].

4.7   Exercise

4.1 Consider the problem of the first-order hyperbolic equation

$$
\begin{cases}
\dfrac{\partial u}{\partial t} + a\dfrac{\partial u}{\partial x} = f(x,t), & 0 < x \leqslant L, \quad 0 < t \leqslant T, \\[2mm]
u(x,0) = \varphi(x), & 0 \leqslant x \leqslant L, \\[2mm]
u(0,t) = \alpha(t), & 0 < t \leqslant T,
\end{cases}
$$

where $a > 0$ is a constant and $\varphi(0) = \alpha(0)$. Divide the interval $[0, L]$ into m equal subintervals and $[0, T]$ into n equal subintervals. Denote $h = L/m$, $\tau = T/n$, $x_i = ih$, $0 \leqslant i \leqslant m$, $t_k = k\tau$, $0 \leqslant k \leqslant n$. A difference scheme for the above problem reads

$$
\begin{cases}
\dfrac{u_i^{k+1} - u_i^k}{\tau} + a\dfrac{u_i^k - u_{i-1}^k}{h} = f(x_i, t_k), & 1 \leqslant i \leqslant m, \quad 0 \leqslant k \leqslant n-1, \\[2mm]
u_i^0 = \varphi(x_i), & 0 \leqslant i \leqslant m, \\[2mm]
u_0^k = \alpha(t_k), & 1 \leqslant k \leqslant n.
\end{cases}
$$

(1) Show the expression of the local truncation error of the difference scheme.
(2) If $\alpha(t) \equiv 0$, prove that the difference solution has the following priori estimate:

$$\left\| u^k \right\|_\infty \leqslant \left\| u^0 \right\|_\infty + \tau \sum_{l=0}^{k-1} \left\| f^l \right\|_\infty, \qquad 1 \leqslant k \leqslant n$$

when $s = a\tau/h \leqslant 1$, where

$$\| u^k \|_\infty = \max_{1 \leqslant i \leqslant m} |u_i^k|, \qquad \| f^k \|_\infty = \max_{1 \leqslant i \leqslant m} |f(x_i, t_k)|.$$

(3) Show that when $s \leqslant 1$, the difference solution is convergent with the convergence order one in the norm $\|\cdot\|_\infty$.
(4) How to solve the difference scheme?

4.2 For the problem given in Exercise 4.1, the following difference scheme can be established

$$
\begin{cases}
\dfrac{1}{2}\left(\dfrac{u_i^{k+1} - u_i^k}{\tau} + \dfrac{u_{i-1}^{k+1} - u_{i-1}^k}{\tau}\right) + \dfrac{a}{2}\left(\dfrac{u_i^k - u_{i-1}^k}{h} + \dfrac{u_i^{k+1} - u_{i-1}^{k+1}}{h}\right) \\
\qquad = f(x_{i-\frac{1}{2}}, t_{k+\frac{1}{2}}), \quad 1 \leqslant i \leqslant m, \quad 0 \leqslant k \leqslant n-1, \\
u_i^0 = \varphi(x_i), \quad 1 \leqslant i \leqslant m, \\
u_0^k = \alpha(t_k), \quad 0 \leqslant k \leqslant n.
\end{cases}
$$

(1) Show the expression of the local truncation error of the difference scheme.
(2) If $\alpha(t) \equiv 0$, then show that for an arbitrary step ratio s, we have

$$\left\| u^k \right\|_A^2 \leqslant e^{\frac{3}{2}k\tau}\left(\left\| u^0 \right\|_A^2 + \frac{3}{2}\tau \sum_{l=0}^{k-1} \left\| f^{l+\frac{1}{2}} \right\|^2\right), \qquad 1 \leqslant k \leqslant n,$$

where

$$\left\| u^k \right\|_A^2 = h \sum_{i=1}^{m}\left(\frac{u_i^k + u_{i-1}^k}{2}\right)^2, \qquad \left\| f^{l+\frac{1}{2}} \right\|^2 = h \sum_{l=1}^{m}\left[f(x_{i-\frac{1}{2}}, t_{l+\frac{1}{2}})\right]^2.$$

(Hints: Multiplying both sides of the difference scheme by $\frac{1}{2}(u_i^{k+1} + u_{i-1}^{k+1} + u_i^k + u_{i-1}^k)$, summing over i, and then using the Gronwall inequality.)
(3) Show that the difference scheme is convergent with the convergence order two in the norm $\|\cdot\|_A$.
(4) How to solve the difference scheme?

4.3 Consider the following difference scheme:

$$
\begin{cases}
\dfrac{1}{2}\left(\dfrac{u_i^{k+1}-u_i^k}{\tau}+\dfrac{u_{i-1}^{k+1}-u_{i-1}^k}{\tau}\right)+\dfrac{a}{2}\left(\dfrac{u_i^k-u_{i-1}^k}{h}+\dfrac{u_i^{k+1}-u_{i-1}^{k+1}}{h}\right)=0, \\[3mm]
\qquad\qquad\qquad\qquad 1\leqslant i\leqslant m,\quad 0\leqslant k\leqslant n-1, \\[2mm]
u_i^0=\varphi(x_i),\quad 1\leqslant i\leqslant m, \\[2mm]
u_0^k=0,\quad 0\leqslant k\leqslant n,
\end{cases}
$$

where $a>0$, $mh=1$. Try to prove

$$
h\sum_{i=1}^{m}\left(\frac{u_i^k-u_{i-1}^k}{h}\right)^2\leqslant h\sum_{i=1}^{m}\left(\frac{u_i^0-u_{i-1}^0}{h}\right)^2,\quad 1\leqslant k\leqslant n.
$$

(Hints: Multiplying both sides of the difference scheme by $\delta_x\delta_t u_{i-\frac{1}{2}}^{k+\frac{1}{2}}$, noticing that it has two equivalent forms: $\frac{1}{\tau}\left(\delta_x u_{i-\frac{1}{2}}^{k+1}-\delta_x u_{i-\frac{1}{2}}^{k}\right)$, $\frac{1}{h}\left(\delta_t u_i^{k+\frac{1}{2}}-\delta_t u_{i-1}^{k+\frac{1}{2}}\right)$, and then summing over i from 1 to m.)

4.4 Let $\{v_i^k\,|\,0\leqslant i\leqslant m,0\leqslant k\leqslant n\}$ be the solution of the difference scheme (4.38), where $g_i^k\equiv 0$. For any step ratio s, try to prove

$$
2\|\delta_t v^{\frac{1}{2}}\|^2+\frac{c^2}{2}\left(|v^1|_1^2+|v^0|_1^2\right)=|v^0|_1^2,
$$

$$
h\sum_{i=1}^{m-1}(\delta_t v_i^{k+\frac{1}{2}})^2+\frac{c^2}{2}\left(|v^{k+1}|_1^2+|v^k|_1^2\right)
$$

$$
=\|\delta_t v^{\frac{1}{2}}\|^2+\frac{c^2}{2}\left(|v^1|_1^2+|v^0|_1^2\right),\quad 0\leqslant k\leqslant n-1.
$$

4.5 For the problem

$$
\begin{cases}
\dfrac{\partial^2 u}{\partial t^2}-c^2\dfrac{\partial^2 u}{\partial x^2}=f(x,t),\quad 0<x<L,\quad 0<t\leqslant T, \\[3mm]
u(x,0)=\varphi(x),\quad \dfrac{\partial u(x,0)}{\partial t}=\psi(x),\quad 0\leqslant x\leqslant L, \\[3mm]
u(0,t)=0,\quad u(L,t)=0,\quad 0<t\leqslant T,
\end{cases}
$$

where $\varphi(0) = \varphi(L) = \psi(0) = \psi(L) = 0$. Derive the difference scheme

$$
\begin{cases}
\delta_t^2 u_i^k - \dfrac{c^2}{4}\left(\delta_x^2 u_i^{k-1} + 2\delta_x^2 u_i^k + \delta_x^2 u_i^{k+1}\right) = f(x_i, t_k), \\[2mm]
\qquad 1 \leqslant i \leqslant m-1, \quad 1 \leqslant k \leqslant n-1, \\[2mm]
u_i^0 = \varphi(x_i), \quad u_i^1 = \varphi(x_i) + \tau\psi(x_i), \quad 1 \leqslant i \leqslant m-1, \\[2mm]
u_0^k = 0, \quad u_m^k = 0, \quad 0 \leqslant k \leqslant n.
\end{cases}
$$

(1) Show the local truncation error of the difference scheme.
(2) Show the priori estimate of the difference solution.

4.6 Consider the problem

$$
\begin{cases}
\dfrac{\partial^2 u}{\partial t^2} - c^2 \dfrac{\partial^2 u}{\partial x^2} = 0, & 0 < x < L, \quad 0 < t \leqslant T, \\[3mm]
u(x,0) = \varphi(x), \quad \dfrac{\partial u(x,0)}{\partial t} = \psi(x), & 0 \leqslant x \leqslant L, \\[3mm]
u(0,t) = \alpha(t), \quad u(L,t) = \beta(t), & 0 < t \leqslant T,
\end{cases}
\tag{4.69}
$$

where c is a constant, $\varphi(x), \psi(x), \alpha(t)$, and $\beta(t)$ are known functions, and $\varphi(0) = \alpha(0)$, $\varphi(L) = \beta(0)$, $\psi(0) = \alpha'(0)$, and $\psi(L) = \beta'(0)$. Let

$$
v = \frac{\partial u}{\partial t}, \quad w = c\frac{\partial u}{\partial x}.
$$

Then (4.69) is equivalent to

$$
\begin{cases}
\dfrac{\partial v}{\partial t} - c\dfrac{\partial w}{\partial x} = 0, & 0 < x < L, \quad 0 < t \leqslant T, \\[3mm]
\dfrac{\partial w}{\partial t} - c\dfrac{\partial v}{\partial x} = 0, & 0 < x < L, \quad 0 < t \leqslant T, \\[3mm]
v(x,0) = \psi(x), \quad w(x,0) = c\varphi'(x), & 0 \leqslant x \leqslant L, \\[3mm]
v(0,t) = \alpha'(t), \quad v(L,t) = \beta'(t), & 0 < t \leqslant T.
\end{cases}
\tag{4.70}
$$

Define the following notation:

$$
\delta_t u_i^{k+\frac{1}{2}} = \frac{1}{\tau}\left(u_i^{k+1} - u_i^k\right), \quad \delta_t u_{i-\frac{1}{2}}^{k+\frac{1}{2}} = \frac{1}{2}\left(\delta_t u_{i-1}^{k+\frac{1}{2}} + \delta_t u_i^{k+\frac{1}{2}}\right),
$$

$$
\delta_x u_{i-\frac{1}{2}}^k = \frac{1}{h}\left(u_i^k - u_{i-1}^k\right), \quad \delta_x u_{i-\frac{1}{2}}^{k+\frac{1}{2}} = \frac{1}{2}\left(\delta_x u_{i-\frac{1}{2}}^k + \delta_x u_{i-\frac{1}{2}}^{k+1}\right).
$$

A difference scheme for (4.70) reads

$$
\begin{cases}
\delta_t v^{k+\frac{1}{2}}_{i-\frac{1}{2}} - c\delta_x w^{k+\frac{1}{2}}_{i-\frac{1}{2}} = 0, & 1 \leqslant i \leqslant m, \quad 0 \leqslant k \leqslant n-1, & \text{(4.71a)} \\[2ex]
\delta_t w^{k+\frac{1}{2}}_{i-\frac{1}{2}} - c\delta_x v^{k+\frac{1}{2}}_{i-\frac{1}{2}} = 0, & 1 \leqslant i \leqslant m, \quad 0 \leqslant k \leqslant n-1, & \text{(4.71b)} \\[2ex]
v^0_i = \psi(x_i), \quad w^0_i = c\varphi'(x_i), & 0 \leqslant i \leqslant m, & \text{(4.71c)} \\[2ex]
v^k_0 = \alpha'(t_k), \quad v^k_m = \beta'(t_k), & 1 \leqslant k \leqslant n. & \text{(4.71d)}
\end{cases}
$$

When $\{w^k_i \mid 0 \leqslant i \leqslant m, 1 \leqslant k \leqslant n\}$ has been solved, using

$$
u^k_0 = \alpha(t_k), \quad c\delta_x u^k_{i-\frac{1}{2}} = w^k_{i-\frac{1}{2}}, \qquad 1 \leqslant i \leqslant m, \quad 1 \leqslant k \leqslant n \qquad \text{(4.72)}
$$

solve $\{u^k_i \mid 0 \leqslant i \leqslant m, 1 \leqslant k \leqslant n\}$.

(1) Show the local truncation error of the difference scheme (4.71) and (4.72).
(2) Suppose $\{v^k_0, w^k_0, v^k_1, w^k_1, \ldots, v^k_m, w^k_m\}$ are known. Write (4.71) as a system of linear equations in

$$
\left\{ v^{k+1}_0, w^{k+1}_0, v^{k+1}_1, w^{k+1}_1, \ldots, v^{k+1}_m, w^{k+1}_m \right\}
$$

(the matrix-vector form).
(3) Suppose $\alpha(t) \equiv 0$, $\beta(t) \equiv 0$. Multiplying both sides of (4.71a) by $2v^{k+\frac{1}{2}}_{i-\frac{1}{2}}$, (4.71b) by $2w^{k+\frac{1}{2}}_{i-\frac{1}{2}}$, adding up the results, and then summing over i from 1 to m and utilizing (4.71d), try to obtain a priori estimate with respect to

$$
h \sum_{i=1}^{m} \left[(v^k_{i-\frac{1}{2}})^2 + (w^k_{i-\frac{1}{2}})^2 \right].
$$

Then using (4.72), try to give a priori estimate for $h \sum_{i=1}^{m} (\delta_x u^k_{i-\frac{1}{2}})^2$. Finally, applying Lemma 1.4, try to get a priori estimate for $\max_{0 \leqslant i \leqslant m} |u^k_i|$.

(4) Consider the system of error equations of the difference scheme (4.71)–(4.72), and prove that the approximate solution $\{u^k_i \mid 0 \leqslant i \leqslant m, 1 \leqslant k \leqslant n\}$ is convergent with the convergence order two in the maximum norm.

4.7 Let $\Omega = (0, L_1) \times (0, L_2)$. Take two positive integers m_1, m_2 and denote $h_1 = \frac{L_1}{m_1}$, $h_2 = \frac{L_2}{m_2}$, $x_i = ih_1$, $y_j = jh_2$, $\Omega_h = \{(x_i, y_j) \mid 0 \leqslant i \leqslant m_1, 0 \leqslant j \leqslant m_2\}$, $\omega = \{(i, j) \mid (x_i, y_j) \in \Omega\}$, $\gamma = \{(i, j) \mid (x_i, y_j) \in \partial\Omega\}$.

Suppose $f = \{f(x_i, y_j) \mid 0 \leqslant i \leqslant m_1, 0 \leqslant j \leqslant m_2\}$ and $f(x_i, y_j) = 0$ when $(i, j) \in \gamma$. Denote

$$\|f\|^2 = h_1 h_2 \sum_{i=1}^{m_1-1} \sum_{j=1}^{m_2-1} f^2(x_i, y_j).$$

Assume that f can be expanded into the following finite Fourier series:

$$f(x_i, y_j) = \sum_{l_1=1}^{m_1-1} \sum_{l_2=1}^{m_2-1} A_{l_1, l_2}\left(\sin \frac{l_1 \pi x_i}{L_1}\right)\left(\sin \frac{l_2 \pi y_j}{L_2}\right).$$

Show

(1)

$$A_{l_1, l_2} = \frac{4}{L_1 L_2} \cdot h_1 h_2 \sum_{i=1}^{m_1-1} \sum_{j=1}^{m_2-1} f(x_i, y_j)\left(\sin \frac{l_1 \pi x_i}{L_1}\right)$$

$$\times \left(\sin \frac{l_2 \pi y_j}{L_2}\right), \quad (l_1, l_2) \in \omega.$$

(2)

$$\|f\|^2 = \frac{L_1 L_2}{4} \sum_{l_1=1}^{m_1-1} \sum_{l_2=1}^{m_2-1} A_{l_1, l_2}^2.$$

4.8 Let $u = \{u_{ij} \mid 0 \leqslant i \leqslant m_1, 0 \leqslant j \leqslant m_2\}$ be the solution of the following difference scheme:

$$\begin{cases} -(\delta_x^2 u_{ij} + \delta_y^2 u_{ij}) = f(x_i, y_j), & (i, j) \in \omega, \\ u_{ij} = 0, & (i, j) \in \gamma. \end{cases}$$

Applying the theory of finite Fourier series in Exercise 4.7, try to prove

$$\|u\| \leqslant \frac{1}{\dfrac{4}{L_1^2} + \dfrac{4}{L_2^2}} \|f\|.$$

Table 4.11 (Exercise 4.9)
Numerical results at some
node points
($h = 1/100$, $\tau = 1/200$)

(x,t)	NS	ES	$|ES - NS|$
$(0.5, 0.1)$			
$(0.5, 0.2)$			
$(0.5, 0.3)$			
$(0.5, 0.4)$			
$(0.5, 0.5)$			
$(0.5, 0.6)$			
$(0.5, 0.7)$			
$(0.5, 0.8)$			
$(0.5, 0.9)$			
$(0.5, 1.0)$			

Table 4.12 (Exercise 4.9)
Numerical results at some
node points
($h = 1/200$, $\tau = 1/100$)

k	(x,t)	NS	ES	$|ES - NS|$
1	$(0.5, 0.01)$			
2	$(0.5, 0.02)$			
3	$(0.5, 0.03)$			
4	$(0.5, 0.04)$			
5	$(0.5, 0.05)$			
6	$(0.5, 0.06)$			
7	$(0.5, 0.07)$			
8	$(0.5, 0.08)$			
9	$(0.5, 0.09)$			
10	$(0.5, 0.10)$			

4.9 Apply the explicit difference scheme (4.14) to compute the problem

$$\begin{cases} \dfrac{\partial^2 u}{\partial t^2} - \dfrac{\partial^2 u}{\partial x^2} = (t^2 - x^2)\sin(xt), & 0 < x < 1, \quad 0 < t \leqslant 1, \\ u(x,0) = 0, \quad \dfrac{\partial u(x,0)}{\partial t} = x, & 0 \leqslant x \leqslant 1, \\ u(0,t) = 0, \quad u(1,t) = \sin t, & 0 < t \leqslant 1. \end{cases}$$

The exact solution of the problem is $u(x,t) = \sin(xt)$.

(1) Fill in Tables 4.11, 4.12, and 4.13, respectively.
(2) Draw the surfaces of the exact solution and the numerical solution with $h = 1/10$ and $\tau = 1/20$.
(3) Draw the error surfaces of the numerical solution with $(h, \tau) = (1/10, 1/20)$, $(1/20, 1/40)$, $(1/40, 1/80)$ in the same coordinate system.

4.10 Apply the implicit difference scheme (4.36) to compute the problem in Exercise 4.9.

(1) Fill in Tables 4.14, 4.15, and 4.16, respectively.

Table 4.13 (Exercise 4.9) The maximum errors of numerical solutions with different step sizes ($s = 1/2$)

h	τ	$E_\infty(h, \tau)$	$E_\infty(2h, 2\tau)/E_\infty(h, \tau)$
1/10	1/20		
1/20	1/40		
1/40	1/80		
1/80	1/160		
1/160	1/320		
1/320	1/640		
1/640	1/1280		

Table 4.14 (Exercise 4.10) Numerical results at some node points ($h = 1/100, \tau = 1/200$)

(x, t)	NS	ES	$|ES - NS|$
(0.5, 0.1)			
(0.5, 0.2)			
(0.5, 0.3)			
(0.5, 0.4)			
(0.5, 0.5)			
(0.5, 0.6)			
(0.5, 0.7)			
(0.5, 0.8)			
(0.5, 0.9)			
(0.5, 1.0)			

Table 4.15 (Exercise 4.10) Numerical results at some node points ($h = 1/200, \tau = 1/100$)

(x, t)	NS	ES	$|ES - NS|$
(0.5, 0.1)			
(0.5, 0.2)			
(0.5, 0.3)			
(0.5, 0.4)			
(0.5, 0.5)			
(0.5, 0.6)			
(0.5, 0.7)			
(0.5, 0.8)			
(0.5, 0.9)			
(0.5, 1.0)			

(2) Draw the surfaces of the exact and numerical solutions with $h = 1/10, \tau = 1/10$.

(3) Draw the error surfaces of numerical solutions with $(h, \tau) = (1/10, 1/10)$, $(1/20, 1/20)$, $(1/40, 1/40)$ in the same coordinate system.

4.11 Apply the compact difference scheme (4.47) to compute the problem in Exercise 4.9.

(1) Fill in Tables 4.17, 4.18, and 4.19, respectively.

(2) Draw the surfaces of the exact and numerical solutions with $h = 1/10$ and $\tau = 1/100$.

Table 4.16 (Exercise 4.10) The maximum errors of numerical solutions with different step sizes ($s = 1$)

h	τ	$E_\infty(h, \tau)$	$E_\infty(2h, 2\tau)/E_\infty(h, \tau)$
1/10	1/10		
1/20	1/20		
1/40	1/40		
1/80	1/80		
1/160	1/160		
1/320	1/320		

Table 4.17 (Exercise 4.11) Numerical results at some node points ($h = 1/10$, $\tau = 1/100$)

(x, t)	NS	ES	$|ES - NS|$
(0.5, 0.1)			
(0.5, 0.2)			
(0.5, 0.3)			
(0.5, 0.4)			
(0.5, 0.5)			
(0.5, 0.6)			
(0.5, 0.7)			
(0.5, 0.8)			
(0.5, 0.9)			
(0.5, 1.0)			

Table 4.18 (Exercise 4.11) Numerical results at some node points ($h = 1/20$, $\tau = 1/400$)

(x, t)	NS	ES	$|ES - NS|$
(0.5, 0.1)			
(0.5, 0.2)			
(0.5, 0.3)			
(0.5, 0.4)			
(0.5, 0.5)			
(0.5, 0.6)			
(0.5, 0.7)			
(0.5, 0.8)			
(0.5, 0.9)			
(0.5, 1.0)			

Table 4.19 (Exercise 4.11) The maximum errors of numerical solutions with different step sizes ($\tau = h^2$)

h	τ	$E_\infty(h, \tau)$	$E_\infty(2h, 4\tau)/E_\infty(h, \tau)$
1/10	1/100		
1/20	1/400		
1/40	1/1600		
1/80	1/6400		

(3) Draw the error surfaces of numerical solutions with $(h, \tau) = (1/10, 1/100)$, $(1/20, 1/400)$ in the same coordinate system.

References

1. Hu, J.W., Tang, H.M.: Numerical Methods for Differential Equations, 2nd edn. Science Press, Beijing (2020)
2. Li, L.K., Yu, C.H., Zhu, Z.H.: Numerical Methods for Differential Equations. Fudan University Press, Shanghai (1999)
3. Li, R.H., Liu, B.: Numerical Methods for Differential Equations, 4th edn. Higher Education Press, Beijing (2009)
4. Lu, J.F., Guan, Z.: Numerical Methods for Partial Differential Equations, 3rd edn. Tsinghua University Press, Beijing (2016)
5. Richtmyer, R.D., Morton, K.W.: Difference Methods for Initial Value Problems, 2nd edn. Wiley, London (1967)

Chapter 5
Alternating Direction Implicit Methods for High-Dimensional Evolution Equations

For one-dimensional parabolic equations, the forward Euler scheme is simple to implement; however, its step ratio is restricted. The backward Euler scheme and the Crank-Nicolson scheme are both unconditionally stable, and the computational cost is not large when the double sweep method is applied. For high-dimensional parabolic equations, the forward Euler scheme is easy to implement, but its stability condition is more stringent than that in the one-dimensional case. Although the backward Euler and Crank-Nicolson schemes are unconditionally stable, the size of the system of difference equations at each time level becomes larger, and the system is no longer a tridiagonal system of linear equations. Solving such a system requires considerable CPU time. Similar challenges can be found for hyperbolic equations. Therefore, it is necessary to explore new unconditionally stable difference schemes with reduced computational complexity. The alternating direction implicit (ADI) scheme introduced in this chapter is unconditionally stable and can be solved efficiently using the double sweep method.

5.1 ADI Schemes for Two-Dimensional Parabolic Equations

As a typical model, we consider the initial-boundary value problem of the two-dimensional heat conduction equation of the form

$$
\begin{cases}
\dfrac{\partial u}{\partial t} - a\left(\dfrac{\partial^2 u}{\partial x^2} + \dfrac{\partial^2 u}{\partial y^2}\right) = f(x, y, t), & (x, y) \in \Omega, \quad 0 < t \leqslant T, & (5.1a) \\[3mm]
u(x, y, 0) = \varphi(x, y), & (x, y) \in \bar{\Omega}, & (5.1b) \\[3mm]
u(x, y, t) = \alpha(x, y, t), & (x, y) \in \Gamma, \quad 0 < t \leqslant T, & (5.1c)
\end{cases}
$$

© Science Press 2026

Z.-Z. Sun et al., *Numerical Solutions to Partial Differential Equations with Finite Difference Methods*, Springer Asia Pacific Mathematics Series 9, https://doi.org/10.1007/978-981-95-5563-5_5

where a is a positive constant (commonly referred to as the heat conduction coefficient), $\Omega = (0, L_1) \times (0, L_2)$, Γ is the boundary of Ω and $\alpha(x, y, 0) = \varphi(x, y)$ when $(x, y) \in \Gamma$.

Take three positive integers m_1, m_2 and n. Denote $h_1 = L_1/m_1$, $h_2 = L_2/m_2$, $\tau = T/n$; $x_i = ih_1$, $0 \leqslant i \leqslant m_1$; $y_j = jh_2$, $0 \leqslant j \leqslant m_2$; $t_k = k\tau$, $0 \leqslant k \leqslant n$; $\Omega_h = \{(x_i, y_j) \,|\, 0 \leqslant i \leqslant m_1, 0 \leqslant j \leqslant m_2\}$, $\Omega_\tau = \{t_k \,|\, 0 \leqslant k \leqslant n\}$, $\omega = \{(i, j) \,|\, (x_i, y_j) \in \Omega\}$, $\gamma = \{(i, j) \,|\, (x_i, y_j) \in \Gamma\}$ and $\bar{\omega} = \omega \cup \gamma$. Besides, denote $r_1 = a\tau/h_1^2$, $r_2 = a\tau/h_2^2$, $t_{k+\frac{1}{2}} = \dfrac{1}{2}(t_k + t_{k+1})$, $f_{ij}^{k+\frac{1}{2}} = f(x_i, y_j, t_{k+\frac{1}{2}})$. We call (x_i, y_j, t_k) a grid point and r_1, r_2 the grid ratios.

For a grid function $v = \{v_{ij}^k \,|\, 0 \leqslant i \leqslant m_1, 0 \leqslant j \leqslant m_2,\ 0 \leqslant k \leqslant n\}$ on $\Omega_h \times \Omega_\tau$, introduce the following notation:

$$v_{ij}^{k+\frac{1}{2}} = \frac{1}{2}\left(v_{ij}^k + v_{ij}^{k+1}\right), \quad \delta_t v_{ij}^{k+\frac{1}{2}} = \frac{1}{\tau}\left(v_{ij}^{k+1} - v_{ij}^k\right),$$

$$\delta_x v_{i-\frac{1}{2},j}^k = \frac{1}{h_1}\left(v_{ij}^k - v_{i-1,j}^k\right), \quad \delta_x^2 v_{ij}^k = \frac{1}{h_1}\left(\delta_x v_{i+\frac{1}{2},j}^k - \delta_x v_{i-\frac{1}{2},j}^k\right),$$

$$\delta_y v_{i,j-\frac{1}{2}}^k = \frac{1}{h_2}\left(v_{ij}^k - v_{i,j-1}^k\right), \quad \delta_y^2 v_{ij}^k = \frac{1}{h_2}\left(\delta_y v_{i,j+\frac{1}{2}}^k - \delta_y v_{i,j-\frac{1}{2}}^k\right),$$

$$\Delta_t v_{ij}^k = \frac{1}{2\tau}\left(v_{ij}^{k+1} - v_{ij}^{k-1}\right), \quad \Delta_h v_{ij}^k = \delta_x^2 v_{ij}^k + \delta_y^2 v_{ij}^k.$$

It is easy to know that

$$\delta_x^2 v_{ij}^k = \frac{1}{h_1^2}\left(v_{i-1,j}^k - 2v_{ij}^k + v_{i+1,j}^k\right), \quad \delta_y^2 v_{ij}^k = \frac{1}{h_2^2}\left(v_{i,j-1}^k - 2v_{ij}^k + v_{i,j+1}^k\right).$$

Let

$$v^k = \{v_{ij}^k \,|\, (i, j) \in \bar{\omega}\}.$$

Then v^k is a grid function on Ω_h.

Denote

$$\mathcal{V}_h = \left\{w \,|\, w = \{w_{ij} \,|\, (i, j) \in \bar{\omega}\} \text{ is a grid function on } \Omega_h\right\},$$

$$\overset{\circ}{\mathcal{V}}_h = \left\{w \,|\, w = \{w_{ij} \,|\, (i, j) \in \bar{\omega}\} \in \mathcal{V}_h \text{ and } w_{ij} = 0 \text{ when } (i, j) \in \gamma\right\}.$$

Introduce the notation of inner products and norms as those in Chap. 2.

5.1.1   *Derivation of the Difference Scheme*

Define the grid function on $\Omega_h \times \Omega_\tau$ by

$$U = \{U_{ij}^k \mid (i, j) \in \bar{\omega}, 0 \leqslant k \leqslant n\},$$

where

$$U_{ij}^k = u(x_i, y_j, t_k), \quad (i, j) \in \bar{\omega}, \quad 0 \leqslant k \leqslant n.$$

Considering the differential equation (5.1a) at the point $(x_i, y_j, t_{k+\frac{1}{2}})$, we have

$$\frac{\partial u(x_i, y_j, t_{k+\frac{1}{2}})}{\partial t} - a\left[\frac{\partial^2 u(x_i, y_j, t_{k+\frac{1}{2}})}{\partial x^2} + \frac{\partial^2 u(x_i, y_j, t_{k+\frac{1}{2}})}{\partial y^2}\right] = f(x_i, y_j, t_{k+\frac{1}{2}}),$$

$$(i, j) \in \omega, \quad 0 \leqslant k \leqslant n - 1. \tag{5.2}$$

It follows from the Taylor expansion with the integral remainder that

$$\frac{\partial u(x_i, y_j, t_{k+\frac{1}{2}})}{\partial t}$$

$$= \delta_t U_{ij}^{k+\frac{1}{2}} - \frac{\tau^2}{16} \int_0^1 \left[u_{ttt}\left(x_i, y_j, t_{k+\frac{1}{2}} - \frac{s\tau}{2}\right) + u_{ttt}\left(x_i, y_j, t_{k+\frac{1}{2}} + \frac{s\tau}{2}\right)\right](1 - s)^2 ds;$$

$$\frac{\partial^2 u(x_i, y_j, t_{k+\frac{1}{2}})}{\partial x^2}$$

$$= \frac{1}{2}\left[\frac{\partial^2 u(x_i, y_j, t_k)}{\partial x^2} + \frac{\partial^2 u(x_i, y_j, t_{k+1})}{\partial x^2}\right]$$

$$- \frac{\tau^2}{8} \int_0^1 \left[u_{xxtt}\left(x_i, y_j, t_{k+\frac{1}{2}} - \frac{s\tau}{2}\right) + u_{xxtt}\left(x_i, y_j, t_{k+\frac{1}{2}} + \frac{s\tau}{2}\right)\right](1 - s)ds$$

$$= \frac{1}{2}\left[\delta_x^2 U_{ij}^k - \frac{h_1^2}{6} \int_0^1 \left(u_{xxxx}(x_i - sh_1, y_j, t_k) + u_{xxxx}(x_i + sh_1, y_j, t_k)\right)(1 - s)^3 ds\right.$$

$$+ \delta_x^2 U_{ij}^{k+1} - \frac{h_1^2}{6} \int_0^1 \left(u_{xxxx}(x_i - sh_1, y_j, t_{k+1})\right.$$

$$\left.\left. + u_{xxxx}(x_i + sh_1, y_j, t_{k+1})\right)(1 - s)^3 ds\right]$$

$$- \frac{\tau^2}{8} \int_0^1 \left[u_{xxtt}\left(x_i, y_j, t_{k+\frac{1}{2}} - \frac{s\tau}{2}\right) + u_{xxtt}\left(x_i, y_j, t_{k+\frac{1}{2}} + \frac{s\tau}{2}\right)\right](1 - s)ds;$$

$$\frac{\partial^2 u(x_i, y_j, t_{k+\frac{1}{2}})}{\partial y^2}$$

$$= \frac{1}{2}\left[\frac{\partial^2 u(x_i, y_j, t_k)}{\partial y^2} + \frac{\partial^2 u(x_i, y_j, t_{k+1})}{\partial y^2}\right]$$

$$- \frac{\tau^2}{8}\int_0^1\left[u_{yytt}\left(x_i, y_j, t_{k+\frac{1}{2}} - \frac{s\tau}{2}\right) + u_{yytt}\left(x_i, y_j, t_{k+\frac{1}{2}} + \frac{s\tau}{2}\right)\right](1-s)ds$$

$$= \frac{1}{2}\left[\delta_y^2 U_{ij}^k - \frac{h_2^2}{6}\int_0^1\left(u_{yyyy}(x_i, y_j - sh_2, t_k) + u_{yyyy}(x_i, y_j + sh_2, t_k)\right)(1-s)^3 ds\right.$$

$$+ \delta_y^2 U_{ij}^{k+1} - \frac{h_2^2}{6}\int_0^1\left(u_{yyyy}(x_i, y_j - sh_2, t_{k+1})\right.$$

$$\left.\left. + u_{yyyy}(x_i, y_j + sh_2, t_{k+1})\right)(1-s)^3 ds\right]$$

$$- \frac{\tau^2}{8}\int_0^1\left[u_{yytt}\left(x_i, y_j, t_{k+\frac{1}{2}} - \frac{s\tau}{2}\right) + u_{yytt}\left(x_i, y_j, t_{k+\frac{1}{2}} + \frac{s\tau}{2}\right)\right](1-s)ds.$$

Substituting the above three equalities into (5.2), we have

$$\delta_t U_{ij}^{k+\frac{1}{2}} - a\left(\delta_x^2 U_{ij}^{k+\frac{1}{2}} + \delta_y^2 U_{ij}^{k+\frac{1}{2}}\right) = f_{ij}^{k+\frac{1}{2}} + (R_1)_{ij}^k, \quad (i, j) \in \omega, \ \ 0 \leqslant k \leqslant n-1,$$

$$(5.3)$$

where

$$(R_1)_{ij}^k = \frac{1}{8}\tau^2\left\{\frac{1}{2}\int_0^1\left[u_{ttt}\left(x_i, y_j, t_{k+\frac{1}{2}} - \frac{s\tau}{2}\right) + u_{ttt}\left(x_i, y_j, t_{k+\frac{1}{2}} + \frac{s\tau}{2}\right)\right](1-s)^2 ds\right.$$

$$- a\int_0^1\left[u_{xxtt}\left(x_i, y_j, t_{k+\frac{1}{2}} - \frac{s\tau}{2}\right) + u_{xxtt}\left(x_i, y_j, t_{k+\frac{1}{2}} + \frac{s\tau}{2}\right)\right.$$

$$\left.\left. + u_{yytt}\left(x_i, y_j, t_{k+\frac{1}{2}} - \frac{s\tau}{2}\right) + u_{yytt}\left(x_i, y_j, t_{k+\frac{1}{2}} + \frac{s\tau}{2}\right)\right](1-s)ds\right\}$$

$$- \frac{a}{12}h_1^2\int_0^1\left[u_{xxxx}(x_i - sh_1, y_j, t_k) + u_{xxxx}(x_i + sh_1, y_j, t_k)\right.$$

$$\left. + u_{xxxx}(x_i - sh_1, y_j, t_{k+1}) + u_{xxxx}(x_i + sh_1, y_j, t_{k+1})\right](1-s)^3 ds$$

$$- \frac{a}{12}h_2^2\int_0^1\left[u_{yyyy}(x_i, y_j - sh_2, t_k) + u_{yyyy}(x_i, y_j + sh_2, t_k)\right.$$

$$\left. + u_{yyyy}(x_i, y_j - sh_2, t_{k+1}) + u_{yyyy}(x_i, y_j + sh_2, t_{k+1})\right](1-s)^3 ds.$$

It is easy to see that there is a positive constant c_1 such that

$$
\begin{cases}
\left|(R_1)_{ij}^k\right| \leqslant c_1(\tau^2 + h_1^2 + h_2^2), & (i, j) \in \omega, \quad 0 \leqslant k \leqslant n - 1, \\
\left|\delta_t(R_1)_{ij}^{k+\frac{1}{2}}\right| \leqslant c_1(\tau^2 + h_1^2 + h_2^2), & (i, j) \in \omega, \quad 0 \leqslant k \leqslant n - 2,
\end{cases}
\tag{5.4}
$$

where

$$
\delta_t(R_1)_{ij}^{k+\frac{1}{2}} = \frac{1}{\tau}\left[(R_1)_{ij}^{k+1} - (R_1)_{ij}^k\right].
$$

Noticing the initial-boundary value conditions (5.1b) and (5.1c), we have

$$
\begin{cases}
U_{ij}^0 = \varphi(x_i, y_j), & (i, j) \in \omega, \\
U_{ij}^k = \alpha(x_i, y_j, t_k), & (i, j) \in \gamma, \quad 0 \leqslant k \leqslant n.
\end{cases}
\tag{5.5}
$$

Omitting the small term $(R_1)_{ij}^k$ in (5.3) and replacing U_{ij}^k with u_{ij}^k, a difference scheme reads

$$
\begin{cases}
\delta_t u_{ij}^{k+\frac{1}{2}} - a\Delta_h u_{ij}^{k+\frac{1}{2}} = f_{ij}^{k+\frac{1}{2}}, & (i, j) \in \omega, \quad 0 \leqslant k \leqslant n - 1, & (5.6a) \\
u_{ij}^0 = \varphi(x_i, y_j), & (i, j) \in \omega, & (5.6b) \\
u_{ij}^k = \alpha(x_i, y_j, t_k), & (i, j) \in \gamma, \quad 0 \leqslant k \leqslant n. & (5.6c)
\end{cases}
$$

The stencil of the difference scheme (5.6) is shown in Fig. 5.1. It is a two-time-level implicit difference scheme. The local truncation error of the difference scheme (5.6) is $(R_1)_{ij}^k$.

The solve the difference scheme (5.6), it is necessary to solve a large sparse system of linear equations at each time level.

Fig. 5.1 The stencil of the difference scheme (5.6)

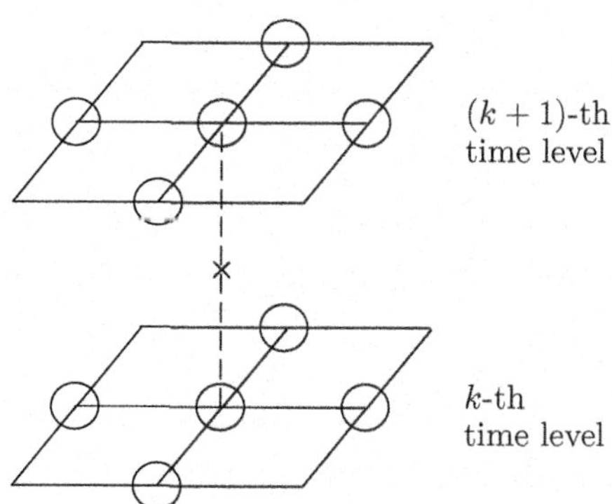

To construct an ADI difference scheme, adding a small term $\frac{1}{4}a^2\tau^2\delta_x^2\delta_y^2\delta_t U_{ij}^{k+\frac{1}{2}}$ on both sides of (5.3), we obtain

$$\delta_t U_{ij}^{k+\frac{1}{2}} - a\Delta_h U_{ij}^{k+\frac{1}{2}} + \frac{1}{4}a^2\tau^2\delta_x^2\delta_y^2\delta_t U_{ij}^{k+\frac{1}{2}} = f_{ij}^{k+\frac{1}{2}} + (R_2)_{ij}^k,$$

$$(i, j) \in \omega, \quad 0 \leqslant k \leqslant n - 1, \tag{5.7}$$

where

$$(R_2)_{ij}^k = (R_1)_{ij}^k + \frac{1}{4}a^2\tau^2\delta_x^2\delta_y^2\delta_t U_{ij}^{k+\frac{1}{2}}.$$

Applying the results in Exercise 1.1, we have

$$\delta_x^2\delta_y^2\delta_t U_{ij}^{k+\frac{1}{2}}$$

$$= \frac{1}{2}\delta_x^2\delta_y^2 \int_0^1 \left[u_t\left(x_i, y_j, t_{k+\frac{1}{2}} - \frac{s\tau}{2}\right) + u_t\left(x_i, y_j, t_{k+\frac{1}{2}} + \frac{s\tau}{2}\right) \right] \mathrm{d}s$$

$$= \frac{1}{2}\delta_y^2 \int_0^1 \left[\delta_x^2\left(u_t\left(x_i, y_j, t_{k+\frac{1}{2}} - \frac{s\tau}{2}\right) + u_t\left(x_i, y_j, t_{k+\frac{1}{2}} + \frac{s\tau}{2}\right)\right) \right] \mathrm{d}s$$

$$= \frac{1}{2}\delta_y^2 \int_0^1 \left[\int_0^1 \left(u_{xxt}\left(x_i - s_1 h_1, y_j, t_{k+\frac{1}{2}} - \frac{s\tau}{2}\right) + u_{xxt}\left(x_i - s_1 h_1, y_j, t_{k+\frac{1}{2}} + \frac{s\tau}{2}\right) \right.\right.$$

$$+ u_{xxt}\left(x_i + s_1 h_1, y_j, t_{k+\frac{1}{2}} - \frac{s\tau}{2}\right)$$

$$\left.\left. + u_{xxt}\left(x_i + s_1 h_1, y_j, t_{k+\frac{1}{2}} + \frac{s\tau}{2}\right) \right)(1 - s_1)\mathrm{d}s_1 \right] \mathrm{d}s$$

$$= \frac{1}{2}\int_0^1 \left[\int_0^1 \delta_y^2\left(u_{xxt}\left(x_i - s_1 h_1, y_j, t_{k+\frac{1}{2}} - \frac{s\tau}{2}\right) + u_{xxt}\left(x_i - s_1 h_1, y_j, t_{k+\frac{1}{2}} + \frac{s\tau}{2}\right) \right.\right.$$

$$+ u_{xxt}\left(x_i + s_1 h_1, y_j, t_{k+\frac{1}{2}} - \frac{s\tau}{2}\right)$$

$$\left.\left. + u_{xxt}\left(x_i + s_1 h_1, y_j, t_{k+\frac{1}{2}} + \frac{s\tau}{2}\right) \right)(1 - s_1)\mathrm{d}s_1 \right] \mathrm{d}s$$

$$= \frac{1}{2}\int_0^1 \left\{ \int_0^1 \left[\int_0^1 \left(u_{xxyyt}\left(x_i - s_1 h_1, y_j - s_2 h_2, t_{k+\frac{1}{2}} - \frac{s\tau}{2}\right) \right.\right.\right.$$

$$+ u_{xxyyt}\left(x_i - s_1 h_1, y_j - s_2 h_2, t_{k+\frac{1}{2}} + \frac{s\tau}{2}\right)$$

$$+ u_{xxyyt}\left(x_i + s_1 h_1, y_j - s_2 h_2, t_{k+\frac{1}{2}} - \frac{s\tau}{2}\right)$$

$$+ u_{xxyyt}\left(x_i + s_1 h_1, y_j - s_2 h_2, t_{k+\frac{1}{2}} + \frac{s\tau}{2}\right)$$

$$+ u_{xxyyt}\left(x_i - s_1 h_1, \, y_j + s_2 h_2, \, t_{k+\frac{1}{2}} - \frac{s\tau}{2}\right)$$

$$+ u_{xxyyt}\left(x_i - s_1 h_1, \, y_j + s_2 h_2, \, t_{k+\frac{1}{2}} + \frac{s\tau}{2}\right)$$

$$+ u_{xxyyt}\left(x_i + s_1 h_1, \, y_j + s_2 h_2, \, t_{k+\frac{1}{2}} - \frac{s\tau}{2}\right)$$

$$+ u_{xxyyt}\left(x_i + s_1 h_1, \, y_j + s_2 h_2, \, t_{k+\frac{1}{2}} + \frac{s\tau}{2}\right)\bigg)(1 - s_2)\mathrm{d}s_2\bigg](1 - s_1)\mathrm{d}s_1\bigg\}\,\mathrm{d}s.$$

Noticing (5.4), one can show that there is a positive constant c_2 such that

$$\begin{cases} \left|(R_2)_{ij}^k\right| \leqslant c_2(\tau^2 + h_1^2 + h_2^2), & (i, j) \in \omega, \quad 0 \leqslant k \leqslant n - 1, \\[2mm] \left|\delta_t(R_2)_{ij}^{k+\frac{1}{2}}\right| \leqslant c_2(\tau^2 + h_1^2 + h_2^2), & (i, j) \in \omega, \quad 0 \leqslant k \leqslant n - 2, \end{cases} \tag{5.8}$$

where

$$\delta_t(R_2)_{ij}^{k+\frac{1}{2}} = \frac{1}{\tau}\left[(R_2)_{ij}^{k+1} - (R_2)_{ij}^k\right].$$

Omitting the small term $(R_2)_{ij}^k$ in (5.7), noticing (5.5) and replacing U_{ij}^k with u_{ij}^k, a difference scheme reads

$$\begin{cases} \delta_t u_{ij}^{k+\frac{1}{2}} - a\Delta_h u_{ij}^{k+\frac{1}{2}} + \frac{1}{4}a^2\tau^2\delta_x^2\delta_y^2\delta_t u_{ij}^{k+\frac{1}{2}} = f_{ij}^{k+\frac{1}{2}}, \\ \hspace{4cm} (i, j) \in \omega, \quad 0 \leqslant k \leqslant n - 1, & (5.9\mathrm{a}) \\[2mm] u_{ij}^0 = \varphi(x_i, y_j), \quad (i, j) \in \omega, & (5.9\mathrm{b}) \\[2mm] u_{ij}^k = \alpha(x_i, y_j, t_k), \quad (i, j) \in \gamma, \quad 0 \leqslant k \leqslant n. & (5.9\mathrm{c}) \end{cases}$$

The stencil of the difference scheme (5.9) is shown in Fig. 5.2.

Fig. 5.2 The stencil of the difference scheme (5.9)

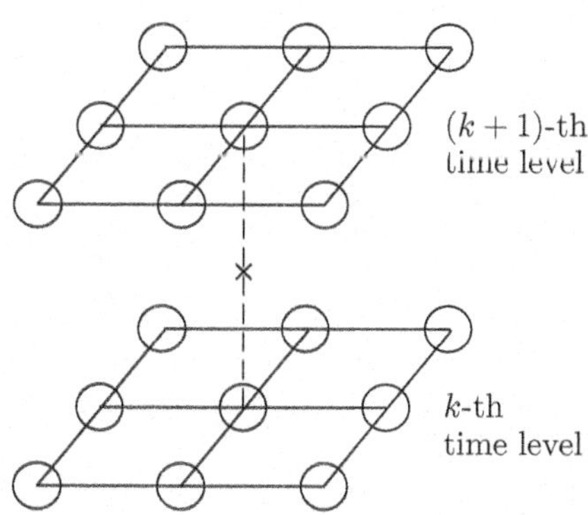

5.1.2   Existence of the Difference Solution

Theorem 5.1 *The difference scheme (5.9) is uniquely solvable.*

Proof Denote

$$u^k = \{u_{ij}^k \mid (i, j) \in \bar{\omega}\}.$$

It follows from (5.9b)–(5.9c) that u^0 has been determined.

Now suppose that u^k has been obtained, and then the difference scheme in u^{k+1} is written as

$$\begin{cases} \delta_t u_{ij}^{k+\frac{1}{2}} - a\Delta_h u_{ij}^{k+\frac{1}{2}} + \dfrac{1}{4}a^2\tau^2\delta_x^2\delta_y^2\delta_t u_{ij}^{k+\frac{1}{2}} = f_{ij}^{k+\frac{1}{2}}, & (i, j) \in \omega, \\ u_{ij}^{k+1} = \alpha(x_i, y_j, t_{k+1}), & (i, j) \in \gamma. \end{cases}$$

Consider its homogeneous one

$$\begin{cases} \dfrac{1}{\tau}u_{ij}^{k+1} - \dfrac{1}{2}a\Delta_h u_{ij}^{k+1} + \dfrac{a^2}{4}\tau\delta_x^2\delta_y^2 u_{ij}^{k+1} = 0, & (i, j) \in \omega, & (5.10a) \\ u_{ij}^{k+1} = 0, & (i, j) \in \gamma. & (5.10b) \end{cases}$$

Taking an inner product of (5.10a) with u^{k+1}, we have

$$\frac{1}{\tau}\|u^{k+1}\|^2 - \frac{1}{2}a\big(\Delta_h u^{k+1}, u^{k+1}\big) + \frac{a^2}{4}\tau\big(\delta_x^2\delta_y^2 u^{k+1}, u^{k+1}\big) = 0. \tag{5.11}$$

Noticing (5.10b) and using the summation by parts, we have

$$-\big(\Delta_h u^{k+1}, u^{k+1}\big) = |u^{k+1}|_1^2, \quad \big(\delta_x^2\delta_y^2 u^{k+1}, u^{k+1}\big) = \|\delta_x\delta_y u^{k+1}\|^2.$$

Substituting the above two equalities into (5.11), we have

$$\frac{1}{\tau}\|u^{k+1}\|^2 + \frac{1}{2}a|u^{k+1}|_1^2 + \frac{a^2}{4}\tau\|\delta_x\delta_y u^{k+1}\|^2 = 0.$$

It is easy to know that

$$u_{ij}^{k+1} = 0, \quad (i, j) \in \omega.$$

By induction, the difference scheme (5.9) is uniquely solvable. $\square$

5.1.3 Implementation of the Difference Scheme and Numerical Examples

The difference equation (5.9a) can be rewritten as

$$u_{ij}^{k+1} - \frac{a\tau}{2}\delta_x^2 u_{ij}^{k+1} - \frac{a\tau}{2}\delta_y^2 u_{ij}^{k+1} + \frac{a^2\tau^2}{4}\delta_x^2\delta_y^2 u_{ij}^{k+1}$$

$$= u_{ij}^k + \frac{a\tau}{2}\delta_x^2 u_{ij}^k + \frac{a\tau}{2}\delta_y^2 u_{ij}^k + \frac{a^2\tau^2}{4}\delta_x^2\delta_y^2 u_{ij}^k + \tau f_{ij}^{k+\frac{1}{2}}, \quad (i, j) \in \omega, \ 0 \leqslant k \leqslant n-1,$$

or

$$\left(\mathcal{I} - \frac{a\tau}{2}\delta_x^2\right)\left(\mathcal{I} - \frac{a\tau}{2}\delta_y^2\right) u_{ij}^{k+1} = \left(\mathcal{I} + \frac{a\tau}{2}\delta_x^2\right)\left(\mathcal{I} + \frac{a\tau}{2}\delta_y^2\right) u_{ij}^k + \tau f_{ij}^{k+\frac{1}{2}},$$

$$(i, j) \in \omega, \quad 0 \leqslant k \leqslant n-1, \tag{5.12}$$

where $\mathcal{I}u_{ij}^k = u_{ij}^k$, and $\mathcal{I}$ is called the identity operator.

Next, two ADI schemes are introduced.

P-R ADI Scheme (1955) [6]

Introduce the intermediate variable $\{\bar{u}_{ij}\}$ and rewrite (5.12) as follows:

$$\begin{cases} \left(\mathcal{I} - \frac{a\tau}{2}\delta_x^2\right)\bar{u}_{ij} = \left(\mathcal{I} + \frac{a\tau}{2}\delta_y^2\right) u_{ij}^k + \frac{\tau}{2} f_{ij}^{k+\frac{1}{2}}, & (i, j) \in \omega; \quad (5.13a) \\[2ex] \left(\mathcal{I} - \frac{a\tau}{2}\delta_y^2\right) u_{ij}^{k+1} = \left(\mathcal{I} + \frac{a\tau}{2}\delta_x^2\right)\bar{u}_{ij} + \frac{\tau}{2} f_{ij}^{k+\frac{1}{2}}, & (i, j) \in \omega. \ (5.13b) \end{cases}$$

Conversely, eliminating the intermediate variable $\{\bar{u}_{ij}\}$ from (5.13), we have

$$\left(\mathcal{I} - \frac{a\tau}{2}\delta_x^2\right)\left(\mathcal{I} - \frac{a\tau}{2}\delta_y^2\right) u_{ij}^{k+1}$$

$$= \left(\mathcal{I} - \frac{a\tau}{2}\delta_x^2\right)\left[\left(\mathcal{I} + \frac{a\tau}{2}\delta_x^2\right)\bar{u}_{ij} + \frac{\tau}{2} f_{ij}^{k+\frac{1}{2}}\right]$$

$$= \left(\mathcal{I} - \frac{a\tau}{2}\delta_x^2\right)\left(\mathcal{I} + \frac{a\tau}{2}\delta_x^2\right)\bar{u}_{ij} + \frac{\tau}{2}\left(\mathcal{I} - \frac{a\tau}{2}\delta_x^2\right) f_{ij}^{k+\frac{1}{2}}$$

$$= \left(\mathcal{I} + \frac{a\tau}{2}\delta_x^2\right)\left(\mathcal{I} - \frac{a\tau}{2}\delta_x^2\right)\bar{u}_{ij} + \frac{\tau}{2}\left(\mathcal{I} - \frac{a\tau}{2}\delta_x^2\right) f_{ij}^{k+\frac{1}{2}}$$

$$= \left(\mathcal{I} + \frac{a\tau}{2}\delta_x^2\right)\left[\left(\mathcal{I} + \frac{a\tau}{2}\delta_y^2\right) u_{ij}^k + \frac{\tau}{2} f_{ij}^{k+\frac{1}{2}}\right] + \frac{\tau}{2}\left(\mathcal{I} - \frac{a\tau}{2}\delta_x^2\right) f_{ij}^{k+\frac{1}{2}}$$

$$= \left(\mathcal{I} + \frac{a\tau}{2}\delta_x^2\right)\left(\mathcal{I} + \frac{a\tau}{2}\delta_y^2\right) u_{ij}^k + \tau f_{ij}^{k+\frac{1}{2}}, \quad (i, j) \in \omega,$$

which is exactly (5.12).

In addition, it follows from (5.13b) that

$$\left(\mathcal{I} + \frac{a\tau}{2}\delta_x^2\right)\bar{u}_{ij} = \left(\mathcal{I} - \frac{a\tau}{2}\delta_y^2\right)u_{ij}^{k+1} - \frac{\tau}{2}f_{ij}^{k+\frac{1}{2}}.$$

Adding the above equality and (5.13a) together, we get

$$\bar{u}_{ij} = \frac{1}{2}(u_{ij}^k + u_{ij}^{k+1}) - \frac{a\tau}{4}\left(\delta_y^2 u_{ij}^{k+1} - \delta_y^2 u_{ij}^k\right) = u_{ij}^{k+\frac{1}{2}} - \frac{a\tau^2}{4}\delta_y^2\delta_t u_{ij}^{k+\frac{1}{2}}.$$

Since $\{u_{0j}^k\}$ and $\{u_{m_1,j}^k\}$ are known, the intermediate variable should satisfy

$$\bar{u}_{0j} = u_{0j}^{k+\frac{1}{2}} - \frac{a\tau^2}{4}\delta_y^2\delta_t u_{0j}^{k+\frac{1}{2}}, \quad \bar{u}_{m_1,j} = u_{m_1,j}^{k+\frac{1}{2}} - \frac{a\tau^2}{4}\delta_y^2\delta_t u_{m_1,j}^{k+\frac{1}{2}}, \quad 1 \leqslant j \leqslant m_2 - 1.$$

When the value $\{u_{ij}^k \mid 0 \leqslant i \leqslant m_1, 0 \leqslant j \leqslant m_2\}$ at the k-th time level is known, we can obtain the intermediate variable $\{\bar{u}_{ij} \mid (i, j) \in \omega\}$ from (5.13a): For an arbitrary fixed j ($1 \leqslant j \leqslant m_2 - 1$), we take the boundary value conditions

$$\bar{u}_{0j} = u_{0j}^{k+\frac{1}{2}} - \frac{a\tau^2}{4}\delta_y^2\delta_t u_{0j}^{k+\frac{1}{2}}, \quad \bar{u}_{m_1,j} = u_{m_1,j}^{k+\frac{1}{2}} - \frac{a\tau^2}{4}\delta_y^2\delta_t u_{m_1,j}^{k+\frac{1}{2}}, \tag{5.14}$$

and solve

$$\left(\mathcal{I} - \frac{a\tau}{2}\delta_x^2\right)\bar{u}_{ij} = \left(\mathcal{I} + \frac{a\tau}{2}\delta_y^2\right)u_{ij}^k + \frac{\tau}{2}f_{ij}^{k+\frac{1}{2}}, \quad 1 \leqslant i \leqslant m_1 - 1, \tag{5.15}$$

and then one obtains $\{\bar{u}_{ij} \mid 1 \leqslant i \leqslant m_1 - 1\}$.

When $\{\bar{u}_{ij} \mid (i, j) \in \omega\}$ has been determined, it remains to solve the value $\{u_{ij}^{k+1} \mid (i, j) \in \omega\}$ at the $(k + 1)$-th time level by using (5.13b): For an arbitrary fixed i ($1 \leqslant i \leqslant m_1 - 1$), we take the boundary value conditions

$$u_{i0}^{k+1} = \alpha(x_i, y_0, t_{k+1}), \quad u_{i,m_2}^{k+1} = \alpha(x_i, y_{m_2}, t_{k+1}), \tag{5.16}$$

and solve

$$\left(\mathcal{I} - \frac{a\tau}{2}\delta_y^2\right)u_{ij}^{k+1} = \left(\mathcal{I} + \frac{a\tau}{2}\delta_x^2\right)\bar{u}_{ij} + \frac{\tau}{2}f_{ij}^{k+\frac{1}{2}}, \quad 1 \leqslant j \leqslant m_2 - 1, \tag{5.17}$$

and then it yields $\{u_{ij}^{k+1} \mid 1 \leqslant j \leqslant m_2 - 1\}$.

Note that (5.14)–(5.15) is an implicit scheme in the x direction, and (5.16)–(5.17) is an implicit scheme in the y direction. Both are tridiagonal systems of linear equations, which can be solved using the double sweep method. In this sense, (5.13) is referred to as an **ADI scheme**, also known as the Peaceman-Rachford ADI (P-R ADI) method, first proposed by Peaceman and Rachford [6].

Equations (5.13a) and (5.13b) can be rewritten as

$$\frac{\bar{u}_{ij} - u_{ij}^k}{\tau/2} - a\big(\delta_x^2 \bar{u}_{ij} + \delta_y^2 u_{ij}^k\big) = f_{ij}^{k+\frac{1}{2}}, \quad (i, j) \in \omega,$$

$$\frac{u_{ij}^{k+1} - \bar{u}_{ij}}{\tau/2} - a\big(\delta_x^2 \bar{u}_{ij} + \delta_y^2 u_{ij}^{k+1}\big) = f_{ij}^{k+\frac{1}{2}}, \quad (i, j) \in \omega.$$

Therefore, the P-R ADI method can also be interpreted as follows:

First, discretize (5.1a) over the interval $[t_k, t_{k+\frac{1}{2}}]$, approximating u_{xx} using the second-order centered difference quotient with respect to x at $t = t_{k+\frac{1}{2}}$, and approximating u_{yy} using the second-order centered difference quotient with respect to y at $t = t_k$. Next, discretize (5.1a) over the interval $[t_{k+\frac{1}{2}}, t_{k+1}]$, approximating u_{xx} using the second-order centered difference quotient with respect to x at $t = t_{k+\frac{1}{2}}$, and approximating u_{yy} using the second-order centered difference quotient with respect to y at $t = t_{k+1}$.

D'Yakonov ADI Scheme (1964) [1]

Let $u_{ij}^* = \left(\mathcal{I} - \dfrac{a\tau}{2}\delta_y^2\right) u_{ij}^{k+1}$. Then (5.12) is equivalent to

$$\begin{cases} \left(\mathcal{I} - \dfrac{a\tau}{2}\delta_x^2\right) u_{ij}^* = \left(\mathcal{I} + \dfrac{a\tau}{2}\delta_x^2\right)\left(\mathcal{I} + \dfrac{a\tau}{2}\delta_y^2\right) u_{ij}^k + \tau f_{ij}^{k+\frac{1}{2}}, \quad (i, j) \in \omega, \\[4mm] \hspace{8cm} (5.18a) \\[4mm] \left(\mathcal{I} - \dfrac{a\tau}{2}\delta_y^2\right) u_{ij}^{k+1} = u_{ij}^*, \quad (i, j) \in \omega. \hspace{3cm} (5.18b) \end{cases}$$

Conversely, eliminating the intermediate variable $\{u_{ij}^*\}$, we have

$$\left(\mathcal{I} - \frac{a\tau}{2}\delta_x^2\right)\left(\mathcal{I} - \frac{a\tau}{2}\delta_y^2\right) u_{ij}^{k+1} = \left(\mathcal{I} - \frac{a\tau}{2}\delta_x^2\right) u_{ij}^*$$

$$= \left(\mathcal{I} + \frac{a\tau}{2}\delta_x^2\right)\left(\mathcal{I} + \frac{a\tau}{2}\delta_y^2\right) u_{ij}^k + \tau f_{ij}^{k+\frac{1}{2}}, \quad (i, j) \in \omega,$$

which is exactly (5.12).

When the value $\{u_{ij}^k \,|\, (i, j) \in \bar{\omega}\}$ is known, the intermediate variable $\{u_{ij}^* \,|\, (i, j) \in \omega\}$ can be solved by using (5.18a): In detail, for an arbitrary fixed j $(1 \leqslant j \leqslant m_2 - 1)$, taking the boundary value conditions

$$u_{0j}^* = \left(\mathcal{I} - \frac{a\tau}{2}\delta_y^2\right) u_{0j}^{k+1}, \quad u_{m_1, j}^* = \left(\mathcal{I} - \frac{a\tau}{2}\delta_y^2\right) u_{m_1, j}^{k+1}, \tag{5.19}$$

and solving

$$\left(\mathcal{I} - \frac{a\tau}{2}\delta_x^2\right) u_{ij}^* = \left(\mathcal{I} + \frac{a\tau}{2}\delta_x^2\right)\left(\mathcal{I} + \frac{a\tau}{2}\delta_y^2\right) u_{ij}^k + \tau f_{ij}^{k+\frac{1}{2}}, \quad 1 \leqslant i \leqslant m_1 - 1,$$

$$(5.20)$$

we obtain $\{u_{ij}^* \mid 1 \leqslant i \leqslant m_1 - 1\}$.

When $\{u_{ij}^* \mid (i, j) \in \omega\}$ has been solved, the value $\{u_{ij}^{k+1} \mid (i, j) \in \omega\}$ can be solved by using (5.18b): For an arbitrary fixed i ($1 \leqslant i \leqslant m_1 - 1$), taking the boundary value conditions

$$u_{i0}^{k+1} = \alpha(x_i, y_0, t_{k+1}), \quad u_{i,m_2}^{k+1} = \alpha(x_i, y_{m_2}, t_{k+1}), \tag{5.21}$$

and solving

$$\left(\mathcal{I} - \frac{a\tau}{2}\delta_y^2\right) u_{ij}^{k+1} = u_{ij}^*, \quad 1 \leqslant j \leqslant m_2 - 1, \tag{5.22}$$

we obtain $\{u_{ij}^{k+1} \mid 1 \leqslant j \leqslant m_2 - 1\}$.

Note that (5.19)–(5.20) is an implicit scheme in x direction, and (5.21)–(5.22) is an implicit one in y direction. Both schemes are the tridiagonal systems of linear equations, which can be solved by the double sweep method. We call (5.18) D'Yakonov ADI scheme.

It is easy to extend D'Yakonov ADI scheme to the three-dimensional case.

Example 5.1 Apply D'yakonov ADI scheme to compute the problem

$$\begin{cases} \dfrac{\partial u}{\partial t} - \left(\dfrac{\partial^2 u}{\partial x^2} + \dfrac{\partial^2 u}{\partial y^2}\right) = -\dfrac{3}{2}e^{\frac{1}{2}(x+y)-t}, & 0 < x, y < 1, \quad 0 < t \leqslant 1, \\ u(x, y, 0) = e^{\frac{1}{2}(x+y)}, & 0 \leqslant x, y \leqslant 1, \\ u(0, y, t) = e^{\frac{1}{2}y-t}, \quad u(1, y, t) = e^{\frac{1}{2}(1+y)-t}, & 0 \leqslant y \leqslant 1, \quad 0 < t \leqslant 1, \\ u(x, 0, t) = e^{\frac{1}{2}x-t}, \quad u(x, 1, t) = e^{\frac{1}{2}(1+x)-t}, & 0 < x < 1, \quad 0 < t \leqslant 1. \end{cases}$$

$$(5.23)$$

The exact solution of the problem is $u(x, y, t) = e^{\frac{1}{2}(x+y)-t}$.

Take $h_1 = h_2 = h = 1/m$ and denote the maximum error

$$E_\infty(h, \tau) = \max_{0 \leqslant i, j \leqslant m, 0 \leqslant k \leqslant n} \left| u(x_i, y_j, t_k) - u_{ij}^k \right|.$$

Table 5.1 presents a subset of the numerical results computed with step sizes $h = 1/100$ and $\tau = 1/100$. Table 5.2 provides the maximum errors $E_\infty(h, \tau)$ of the numerical solutions for various step sizes.

From Table 5.2, it is observed that when both the spatial and temporal step sizes are halved, the maximum errors decrease to one-quarter of their original values.

Table 5.1 (Example 5.1) Part of numerical solutions, exact solutions, and absolute values of the errors ($h = 1/100$, $\tau = 1/100$)

$(x,\ y,\ t)$	NS	ES	|ES−NS|
$(0.25, 0.25, 0.25)$	0.9999997	1.0000000	3.077e−7
$(0.75, 0.25, 0.25)$	1.2840250	1.2840250	3.494e−7
$(0.25, 0.75, 0.25)$	1.2840250	1.2840250	3.494e−7
$(0.75, 0.75, 0.25)$	1.6487210	1.6487210	3.998e−7
$(0.25, 0.25, 0.50)$	0.7788005	0.7788008	2.419e−7
$(0.75, 0.25, 0.50)$	0.9999997	1.0000000	2.744e−7
$(0.25, 0.75, 0.50)$	0.9999997	1.0000000	2.744e−7
$(0.75, 0.75, 0.50)$	1.2840250	1.2840250	3.137e−7
$(0.25, 0.25, 0.75)$	0.6065305	0.6065307	1.884e−7
$(0.75, 0.25, 0.75)$	0.7788006	0.7788008	2.137e−7
$(0.25, 0.75, 0.75)$	0.7788006	0.7788008	2.137e−7
$(0.75, 0.75, 0.75)$	0.9999998	1.0000000	2.443e−7
$(0.25, 0.25, 1.00)$	0.4723664	0.4723666	1.467e−7
$(0.75, 0.25, 1.00)$	0.6065305	0.6065307	1.665e−7
$(0.25, 0.75, 1.00)$	0.6065305	0.6065307	1.665e−7
$(0.75, 0.75, 1.00)$	0.7788006	0.7788008	1.903e−7

Table 5.2 (Example 5.1) The maximum errors $E_\infty(h, \tau)$ of numerical solutions with different step sizes

h	τ	$E_\infty(h, \tau)$	$E_\infty(2h, 2\tau)/E_\infty(h, \tau)$
1/10	1/10	6.020e−5	
1/20	1/20	1.513e−5	3.979
1/40	1/40	3.773e−6	4.010
1/80	1/80	9.447e−7	3.994
1/160	1/160	2.361e−7	4.001

Fig. 5.3 (Example 5.1) The surface of the exact solution at $t = 1$

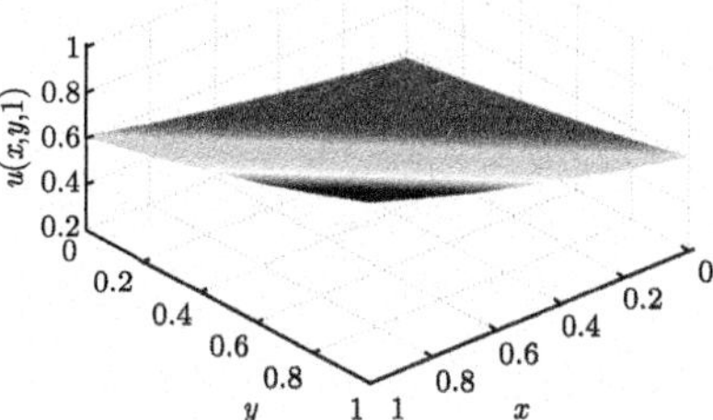

Figures 5.3 and 5.4 display the surfaces of the exact and numerical solutions at $t = 1$ with step sizes $h = 1/10$ and $\tau = 1/10$, respectively. Figure 5.5 illustrates the error surfaces of the numerical solutions for different step sizes at $t = 1$.

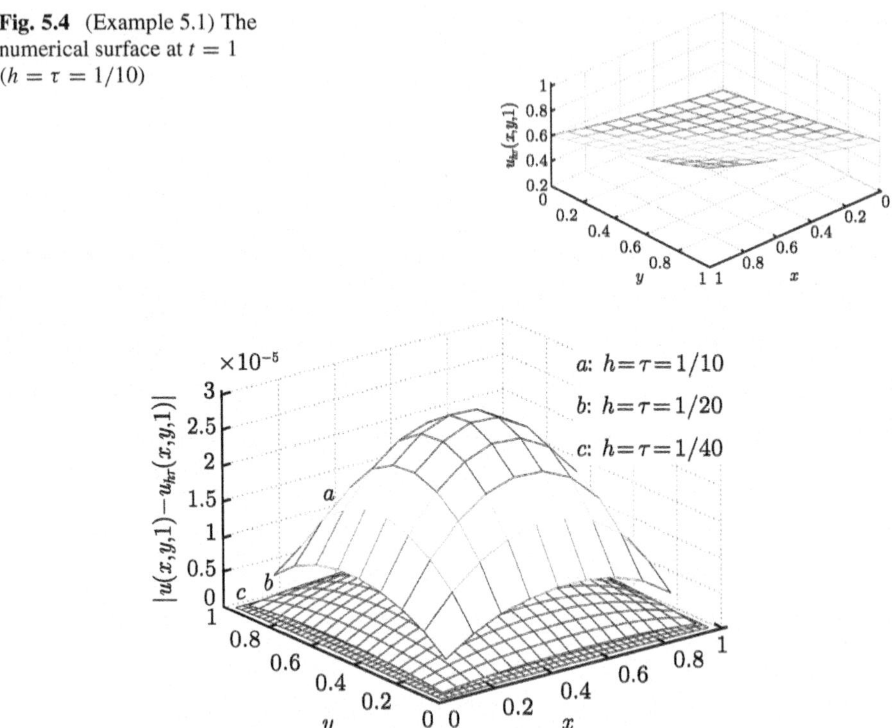

Fig. 5.4 (Example 5.1) The numerical surface at $t = 1$ ($h = \tau = 1/10$)

Fig. 5.5 (Example 5.1) The numerical error surfaces at $t = 1$

5.1.4   A Priori Estimate of the Difference Solution

We begin by presenting two fundamental discrete Gronwall inequalities.

Lemma 5.1 (Gronwall Inequalities)

(a) *Suppose $\{F^k\}_{k=0}^{\infty}$ is a nonnegative sequence, c and g are both nonnegative constants, satisfying*

$$F^k \leqslant c\tau \sum_{l=0}^{k-1} F^l + g, \quad k = 0, 1, 2, \ldots,$$

then we have

$$F^k \leqslant \mathrm{e}^{ck\tau} g, \quad k = 0, 1, 2, \ldots.$$

(b) *Suppose $\{F^k\}_{k=0}^{\infty}$ is a nonnegative sequence, $\{g^k\}_{k=0}^{\infty}$ is a nonnegative monotone non-decreasing sequence, satisfying*

$$F^k \leqslant c\tau \sum_{l=0}^{k-1} F^l + g^k, \quad k = 0, 1, 2, \ldots,$$

then we have

$$F^k \leqslant e^{ck\tau} g^k, \quad k = 0, 1, 2, \ldots.$$

Proof

(a) It is easy to know that

$$F^0 \leqslant g.$$

Let

$$G^k = c\tau \sum_{l=0}^{k-1} F^l + g, \quad k = 0, 1, 2, \ldots.$$

Then we have

$$G^0 = g,$$

$$F^k \leqslant G^k, \quad k = 0, 1, 2, \ldots,$$

$$G^k = G^{k-1} + c\tau F^{k-1} \leqslant G^{k-1} + c\tau G^{k-1} = (1 + c\tau)G^{k-1}, \quad k = 1, 2, 3, \ldots.$$

It follows from recursion that

$$G^k \leqslant (1 + c\tau)^k G^0 \leqslant e^{ck\tau} g, \quad k = 0, 1, 2, \ldots.$$

Therefore,

$$F^k \leqslant G^k \leqslant e^{ck\tau} g, \quad k = 0, 1, 2, \ldots.$$

(b) It is easy to know that

$$F^0 \leqslant g^0.$$

Let

$$G^k = c\tau \sum_{l=0}^{k-1} F^l + g^k, \quad k = 0, 1, 2, \dots.$$

Then we have

$$G^0 = g^0,$$

$$F^k \leqslant G^k, \quad k = 0, 1, 2, \dots,$$

$$\begin{aligned}
G^k &= c\tau \sum_{l=0}^{k-2} F^l + g^{k-1} + c\tau F^{k-1} + (g^k - g^{k-1}) \\
&= G^{k-1} + c\tau F^{k-1} + (g^k - g^{k-1}) \\
&\leqslant (1 + c\tau)G^{k-1} + (g^k - g^{k-1}), \quad k = 1, 2, \dots.
\end{aligned}$$

With the help of Lemma 3.3, we have

$$F^k \leqslant G^k \leqslant e^{ck\tau}\left[G^0 + \sum_{l=1}^{k}(g^l - g^{l-1})\right] = e^{ck\tau}g^k, \quad k = 0, 1, 2, \dots.$$

$\square$

Theorem 5.2 *Let $\{v_{ij}^k \mid (i, j) \in \bar{\omega}, \ 0 \leqslant k \leqslant n\}$ be the solution of the difference scheme*

$$\begin{cases}
\delta_t v_{ij}^{k+\frac{1}{2}} - a\Delta_h v_{ij}^{k+\frac{1}{2}} + \dfrac{1}{4}a^2\tau^2\delta_x^2\delta_y^2\delta_t v_{ij}^{k+\frac{1}{2}} = g_{ij}^k, & \\
\hspace{4cm} (i, j) \in \omega, \quad 0 \leqslant k \leqslant n - 1, & (5.24\text{a}) \\
v_{ij}^0 = \varphi(x_i, y_j), \quad (i, j) \in \omega, & (5.24\text{b}) \\
v_{ij}^k = 0, \quad (i, j) \in \gamma, \quad 0 \leqslant k \leqslant n. & (5.24\text{c})
\end{cases}$$

Then we have

$$\|\Delta_h v^{k+1}\|^2$$

$$\leqslant e^{2k\tau}\left[4\|\Delta_h v^0\|^2 + 2a^{-2}\left(\|g^0\|^2 + 2\max_{0\leqslant l\leqslant k}\|g^l\|^2 + \tau\sum_{l=1}^{k}\|\delta_t g^{l-\frac{1}{2}}\|^2\right)\right],$$

$$0 \leqslant k \leqslant n - 1,$$

where

$$\|g^k\|^2 = h_1 h_2 \sum_{i=1}^{m_1-1} \sum_{j=1}^{m_2-1} (g_{ij}^k)^2, \quad \|\delta_t g^{l-\frac{1}{2}}\|^2 = h_1 h_2 \sum_{i=1}^{m_1-1} \sum_{j=1}^{m_2-1} (\delta_t g_{ij}^{l-\frac{1}{2}})^2.$$

Proof Taking an inner product of (5.24a) with $-\Delta_h \delta_t v^{k+\frac{1}{2}}$, we have

$$-\left(\delta_t v^{k+\frac{1}{2}}, \Delta_h \delta_t v^{k+\frac{1}{2}}\right) + a\left(\Delta_h v^{k+\frac{1}{2}}, \Delta_h \delta_t v^{k+\frac{1}{2}}\right)$$

$$-\frac{1}{4}a^2\tau^2\left(\delta_x^2 \delta_y^2 \delta_t v^{k+\frac{1}{2}}, \Delta_h \delta_t v^{k+\frac{1}{2}}\right) = -\left(g^k, \Delta_h \delta_t v^{k+\frac{1}{2}}\right), \quad 0 \leqslant k \leqslant n-1.$$

Noticing

$$-\left(\delta_t v^{k+\frac{1}{2}}, \Delta_h \delta_t v^{k+\frac{1}{2}}\right) = \left|\delta_t v^{k+\frac{1}{2}}\right|_1^2,$$

$$\left(\Delta_h v^{k+\frac{1}{2}}, \Delta_h \delta_t v^{k+\frac{1}{2}}\right) = \frac{1}{2\tau}\left(\|\Delta_h v^{k+1}\|^2 - \|\Delta_h v^k\|^2\right),$$

$$-\left(\delta_x^2 \delta_y^2 \delta_t v^{k+\frac{1}{2}}, \Delta_h \delta_t v^{k+\frac{1}{2}}\right) = \left|\delta_x \delta_y \delta_t v^{k+\frac{1}{2}}\right|_1^2,$$

we have

$$\frac{a}{2\tau}\left(\|\Delta_h v^{k+1}\|^2 - \|\Delta_h v^k\|^2\right) \leqslant -\left(g^k, \Delta_h \delta_t v^{k+\frac{1}{2}}\right), \quad 0 \leqslant k \leqslant n-1.$$

Replacing k with l and summing over l from 0 to k, we have

$$\frac{a}{2\tau}\left(\|\Delta_h v^{k+1}\|^2 - \|\Delta_h v^0\|^2\right)$$

$$\leqslant -\sum_{l=0}^{k}\left(g^l, \Delta_h \delta_t v^{l+\frac{1}{2}}\right)$$

$$= \frac{1}{\tau}\left[(g^0, \Delta_h v^0) - (g^k, \Delta_h v^{k+1})\right] + \sum_{l=1}^{k}(\delta_t g^{l-\frac{1}{2}}, \Delta_h v^l), \quad 0 \leqslant k \leqslant n-1.$$

Multiplying both sides of the above inequality by $\frac{2\tau}{a}$ and rearranging the resulting expression, we obtain

$$\|\Delta_h v^{k+1}\|^2 \leqslant \|\Delta_h v^0\|^2 + \frac{2}{a}\left[(g^0, \Delta_h v^0) - (g^k, \Delta_h v^{k+1})\right]$$

$$+ \frac{2}{a}\tau \sum_{l=1}^{k}(\delta_t g^{l-\frac{1}{2}}, \Delta_h v^l)$$

$$\leqslant \|\Delta_h v^0\|^2 + \left(a^{-2}\|g^0\|^2 + \|\Delta_h v^0\|^2\right) + \left(2a^{-2}\|g^k\|^2 + \frac{1}{2}\|\Delta_h v^{k+1}\|^2\right)$$

$$+ \tau \sum_{l=1}^{k} \left(a^{-2}\|\delta_t g^{l-\frac{1}{2}}\|^2 + \|\Delta_h v^l\|^2\right), \quad 0 \leqslant k \leqslant n-1.$$

Therefore,

$$\|\Delta_h v^{k+1}\|^2 \leqslant 4\|\Delta_h v^0\|^2 + 2a^{-2}\left(\|g^0\|^2 + 2\|g^k\|^2\right)$$

$$+ 2a^{-2}\tau \sum_{l=1}^{k} \|\delta_t g^{l-\frac{1}{2}}\|^2 + 2\tau \sum_{l=1}^{k} \|\Delta_h v^l\|^2, \quad 0 \leqslant k \leqslant n-1.$$

It follows from Lemma 5.1 (Gronwall inequality) that

$$\|\Delta_h v^{k+1}\|^2 \leqslant e^{2k\tau} \left[4\|\Delta_h v^0\|^2 + 2a^{-2}\left(\|g^0\|^2 + 2 \max_{0 \leqslant l \leqslant k} \|g^l\|^2 \right.\right.$$

$$\left.\left. + \tau \sum_{l=1}^{k} \|\delta_t g^{l-\frac{1}{2}}\|^2\right)\right], \quad 0 \leqslant k \leqslant n-1,$$

which completes the proof. $\qquad\square$

5.1.5   Convergence and Stability of the Difference Solution

Convergence

Theorem 5.3 *Let $\{u(x, y, t) \mid (x, y) \in \bar{\Omega}, 0 \leqslant t \leqslant T\}$ be the solution of the problem (5.1) and $\{u_{ij}^k \mid (i, j) \in \bar{\omega}, 0 \leqslant k \leqslant n\}$ be the solution of the difference scheme (5.9). Denote*

$$e_{ij}^k = u(x_i, y_j, t_k) - u_{ij}^k, \quad (i, j) \in \bar{\omega}, \quad 0 \leqslant k \leqslant n.$$

Then we have

$$\|e^k\|_\infty \leqslant \frac{1}{12a}\sqrt{6(\sqrt{2}+1)(3+T)e^T \, L_1 L_2 c_2 (\tau^2 + h_1^2 + h_2^2)}, \quad 1 \leqslant k \leqslant n.$$

Proof Subtracting (5.9) from (5.7) and (5.5), we have the system of error equations

$$\begin{cases} \delta_t e_{ij}^{k+\frac{1}{2}} - a\Delta_h e_{ij}^{k+\frac{1}{2}} + \frac{1}{4}a^2\tau^2\delta_x^2\delta_y^2\delta_t e_{ij}^{k+\frac{1}{2}} = (R_2)_{ij}^k, \\ \qquad\qquad\qquad (i,j)\in\omega, \quad 0\leqslant k\leqslant n-1, \\ e_{ij}^0 = 0, \quad (i,j)\in\omega, \\ e_{ij}^k = 0, \quad (i,j)\in\gamma, \quad 0\leqslant k\leqslant n. \end{cases}$$

Applying Theorem 5.2 and noticing (5.8), we have

$$\|\Delta_h e^{k+1}\|^2 \leqslant e^{2k\tau}\left[4\|\Delta_h e^0\|^2 + 2a^{-2}\Big(\|(R_2)^0\|^2\right.$$

$$\left.+ 2\max_{0\leqslant l\leqslant k}\|(R_2)^l\|^2 + \tau\sum_{l=1}^{k}\|\delta_t(R_2)^{l-\frac{1}{2}}\|^2\Big)\right]$$

$$\leqslant 2e^{2T}L_1L_2a^{-2}(3+T)c_2^2(\tau^2+h_1^2+h_2^2)^2, \quad 0\leqslant k\leqslant n-1.$$

Taking the square root on both sides of the above inequality, we obtain

$$\|\Delta_h e^k\| \leqslant e^T a^{-1}\sqrt{2L_1L_2(3+T)}c_2(\tau^2+h_1^2+h_2^2), \quad 1\leqslant k\leqslant n.$$

It follows from Lemma 2.8 again that

$$\|e^k\|_\infty \leqslant \frac{1}{12}\sqrt{3(\sqrt{2}+1)L_1L_2}\,\|\Delta_h e^k\|$$

$$\leqslant \frac{1}{12}\sqrt{3(\sqrt{2}+1)L_1L_2}\,e^T a^{-1}\sqrt{2L_1L_2(3+T)}c_2(\tau^2+h_1^2+h_2^2)$$

$$= \frac{1}{12a}\sqrt{6(\sqrt{2}+1)(3+T)}e^T L_1L_2c_2(\tau^2+h_1^2+h_2^2), \quad 1\leqslant k\leqslant n,$$

which completes the proof. □

Stability

Theorem 5.4 *For arbitrary grid ratios $r_1 \equiv a\frac{\tau}{h_1^2}$ and $r_2 \equiv a\frac{\tau}{h_2^2}$, the solution of the difference scheme (5.9) is stable with respect to the initial value and the right-hand side function in the following sense: Let $\{u_{ij}^k \mid (i,j)\in\bar\omega, \ 0\leqslant k\leqslant n\}$ be the solution of the system of difference equations*

$$\begin{cases} \delta_t u_{ij}^{k+\frac{1}{2}} - a\Delta_h u_{ij}^{k+\frac{1}{2}} + \frac{1}{4}a^2\tau^2\delta_x^2\delta_y^2\delta_t u_{ij}^{k+\frac{1}{2}} = f_{ij}^k, \\ \qquad\qquad\qquad (i,j)\in\omega, \quad 0\leqslant k\leqslant n-1, \\ u_{ij}^0 = \varphi_{ij}, \quad (i,j)\in\omega, \\ u_{ij}^k = 0, \quad (i,j)\in\gamma, \quad 0\leqslant k\leqslant n. \end{cases}$$

Then we have

$$\|\Delta_h u^{k+1}\|^2 \leqslant e^{2k\tau}\left[4\|\Delta_h u^0\|^2 + 2a^{-2}\left(\|f^0\|^2 + 2\max_{0\leqslant l\leqslant k}\|f^l\|^2\right)\right.$$

$$\left. + 2a^{-2}\tau\sum_{l=1}^{k}\|\delta_t f^{l-\frac{1}{2}}\|^2\right], \quad 0\leqslant k\leqslant n-1.$$

Proof A direct application of Theorem 5.2 gives the result.   □

5.2 The Compact ADI Scheme for the Two-Dimensional Parabolic Equation

In this section, we use the initial-boundary value problem of the two-dimensional heat conduction equation (5.1) as an example to introduce the compact ADI scheme [3, 8].

Let $v = \{v_{ij} \,|\, (i, j) \in \bar{\omega}\}$ be the grid function on Ω_h. Introduce the following notation:

$$\mathcal{A}v_{ij} = \begin{cases} \dfrac{1}{12}\left(v_{i-1,j} + 10v_{ij} + v_{i+1,j}\right), & 1\leqslant i\leqslant m_1-1, \quad 0\leqslant j\leqslant m_2, \\ v_{ij}, & i = 0, m_1, \quad 0\leqslant j\leqslant m_2, \end{cases}$$

$$\mathcal{B}v_{ij} = \begin{cases} \dfrac{1}{12}\left(v_{i,j-1} + 10v_{ij} + v_{i,j+1}\right), & 1\leqslant j\leqslant m_2-1, \quad 0\leqslant i\leqslant m_1, \\ v_{ij}, & j = 0, m_2, \quad 0\leqslant i\leqslant m_1. \end{cases}$$

It is easy to know that

$$\mathcal{A}v_{ij} = \left(\mathcal{I} + \frac{h_1^2}{12}\delta_x^2\right)v_{ij}, \quad \mathcal{B}v_{ij} = \left(\mathcal{I} + \frac{h_2^2}{12}\delta_y^2\right)v_{ij}, \quad (i, j) \in \omega.$$

5.2.1 The Derivation of the Difference Scheme

Let

$$v = \frac{\partial^2 u}{\partial x^2}, \quad w = \frac{\partial^2 u}{\partial y^2}.$$

Then (5.1a) is equivalent to

$$
\begin{cases}
\dfrac{\partial u}{\partial t} - av - aw = f(x, y, t), & (5.25a) \\[3mm]
v = \dfrac{\partial^2 u}{\partial x^2}, & (5.25b) \\[3mm]
w = \dfrac{\partial^2 u}{\partial y^2}. & (5.25c)
\end{cases}
$$

Define the grid functions by

$$
U_{ij}^k = u(x_i, y_j, t_k), \quad V_{ij}^k = v(x_i, y_j, t_k), \quad W_{ij}^k = w(x_i, y_j, t_k),
$$

$$
(i, j) \in \bar{\omega}, \quad 0 \leqslant k \leqslant n.
$$

Considering Eq. (5.25a) at the point $(x_i, y_j, t_{k+\frac{1}{2}})$ gives

$$
\frac{\partial u}{\partial t}(x_i, y_j, t_{k+\frac{1}{2}}) - av(x_i, y_j, t_{k+\frac{1}{2}}) - aw(x_i, y_j, t_{k+\frac{1}{2}}) = f_{ij}^{k+\frac{1}{2}},
$$

$$
(i, j) \in \bar{\omega}, \quad 0 \leqslant k \leqslant n - 1. \quad (5.26)
$$

By the Taylor expansion with an integral remainder, we have

$$
\frac{\partial u(x_i, y_j, t_{k+\frac{1}{2}})}{\partial t} = \delta_t U_{ij}^{k+\frac{1}{2}} - \frac{\tau^2}{16} \int_0^1 \left(u_{ttt}\left(x_i, y_j, t_{k+\frac{1}{2}} - \frac{s\tau}{2}\right) \right.
$$

$$
\left. + u_{ttt}\left(x_i, y_j, t_{k+\frac{1}{2}} + \frac{s\tau}{2}\right) \right)(1 - s)^2 ds,
$$

$$
v(x_i, y_j, t_{k+\frac{1}{2}}) = V_{ij}^{k+\frac{1}{2}} - \frac{\tau^2}{8} \int_0^1 \left(v_{tt}\left(x_i, y_j, t_{k+\frac{1}{2}} - \frac{s\tau}{2}\right) \right.
$$

$$
\left. + v_{tt}\left(x_i, y_j, t_{k+\frac{1}{2}} + \frac{s\tau}{2}\right) \right)(1 - s)ds
$$

$$
w(x_i, y_j, t_{k+\frac{1}{2}}) = W_{ij}^{k+\frac{1}{2}} - \frac{\tau^2}{8} \int_0^1 \left(w_{tt}\left(x_i, y_j, t_{k+\frac{1}{2}} - \frac{s\tau}{2}\right) \right.
$$

$$
\left. + w_{tt}\left(x_i, y_j, t_{k+\frac{1}{2}} + \frac{s\tau}{2}\right) \right)(1 - s)ds.
$$

Substituting the above three equalities into (5.26), we obtain

$$
\delta_t U_{ij}^{k+\frac{1}{2}} - aV_{ij}^{k+\frac{1}{2}} - aW_{ij}^{k+\frac{1}{2}} = f_{ij}^{k+\frac{1}{2}} + \tau^2 g_{ij}^{k+\frac{1}{2}},
$$

$$
(i, j) \in \bar{\omega}, \quad 0 \leqslant k \leqslant n - 1, \quad (5.27)
$$

where

$$
g_{ij}^{k+\frac{1}{2}} = \frac{1}{16} \int_0^1 \left(u_{ttt}\left(x_i, y_j, t_{k+\frac{1}{2}} - \frac{s\tau}{2}\right) + u_{ttt}\left(x_i, y_j, t_{k+\frac{1}{2}} + \frac{s\tau}{2}\right) \right)(1-s)^2 ds
$$

$$
- \frac{a}{8} \int_0^1 \left[v_{tt}\left(x_i, y_j, t_{k+\frac{1}{2}} - \frac{s\tau}{2}\right) + v_{tt}\left(x_i, y_j, t_{k+\frac{1}{2}} + \frac{s\tau}{2}\right) \right.
$$

$$
\left. + w_{tt}\left(x_i, y_j, t_{k+\frac{1}{2}} - \frac{s\tau}{2}\right) + w_{tt}\left(x_i, y_j, t_{k+\frac{1}{2}} + \frac{s\tau}{2}\right) \right](1-s)ds.
$$

Performing the operator $\mathcal{AB}$ on both sides of (5.27) produces

$$
\mathcal{AB}\delta_t U_{ij}^{k+\frac{1}{2}} - a\,\mathcal{AB}V_{ij}^{k+\frac{1}{2}} - a\,\mathcal{AB}W_{ij}^{k+\frac{1}{2}}
$$

$$
= \mathcal{AB}f_{ij}^{k+\frac{1}{2}} + \tau^2 \mathcal{AB}g_{ij}^{k+\frac{1}{2}}, \quad (i, j) \in \omega, \quad 0 \leqslant k \leqslant n-1. \tag{5.28}
$$

Considering Eq. (5.25b) at the grid point (x_i, y_j, t_k) gives

$$
v(x_i, y_j, t_k) = \frac{\partial^2 u}{\partial x^2}(x_i, y_j, t_k), \quad (i, j) \in \bar{\omega}, \quad 0 \leqslant k \leqslant n.
$$

According to Lemma 1.2, we have

$$
\mathcal{A}V_{ij}^k = \delta_x^2 U_{ij}^k + \frac{h_1^4}{360} \int_0^1 \left[u_{xxxxxx}(x_i - sh_1, y_j, t_k) \right.
$$

$$
\left. + u_{xxxxxx}(x_i + sh_1, y_j, t_k) \right](1-s)^3 [5 - 3(1-s)^2] ds,
$$

$$
1 \leqslant i \leqslant m_1 - 1, \quad 0 \leqslant j \leqslant m_2, \quad 0 \leqslant k \leqslant n.
$$

Averaging the above equality with the superscripts k and $k+1$ yields

$$
\mathcal{A}V_{ij}^{k+\frac{1}{2}} = \delta_x^2 U_{ij}^{k+\frac{1}{2}} + \frac{h_1^4}{720} \int_0^1 \left[u_{xxxxxx}(x_i - sh_1, y_j, t_k) \right.
$$

$$
+ u_{xxxxxx}(x_i + sh_1, y_j, t_k) + u_{xxxxxx}(x_i - sh_1, y_j, t_{k+1})
$$

$$
\left. + u_{xxxxxx}(x_i + sh_1, y_j, t_{k+1}) \right](1-s)^3 [5 - 3(1-s)^2] ds,
$$

$$
1 \leqslant i \leqslant m_1 - 1, \quad 0 \leqslant j \leqslant m_2, \quad 0 \leqslant k \leqslant n-1.
$$

Performing the operator $\mathcal{B}$ on both sides of the above equality arrives at

$$
\mathcal{AB}V_{ij}^{k+\frac{1}{2}} = \mathcal{B}\delta_x^2 U_{ij}^{k+\frac{1}{2}} + \frac{h_1^4}{720}\mathcal{B}\left\{ \int_0^1 \left[u_{xxxxxx}(x_i - sh_1, y_j, t_k) \right. \right.
$$

$$+u_{xxxxxx}(x_i + sh_1, y_j, t_k) + u_{xxxxxx}(x_i - sh_1, y_j, t_{k+1})$$

$$+u_{xxxxxx}(x_i + sh_1, y_j, t_{k+1})\big](1 - s)^3[5 - 3(1 - s)^2]ds\bigg\},$$

$$(i, j) \in \omega, \quad 0 \leqslant k \leqslant n - 1. \tag{5.29}$$

In a similar way, considering (5.25c) at the grid point (x_i, y_j, t_k) gives

$$\mathcal{A}\mathcal{B}W_{ij}^{k+\frac{1}{2}} = \mathcal{A}\delta_y^2 U_{ij}^{k+\frac{1}{2}} + \frac{h_2^4}{720}\mathcal{A}\bigg\{ \int_0^1 \big[u_{yyyyyy}(x_i, y_j - sh_2, t_k)$$

$$+u_{yyyyyy}(x_i, y_j + sh_2, t_k) + u_{yyyyyy}(x_i, y_j - sh_2, t_{k+1})$$

$$+u_{yyyyyy}(x_i, y_j + sh_2, t_{k+1})\big](1 - s)^3[5 - 3(1 - s)^2]ds\bigg\},$$

$$(i, j) \in \omega, \quad 0 \leqslant k \leqslant n - 1. \tag{5.30}$$

Substituting (5.29) and (5.30) into (5.28) and adding a small term $\frac{1}{4}a^2\tau^2\delta_x^2\delta_y^2\delta_t$ $U_{ij}^{k+\frac{1}{2}}$ on both sides produce

$$\mathcal{A}\mathcal{B}\delta_t U_{ij}^{k+\frac{1}{2}} - a\Big(\mathcal{B}\delta_x^2 U_{ij}^{k+\frac{1}{2}} + \mathcal{A}\delta_y^2 U_{ij}^{k+\frac{1}{2}}\Big) + \frac{1}{4}a^2\tau^2\delta_x^2\delta_y^2\delta_t U_{ij}^{k+\frac{1}{2}}$$

$$= \mathcal{A}\mathcal{B}f_{ij}^{k+\frac{1}{2}} + (R_3)_{ij}^k, \quad (i, j) \in \omega, \quad 0 \leqslant k \leqslant n - 1, \tag{5.31}$$

where

$$(R_3)_{ij}^k = \tau^2\bigg\{\mathcal{A}\mathcal{B}g_{ij}^{k+\frac{1}{2}} + \frac{1}{4}a^2\delta_x^2\delta_y^2\delta_t U_{ij}^{k+\frac{1}{2}}\bigg\}$$

$$+ a\frac{h_1^4}{720}\mathcal{B}\bigg\{ \int_0^1 \big[u_{xxxxxx}(x_i - sh_1, y_j, t_k)$$

$$+ u_{xxxxxx}(x_i + sh_1, y_j, t_k) + u_{xxxxxx}(x_i - sh_1, y_j, t_{k+1})$$

$$+ u_{xxxxxx}(x_i + sh_1, y_j, t_{k+1})\big](1 - s)^3[5 - 3(1 - s)^2]ds\bigg\}$$

$$+ a\frac{h_2^4}{720}\mathcal{A}\bigg\{ \int_0^1 \big[u_{yyyyyy}(x_i, y_j - sh_2, t_k)$$

$$+ u_{yyyyyy}(x_i, y_j + sh_2, t_k) + u_{yyyyyy}(x_i, y_j - sh_2, t_{k+1})$$

$$+ u_{yyyyyy}(x_i, y_j + sh_2, t_{k+1})\big](1 - s)^3[5 - 3(1 - s)^2]ds\bigg\},$$

$$(i, j) \in \omega, \quad 0 \leqslant k \leqslant n - 1.$$

There is a positive constant c_3 such that

$$\begin{cases} \left|(R_3)_{ij}^k\right| \leqslant c_3(\tau^2 + h_1^4 + h_2^4), & (i, j) \in \omega, \quad 0 \leqslant k \leqslant n - 1, \\ \left|\delta_t (R_3)_{ij}^{k+\frac{1}{2}}\right| \leqslant c_3(\tau^2 + h_1^4 + h_2^4), & (i, j) \in \omega, \quad 0 \leqslant k \leqslant n - 2, \end{cases} \tag{5.32}$$

where

$$\delta_t (R_3)_{ij}^{k+\frac{1}{2}} = \frac{1}{\tau}\left[(R_3)_{ij}^{k+1} - (R_3)_{ij}^k\right].$$

Noticing the initial-boundary value conditions (5.1b) and (5.1c), we have

$$\begin{cases} U_{ij}^0 = \varphi(x_i, y_j), & (i, j) \in \omega, \\ U_{ij}^k = \alpha(x_i, y_j, t_k), & (i, j) \in \gamma, \quad 0 \leqslant k \leqslant n. \end{cases} \tag{5.33}$$

Omitting the small term $(R_3)_{ij}^k$ in (5.31), a compact difference scheme for (5.1) reads

$$\begin{cases} \mathcal{A}\mathcal{B}\delta_t u_{ij}^{k+\frac{1}{2}} - a(\mathcal{B}\delta_x^2 u_{ij}^{k+\frac{1}{2}} + \mathcal{A}\delta_y^2 u_{ij}^{k+\frac{1}{2}}) + \frac{1}{4}a^2\tau^2\delta_x^2\delta_y^2\delta_t u_{ij}^{k+\frac{1}{2}} = \mathcal{A}\mathcal{B}f_{ij}^{k+\frac{1}{2}}, \\ \hspace{8cm} (i, j) \in \omega, \quad 0 \leqslant k \leqslant n - 1, \hspace{1cm} (5.34a) \\ u_{ij}^0 = \varphi(x_i, y_j), \quad (i, j) \in \omega, \hspace{4.5cm} (5.34b) \\ u_{ij}^k = \alpha(x_i, y_j, t_k), \quad (i, j) \in \gamma, \quad 0 \leqslant k \leqslant n. \hspace{2cm} (5.34c) \end{cases}$$

The computational stencil of the compact difference scheme (5.34) is the same to that in Fig. 5.2.

5.2.2   Existence of the Difference Solution

Lemma 5.2 *For $w \in \overset{\circ}{\mathcal{V}}_h$, we have*

$$(\mathcal{A}\mathcal{B}w, w) \geqslant \frac{1}{3}\|w\|^2.$$

Proof A direct calculation gives

$$(\mathcal{A}\mathcal{B}w, w)$$

$$= \left(\left(\mathcal{I} + \frac{h_1^2}{12}\delta_x^2\right)\left(\mathcal{I} + \frac{h_2^2}{12}\delta_y^2\right)w, w\right)$$

$$= \left(\left(\mathcal{I} + \frac{h_1^2}{12}\delta_x^2 + \frac{h_2^2}{12}\delta_y^2 + \frac{1}{144}h_1^2 h_2^2 \delta_x^2 \delta_y^2 \right) w, \, w \right)$$

$$= \left(w, \, w \right) + \frac{h_1^2}{12}\left(\delta_x^2 w, \, w \right) + \frac{h_2^2}{12}\left(\delta_y^2 w, \, w \right) + \frac{1}{144}h_1^2 h_2^2 \left(\delta_x^2 \delta_y^2 w, \, w \right)$$

$$= \|w\|^2 - \frac{h_1^2}{12}\|\delta_x w\|^2 - \frac{h_2^2}{12}\|\delta_y w\|^2 + \frac{1}{144}h_1^2 h_2^2 \|\delta_x \delta_y w\|^2$$

$$\geqslant \|w\|^2 - \frac{h_1^2}{12}\|\delta_x w\|^2 - \frac{h_2^2}{12}\|\delta_y w\|^2$$

$$\geqslant \|w\|^2 - \frac{h_1^2}{12}\cdot\frac{4}{h_1^2}\|w\|^2 - \frac{h_2^2}{12}\cdot\frac{4}{h_2^2}\|w\|^2$$

$$= \frac{1}{3}\|w\|^2.$$

$\square$

Theorem 5.5 *The compact difference scheme (5.34) is uniquely solvable.*

Proof Denote

$$u^k = \{u_{ij}^k \mid (i, j) \in \bar{\omega}\}.$$

It follows from (5.34b) and (5.34c) that u^0 has been determined. Now suppose u^k has been determined, the compact difference scheme in u^{k+1} is

$$\begin{cases} \mathcal{A}\mathcal{B}\delta_t u_{ij}^{k+\frac{1}{2}} - a\left(\mathcal{B}\delta_x^2 u_{ij}^{k+\frac{1}{2}} + \mathcal{A}\delta_y^2 u_{ij}^{k+\frac{1}{2}} \right) + \frac{1}{4}a^2\tau^2\delta_x^2\delta_y^2\delta_t u_{ij}^{k+\frac{1}{2}} \\ \qquad\qquad\qquad\qquad\qquad\qquad = \mathcal{A}\mathcal{B}f_{ij}^{k+\frac{1}{2}}, \quad (i, j) \in \omega, \\ u_{ij}^{k+1} = \alpha(x_i, y_j, t_{k+1}), \quad (i, j) \in \gamma. \end{cases}$$

The corresponding homogeneous system is

$$\begin{cases} \dfrac{1}{\tau}\mathcal{A}\mathcal{B}u_{ij}^{k+1} - \dfrac{1}{2}a\left(\mathcal{B}\delta_x^2 u_{ij}^{k+1} + \mathcal{A}\delta_y^2 u_{ij}^{k+1} \right) \\ \qquad\qquad\qquad + \dfrac{1}{4}a^2\tau\delta_x^2\delta_y^2 u_{ij}^{k+1} = 0, \quad (i, j) \in \omega, \qquad\qquad (5.35a) \\ u_{ij}^{k+1} = 0, \quad (i, j) \in \gamma. \qquad\qquad\qquad\qquad\qquad\qquad\qquad (5.35b) \end{cases}$$

Taking an inner product of (5.35a) with u^{k+1} produces

$$
\frac{1}{\tau}\left(\mathcal{A}\mathcal{B}u^{k+1},u^{k+1}\right)-\frac{1}{2}a\left(\mathcal{B}\delta_x^2 u^{k+1}+\mathcal{A}\delta_y^2 u^{k+1},u^{k+1}\right)
$$

$$
+\frac{1}{4}a^2\tau\left(\delta_x^2\delta_y^2 u^{k+1},u^{k+1}\right)=0. \tag{5.36}
$$

Below, we estimate each term in (5.36).
By Lemma 5.2, we obtain

$$
\left(\mathcal{A}\mathcal{B}u^{k+1},u^{k+1}\right)\geqslant\frac{1}{3}\|u^{k+1}\|^2. \tag{5.37}
$$

It follows from Lemma 2.4 that

$$
-\left(\mathcal{B}\delta_x^2 u^{k+1}+\mathcal{A}\delta_y^2 u^{k+1},u^{k+1}\right)\geqslant\frac{2}{3}|u^{k+1}|_1^2. \tag{5.38}
$$

Noticing (5.35b) and with the help of the summation by parts, we have

$$
\left(\delta_x^2\delta_y^2 u^{k+1},u^{k+1}\right)=\|\delta_x\delta_y u^{k+1}\|^2. \tag{5.39}
$$

Substituting (5.37)–(5.39) into (5.36), we have

$$
\frac{1}{3\tau}\|u^{k+1}\|^2+\frac{a}{3}|u^{k+1}|_1^2+\frac{1}{4}a^2\tau\|\delta_x\delta_y u^{k+1}\|^2\leqslant 0,
$$

which implies

$$
u_{ij}^{k+1}=0,\quad (i,j)\in\bar{\omega}.
$$

By induction, the compact difference scheme (5.34) has a unique solution. $\square$

5.2.3   *Implementation of the Difference Scheme and Numerical Examples*

We can rewrite (5.34a) as

$$
\mathcal{A}\mathcal{B}u_{ij}^{k+1}-\frac{a\tau}{2}\mathcal{B}\delta_x^2 u_{ij}^{k+1}-\frac{a\tau}{2}\mathcal{A}\delta_y^2 u_{ij}^{k+1}+\frac{a^2\tau^2}{4}\delta_x^2\delta_y^2 u_{ij}^{k+1}
$$

$$
=\mathcal{A}\mathcal{B}u_{ij}^{k}+\frac{a\tau}{2}\mathcal{B}\delta_x^2 u_{ij}^{k}+\frac{a\tau}{2}\mathcal{A}\delta_y^2 u_{ij}^{k}+\frac{a^2\tau^2}{4}\delta_x^2\delta_y^2 u_{ij}^{k}+\tau\mathcal{A}\mathcal{B}f_{ij}^{k+\frac{1}{2}},
$$

or

$$\left(\mathcal{A} - \frac{a\tau}{2}\delta_x^2\right)\left(\mathcal{B} - \frac{a\tau}{2}\delta_y^2\right)u_{ij}^{k+1}$$
$$= \left(\mathcal{A} + \frac{a\tau}{2}\delta_x^2\right)\left(\mathcal{B} + \frac{a\tau}{2}\delta_y^2\right)u_{ij}^{k} + \tau\mathcal{A}\mathcal{B}f_{ij}^{k+\frac{1}{2}}. \tag{5.40}$$

Let

$$\bar{u}_{ij} = \left(\mathcal{B} - \frac{a\tau}{2}\delta_y^2\right)u_{ij}^{k+1}.$$

Then we have

$$\begin{cases} \left(\mathcal{A} - \frac{a\tau}{2}\delta_x^2\right)\bar{u}_{ij} = \left(\mathcal{A} + \frac{a\tau}{2}\delta_x^2\right)\left(\mathcal{B} + \frac{a\tau}{2}\delta_y^2\right)u_{ij}^{k} + \tau\mathcal{A}\mathcal{B}f_{ij}^{k+\frac{1}{2}}, & \text{(5.41a)} \\ \left(\mathcal{B} - \frac{a\tau}{2}\delta_y^2\right)u_{ij}^{k+1} = \bar{u}_{ij}. & \text{(5.41b)} \end{cases}$$

When the value $\{u_{ij}^{k} \mid (i, j) \in \bar{\omega}\}$ is known, we can determine the intermediate variable $\{\bar{u}_{ij} \mid (i, j) \in \omega\}$ by (5.41a). In detail, for an arbitrary fixed j ($1 \leqslant j \leqslant m_2 - 1$), taking the boundary value conditions

$$\bar{u}_{0j} = \left(\mathcal{B} - \frac{a\tau}{2}\delta_y^2\right)u_{0j}^{k+1}, \quad \bar{u}_{m_1,j} = \left(\mathcal{B} - \frac{a\tau}{2}\delta_y^2\right)u_{m_1,j}^{k+1}, \tag{5.42}$$

and solving

$$\left(\mathcal{A} - \frac{a\tau}{2}\delta_x^2\right)\bar{u}_{ij} = \left(\mathcal{A} + \frac{a\tau}{2}\delta_x^2\right)\left(\mathcal{B} + \frac{a\tau}{2}\delta_y^2\right)u_{ij}^{k} + \tau\mathcal{A}\mathcal{B}f_{ij}^{k+\frac{1}{2}},$$
$$1 \leqslant i \leqslant m_1 - 1, \tag{5.43}$$

we have $\{\bar{u}_{ij} \mid 1 \leqslant i \leqslant m_1 - 1\}$.

When $\{\bar{u}_{ij} \mid (i, j) \in \omega\}$ has been solved, the value $\{u_{ij}^{k+1} \mid (i, j) \in \omega\}$ can be solved by (5.41b): For a fixed i ($1 \leqslant i \leqslant m_1 - 1$), taking the boundary value conditions

$$u_{i0}^{k+1} = \alpha(x_i, y_0, t_{k+1}), \quad u_{i,m_2}^{k+1} = \alpha(x_i, y_{m_2}, t_{k+1}), \tag{5.44}$$

and solving

$$\left(\mathcal{B} - \frac{a\tau}{2}\delta_y^2\right)u_{ij}^{k+1} = \bar{u}_{ij}, \quad 1 \leqslant j \leqslant m_2 - 1, \tag{5.45}$$

it yields $\{u_{ij}^{k+1} \mid 1 \leqslant j \leqslant m_2 - 1\}$.

The difference scheme (5.42)–(5.43) is implicit in the x direction, while (5.44)–(5.45) is implicit in the y direction. Both result in tridiagonal systems of linear equations, which can be solved using the double sweep method. We refer to (5.41) as the **compact ADI scheme**. The following P-R compact ADI scheme for (5.40) can also be constructed as

$$
\begin{cases}
\left(\mathcal{A} - \dfrac{a\tau}{2}\delta_x^2\right)\bar{u}_{ij} = \left(\mathcal{B} + \dfrac{a\tau}{2}\delta_y^2\right)u_{ij}^k + \dfrac{\tau}{2}\mathcal{B}f_{ij}^{k+\frac{1}{2}}, \\
\left(\mathcal{B} - \dfrac{a\tau}{2}\delta_y^2\right)u_{ij}^{k+1} = \left(\mathcal{A} + \dfrac{a\tau}{2}\delta_x^2\right)\bar{u}_{ij} + \dfrac{\tau}{2}\mathcal{B}f_{ij}^{k+\frac{1}{2}}.
\end{cases}
$$

Example 5.2 Apply the compact ADI scheme (5.41) to compute the problem (5.23) in Example 5.1.

Taking $h_1 = h_2 = h = 1/m$, denote the maximum error by

$$
E_\infty(h, \tau) = \max_{0 \leqslant i, j \leqslant m, 0 \leqslant k \leqslant n} \left|u(x_i, y_j, t_k) - u_{ij}^k\right|.
$$

Table 5.3 shows some numerical results with the step sizes $h = 1/10$ and $\tau = 1/100$. Table 5.4 gives the maximum errors $E_\infty(h, \tau)$ of numerical solutions with different step sizes.

From Table 5.4, we observe that when the spatial step size is halved and the temporal step size is reduced to a quarter of its original value, the maximum error

Table 5.3 (Example 5.2) Part of numerical solutions, the exact solutions, and the absolute values of the errors ($h = 1/10$, $\tau = 1/100$)

(x, y, t)	NS	ES	|ES−NS|
$(0.4, 0.4, 0.25)$	1.161834	1.161834	6.020e−7
$(0.8, 0.4, 0.25)$	1.419067	1.419068	4.617e−7
$(0.4, 0.8, 0.25)$	1.419067	1.419068	4.617e−7
$(0.8, 0.8, 0.25)$	1.733253	1.733253	3.728e−7
$(0.4, 0.4, 0.50)$	0.904837	0.904837	4.738e−7
$(0.8, 0.4, 0.50)$	1.105171	1.105171	3.626e−7
$(0.4, 0.8, 0.50)$	1.105171	1.105171	3.626e−7
$(0.8, 0.8, 0.50)$	1.349859	1.349859	2.922e−7
$(0.4, 0.4, 0.75)$	0.7046877	0.7046881	3.690e−7
$(0.8, 0.4, 0.75)$	0.8607077	0.860708	2.824e−7
$(0.4, 0.8, 0.75)$	0.8607077	0.860708	2.824e−7
$(0.8, 0.8, 0.75)$	1.0512710	1.0512710	2.276e−7
$(0.4, 0.4, 1.00)$	0.5488113	0.5488116	2.874e−7
$(0.8, 0.4, 1.00)$	0.6703198	0.6703200	2.199e−7
$(0.4, 0.8, 1.00)$	0.6703198	0.6703200	2.199e−7
$(0.8, 0.8, 1.00)$	0.8187306	0.8187308	1.772e−7

Table 5.4 (Example 5.2) The maximum errors of numerical solutions with different step sizes

h	τ	$E_\infty(h, \tau)$	$E_\infty(2h, 4\tau)/E_\infty(h, \tau)$
1/10	1/100	7.087e−7	
1/20	1/400	4.469e−8	15.86
1/40	1/1600	2.793e−9	16.00

Fig. 5.6 (Example 5.2) The surface of the exact solution at $t = 1$

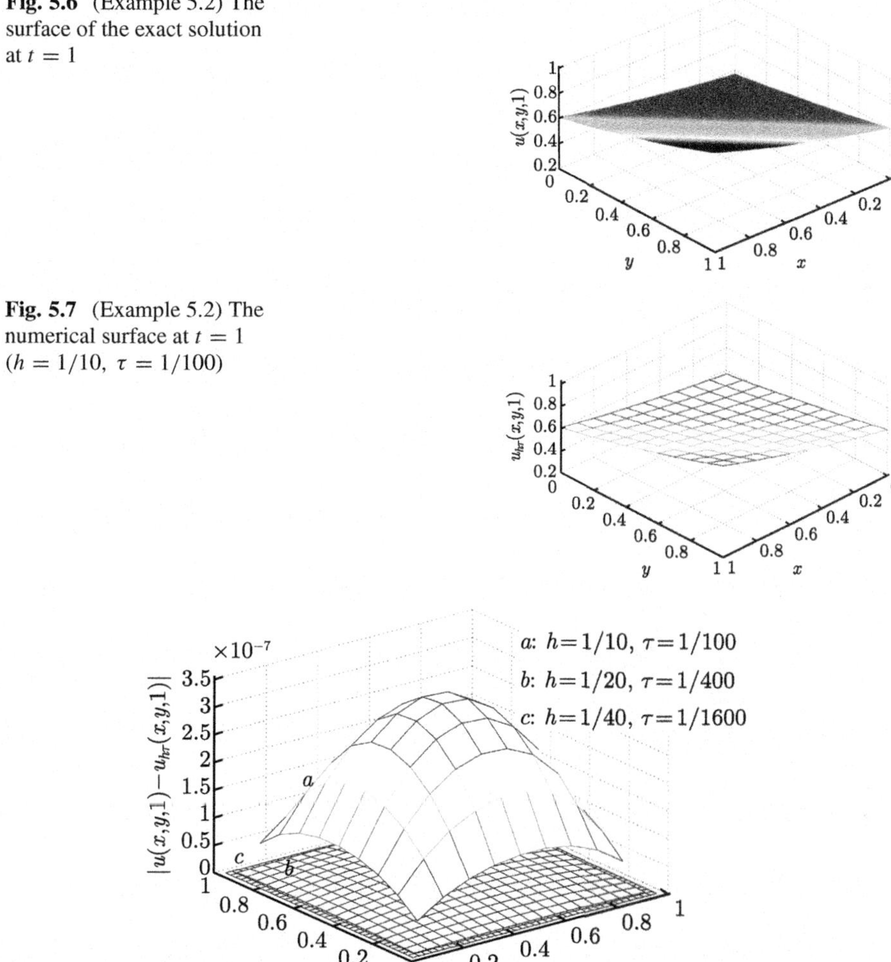

Fig. 5.7 (Example 5.2) The numerical surface at $t = 1$ ($h = 1/10$, $\tau = 1/100$)

Fig. 5.8 (Example 5.2) The numerical error surfaces at $t = 1$ with different step sizes

decreases to one-sixteenth of its original value. Figures 5.6 and 5.7 display the surface of the exact solution and the surface of the numerical solution at $t = 1$ with $h = 1/10$ and $\tau = 1/100$, respectively. Figure 5.8 illustrates the error surfaces at $t = 1$ for different step sizes.

5.2.4   A Priori Estimate of the Difference Solution

Lemma 5.3 *For $w \in \overset{\circ}{\mathcal{V}}_h$, we have*

$$-(\mathcal{ABw}, \Delta_h w) \geqslant \frac{1}{3}|w|_1^2.$$

Proof

$$-\left(\mathcal{ABw}, \Delta_h w\right)$$

$$= -\left(\left(\mathcal{I} + \frac{h_1^2}{12}\delta_x^2\right)\left(\mathcal{I} + \frac{h_2^2}{12}\delta_y^2\right)w, (\delta_x^2 + \delta_y^2)w\right)$$

$$= -\left(\left(\mathcal{I} + \frac{h_1^2}{12}\delta_x^2 + \frac{h_2^2}{12}\delta_y^2 + \frac{1}{144}h_1^2 h_2^2 \delta_x^2 \delta_y^2\right)w, \delta_x^2 w\right)$$

$$\quad - \left(\left(\mathcal{I} + \frac{h_1^2}{12}\delta_x^2 + \frac{h_2^2}{12}\delta_y^2 + \frac{1}{144}h_1^2 h_2^2 \delta_x^2 \delta_y^2\right)w, \delta_y^2 w\right)$$

$$= \|\delta_x w\|^2 - \frac{h_1^2}{12}\|\delta_x^2 w\|^2 - \frac{h_2^2}{12}\|\delta_y \delta_x w\|^2 + \frac{1}{144}h_1^2 h_2^2 \|\delta_x^2 \delta_y w\|^2$$

$$\quad + \|\delta_y w\|^2 - \frac{h_1^2}{12}\|\delta_x \delta_y w\|^2 - \frac{h_2^2}{12}\|\delta_y^2 w\|^2 + \frac{1}{144}h_1^2 h_2^2 \|\delta_x \delta_y^2 w\|^2$$

$$\geqslant \|\delta_x w\|^2 - \frac{h_1^2}{12}\cdot\frac{4}{h_1^2}\|\delta_x w\|^2 - \frac{h_2^2}{12}\cdot\frac{4}{h_2^2}\|\delta_x w\|^2$$

$$\quad + \|\delta_y w\|^2 - \frac{h_1^2}{12}\cdot\frac{4}{h_1^2}\|\delta_y w\|^2 - \frac{h_2^2}{12}\cdot\frac{4}{h_2^2}\|\delta_y w\|^2$$

$$= \frac{1}{3}|w|_1^2.$$

$\square$

Theorem 5.6 *Let $v = \{v_{ij}^k \mid (i, j) \in \bar{\omega}, 0 \leqslant k \leqslant n\}$ be the solution of the compact ADI scheme*

$$\begin{cases} \mathcal{AB}\delta_t v_{ij}^{k+\frac{1}{2}} - a\left(\mathcal{B}\delta_x^2 v_{ij}^{k+\frac{1}{2}} + \mathcal{A}\delta_y^2 v_{ij}^{k+\frac{1}{2}}\right) + \frac{1}{4}a^2\tau^2\delta_x^2\delta_y^2\delta_t v_{ij}^{k+\frac{1}{2}} = g_{ij}^k, \\ \qquad\qquad\qquad\qquad (i, j) \in \omega, \quad 0 \leqslant k \leqslant n - 1, & \text{(5.46a)} \\ v_{ij}^0 = \varphi(x_i, y_j), \quad (i, j) \in \omega, & \text{(5.46b)} \\ v_{ij}^k = 0, \quad (i, j) \in \gamma, \quad 0 \leqslant k \leqslant n. & \text{(5.46c)} \end{cases}$$

Then we have

$$\|\Delta_h v^{k+1}\|^2 \leqslant e^{k\tau}\left[6\|\Delta_h v^0\|^2 + 3a^{-2}\left(\|g^0\|^2 + 3\max_{0\leqslant l\leqslant k}\|g^l\|^2\right.\right.$$

$$\left.\left. +3\tau\sum_{l=1}^{k}\|\delta_t g^{l-\frac{1}{2}}\|^2\right)\right],$$

$$0\leqslant k\leqslant n-1,$$

where $\|g^l\|$ and $\|\delta_t g^{l-\frac{1}{2}}\|$ have the same definitions as those in Theorem 5.2.

Proof Taking an inner product of (5.46a) with $-\Delta_h\delta_t v^{k+\frac{1}{2}}$, we have

$$-\left(\mathcal{A}\mathcal{B}\delta_t v^{k+\frac{1}{2}}, \Delta_h\delta_t v^{k+\frac{1}{2}}\right) + a\left(\mathcal{B}\delta_x^2 v^{k+\frac{1}{2}} + \mathcal{A}\delta_y^2 v^{k+\frac{1}{2}}, \Delta_h\delta_t v^{k+\frac{1}{2}}\right)$$

$$-\frac{1}{4}a^2\tau^2\left(\delta_x^2\delta_y^2\delta_t v^{k+\frac{1}{2}}, \Delta_h\delta_t v^{k+\frac{1}{2}}\right) = -\left(g^k, \Delta_h\delta_t v^{k+\frac{1}{2}}\right), \quad 0\leqslant k\leqslant n-1.$$

$$(5.47)$$

Next, we estimate each term in (5.47).

With the help of Lemma 5.3, for the first term on the left-hand side, we have

$$-\left(\mathcal{A}\mathcal{B}\delta_t v^{k+\frac{1}{2}}, \Delta_h\delta_t v^{k+\frac{1}{2}}\right) \geqslant \frac{1}{3}\left|\delta_t v^{k+\frac{1}{2}}\right|_1^2. \tag{5.48}$$

Denote

$$E^k = \left(\mathcal{B}\delta_x^2 v^k + \mathcal{A}\delta_y^2 v^k, \Delta_h v^k\right).$$

Noticing that

$$\left(\mathcal{B}\delta_x^2 v^k + \mathcal{A}\delta_y^2 v^k, \Delta_h v^{k+1}\right) = \left(\mathcal{B}\delta_x^2 v^{k+1} + \mathcal{A}\delta_y^2 v^{k+1}, \Delta_h v^k\right),$$

the second term on the left-hand side becomes

$$\left(\mathcal{B}\delta_x^2 v^{k+\frac{1}{2}} + \mathcal{A}\delta_y^2 v^{k+\frac{1}{2}}, \Delta_h\delta_t v^{k+\frac{1}{2}}\right) = \frac{1}{2\tau}\left(E^{k+1} - E^k\right). \tag{5.49}$$

The third term has the estimate of the form

$$-\left(\delta_x^2\delta_y^2\delta_t v^{k+\frac{1}{2}}, \Delta_h\delta_t v^{k+\frac{1}{2}}\right) = -\left(\delta_x^2\delta_y^2\delta_t v^{k+\frac{1}{2}}, (\delta_x^2 + \delta_y^2)\delta_t v^{k+\frac{1}{2}}\right)$$

$$= \|\delta_x^2\delta_y\delta_t v^{k+\frac{1}{2}}\|^2 + \|\delta_x\delta_y^2\delta_t v^{k+\frac{1}{2}}\|^2 = \left|\delta_x\delta_y\delta_t v^{k+\frac{1}{2}}\right|_1^2. \tag{5.50}$$

Substituting (5.48)–(5.50) into (5.47), we have

$$\frac{a}{2\tau}\left(E^{k+1} - E^k\right) \leqslant -\left(g^k, \Delta_h \delta_t v^{k+\frac{1}{2}}\right), \quad 0 \leqslant k \leqslant n - 1.$$

Replacing k with l and summing over l from 0 to k, we have

$$\frac{a}{2\tau}\left(E^{k+1} - E^0\right)$$

$$\leqslant -\sum_{l=0}^{k}\left(g^l, \Delta_h \delta_t v^{l+\frac{1}{2}}\right)$$

$$= \frac{1}{\tau}\left[\left(g^0, \Delta_h v^0\right) - \left(g^k, \Delta_h v^{k+1}\right)\right] + \sum_{l=1}^{k}\left(\delta_t g^{l-\frac{1}{2}}, \Delta_h v^l\right), \quad 0 \leqslant k \leqslant n - 1,$$

namely,

$$E^{k+1} \leqslant E^0 + 2a^{-1}\left(g^0, \Delta_h v^0\right) - 2a^{-1}\left(g^k, \Delta_h v^{k+1}\right)$$

$$+ 2a^{-1}\tau \sum_{l=1}^{k}\left(\delta_t g^{l-\frac{1}{2}}, \Delta_h v^l\right), \quad 0 \leqslant k \leqslant n - 1. \tag{5.51}$$

It follows from Lemma 2.6 that

$$\frac{2}{3}\|\Delta_h v^k\|^2 \leqslant E^k \leqslant \|\Delta_h v^k\|^2.$$

By using (5.51), it yields

$$\frac{2}{3}\|\Delta_h v^{k+1}\|^2$$

$$\leqslant \|\Delta_h v^0\|^2 + 2a^{-1}\left(g^0, \Delta_h v^0\right) - 2a^{-1}\left(g^k, \Delta_h v^{k+1}\right) + 2a^{-1}\tau \sum_{l=1}^{k}\left(\delta_t g^{l-\frac{1}{2}}, \Delta_h v^l\right)$$

$$\leqslant \|\Delta_h v^0\|^2 + \left(a^{-2}\|g^0\|^2 + \|\Delta_h v^0\|^2\right) + \left(3a^{-2}\|g^k\|^2 + \frac{1}{3}\|\Delta_h v^{k+1}\|^2\right)$$

$$+ \tau \sum_{l=1}^{k}\left(3a^{-2}\|\delta_t g^{l-\frac{1}{2}}\|^2 + \frac{1}{3}\|\Delta_h v^l\|^2\right), \quad 0 \leqslant k \leqslant n - 1.$$

Multiplying both sides of the above inequality by 3 and rearranging the corresponding result, we have

$$\|\Delta_h v^{k+1}\|^2 \leqslant 6\|\Delta_h v^0\|^2 + 3a^{-2}\|g^0\|^2 + 9a^{-2}\|g^k\|^2$$

$$+ 9a^{-2}\tau \sum_{l=1}^{k} \|\delta_t g^{l-\frac{1}{2}}\|^2 + \tau \sum_{l=1}^{k} \|\Delta_h v^l\|^2, \quad 0 \leqslant k \leqslant n-1.$$

It follows from Lemma 5.1 (Gronwall inequality) that

$$\|\Delta_h v^{k+1}\|^2 \leqslant e^{k\tau}\left[6\|\Delta_h v^0\|^2 + 3a^{-2}\left(\|g^0\|^2 + 3\max_{0\leqslant l\leqslant k}\|g^l\|^2\right.\right.$$

$$\left.\left.+3\tau \sum_{l=1}^{k}\|\delta_t g^{l-\frac{1}{2}}\|^2\right)\right], \quad 0 \leqslant k \leqslant n-1.$$

$\square$

5.2.5   *Convergence and Stability of the Difference Solution*

Convergence

Now we will discuss the convergence.

Theorem 5.7 *Let $\{u(x, y, t) \mid (x, y) \in \bar{\Omega}, 0 \leqslant t \leqslant T\}$ be the solution of the problem (5.1) and $\{u_{ij}^k \mid (i, j) \in \bar{\omega}, 0 \leqslant k \leqslant n\}$ be the solution of the difference scheme (5.34). Denote*

$$e_{ij}^k = u(x_i, y_j, t_k) - u_{ij}^k, \quad (i, j) \in \bar{\omega}, \quad 0 \leqslant k \leqslant n.$$

Then we have

$$\|\Delta_h e^{k+1}\| \leqslant \sqrt{3(4+3T)e^T L_1 L_2}\, a^{-1} c_3\left(\tau^2 + h_1^4 + h_2^4\right), \quad 0 \leqslant k \leqslant n-1.$$

Proof Subtracting (5.34) from (5.31) and (5.33), respectively, we have the system of error equations

$$\begin{cases} \mathcal{A}\mathcal{B}\delta_t e_{ij}^{k+\frac{1}{2}} - a\left(\mathcal{B}\delta_x^2 e_{ij}^{k+\frac{1}{2}} + \mathcal{A}\delta_y^2 e_{ij}^{k+\frac{1}{2}}\right) + \frac{1}{4}a^2\tau^2\delta_x^2\delta_y^2\delta_t e_{ij}^{k+\frac{1}{2}} = (R_3)_{ij}^k, \\ \qquad\qquad\qquad (i, j) \in \omega, \quad 0 \leqslant k \leqslant n-1, \\ e_{ij}^0 = 0, \quad (i, j) \in \omega, \\ e_{ij}^k = 0, \quad (i, j) \in \gamma, \quad 0 \leqslant k \leqslant n. \end{cases}$$

Utilizing Theorem 5.6 and noticing (5.32), we have

$$\|\Delta_h e^{k+1}\|^2$$

$$\leqslant e^{k\tau}\left[6\|\Delta_h e^0\|^2 + 3a^{-2}\left(\|(R_3)^0\|^2 + 3\max_{0\leqslant l\leqslant k}\|(R_3)^l\|^2 + 3\tau\sum_{l=1}^{k}\|\delta_t(R_3)^{l-\frac{1}{2}}\|^2\right)\right]$$

$$\leqslant e^{k\tau}\cdot 3a^{-2}(4+3k\tau)L_1L_2c_3^2(\tau^2+h_1^4+h_2^4)^2$$

$$\leqslant e^T\cdot 3a^{-2}(4+3T)L_1L_2c_3^2(\tau^2+h_1^4+h_2^4)^2,\quad 0\leqslant k\leqslant n-1.$$

Taking the square root on both sides of the above inequality produces

$$\|\Delta_h e^{k+1}\| \leqslant \sqrt{3(4+3T)e^TL_1L_2}\,a^{-1}c_3(\tau^2+h_1^4+h_2^4),\quad 0\leqslant k\leqslant n-1.$$

$\square$

The convergence in the maximum norm is easily obtained with the help of Lemma 2.8.

Stability

Theorem 5.8 *The solution of the difference scheme (5.34) is stable with respect to the initial value and the nonhomogeneous term in the following sense: Let $\{u_{ij}^k \mid (i, j) \in \bar{\omega}, 0 \leqslant k \leqslant n\}$ be the solution of the difference scheme*

$$\begin{cases}
\mathcal{A}\mathcal{B}\delta_t u_{ij}^{k+\frac{1}{2}} = a\left(\mathcal{B}\delta_x^2 u_{ij}^{k+\frac{1}{2}} + \mathcal{A}\delta_y^2 u_{ij}^{k+\frac{1}{2}}\right) - \frac{a^2}{4}\tau^2\delta_x^2\delta_y^2\delta_t u_{ij}^{k+\frac{1}{2}} + g_{ij}^k, \\
\qquad\qquad\qquad (i, j) \in \omega,\quad 0\leqslant k\leqslant n-1, \\
u_{ij}^0 = \varphi(x_i, y_j),\quad (i, j) \in \omega, \\
u_{ij}^k = 0,\quad (i, j) \in \gamma,\quad 0\leqslant k\leqslant n.
\end{cases}$$

Then we have

$$\|\Delta_h u^{k+1}\|^2 \leqslant e^{k\tau}\left[6\|\Delta_h u^0\|^2 + 3a^{-2}\left(\|g^0\|^2 + 3\max_{0\leqslant l\leqslant k}\|g^l\|^2\right.\right.$$

$$\left.\left.+3\tau\sum_{l=1}^{k}\|\delta_t g^{l-\frac{1}{2}}\|^2\right)\right],$$

$$0\leqslant k\leqslant n-1.$$

Proof A direct application of Theorem 5.6 arrives at the result. $\square$

5.3   The ADI Scheme for Two-Dimensional Hyperbolic Equations

As a model, we consider the initial-boundary value problem of the two-dimensional wave equation

$$
\begin{cases}
\dfrac{\partial^2 u}{\partial t^2} - c^2 \left(\dfrac{\partial^2 u}{\partial x^2} + \dfrac{\partial^2 u}{\partial y^2} \right) = f(x, y, t), & (x, y, t) \in \Omega \times (0, T], \quad (5.52a) \\[4mm]
u(x, y, 0) = \varphi(x, y), \quad \dfrac{\partial u(x, y, 0)}{\partial t} = \psi(x, y), & (x, y) \in \bar{\Omega}, \quad (5.52b) \\[4mm]
u(x, y, t) = \alpha(x, y, t), \quad (x, y) \in \Gamma, \quad 0 < t \leqslant T, & (5.52c)
\end{cases}
$$

where c is a positive constant (generally denotes the wave speed), $\Omega = (0, L_1) \times (0, L_2)$, Γ is the boundary of Ω, and $\alpha(x, y, 0) = \varphi(x, y)$ and $\partial \alpha(x, y, 0)/\partial t = \psi(x, y)$ when $(x, y) \in \Gamma$.

5.3.1   Derivation of the Difference Scheme

Lemma 5.4 *Let $g \in C^4[0, \tau]$. Then we have*

$$
g''(\tau) = \frac{2}{\tau} \left[\frac{g(\tau) - g(0)}{\tau} - g'(0) + \frac{\tau^2}{3} g'''(0) \right] + \frac{5}{12} \tau^2 g^{(4)}(\xi \tau), \quad \xi \in (0, 1).
$$

Proof It follows from the Taylor expansion that

$$
g(\tau) = g(0) + \tau g'(0) + \frac{\tau^2}{2} g''(0) + \frac{\tau^3}{6} g'''(0) + \frac{\tau^4}{6} \int_0^1 g^{(4)}(s\tau)(1-s)^3 ds \quad (5.53)
$$

and

$$
g''(\tau) = g''(0) + \tau g'''(0) + \tau^2 \int_0^1 g^{(4)}(s\tau)(1 - s) ds. \quad (5.54)
$$

By means of (5.53), we have

$$
g''(0) = \frac{2}{\tau} \left[\frac{g(\tau) - g(0)}{\tau} - g'(0) \right] - \frac{\tau}{3} g'''(0) - \frac{\tau^2}{3} \int_0^1 g^{(4)}(s\tau)(1 - s)^3 ds.
$$

$$
(5.55)
$$

Substituting (5.55) into (5.54), we have

$$
\begin{aligned}
g''(\tau) &= \frac{2}{\tau}\left[\frac{g(\tau)-g(0)}{\tau} - g'(0)\right] + \frac{2\tau}{3}g'''(0) \\
&\quad + \tau^2 \int_0^1 g^{(4)}(s\tau)(1-s)\left[1 - \frac{1}{3}(1-s)^2\right]ds \\
&= \frac{2}{\tau}\left[\frac{g(\tau)-g(0)}{\tau} - g'(0)\right] + \frac{2\tau}{3}g'''(0) \\
&\quad + \tau^2 g^{(4)}(\xi\tau) \int_0^1 (1-s)\left[1 - \frac{1}{3}(1-s)^2\right]ds \\
&= \frac{2}{\tau}\left[\frac{g(\tau)-g(0)}{\tau} - g'(0) + \frac{\tau^2}{3}g'''(0)\right] + \frac{5}{12}\tau^2 g^{(4)}(\xi\tau), \quad \xi \in (0,1).
\end{aligned}
$$

$\square$

(I) Considering Eq. (5.52a) at the grid point (x_i, y_j, t_1), we have

$$
\frac{\partial^2 u}{\partial t^2}(x_i, y_j, t_1) - c^2\left(\frac{\partial^2 u}{\partial x^2}(x_i, y_j, t_1) + \frac{\partial^2 u}{\partial y^2}(x_i, y_j, t_1)\right) = f_{ij}^1, \quad (i, j) \in \omega.
\tag{5.56}
$$

Combining (5.52a) with (5.52b), it yields

$$
u_{ttt}(x, y, 0) = c^2\big(\psi_{xx}(x, y) + \psi_{yy}(x, y)\big) + f_t(x, y, 0) \equiv \rho(x, y).
$$

In light of Lemma 5.4, we have

$$
\begin{aligned}
\frac{\partial^2 u}{\partial t^2}(x_i, y_j, t_1) &= \frac{2}{\tau}\left(\delta_t U_{ij}^{\frac{1}{2}} - u_t(x_i, y_j, t_0) + \frac{\tau^2}{3}u_{ttt}(x_i, y_j, 0)\right) \\
&\quad + \frac{5}{12}\tau^2 u_{tttt}(x_i, y_j, \xi_{ij}\tau) \\
&= \frac{2}{\tau}\left(\delta_t U_{ij}^{\frac{1}{2}} - \psi_{ij} + \frac{\tau^2}{3}\rho_{ij}\right) + \frac{5}{12}\tau^2 u_{tttt}(x_i, y_j, \xi_{ij}\tau),
\end{aligned}
$$

where $\psi_{ij} = \psi(x_i, y_j)$, $\rho_{ij} = \rho(x_i, y_j)$, and $\xi_{ij} \in (0, 1)$. It follows from Lemma 1.2 that

$$
\frac{\partial^2 u(x_i, y_j, t_1)}{\partial x^2} = \delta_x^2 U_{ij}^1 - \frac{h_1^2}{12}u_{xxxx}(x_i + \eta_{ij}h_1, y_j, t_1), \quad \eta_{ij} \in (-1, 1);
$$

$$
\frac{\partial^2 u(x_i, y_j, t_1)}{\partial y^2} = \delta_y^2 U_{ij}^1 - \frac{h_2^2}{12}u_{yyyy}(x_i, y_j + \varsigma_{ij}h_2, t_1), \quad \varsigma_{ij} \in (-1, 1).
$$

Substituting the above three equalities into (5.56) and adding a small term $\frac{c^4\tau^2}{2}\delta_x^2\delta_y^2 U_{ij}^1$ on both sides of the corresponding result produce

$$\frac{2}{\tau}\left(\delta_t U_{ij}^{\frac{1}{2}} - \psi_{ij} + \frac{\tau^2}{3}\rho_{ij}\right) - c^2\Delta_h U_{ij}^1 + \frac{c^4\tau^2}{2}\delta_x^2\delta_y^2 U_{ij}^1 = f_{ij}^1 + (R_4)_{ij}^0, \quad (i,j)\in\omega,$$

(5.57)

where

$$(R_4)_{ij}^0 = -\frac{5}{12}\tau^2 u_{tttt}(x_i, y_j, \xi_{ij}\tau) - \frac{c^2 h_1^2}{12} u_{xxxx}(x_i + \eta_{ij}h_1, y_j, t_1)$$

$$- \frac{c^2 h_2^2}{12} u_{yyyy}(x_i, y_j + \varsigma_{ij}h_2, t_1) + \frac{c^4\tau^2}{2}\delta_x^2\delta_y^2 U_{ij}^1, \quad (i,j)\in\omega.$$

There is a constant c_4 such that

$$\left|(R_4)_{ij}^0\right| \leqslant c_4(\tau^2 + h_1^2 + h_2^2), \quad (i,j)\in\omega.$$

(5.58)

(II) Considering Eq. (5.52a) at the grid point (x_i, y_j, t_k), we have

$$\frac{\partial^2 u}{\partial t^2}(x_i, y_j, t_k) - c^2\left(\frac{\partial^2 u}{\partial x^2}(x_i, y_j, t_k) + \frac{\partial^2 u}{\partial y^2}(x_i, y_j, t_k)\right) = f_{ij}^k,$$

$$(i,j)\in\omega, \quad 1\leqslant k \leqslant n-1.$$

(5.59)

Using Lemma 1.2, we have

$$\frac{\partial^2 u}{\partial t^2}(x_i, y_j, t_k)$$

$$= \delta_t^2 U_{ij}^k - \frac{\tau^2}{6}\int_0^1\left(u_{tttt}(x_i, y_j, t_k - s\tau) + u_{tttt}(x_i, y_j, t_k + s\tau)\right)(1-s)^3 ds;$$

$$\frac{\partial^2 u(x_i, y_j, t_k)}{\partial x^2}$$

$$= \frac{1}{2}\left[\frac{\partial^2 u(x_i, y_j, t_{k-1})}{\partial x^2} + \frac{\partial^2 u(x_i, y_j, t_{k+1})}{\partial x^2}\right]$$

$$- \frac{\tau^2}{2}\int_0^1\left(u_{xxtt}(x_i, y_j, t_k - s\tau) + u_{xxtt}(x_i, y_j, t_k + s\tau)\right)(1-s)ds$$

$$= \frac{1}{2}\left[\delta_x^2 U_{ij}^{k-1} - \frac{h_1^2}{6}\int_0^1\left(u_{xxxx}(x_i - sh_1, y_j, t_{k-1})\right.\right.$$

$$\left.\left. + u_{xxxx}(x_i + sh_1, y_j, t_{k-1})\right)(1-s)^3 ds\right.$$

$$+ \delta_x^2 U_{ij}^{k+1} - \frac{h_1^2}{6} \int_0^1 \Big(u_{xxxx}(x_i - sh_1, y_j, t_{k+1})$$

$$+ u_{xxxx}(x_i + sh_1, y_j, t_{k+1}) \Big)(1 - s)^3 ds \Bigg]$$

$$- \frac{\tau^2}{2} \int_0^1 \Big(u_{xxtt}(x_i, y_j, t_k - s\tau) + u_{xxtt}(x_i, y_j, t_k + s\tau) \Big)(1 - s) ds;$$

$$\frac{\partial^2 u(x_i, y_j, t_k)}{\partial y^2}$$

$$= \frac{1}{2} \left[\frac{\partial^2 u(x_i, y_j, t_{k-1})}{\partial y^2} + \frac{\partial^2 u(x_i, y_j, t_{k+1})}{\partial y^2} \right]$$

$$- \frac{\tau^2}{2} \int_0^1 \Big(u_{yytt}(x_i, y_j, t_k - s\tau) + u_{yytt}(x_i, y_j, t_k + s\tau) \Big)(1 - s) ds$$

$$= \frac{1}{2} \Bigg[\delta_y^2 U_{ij}^{k-1} - \frac{h_2^2}{6} \int_0^1 \Big(u_{yyyy}(x_i, y_j - sh_2, t_{k-1})$$

$$+ u_{yyyy}(x_i, y_j + sh_2, t_{k-1}) \Big)(1 - s)^3 ds$$

$$+ \delta_y^2 U_{ij}^{k+1} - \frac{h_2^2}{6} \int_0^1 \Big(u_{yyyy}(x_i, y_j - sh_2, t_{k+1})$$

$$+ u_{yyyy}(x_i, y_j + sh_2, t_{k+1}) \Big)(1 - s)^3 ds \Bigg]$$

$$- \frac{\tau^2}{2} \int_0^1 \Big(u_{yytt}(x_i, y_j, t_k - s\tau) + u_{yytt}(x_i, y_j, t_k + s\tau) \Big)(1 - s) ds.$$

Substituting the above three equalities into (5.59) and adding a small term $\frac{c^4 \tau^2}{2} \delta_x^2 \delta_y^2 U_{ij}^{\bar{k}}$ on both sides of the resultant equality, we have

$$\delta_t^2 U_{ij}^k - c^2 \Delta_h U_{ij}^{\bar{k}} + \frac{c^4 \tau^2}{2} \delta_x^2 \delta_y^2 U_{ij}^{\bar{k}} = f_{ij}^k + (R_4)_{ij}^k, \quad (i, j) \in \omega, \quad 1 \leqslant k \leqslant n - 1,$$
$$(5.60)$$

where

$$(R_4)_{ij}^k$$

$$= \tau^2 \Bigg[\frac{1}{6} \int_0^1 \Big(u_{tttt}(x_i, y_j, t_k - s\tau) + u_{tttt}(x_i, y_j, t_k + s\tau) \Big)(1 - s)^3 ds$$

$$- \frac{c^2}{2} \int_0^1 \Big(u_{xxtt}(x_i, y_j, t_k - s\tau) + u_{xxtt}(x_i, y_j, t_k + s\tau) \Big)(1 - s) ds$$

$$-\frac{c^2}{2}\int_0^1 \Big(u_{yytt}(x_i, y_j, t_k - s\tau) + u_{yytt}(x_i, y_j, t_k + s\tau)\Big)(1 - s)\,ds$$

$$+\frac{c^4}{2}\delta_x^2\delta_y^2 U_{ij}^{\bar{k}}\Big]$$

$$-c^2\frac{h_1^2}{12}\int_0^1 \Big[u_{xxxx}(x_i - sh_1, y_j, t_{k-1}) + u_{xxxx}(x_i + sh_1, y_j, t_{k-1})$$

$$+ u_{xxxx}(x_i - sh_1, y_j, t_{k+1}) + u_{xxxx}(x_i + sh_1, y_j, t_{k+1})\Big](1 - s)^3 ds$$

$$-c^2\frac{h_2^2}{12}\int_0^1 \Big[u_{yyyy}(x_i, y_j - sh_2, t_{k-1}) + u_{yyyy}(x_i, y_j + sh_2, t_{k-1})$$

$$+ u_{yyyy}(x_i, y_j - sh_2, t_{k+1}) + u_{yyyy}(x_i, y_j + sh_2, t_{k+1})\Big](1 - s)^3 ds,$$

$$(i, j) \in \omega, \quad 1 \leqslant k \leqslant n - 1.$$

There is a constant c_5 such that

$$\begin{cases} \big|(R_4)_{ij}^k\big| \leqslant c_5(\tau^2 + h_1^2 + h_2^2), & (i, j) \in \omega, \quad 1 \leqslant k \leqslant n - 1, \\ \big|\Delta_t(R_4)_{ij}^k\big| \leqslant c_5(\tau^2 + h_1^2 + h_2^2), & (i, j) \in \omega, \quad 2 \leqslant k \leqslant n - 2, \end{cases} \tag{5.61}$$

where

$$\Delta_t(R_4)_{ij}^k = \frac{1}{2\tau}\Big[(R_4)_{ij}^{k+1} - (R_4)_{ij}^{k-1}\Big].$$

Noticing the initial-boundary value conditions (5.52b)–(5.52c), we have

$$\begin{cases} U_{ij}^0 = \varphi(x_i, y_j), & (i, j) \in \omega, \\ U_{ij}^k = \alpha(x_i, y_j, t_k), & (i, j) \in \gamma, \quad 0 \leqslant k \leqslant n. \end{cases} \tag{5.62}$$

Omitting the small terms $(R_4)_{ij}^0$ in (5.57) and $(R_4)_{ij}^k$ in (5.60), a difference scheme for the problem (5.52) reads

$$\begin{cases} \dfrac{2}{\tau}\Big(\delta_t u_{ij}^{\frac{1}{2}} - \psi_{ij} + \dfrac{\tau^2}{3}\rho_{ij}\Big) - c^2\Delta_h u_{ij}^1 + \dfrac{c^4\tau^2}{2}\delta_x^2\delta_y^2 u_{ij}^1 = f_{ij}^1, & (i, j) \in \omega, \quad (5.63a)\\[3mm] \delta_t^2 u_{ij}^k - c^2\Delta_h u_{ij}^{\bar{k}} + \dfrac{c^4\tau^2}{2}\delta_x^2\delta_y^2 u_{ij}^{\bar{k}} = f_{ij}^k, & (i, j) \in \omega, \quad 1 \leqslant k \leqslant n - 1, \quad (5.63b)\\[3mm] u_{ij}^0 = \varphi(x_i, y_j), & (i, j) \in \omega, \quad (5.63c)\\[3mm] u_{ij}^k = \alpha(x_i, y_j, t_k), & (i, j) \in \gamma, \quad 0 \leqslant k \leqslant n. \quad (5.63d) \end{cases}$$

Fig. 5.9 The stencil of the difference scheme (5.63)

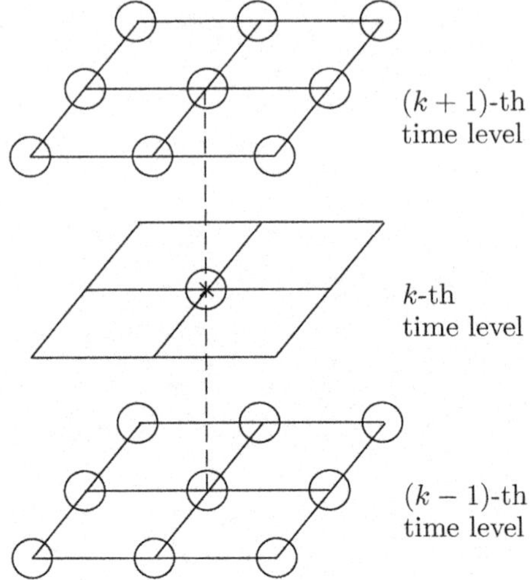

The stencil of the difference scheme (5.63) is shown in Fig. 5.9.

5.3.2 Existence of the Difference Solution

Theorem 5.9 *The difference scheme (5.63) is uniquely solvable.*

Proof Denote

$$u^k = \{u_{ij}^k \mid (i, j) \in \bar{\omega}\}.$$

It follows from (5.63c) and (5.63d) that u^0 has been determined.
The difference scheme in u^1 reads

$$\begin{cases} \dfrac{2}{\tau}\left(\delta_t u_{ij}^{\frac{1}{2}} - \psi_{ij} + \dfrac{1}{3}\tau^2 \rho_{ij}\right) - c^2 \Delta_h u_{ij}^1 + \dfrac{c^4 \tau^2}{2}\delta_x^2 \delta_y^2 u_{ij}^1 = f_{ij}^1, & (i, j) \in \omega, \\ u_{ij}^1 = \alpha(x_i, y_j, t_1), & (i, j) \in \gamma. \end{cases}$$

Consider the corresponding homogeneous one

$$\begin{cases} \dfrac{2}{\tau^2}u_{ij}^1 - c^2 \Delta_h u_{ij}^1 + \dfrac{c^4 \tau^2}{2}\delta_x^2 \delta_y^2 u_{ij}^1 = 0, & (i, j) \in \omega, & \text{(5.64a)} \\ u_{ij}^1 = 0, & (i, j) \in \gamma. & \text{(5.64b)} \end{cases}$$

Taking an inner product of (5.64a) with u^1, it yields

$$\frac{2}{\tau^2}\|u^1\|^2 - c^2(\Delta_h u^1, u^1) + \frac{c^4\tau^2}{2}(\delta_x^2\delta_y^2 u^1, u^1) = 0.$$

Applying the summation by parts and noticing (5.64b), we have

$$\frac{2}{\tau^2}\|u^1\|^2 + c^2|u^1|_1^2 + \frac{c^4\tau^2}{2}\|\delta_x\delta_y u^1\|^2 = 0.$$

It is easy to obtain

$$u_{ij}^1 = 0, \quad (i, j) \in \bar{\omega}.$$

Suppose u^{k-1} and u^k $(1 \leqslant k \leqslant n-1)$ have been determined. The difference scheme in u^{k+1} reads

$$\begin{cases} \delta_t^2 u_{ij}^k - c^2\Delta_h u_{ij}^{\bar{k}} + \dfrac{c^4\tau^2}{2}\delta_x^2\delta_y^2 u_{ij}^{\bar{k}} = f_{ij}^k, & (i, j) \in \omega, \\ u_{ij}^{k+1} = \alpha(x_i, y_j, t_{k+1}), & (i, j) \in \gamma. \end{cases}$$

Consider its corresponding homogeneous one

$$\begin{cases} \dfrac{1}{\tau^2}u_{ij}^{k+1} - \dfrac{c^2}{2}\Delta_h u_{ij}^{k+1} + \dfrac{c^4\tau^2}{4}\delta_x^2\delta_y^2 u_{ij}^{k+1} = 0, & (i, j) \in \omega, \quad (5.65a) \\ u_{ij}^{k+1} = 0, \quad (i, j) \in \gamma. & \hspace{3.5cm} (5.65b) \end{cases}$$

Taking an inner product on both sides of (5.65a) with u^{k+1} produces

$$\frac{1}{\tau^2}\|u^{k+1}\|^2 - \frac{c^2}{2}(\Delta_h u^{k+1}, u^{k+1}) + \frac{c^4\tau^2}{4}(\delta_x^2\delta_y^2 u^{k+1}, u^{k+1}) = 0.$$

Applying the summation by parts to the second term above and noticing (5.65b), we have

$$\frac{1}{\tau^2}\|u^{k+1}\|^2 + \frac{c^2}{2}|u^{k+1}|_1^2 + \frac{c^4\tau^2}{4}\|\delta_x\delta_y u^{k+1}\|^2 = 0.$$

Consequently,

$$u_{ij}^{k+1} = 0, \quad (i, j) \in \bar{\omega}.$$

By induction, the system of difference equations (5.63) is uniquely solvable. $\square$

5.3.3 *Implementation of the Difference Scheme and Numerical Examples*

It follows from (5.63c) and (5.63d) that $u^0 = \{u_{ij}^0 \mid 0 \leqslant i \leqslant m_1, 0 \leqslant j \leqslant m_2\}$ has been determined.

We first consider the computation of (5.63a).

Equation (5.63a) can be rewritten as

$$u_{ij}^1 - \frac{c^2\tau^2}{2}\Delta_h u_{ij}^1 + \frac{c^4\tau^4}{4}\delta_x^2\delta_y^2 u_{ij}^1 = u_{ij}^0 + \tau\psi_{ij} - \frac{\tau^3}{3}\rho_{ij} + \frac{\tau^2}{2}f_{ij}^1, \quad (i, j) \in \omega.$$

Decompose the above equality as follows:

$$\left(\mathcal{I} - \frac{c^2\tau^2}{2}\delta_x^2\right)\left(\mathcal{I} - \frac{c^2\tau^2}{2}\delta_y^2\right)u_{ij}^1 = u_{ij}^0 + \tau\psi_{ij} - \frac{\tau^3}{3}\rho_{ij} + \frac{\tau^2}{2}f_{ij}^1, \quad (i, j) \in \omega.$$

Let

$$\bar{u}_{ij} = \left(\mathcal{I} - \frac{c^2\tau^2}{2}\delta_y^2\right)u_{ij}^1.$$

Then we have

$$\begin{cases} \left(\mathcal{I} - \dfrac{c^2\tau^2}{2}\delta_x^2\right)\bar{u}_{ij} = u_{ij}^0 + \tau\psi_{ij} - \dfrac{\tau^3}{3}\rho_{ij} + \dfrac{\tau^2}{2}f_{ij}^1, & (5.66a) \\[4mm] \left(\mathcal{I} - \dfrac{c^2\tau^2}{2}\delta_y^2\right)u_{ij}^1 = \bar{u}_{ij}. & (5.66b) \end{cases}$$

Using (5.66a), we can solve the intermediate variable $\{\bar{u}_{ij} \mid (i, j) \in \omega\}$: for an arbitrary fixed j ($1 \leqslant j \leqslant m_2 - 1$), taking the boundary value conditions

$$\bar{u}_{0j} = \left(\mathcal{I} - \frac{c^2\tau^2}{2}\delta_y^2\right)u_{0j}^1, \quad \bar{u}_{m_1,j} = \left(\mathcal{I} - \frac{c^2\tau^2}{2}\delta_y^2\right)u_{m_1,j}^1 \tag{5.67}$$

and solving

$$\left(\mathcal{I} - \frac{c^2\tau^2}{2}\delta_x^2\right)\bar{u}_{ij} = u_{ij}^0 + \tau\psi_{ij} - \frac{\tau^3}{3}\rho_{ij} + \frac{\tau^2}{2}f_{ij}^1, \quad 1 \leqslant i \leqslant m_1 - 1, \tag{5.68}$$

we obtain $\{\bar{u}_{ij} \mid 1 \leqslant i \leqslant m_1 - 1\}$.

When $\{\bar{u}_{ij} \mid (i, j) \in \omega\}$ has been solved, it follows from (5.66b) that the value of $\{u_{ij}^1 \mid (i, j) \in \omega\}$ can be determined: For a fixed i ($1 \leqslant i \leqslant m_1 - 1$), taking the boundary value conditions

$$u_{i0}^1 = \alpha(x_i, y_0, t_1), \quad u_{i,m_2}^1 = \alpha(x_i, y_{m_2}, t_1) \tag{5.69}$$

and solving

$$\left(\mathcal{I} - \frac{c^2\tau^2}{2}\delta_y^2\right) u_{ij}^1 = \bar{u}_{ij}, \quad 1 \leqslant j \leqslant m_2 - 1, \tag{5.70}$$

we obtain $\{u_{ij}^1 \mid 1 \leqslant j \leqslant m_2 - 1\}$.

The difference scheme (5.67)–(5.68) is implicit in the x direction, while (5.69)–(5.70) is implicit in the y direction. Both schemes result in tridiagonal systems of linear equations, which can be solved using the double sweep method. In this context, (5.66) is referred to as the **ADI scheme**.

Considering the computation of (5.63b) and noticing

$$\delta_t^2 u_{ij}^k = \frac{u_{ij}^{k+1} - 2u_{ij}^k + u_{ij}^{k-1}}{\tau^2} = \frac{2u_{ij}^{\bar{k}} - 2u_{ij}^k}{\tau^2} = \frac{2}{\tau^2}\left(u_{ij}^{\bar{k}} - u_{ij}^k\right),$$

Equation (5.63b) can be rewritten as

$$\frac{2}{\tau^2}(u_{ij}^{\bar{k}} - u_{ij}^k) - c^2 \Delta_h u_{ij}^{\bar{k}} + \frac{c^4\tau^2}{2}\delta_x^2\delta_y^2 u_{ij}^{\bar{k}} = f_{ij}^k,$$
$$(i, j) \in \omega, \quad 1 \leqslant k \leqslant n - 1,$$

or

$$u_{ij}^{\bar{k}} - \frac{c^2\tau^2}{2}\Delta_h u_{ij}^{\bar{k}} + \frac{c^4\tau^4}{4}\delta_x^2\delta_y^2 u_{ij}^{\bar{k}} = u_{ij}^k + \frac{\tau^2}{2} f_{ij}^k,$$
$$(i, j) \in \omega, \quad 1 \leqslant k \leqslant n - 1.$$

Decompose the above equality as follows

$$\left(\mathcal{I} - \frac{c^2\tau^2}{2}\delta_x^2\right)\left(\mathcal{I} - \frac{c^2\tau^2}{2}\delta_y^2\right) u_{ij}^{\bar{k}} = u_{ij}^k + \frac{\tau^2}{2} f_{ij}^k, \quad (i, j) \in \omega, \quad 1 \leqslant k \leqslant n - 1.$$

Let

$$\bar{u}_{ij} = \left(\mathcal{I} - \frac{c^2\tau^2}{2}\delta_y^2\right) u_{ij}^{\bar{k}}.$$

Then we have

$$
\begin{cases}
\left(\mathcal{I} - \dfrac{c^2\tau^2}{2}\delta_x^2\right)\bar{u}_{ij} = u_{ij}^k + \dfrac{\tau^2}{2}f_{ij}^k, & \text{(5.71a)}\\[3mm]
\left(\mathcal{I} - \dfrac{c^2\tau^2}{2}\delta_y^2\right)u_{ij}^{\bar{k}} = \bar{u}_{ij}. & \text{(5.71b)}
\end{cases}
$$

When the value of $\{u_{ij}^{k-1},\ u_{ij}^k \mid (i, j) \in \omega\}$ has been known, the intermediate variable $\{\bar{u}_{ij} \mid (i, j) \in \omega\}$ can be solved from (5.71a) as follows: For an arbitrary fixed j $(1 \leqslant j \leqslant m_2 - 1)$, taking the boundary value conditions

$$
\bar{u}_{0j} = \left(\mathcal{I} - \frac{c^2\tau^2}{2}\delta_y^2\right)u_{0j}^{\bar{k}}, \quad \bar{u}_{m_1,j} = \left(\mathcal{I} - \frac{c^2\tau^2}{2}\delta_y^2\right)u_{m_1,j}^{\bar{k}}, \tag{5.72}
$$

and solving

$$
\left(\mathcal{I} - \frac{c^2\tau^2}{2}\delta_x^2\right)\bar{u}_{ij} = u_{ij}^k + \frac{\tau^2}{2}f_{ij}^k, \quad 1 \leqslant i \leqslant m_1 - 1, \tag{5.73}
$$

the value of $\{\bar{u}_{ij} \mid 1 \leqslant i \leqslant m_1 - 1\}$ is obtained.

When $\{\bar{u}_{ij} \mid (i, j) \in \omega\}$ has been solved, the value $\{u_{ij}^{k+1} \mid (i, j) \in \omega\}$ at the $(k + 1)$-th time level can be solved from (5.71b) as follows: For a fixed i $(1 \leqslant i \leqslant m_1 - 1)$, taking the boundary value conditions

$$
\begin{cases}
u_{i0}^{\bar{k}} = \dfrac{1}{2}[\alpha(x_i, y_0, t_{k+1}) + \alpha(x_i, y_0, t_{k-1})],\\[3mm]
u_{i,m_2}^{\bar{k}} = \dfrac{1}{2}[\alpha(x_i, y_{m_2}, t_{k+1}) + \alpha(x_i, y_{m_2}, t_{k-1})],
\end{cases}
\tag{5.74}
$$

and solving

$$
\left(\mathcal{I} - \frac{c^2\tau^2}{2}\delta_y^2\right)u_{ij}^{\bar{k}} = \bar{u}_{ij}, \quad 1 \leqslant j \leqslant m_2 - 1, \tag{5.75}
$$

the value of $\{u_{ij}^{\bar{k}} \mid 1 \leqslant j \leqslant m_2 - 1\}$ is determined.

Finally, it follows from

$$
u_{ij}^{k+1} = 2u_{ij}^{\bar{k}} - u_{ij}^{k-1}, \quad (i, j) \in \omega
$$

that the value of $\{u_{ij}^{k+1} \mid (i, j) \in \omega\}$ is obtained.

The difference scheme (5.72)–(5.73) is implicit in the x direction, while (5.74)–(5.75) is implicit in the y direction. Both schemes result in tridiagonal systems of linear equations, which can be solved using the double sweep method. Therefore, (5.71) is referred to as the **ADI scheme**.

Example 5.3 Apply the ADI scheme (5.66) and (5.71) to solve the problem

$$
\begin{cases}
\dfrac{\partial^2 u}{\partial t^2} - \left(\dfrac{\partial^2 u}{\partial x^2} + \dfrac{\partial^2 u}{\partial y^2}\right) = \dfrac{1}{2}e^{\frac{1}{2}(x+y)-t}, & 0 < x, y < 1, \quad 0 < t \leqslant 1, \\[2mm]
u(x, y, 0) = e^{\frac{1}{2}(x+y)}, \quad \dfrac{\partial u(x, y, 0)}{\partial t} = -e^{\frac{1}{2}(x+y)}, & 0 \leqslant x, y \leqslant 1, \\[2mm]
u(0, y, t) = e^{\frac{1}{2}y-t}, \quad u(1, y, t) = e^{\frac{1}{2}(1+y)-t}, & 0 \leqslant y \leqslant 1, \quad 0 < t \leqslant 1, \\[2mm]
u(x, 0, t) = e^{\frac{1}{2}x-t}, \quad u(x, 1, t) = e^{\frac{1}{2}(1+x)-t}, & 0 < x < 1, \quad 0 < t \leqslant 1.
\end{cases}
\tag{5.76}
$$

The exact solution of the problem is $u(x, y, t) = e^{\frac{1}{2}(x+y)-t}$.

Take $h_1 = h_2 = h = 1/m$ and denote the maximum error by

$$
E_\infty(h, \tau) = \max_{0 \leqslant i, j \leqslant m, 0 \leqslant k \leqslant n} |u(x_i, y_j, t_k) - u_{ij}^k|.
$$

Table 5.5 presents some numerical results for $h = 1/100$ and $\tau = 1/100$. Table 5.6 provides the maximum errors $E_\infty(h, \tau)$ for different step sizes. From Table 5.6, it is observed that when both the spatial and temporal step sizes are halved, the maximum errors are reduced to one-quarter of their original values.

Figures 5.10 and 5.11 show the surfaces of the exact solution and the numerical solution with step sizes $h = 1/10$ and $\tau = 1/10$ at $t = 1$, respectively. Figure 5.12 shows the error surfaces of numerical solutions with different step sizes at $t = 1$.

Table 5.5 (Example 5.3) Some numerical solutions, exact solutions, and absolute values of the errors ($h = 1/100$, $\tau = 1/100$)

(x, y, t)	NS	ES	|ES−NS|
(0.25,0.25,0.25)	1.000001	1.000000	5.749e−7
(0.75,0.25,0.25)	1.284026	1.284025	7.371e−7
(0.25,0.75,0.25)	1.284026	1.284025	7.371e−7
(0.75,0.75,0.25)	1.648722	1.648721	9.451e−7
(0.25,0.25,0.50)	0.778802	0.778801	1.158e−6
(0.75,0.25,0.50)	1.000001	1.000000	1.331e−6
(0.25,0.75,0.50)	1.000001	1.000000	1.331e−6
(0.75,0.75,0.50)	1.284027	1.284025	1.521e−6
(0.25,0.25,0.75)	0.6065321	0.6065307	1.399e−6
(0.75,0.25,0.75)	0.7788022	0.7788008	1.395e−6
(0.25,0.75,0.75)	0.7788022	0.7788008	1.395e−6
(0.75,0.75,0.75)	1.000001	1.000000	1.374e−6
(0.25,0.25,1.00)	0.4723671	0.4723666	5.218e−7
(0.75,0.25,1.00)	0.6065311	0.6065307	4.641e−7
(0.25,0.75,1.00)	0.6065311	0.6065307	4.641e−7
(0.75,0.75,1.00)	0.7788012	0.7788008	4.472e−7

Table 5.6 (Example 5.3) The maximum errors of numerical solutions with different step sizes $E_\infty(h, \tau)$

h	τ	$E_\infty(h, \tau)$	$E_\infty(2h, 2\tau)/E_\infty(h, \tau)$
1/10	1/10	3.259e−4	
1/20	1/20	8.261e−5	3.9447
1/40	1/40	2.052e−5	4.0257
1/80	1/80	5.038e−6	4.0735
1/160	1/160	1.249e−6	4.0347

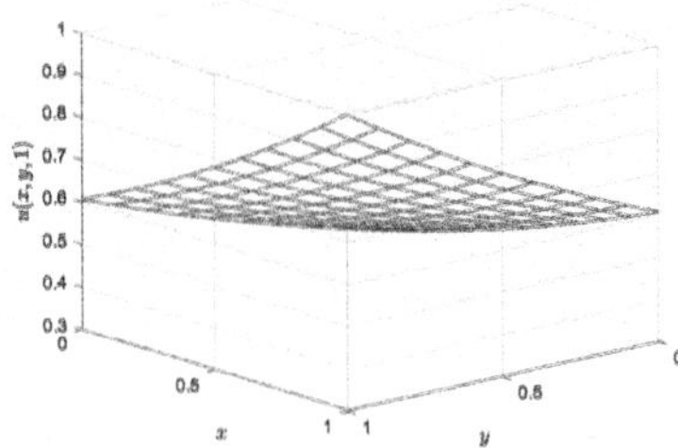

Fig. 5.10 (Example 5.3) The exact solution surface at $t = 1$

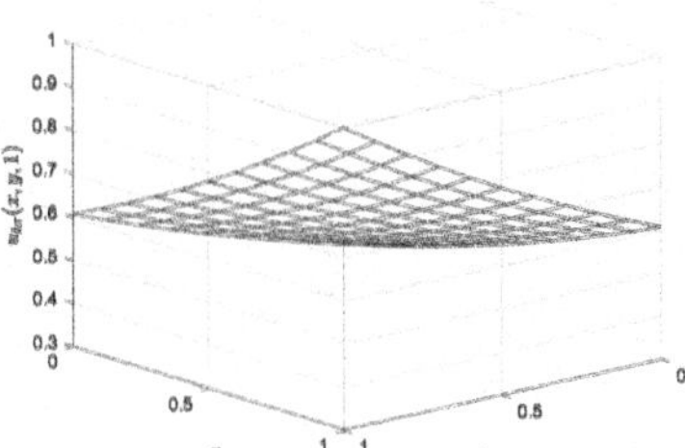

Fig. 5.11 (Example 5.3) The numerical solution surface at $t = 1$ ($h = \tau = 1/10$)

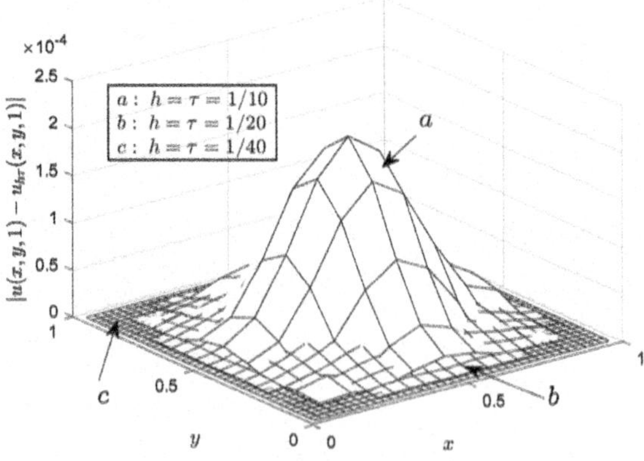

Fig. 5.12 (Example 5.3) The numerical error surfaces at $t = 1$

5.3.4   A Priori Estimate of the Difference Solution

Theorem 5.10 *Let* $\{v_{ij}^{k} \mid (i, j) \in \bar{\omega}, \ 0 \leqslant k \leqslant n\}$ *be the solution of the difference scheme*

$$
\begin{cases}
\dfrac{2}{\tau}\delta_t v_{ij}^{\frac{1}{2}} - c^2 \Delta_h v_{ij}^1 + \dfrac{c^4 \tau^2}{2}\delta_x^2 \delta_y^2 v_{ij}^1 = g_{ij}^0, \quad (i, j) \in \omega, & (5.77\text{a}) \\[2mm]
\delta_t^2 v_{ij}^k - c^2 \Delta_h v_{ij}^{\bar{k}} + \dfrac{c^4 \tau^2}{2}\delta_x^2 \delta_y^2 v_{ij}^{\bar{k}} = g_{ij}^k, \\[1mm]
\qquad\qquad\qquad (i, j) \in \omega, \quad 1 \leqslant k \leqslant n - 1, & (5.77\text{b}) \\[2mm]
v_{ij}^0 = \psi_{ij}, \quad (i, j) \in \omega, & (5.77\text{c}) \\[2mm]
v_{ij}^k = 0, \quad (i, j) \in \gamma, \quad 0 \leqslant k \leqslant n. & (5.77\text{d})
\end{cases}
$$

Then we have

$$
\frac{\|\Delta_h v^{k+1}\|^2 + \|\Delta_h v^k\|^2}{2}
$$

$$
\leqslant e^{2T}\left[6\|\Delta_h v^0\|^2 + 3c^2\tau^2 |\delta_x \delta_y v^0|_1^2 \right.
$$

$$
\left. + 2c^{-4}\left(6\|g^0\|^2 + 4\max_{1 \leqslant l \leqslant k}\|g^l\|^2 + \tau \sum_{l=2}^{k-1}\|\Delta_t g^l\|^2 \right)\right], \quad 0 \leqslant k \leqslant n - 1.
$$

Proof Denote

$$
E^k = |\delta_t v^{k+\frac{1}{2}}|_1^2 + c^2 \frac{\|\Delta_h v^k\|^2 + \|\Delta_h v^{k+1}\|^2}{2} + \frac{c^4 \tau^2}{2}\frac{|\delta_x \delta_y v^k|_1^2 + |\delta_x \delta_y v^{k+1}|_1^2}{2}.
$$

(I) Taking an inner product of (5.77a) with $-\Delta_h \delta_t v^{\frac{1}{2}}$, we have

$$
-\frac{2}{\tau}\left(\delta_t v^{\frac{1}{2}}, \Delta_h \delta_t v^{\frac{1}{2}}\right) + c^2\left(\Delta_h v^1, \Delta_h \delta_t v^{\frac{1}{2}}\right) - \frac{c^4 \tau^2}{2}\left(\delta_x^2 \delta_y^2 v^1, \Delta_h \delta_t v^{\frac{1}{2}}\right)
$$

$$
= -\left(g^0, \Delta_h \delta_t v^{\frac{1}{2}}\right). \tag{5.78}
$$

Next we analyze each term on the left-hand side of (5.78).
The first term:

$$
-\frac{2}{\tau}\left(\delta_t v^{\frac{1}{2}}, \Delta_h \delta_t v^{\frac{1}{2}}\right) = \frac{2}{\tau}|\delta_t v^{\frac{1}{2}}|_1^2.
$$

The second term:

$$\left(\Delta_h v^1, \Delta_h \delta_t v^{\frac{1}{2}}\right) = \frac{1}{2\tau}\left(\|\Delta_h v^1\|^2 - \|\Delta_h v^0\|^2\right) + \frac{\tau}{2}\|\Delta_h \delta_t v^{\frac{1}{2}}\|^2.$$

The third term:

$$-\left(\delta_x^2 \delta_y^2 v^1, \Delta_h \delta_t v^{\frac{1}{2}}\right) = \frac{1}{2\tau}\left(|\delta_x \delta_y v^1|_1^2 - |\delta_x \delta_y v^0|_1^2\right) + \frac{\tau}{2}|\delta_x \delta_y \delta_t v^{\frac{1}{2}}|_1^2.$$

Substituting the above three equalities into (5.78), we have

$$\frac{2}{\tau}|\delta_t v^{\frac{1}{2}}|_1^2 + c^2 \cdot \frac{1}{2\tau}\left(\|\Delta_h v^1\|^2 - \|\Delta_h v^0\|^2\right)$$

$$+ \frac{c^4 \tau^2}{2} \cdot \frac{1}{2\tau}\left(|\delta_x \delta_y v^1|_1^2 - |\delta_x \delta_y v^0|_1^2\right) \leqslant -\left(g^0, \Delta_h \delta_t v^{\frac{1}{2}}\right).$$

It is easy to obtain

$$2|\delta_t v^{\frac{1}{2}}|_1^2 + c^2 \cdot \frac{1}{2}\left(\|\Delta_h v^1\|^2 + \|\Delta_h v^0\|^2\right) + \frac{c^4 \tau^2}{2} \cdot \frac{1}{2}\left(|\delta_x \delta_y v^1|_1^2 + |\delta_x \delta_y v^0|_1^2\right)$$

$$\leqslant c^2 \|\Delta_h v^0\|^2 + \frac{c^4 \tau^2}{2}|\delta_x \delta_y v^0|_1^2 - \left(g^0, \Delta_h v^1 - \Delta_h v^0\right)$$

$$\leqslant c^2 \|\Delta_h v^0\|^2 + \frac{c^4 \tau^2}{2}|\delta_x \delta_y v^0|_1^2 + c^2 \cdot \frac{1}{4}\left(\|\Delta_h v^1\|^2 + \|\Delta_h v^0\|^2\right) + 2c^{-2}\|g^0\|^2,$$

namely,

$$2|\delta_t v^{\frac{1}{2}}|_1^2 + c^2 \cdot \frac{1}{4}\left(\|\Delta_h v^1\|^2 + \|\Delta_h v^0\|^2\right) + \frac{c^4 \tau^2}{2} \cdot \frac{1}{2}\left(|\delta_x \delta_y v^1|_1^2 + |\delta_x \delta_y v^0|_1^2\right)$$

$$\leqslant c^2 \|\Delta_h v^0\|^2 + \frac{c^4 \tau^2}{2}|\delta_x \delta_y v^0|_1^2 + 2c^{-2}\|g^0\|^2.$$

Therefore,

$$E^0 \leqslant 2c^2 \|\Delta_h v^0\|^2 + c^4 \tau^2 |\delta_x \delta_y v^0|_1^2 + 4c^{-2}\|g^0\|^2. \tag{5.79}$$

(II) Taking an inner product of $-\Delta_h \Delta_t v^k$ with (5.77b), we have

$$-\left(\delta_t^2 v^k, \Delta_h \Delta_t v^k\right) + c^2 \left(\Delta_h v^{\bar{k}}, \Delta_h \Delta_t v^k\right) - \frac{c^4 \tau^2}{2}\left(\delta_x^2 \delta_y^2 v^{\bar{k}}, \Delta_h \Delta_t v^k\right)$$

$$= -\left(g^k, \Delta_h \Delta_t v^k\right), \quad 1 \leqslant k \leqslant n - 1. \tag{5.80}$$

Next we analyze each term on the left-hand side of (5.80).

The first term:

$$- \left(\delta_t^2 v^k, \Delta_h \Delta_t v^k\right) = \frac{1}{2\tau}\left(|\delta_t v^{k+\frac{1}{2}}|_1^2 - |\delta_t v^{k-\frac{1}{2}}|_1^2\right).$$

The second term:

$$\left(\Delta_h v^{\bar{k}}, \Delta_h \Delta_t v^k\right) = \frac{1}{4\tau}\left(\|\Delta_h v^{k+1}\|^2 - \|\Delta_h v^{k-1}\|^2\right)$$

$$= \frac{1}{2\tau}\left(\frac{\|\Delta_h v^{k+1}\|^2 + \|\Delta_h v^k\|^2}{2} - \frac{\|\Delta_h v^k\|^2 + \|\Delta_h v^{k-1}\|^2}{2}\right).$$

The third term:

$$- \left(\delta_x^2 \delta_y^2 v^{\bar{k}}, \Delta_h \Delta_t v^k\right) = \frac{1}{4\tau}\left(|\delta_x \delta_y v^{k+1}|_1^2 - |\delta_x \delta_y v^{k-1}|_1^2\right)$$

$$= \frac{1}{2\tau}\left(\frac{|\delta_x \delta_y v^{k+1}|_1^2 + |\delta_x \delta_y v^k|_1^2}{2} - \frac{|\delta_x \delta_y v^k|_1^2 + |\delta_x \delta_y v^{k-1}|_1^2}{2}\right).$$

Substituting the above three equalities into (5.80), we have

$$\frac{1}{2\tau}(E^k - E^{k-1}) = -\left(g^k, \Delta_h \Delta_t v^k\right), \quad 1 \leqslant k \leqslant n - 1. \tag{5.81}$$

When $k = 1$, it yields

$$E^1 = E^0 - 2\tau(g^1, \Delta_h \Delta_t v^1)$$

$$= E^0 - (g^1, \Delta_h v^2 - \Delta_h v^0)$$

$$= E^0 - (g^1, \Delta_h v^2) + (g^1, \Delta_h v^0)$$

$$\leqslant E^0 + 2c^{-2}\|g^1\|^2 + \frac{c^2}{4}(\|\Delta_h v^2\|^2 + \|\Delta_h v^0\|^2).$$

It is easy to see that

$$c^2 \frac{\|\Delta_h v^2\|^2 + \|\Delta_h v^1\|^2}{2} \leqslant 2E^0 + 4c^{-2}\|g^1\|^2 + c^2 \frac{\|\Delta_h v^1\|^2 + \|\Delta_h v^0\|^2}{2},$$

namely,

$$\frac{1}{2}(\|\Delta_h v^2\|^2 + \|\Delta_h v^1\|^2) \leqslant 2c^{-2}E^0 + 4c^{-4}\|g^1\|^2 + \frac{1}{2}(\|\Delta_h v^0\|^2 + \|\Delta_h v^1\|^2). \tag{5.82}$$

When $k \geqslant 2$, replacing k with l in (5.81) and summing over l from 1 to k, we have

$$E^k = E^0 - 2\tau \sum_{l=1}^{k} \left(g^l, \Delta_h \Delta_t v^l\right)$$

$$= E^0 - \sum_{l=1}^{k} \left(g^l, \Delta_h v^{l+1} - \Delta_h v^{l-1}\right)$$

$$= E^0 - \sum_{l=2}^{k+1} \left(g^{l-1}, \Delta_h v^l\right) + \sum_{l=0}^{k-1} \left(g^{l+1}, \Delta_h v^l\right)$$

$$= E^0 + 2\tau \sum_{l=2}^{k-1} \left(\Delta_t g^l, \Delta_h v^l\right) - \left(g^{k-1}, \Delta_h v^k\right)$$

$$- \left(g^k, \Delta_h v^{k+1}\right) + \left(g^1, \Delta_h v^0\right) + \left(g^2, \Delta_h v^1\right), \quad 2 \leqslant k \leqslant n-1.$$

Noticing

$$E^k \geqslant c^2 \frac{\|\Delta_h v^{k+1}\|^2 + \|\Delta_h v^k\|^2}{2},$$

we have

$$c^2 \frac{\|\Delta_h v^{k+1}\|^2 + \|\Delta_h v^k\|^2}{2}$$

$$\leqslant E^0 + 2\tau \sum_{l=2}^{k-1} \left(\Delta_t g^l, \Delta_h v^l\right) - \left(g^{k-1}, \Delta_h v^k\right)$$

$$- \left(g^k, \Delta_h v^{k+1}\right) + \left(g^1, \Delta_h v^0\right) + \left(g^2, \Delta_h v^1\right)$$

$$\leqslant E^0 + \tau \sum_{l=2}^{k-1} \left(c^{-2}\|\Delta_t g^l\|^2 + c^2\|\Delta_h v^l\|^2\right)$$

$$+ \left(c^{-2}\|g^{k-1}\|^2 + \frac{1}{4}c^2\|\Delta_h v^k\|^2\right) + \left(c^{-2}\|g^k\|^2 + \frac{1}{4}c^2\|\Delta_h v^{k+1}\|^2\right)$$

$$+ \left(c^{-2}\|g^1\|^2 + \frac{1}{4}c^2\|\Delta_h v^0\|^2\right) + \left(c^{-2}\|g^2\|^2 + \frac{1}{4}c^2\|\Delta_h v^1\|^2\right), \quad 2 \leqslant k \leqslant n-1.$$

After further calculation, we have

$$
\frac{c^2}{2} \frac{\|\Delta_h v^{k+1}\|^2 + \|\Delta_h v^k\|^2}{2}
$$

$$
\leqslant E^0 + c^{-2}\left(\|g^{k-1}\|^2 + \|g^k\|^2 + \|g^1\|^2 + \|g^2\|^2 + \tau \sum_{l=2}^{k-1} \|\Delta_t g^l\|^2 \right)
$$

$$
+ \frac{c^2}{4}\|\Delta_h v^0\|^2 + \frac{c^2}{4}\|\Delta_h v^1\|^2 + c^2\tau \sum_{l=2}^{k-1} \|\Delta_h v^l\|^2, \quad 2 \leqslant k \leqslant n-1.
$$

Multiplying both sides of the above inequality by $2c^{-2}$, we have

$$
\frac{\|\Delta_h v^{k+1}\|^2 + \|\Delta_h v^k\|^2}{2}
$$

$$
\leqslant 2c^{-2}E^0 + 2c^{-4}\left(\|g^{k-1}\|^2 + \|g^k\|^2 + \|g^1\|^2 + \|g^2\|^2 + \tau \sum_{l=2}^{k-1} \|\Delta_t g^l\|^2 \right)
$$

$$
+ \frac{\|\Delta_h v^0\|^2 + \|\Delta_h v^1\|^2}{2} + 2\tau \sum_{l=1}^{k-1} \frac{\|\Delta_h v^{l+1}\|^2 + \|\Delta_h v^l\|^2}{2}, \quad 2 \leqslant k \leqslant n-1.
$$

In combination of (5.82) and Lemma 5.1 (Gronwall inequality), we have

$$
\frac{\|\Delta_h v^{k+1}\|^2 + \|\Delta_h v^k\|^2}{2}
$$

$$
\leqslant e^{2(k-1)\tau}\left[2c^{-2}E^0 + 2c^{-4}\left(4 \max_{1\leqslant l\leqslant k} \|g^l\|^2 + \tau \sum_{l=2}^{k-1} \|\Delta_t g^l\|^2 \right) \right.
$$

$$
\left. + \frac{\|\Delta_h v^0\|^2 + \|\Delta_h v^1\|^2}{2} \right], \quad 1 \leqslant k \leqslant n-1.
$$

It follows from (5.79) that

$$
\frac{\|\Delta_h v^{k+1}\|^2 + \|\Delta_h v^k\|^2}{2}
$$

$$
\leqslant e^{2(k-1)\tau}\left[3c^{-2}E^0 + 2c^{-4}\left(4 \max_{1\leqslant l\leqslant k} \|g^l\|^2 + \tau \sum_{l=2}^{k-1} \|\Delta_t g^l\|^2 \right) \right]
$$

$$
\leqslant e^{2(k-1)\tau}\left\{ 3c^{-2}\left[2c^2\|\Delta_h v^0\|^2 + c^4\tau^2|\delta_x\delta_y v^0|_1^2 + 4c^{-2}\|g^0\|^2 \right] \right.
$$

$$+ 2c^{-4}\left(4 \max_{1 \leqslant l \leqslant k} \|g^l\|^2 + \tau \sum_{l=2}^{k-1} \|\Delta_t g^l\|^2\right)\Bigg\}$$

$$\leqslant e^{2T}\Bigg[6\|\Delta_h v^0\|^2 + 3c^2\tau^2 |\delta_x \delta_y v^0|_1^2$$

$$+ 2c^{-4}\left(6\|g^0\|^2 + 4 \max_{1 \leqslant l \leqslant k} \|g^l\|^2 + \tau \sum_{l=2}^{k-1} \|\Delta_t g^l\|^2\right)\Bigg], \quad 1 \leqslant k \leqslant n-1.$$

By combining the definition of E^0 with (5.79), it follows that the above inequality also holds for $k = 0$. $\qquad\square$

5.3.5   *Convergence and Stability of the Difference Scheme*

Convergence

Theorem 5.11 *Let* $\{u(x, y, t) \,|\, (x, y) \in \bar{\Omega}, 0 \leqslant t \leqslant T\}$ *be the solution of the problem (5.52) and* $\{u_{ij}^k \,|\, (i, j) \in \bar{\omega}, 0 \leqslant k \leqslant n\}$ *be the solution of the difference scheme (5.63). Denote*

$$e_{ij}^k = u(x_i, y_j, t_k) - u_{ij}^k, \quad (i, j) \in \bar{\omega}, \quad 0 \leqslant k \leqslant n,$$

and then we have

$$\|e^k\|_\infty \leqslant \frac{1}{6} L_1 L_2 c^{-2} e^T \sqrt{3(\sqrt{2}+1)(6c_4^2 + 4c_5^2 + Tc_5^2)}\,(\tau^2 + h_1^2 + h_2^2),\ 1 \leqslant k \leqslant n.$$

Proof Subtracting (5.63) from (5.57), (5.60), and (5.62), we have the system of error equations

$$\begin{cases} \dfrac{2}{\tau}\delta_t e_{ij}^{\frac{1}{2}} - c^2 \Delta_h e_{ij}^1 + \dfrac{c^4 \tau^2}{2}\delta_x^2 \delta_y^2 e_{ij}^1 = (R_4)_{ij}^0, & (i, j) \in \omega, \\[2ex] \delta_t^2 e_{ij}^k - c^2 \Delta_h e_{ij}^{\bar{k}} + \dfrac{c^4 \tau^2}{2}\delta_x^2 \delta_y^2 e_{ij}^{\bar{k}} = (R_4)_{ij}^k, & (i, j) \in \omega, \quad 1 \leqslant k \leqslant n-1, \\[2ex] e_{ij}^0 = 0, & (i, j) \in \omega, \\[1ex] e_{ij}^k = 0, & (i, j) \in \gamma, \quad 0 \leqslant k \leqslant n. \end{cases}$$

Noticing (5.58) and (5.61), it follows from Theorem 5.10 that

$$\frac{1}{2}(\|\Delta_h e^{k+1}\|^2 + \|\Delta_h e^k\|^2)$$

$$\leqslant e^{2T}\Bigg[6\|\Delta_h e^0\|^2 + 3c^2\tau^2 |\delta_x \delta_y e^0|_1^2$$

$$
+ 2c^{-4}\left(6\|(R_4)^0\|^2 + 4\max_{1\leqslant l\leqslant k}\|(R_4)^l\|^2 + \tau\sum_{l=2}^{k-1}\|\Delta_t(R_4)^l\|^2\right)\Big]
$$

$$
= 2c^{-4}e^{2T}\left(6\|(R_4)^0\|^2 + 4\max_{1\leqslant l\leqslant k}\|(R_4)^l\|^2 + \tau\sum_{l=2}^{k-1}\|\Delta_t(R_4)^l\|^2\right)
$$

$$
\leqslant 2c^{-4}e^{2T}\Big[6L_1L_2c_4^2(\tau^2 + h_1^2 + h_2^2)^2 + 4L_1L_2c_5^2(\tau^2 + h_1^2 + h_2^2)^2
$$

$$
+ (k-2)\tau L_1L_2c_5^2(\tau^2 + h_1^2 + h_2^2)^2\Big]
$$

$$
= 2c^{-4}L_1L_2e^{2T}\left(6c_4^2 + 4c_5^2 + Tc_5^2\right)(\tau^2 + h_1^2 + h_2^2)^2, \quad 0\leqslant k\leqslant n-1.
$$

Therefore,

$$
\|\Delta_he^k\|^2 \leqslant 4c^{-4}L_1L_2e^{2T}\left(6c_4^2 + 4c_5^2 + Tc_5^2\right)(\tau^2 + h_1^2 + h_2^2)^2, \quad 1\leqslant k\leqslant n.
$$

Taking the square root on both sides of the above inequality produces

$$
\|\Delta_he^k\| \leqslant 2c^{-2}e^T\sqrt{L_1L_2(6c_4^2 + 4c_5^2 + Tc_5^2)}(\tau^2 + h_1^2 + h_2^2), \quad 1\leqslant k\leqslant n.
$$

It follows again from Lemma 2.8 that

$$
\|e^k\|_\infty \leqslant \frac{1}{12}\sqrt{3(\sqrt{2}+1)L_1L_2}\,\|\Delta_he^k\|
$$

$$
= \frac{1}{6}L_1L_2c^{-2}e^T\sqrt{3(\sqrt{2}+1)(6c_4^2 + 4c_5^2 + Tc_5^2)}\,(\tau^2 + h_1^2 + h_2^2), \; 1\leqslant k\leqslant n.
$$

$$\square$$

Stability

Theorem 5.12 *The solution of the difference scheme (5.63) is stable for arbitrary grid ratios $s_1 \equiv c\frac{\tau}{h_1}$ and $s_2 \equiv c\frac{\tau}{h_2}$ with respect to the initial value and right-hand side function in the following sense: Let $\{u_{ij}^k \,|\, 0\leqslant i\leqslant m_1, 0\leqslant j\leqslant m_2, 0\leqslant k\leqslant n\}$ be the solution of the system of difference equations*

$$
\begin{cases}
\dfrac{2}{\tau}\delta_t u_{ij}^{\frac{1}{2}} - c^2\Delta_h u_{ij}^1 + \dfrac{c^4\tau^2}{2}\delta_x^2\delta_y^2 u_{ij}^1 = f_{ij}^0, & (i,j)\in\omega,\\[2mm]
\delta_t^2 u_{ij}^k - c^2\Delta_h u_{ij}^{\bar k} + \dfrac{c^4\tau^2}{2}\delta_x^2\delta_y^2 u_{ij}^{\bar k} = f_{ij}^k, & (i,j)\in\omega, \quad 1\leqslant k\leqslant n-1,\\[2mm]
u_{ij}^0 = \varphi(x_i, y_j), & (i,j)\in\omega,\\[2mm]
u_{ij}^k = 0, & (i,j)\in\gamma, \quad 0\leqslant k\leqslant n.
\end{cases}
$$

Then we have

$$\frac{\|\Delta_h u^{k+1}\|^2 + \|\Delta_h u^k\|^2}{2}$$

$$\leqslant e^{2T}\left[6\|\Delta_h u^0\|^2 + 3c^2\tau^2|\delta_x\delta_y u^0|_1^2\right.$$

$$\left. + 2c^{-4}\left(6\|f^0\|^2 + 4\max_{1\leqslant l\leqslant k}\|f^l\|^2 + \tau\sum_{l=2}^{k-1}\|\Delta_t f^l\|^2\right)\right], \quad 0\leqslant k\leqslant n-1.$$

Proof A direct application of Theorem 5.10 yields the result. $\square$

5.4 The Compact ADI Scheme for the Two-Dimensional Hyperbolic Equation

With the application of the numerical methods in Sects. 4.4, 5.2, and 5.3, the following compact difference scheme can be established for the initial-boundary value problem of the two-dimensional hyperbolic equation (5.52):

$$\begin{cases} \dfrac{2}{\tau}\mathcal{A}\mathcal{B}\left(\delta_t u_{ij}^{\frac{1}{2}} - \psi_{ij} + \dfrac{\tau^2}{3}\rho_{ij}\right) - c^2(\mathcal{B}\delta_x^2 + \mathcal{A}\delta_y^2)u_{ij}^1 + \dfrac{c^4\tau^2}{2}\delta_x^2\delta_y^2 u_{ij}^1 \\ \quad = \mathcal{A}\mathcal{B}f_{ij}^1, \quad (i,j)\in\omega, & (5.83\text{a}) \\[2ex] \mathcal{A}\mathcal{B}\delta_t^2 u_{ij}^k - c^2\left(\mathcal{B}\delta_x^2 + \mathcal{A}\delta_y^2\right)u_{ij}^{\bar{k}} + c^4\dfrac{\tau^2}{2}\delta_x^2\delta_y^2 u_{ij}^{\bar{k}} = \mathcal{A}\mathcal{B}f_{ij}^k, \\ \qquad\qquad\qquad (i,j)\in\omega, \quad 1\leqslant k\leqslant n-1, & (5.83\text{b}) \\[2ex] u_{ij}^0 = \varphi(x_i, y_j), \quad (i,j)\in\omega, & (5.83\text{c}) \\[1ex] u_{ij}^k = \alpha(x_i, y_j, t_k), \quad (i,j)\in\gamma, \quad 0\leqslant k\leqslant n. & (5.83\text{d}) \end{cases}$$

The computational stencil of the difference scheme (5.83) is illustrated in Fig. 5.13.

It can be shown that the difference scheme (5.83) is uniquely solvable, stable with respect to the initial condition and the right-hand side function, and convergent in the maximum norm with the accuracy of order $O(\tau^2 + h_1^4 + h_2^4)$.

From (5.83c) and (5.83d), it follows that $u^0 = \{u_{ij}^0 \mid (i,j)\in\bar{\omega}\}$ has been determined.

Fig. 5.13 The stencil of the
difference scheme (5.83)

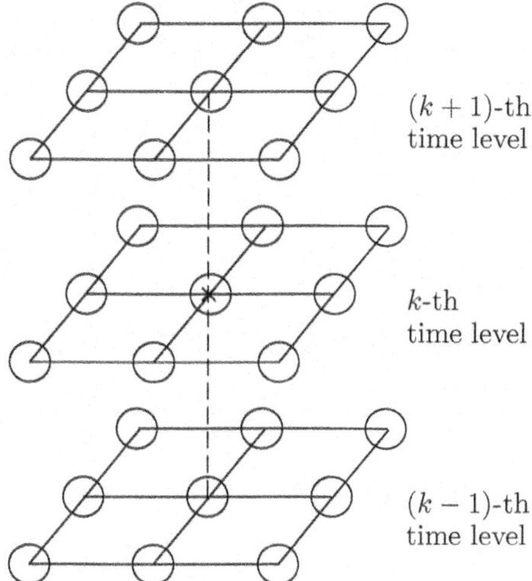

Implementation of (5.83a)
Rewrite (5.83a) as

$$
\mathcal{AB}u_{ij}^1 - \frac{c^2\tau^2}{2}(\mathcal{B}\delta_x^2 + \mathcal{A}\delta_y^2)u_{ij}^1 + \frac{c^4\tau^4}{4}\delta_x^2\delta_y^2 u_{ij}^1
$$

$$
= \mathcal{AB}\left(u_{ij}^0 + \tau\psi_{ij} - \frac{\tau^3}{3}\rho_{ij} + \frac{\tau^2}{2}f_{ij}^1\right), \quad (i, j) \in \omega,
$$

or

$$
\left(\mathcal{A} - \frac{c^2\tau^2}{2}\delta_x^2\right)\left(\mathcal{B} - \frac{c^2\tau^2}{2}\delta_y^2\right)u_{ij}^1 = \mathcal{AB}\left(u_{ij}^0 + \tau\psi_{ij} - \frac{\tau^3}{3}\rho_{ij} + \frac{\tau^2}{2}f_{ij}^1\right), \quad (i, j) \in \omega.
$$

Let

$$
\bar{u}_{ij} = \left(\mathcal{B} - \frac{c^2\tau^2}{2}\delta_y^2\right)u_{ij}^1,
$$

and then (5.83a) can be converted into the compact ADI scheme

$$
\begin{cases}
\left(\mathcal{A} - \dfrac{c^2\tau^2}{2}\delta_x^2\right)\bar{u}_{ij} = \mathcal{AB}\left(u_{ij}^0 + \tau\psi_{ij} - \dfrac{\tau^3}{3}\rho_{ij} + \dfrac{\tau^2}{2}f_{ij}^1\right), & \text{(5.84a)} \\[2ex]
\left(\mathcal{B} - \dfrac{c^2\tau^2}{2}\delta_y^2\right)u_{ij}^1 = \bar{u}_{ij}. & \text{(5.84b)}
\end{cases}
$$

The intermediate variable $\{\bar{u}_{ij} \mid (i, j) \in \omega\}$ can be solved by (5.84a): For an arbitrary fixed j ($1 \leqslant j \leqslant m_2 - 1$), taking the boundary value conditions

$$\bar{u}_{0j} = \left(\mathcal{B} - \frac{c^2\tau^2}{2}\delta_y^2\right) u_{0j}^1, \quad \bar{u}_{m_1,j} = \left(\mathcal{B} - \frac{c^2\tau^2}{2}\delta_y^2\right) u_{m_1,j}^1, \qquad (5.85)$$

and solving

$$\left(\mathcal{A} - \frac{c^2\tau^2}{2}\delta_x^2\right) \bar{u}_{ij} = \mathcal{A}\mathcal{B}\left(u_{ij}^0 + \tau\psi_{ij} - \frac{\tau^3}{3}\rho_{ij} + \frac{\tau^2}{2}f_{ij}^1\right), \quad 1 \leqslant i \leqslant m_1 - 1,$$

$$(5.86)$$

one obtains $\{\bar{u}_{ij} \mid 1 \leqslant i \leqslant m_1 - 1\}$.

When $\{\bar{u}_{ij} \mid (i, j) \in \omega\}$ has been solved, the value of $\{u_{ij}^1 \mid (i, j) \in \omega\}$ can be obtained from (5.84b): For an arbitrary fixed i ($1 \leqslant i \leqslant m_1 - 1$), taking the boundary value conditions

$$u_{i0}^1 = \alpha(x_i, y_0, t_1), \quad u_{i,m_2}^1 = \alpha(x_i, y_{m_2}, t_1), \qquad (5.87)$$

and solving

$$\left(\mathcal{B} - \frac{c^2\tau^2}{2}\delta_y^2\right) u_{ij}^1 = \bar{u}_{ij}, \quad 1 \leqslant j \leqslant m_2 - 1, \qquad (5.88)$$

one gets $\{u_{ij}^1 \mid 1 \leqslant j \leqslant m_2 - 1\}$.

The difference scheme (5.85)–(5.86) is an implicit scheme in the x direction, and (5.87)–(5.88) is an implicit one in the y direction. Both schemes are the tridiagonal systems of linear equations, which can be solved by the double sweep method.

Implementation of (5.83b)

Noticing

$$\delta_t^2 u_{ij}^k = \frac{u_{ij}^{k+1} - 2u_{ij}^k + u_{ij}^{k-1}}{\tau^2} = \frac{2u_{ij}^{\bar{k}} - 2u_{ij}^k}{\tau^2} = \frac{2}{\tau^2}(u_{ij}^{\bar{k}} - u_{ij}^k),$$

Equation (5.83b) can be rewritten as

$$\frac{2}{\tau^2}\mathcal{A}\mathcal{B}(u_{ij}^{\bar{k}} - u_{ij}^k) - c^2(\mathcal{B}\delta_x^2 + \mathcal{A}\delta_y^2)u_{ij}^{\bar{k}} + \frac{c^4\tau^2}{2}\delta_x^2\delta_y^2 u_{ij}^{\bar{k}} = \mathcal{A}\mathcal{B}f_{ij}^k,$$

$$(i, j) \in \omega, \quad 1 \leqslant k \leqslant n - 1,$$

or

$$\left(\mathcal{A} - \frac{c^2\tau^2}{2}\delta_x^2\right)\left(\mathcal{B} - \frac{c^2\tau^2}{2}\delta_y^2\right) u_{ij}^{\bar{k}} = \mathcal{A}\mathcal{B}\left(u_{ij}^k + \frac{\tau^2}{2}f_{ij}^k\right), \quad (i, j) \in \omega, \quad 1 \leqslant k \leqslant n - 1.$$

Let

$$\bar{u}_{ij} = \left(\mathcal{B} - \frac{c^2\tau^2}{2}\delta_y^2 \right) u_{ij}^{\bar{k}}.$$

Then we have

$$\begin{cases} \left(\mathcal{A} - \dfrac{c^2\tau^2}{2}\delta_x^2 \right) \bar{u}_{ij} = \mathcal{A}\mathcal{B}\left(u_{ij}^k + \dfrac{\tau^2}{2}f_{ij}^k \right), & \text{(5.89a)} \\[2em] \left(\mathcal{B} - \dfrac{c^2\tau^2}{2}\delta_y^2 \right) u_{ij}^{\bar{k}} = \bar{u}_{ij}. & \text{(5.89b)} \end{cases}$$

When the value of $\{u_{ij}^{k-1}, u_{ij}^k \mid 0 \leqslant i \leqslant m_1, 0 \leqslant j \leqslant m_2\}$ has been known, the intermediate variable $\{\bar{u}_{ij} \mid (i, j) \in \omega\}$ can be solved by (5.89a): For an arbitrary fixed j ($1 \leqslant j \leqslant m_2 - 1$), taking the boundary value conditions

$$\bar{u}_{0j} = \left(\mathcal{B} - \frac{c^2\tau^2}{2}\delta_y^2 \right) u_{0j}^{\bar{k}}, \quad \bar{u}_{m_1,j} = \left(\mathcal{B} - \frac{c^2\tau^2}{2}\delta_y^2 \right) u_{m_1,j}^{\bar{k}}, \qquad \text{(5.90)}$$

and solving

$$\left(\mathcal{A} - \frac{c^2\tau^2}{2}\delta_x^2 \right) \bar{u}_{ij} = \mathcal{A}\mathcal{B}\left(u_{ij}^k + \frac{c^2\tau^2}{2}f_{ij}^k \right), \qquad 1 \leqslant i \leqslant m_1 - 1, \qquad \text{(5.91)}$$

one obtains $\{\bar{u}_{ij} \mid 1 \leqslant i \leqslant m_1 - 1\}$.

When $\{\bar{u}_{ij} \mid (i, j) \in \omega\}$ has been solved, the value of $\{u_{ij}^{k+1} \mid (i, j) \in \omega\}$ can be solved from (5.89b): For an arbitrary fixed i ($1 \leqslant i \leqslant m_1 - 1$), taking the boundary value conditions

$$u_{i0}^{\bar{k}} = \frac{1}{2}[\alpha(x_i, y_0, t_{k+1}) + \alpha(x_i, y_0, t_{k-1})],$$

$$u_{i,m_2}^{\bar{k}} = \frac{1}{2}[\alpha(x_i, y_{m_2}, t_{k+1}) + \alpha(x_i, y_{m_2}, t_{k-1})], \qquad \text{(5.92)}$$

and solving

$$\left(\mathcal{B} - \frac{c^2\tau^2}{2}\delta_y^2 \right) u_{ij}^{\bar{k}} = \bar{u}_{ij}, \qquad 1 \leqslant j \leqslant m_2 - 1, \qquad \text{(5.93)}$$

one obtains $\{u_{ij}^{\bar{k}} \mid 1 \leqslant j \leqslant m_2 - 1\}$.

Finally, it follows from

$$u_{ij}^{k+1} = 2u_{ij}^{\bar{k}} - u_{ij}^{k-1}, \qquad (i, j) \in \omega$$

that $\{u_{ij}^{k+1} \mid (i, j) \in \omega\}$ is determined.

The difference scheme (5.90)–(5.91) is an implicit scheme in the x direction, and (5.92)–(5.93) is an implicit one in the y direction. Both schemes are the tridiagonal systems of linear equations, which can be solved by the double sweep method.

Example 5.4 Apply the compact ADI scheme (5.84) and (5.89) to compute the problem (5.76) in Example 5.3.

Take $h_1 = h_2 = 1/m$ and denote the maximum numerical error by

$$E_\infty(h, \tau) = \max_{0 \leqslant i, j \leqslant m, 0 \leqslant k \leqslant n} |u(x_i, y_j, t_k) - u_{ij}^k|.$$

Table 5.7 lists some numerical results by taking the step sizes $h = 1/10$ and $\tau = 1/100$. Table 5.8 gives the maximum errors of the numerical solutions with different step sizes. We can see from Table 5.8 that the maximum error decreases to $1/16$ of the original when the spatial step size decreases by half and temporal step size decreases to a quarter of the original, respectively.

Figures 5.14 and 5.15 show the surfaces of the exact solution and numerical solution at $t = 1$ with the step sizes $h = 1/10$ and $\tau = 1/100$, respectively. Figure 5.16 illustrates the error surfaces at $t = 1$ with different step sizes.

Table 5.7 (Example 5.4) Some numerical solutions, exact solutions, and absolute values of the errors ($h = 1/10$, $\tau = 1/100$)

(x, y, t)	NS	ES	|ES-NS|
(0.4,0.4,0.25)	1.1618349	1.1618342	6.302e−7
(0.8,0.4,0.25)	1.4190683	1.4190675	7.172e−7
(0.4,0.8,0.25)	1.4190683	1.4190675	7.172e−7
(0.8,0.8,0.25)	1.7332538	1.7332530	8.115e−7
(0.4,0.4,0.50)	0.9048395	0.9048374	2.087e−6
(0.8,0.4,0.50)	1.1051724	1.1051709	1.482e−6
(0.4,0.8,0.50)	1.1051724	1.1051709	1.482e−6
(0.8,0.8,0.50)	1.3498598	1.3498588	1.019e−6
(0.4,0.4,0.75)	0.7046904	0.7046881	2.298e−6
(0.8,0.4,0.75)	0.8607093	0.8607080	1.372e−6
(0.4,0.8,0.75)	0.8607093	0.8607080	1.372e−6
(0.8,0.8,0.75)	1.0512720	1.0512711	9.060e−7
(0.4,0.4,1.00)	0.5488121	0.5488116	4.579e−7
(0.8,0.4,1.00)	0.6703205	0.6703200	4.373e−7
(0.4,0.8,1.00)	0.6703205	0.6703200	4.373e−7
(0.8,0.8,1.00)	0.8187311	0.8187308	3.533e−7

Table 5.8 (Example 5.4) The maximum numerical errors $E_\infty(h, \tau)$ with different step sizes

h	τ	$E_\infty(h, \tau)$	$E_\infty(2h, 4\tau)/E_\infty(h, \tau)$
1/10	1/100	2.994e−6	
1/20	1/400	1.848e−7	16.1993
1/40	1/1600	1.137e−8	16.2566

Fig. 5.14 (Example 5.4) The exact solution surface at $t = 1$

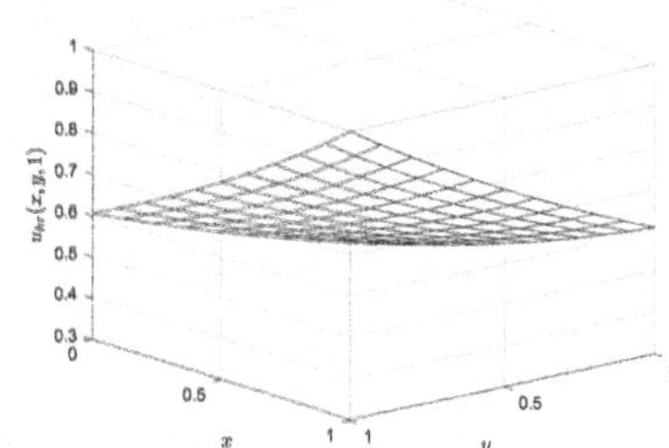

Fig. 5.15 (Example 5.4) The numerical solution surface at $t = 1$ ($h = 1/10$, $\tau = 1/100$)

Fig. 5.16 (Example 5.4) Error surfaces at $t = 1$ with different step sizes

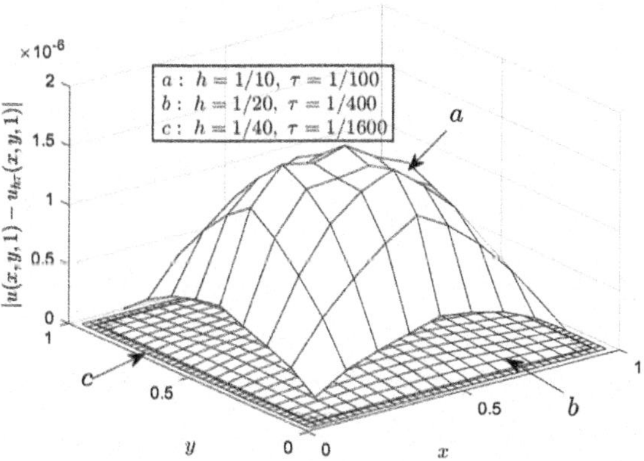

5.5 Summary and Extension

In Sects. 5.1 and 5.2, we introduce the ADI methods and the compact ADI method, respectively, to solve two-dimensional parabolic equations. The existence, uniqueness, convergence, and stability of the solutions to the ADI difference schemes are proved using the energy method. The local truncation error and the H^1-norm error estimate of the compact ADI difference scheme are discussed in [7, 8]. The error analysis in the maximum norm (also known as H^2-analysis) for the ADI and compact ADI difference schemes is presented in [3]. The Richardson extrapolation methods are also covered in these references for the ADI and compact ADI methods. The optimal convergence rate of the forward Euler method for solving linear and nonlinear high-dimensional parabolic equations is discussed in [9]. For high-dimensional parabolic equations with variable coefficients and mixed derivative terms, a second-order difference scheme can be established, and a numerical solution with the fourth-order accuracy in both space and time can be obtained using the Richardson extrapolation method, as seen in [2].

In Sects. 5.3 and 5.4, we introduce the ADI method and the compact ADI method, respectively, for solving two-dimensional hyperbolic equations. The existence, uniqueness, convergence, and stability of the difference solution are proved using the energy method. The error estimate (H^2-analysis) for both the explicit difference scheme and the ADI difference scheme is provided in [4]. For the two-dimensional hyperbolic equation (5.52a), let $v = u_t$, and then (5.52a) is equivalent to [5]

$$\begin{cases} v_t - c^2 \Delta u = f(x, y, t), \\ u_t = v. \end{cases}$$

Then, another difference scheme for (5.52) reads

$$\begin{cases} \delta_t v_{ij}^{k+\frac{1}{2}} - c^2 \Delta_h u_{ij}^{k+\frac{1}{2}} + \dfrac{c^4}{4}\tau^2 \delta_x^2 \delta_y^2 u_{ij}^{k+\frac{1}{2}} = f_{ij}^{k+\frac{1}{2}}, & (i, j) \in \omega, \quad 0 \leqslant k \leqslant n - 1, \\ \delta_t u_{ij}^{k+\frac{1}{2}} = v_{ij}^{k+\frac{1}{2}}, & (i, j) \in \bar{\omega}, \quad 0 \leqslant k \leqslant n - 1, \\ u_{ij}^0 = \varphi_{ij}, \quad v_{ij}^0 = \psi_{ij}, & (i, j) \in \omega, \\ u_{ij}^k = \alpha(x_i, y_j, t_k), & (i, j) \in \gamma, \quad 1 \leqslant k \leqslant n. \end{cases}$$

Eliminating the intermediate variable $\{v_{ij}^k\}$, we have

$$\begin{cases} \dfrac{2}{\tau}\left(\delta_t u_{ij}^{\frac{1}{2}} - \psi_{ij}\right) - c^2 \Delta_h u_{ij}^{\frac{1}{2}} + \dfrac{c^4}{4}\tau^2 \delta_x^2 \delta_y^2 u_{ij}^{\frac{1}{2}} = f_{ij}^{\frac{1}{2}}, & (i, j) \in \omega, \\ \delta_t^2 u_{ij}^k - c^2 \Delta_h \dfrac{u_{ij}^{k-\frac{1}{2}} + u_{ij}^{k+\frac{1}{2}}}{2} + \dfrac{c^4}{4}\tau^2 \delta_x^2 \delta_y^2 \dfrac{u_{ij}^{k-\frac{1}{2}} + u_{ij}^{k+\frac{1}{2}}}{2} = \dfrac{1}{2}\left(f_{ij}^{k-\frac{1}{2}} + f_{ij}^{k+\frac{1}{2}}\right), \\ & (i, j) \in \omega, \quad 1 \leqslant k \leqslant n - 1, \\ u_{ij}^0 = \varphi_{ij}, & (i, j) \in \omega, \\ u_{ij}^k = \alpha(x_i, y_j, t_k), & (i, j) \in \gamma, \quad 1 \leqslant k \leqslant n. \end{cases}$$

Make the operator decomposition as follows:

$$\begin{cases} \left(\mathcal{I} - \dfrac{c^2 \tau^2}{4}\delta_x^2\right)\left(\mathcal{I} - \dfrac{c^2 \tau^2}{4}\delta_y^2\right)u_{ij}^{\frac{1}{2}} = u_{ij}^0 + \dfrac{\tau}{2}\psi_{ij} + \dfrac{\tau^2}{4}f_{ij}^{\frac{1}{2}}, \\ \left(\mathcal{I} - \dfrac{c^2 \tau^2}{4}\delta_x^2\right)\left(\mathcal{I} - \dfrac{c^2 \tau^2}{4}\delta_y^2\right)\dfrac{u_{ij}^{k-\frac{1}{2}} + u_{ij}^{k+\frac{1}{2}}}{2} = u_{ij}^k + \dfrac{\tau^2}{4}\dfrac{f_{ij}^{k-\frac{1}{2}} + f_{ij}^{k+\frac{1}{2}}}{2}. \end{cases}$$

The ADI method is a widely used approach for solving evolution equations in high dimensions. The fundamental idea is to decompose a high-dimensional problem into a series of one-dimensional problems. This method can be easily extended to the three-dimensional case and to problems with derivative boundary conditions, for which the corresponding theoretical results can also be established.

5.6 Exercise

5.1 For the problem (5.1), derive the following ADI scheme

$$\begin{cases} (\mathcal{I} - a\tau\delta_x^2)u_{ij}^* = u_{ij}^k + \tau f_{ij}^{k+1}, \\ (\mathcal{I} - a\tau\delta_y^2)u_{ij}^{k+1} = u_{ij}^*. \end{cases}$$

(1) Eliminate the intermediate variable u_{ij}^* and analyze the local truncation error.

(2) If the boundary value is always 0, prove that the difference solution has a priori estimate

$$\|u^k\|^2 \leqslant \|u^0\|^2 + \tau \sum_{l=1}^{k} \|f^l\|^2, \quad 1 \leqslant k \leqslant n.$$

(3) Show the convergence of the difference scheme.

5.2 Let $\{u_{ij}^k\}$ be the solution of the following difference scheme:

$$\begin{cases} \delta_t u_{ij}^{k+\frac{1}{2}} - a\Delta_h u_{ij}^{k+\frac{1}{2}} = f_{ij}^{k+\frac{1}{2}}, & (i, j) \in \omega, \quad 0 \leqslant k \leqslant n - 1, \\ u_{ij}^0 = \varphi(x_i, y_j), & (i, j) \in \omega, \\ u_{ij}^k = 0, & (i, j) \in \gamma, \quad 0 \leqslant k \leqslant n. \end{cases}$$

Prove that the difference solution has the following priori estimate:

$$|u^k|_1^2 \leqslant |u^0|_1^2 + \frac{1}{2a}\tau \sum_{l=0}^{k-1} \|f^{l+\frac{1}{2}}\|^2, \quad 1 \leqslant k \leqslant n.$$

5.3 Let $\{u_{ij}^k\}$ be the solution of the system of difference equations

$$\begin{cases} \delta_t u_{ij}^{k+\frac{1}{2}} - a\,\Delta_h u_{ij}^{k+\frac{1}{2}} + \frac{1}{4}a^2\tau^2\delta_x^2\delta_y^2\delta_t u_{ij}^{k+\frac{1}{2}} = f_{ij}^{k+\frac{1}{2}}, & (i, j) \in \omega, \quad 0 \leqslant k \leqslant n - 1, \\ u_{ij}^0 = \varphi(x_i, y_j), & (i, j) \in \omega, \\ u_{ij}^k = 0, & (i, j) \in \gamma, \quad 0 \leqslant k \leqslant n. \end{cases}$$

Denote

$$E^k = \|u^k\|^2 + \frac{1}{4}a^2\tau^2\|\delta_x\delta_y u^k\|^2.$$

When $\tau \leqslant \frac{2}{3}$, prove that the following estimate holds:

$$E^k \leqslant e^{\frac{3}{2}k\tau}\left(E^0 + \frac{3}{2}\tau\sum_{l=0}^{k-1}\|f^{l+\frac{1}{2}}\|^2\right), \quad 1 \leqslant k \leqslant n.$$

5.4 Consider the difference scheme (5.9) with $a = 1$. Rewrite the difference equation (5.9a) into

$$\left(\mathcal{I} - \frac{\tau}{2}\delta_x^2\right)\left(\mathcal{I} - \frac{\tau}{2}\delta_y^2\right)\delta_t u_{ij}^{k+\frac{1}{2}} = \Delta_h u_{ij}^k + f_{ij}^{k+\frac{1}{2}},$$

$$(i, j) \in \omega, \quad 0 \leqslant k \leqslant n - 1.$$

Try to derive an ADI algorithm according to the above equality.

5.5 Let $\{u_{ij}^k\}$ be the solution of the difference scheme

$$\begin{cases} \dfrac{2}{\tau}\delta_t u_{ij}^{\frac{1}{2}} - c^2\Delta_h u_{ij}^1 = g_{ij}^0, & (i, j) \in \omega, & & \text{(5.94a)} \\[2mm] \delta_t^2 u_{ij}^k - c^2\Delta_h u_{ij}^{\bar{k}} = g_{ij}^k, & (i, j) \in \omega, & 1 \leqslant k \leqslant n - 1, & \text{(5.94b)} \\[2mm] u_{ij}^0 = \varphi(x_i, y_j), & (i, j) \in \omega, & & \text{(5.94c)} \\[2mm] u_{ij}^k = 0, & (i, j) \in \gamma, & 0 \leqslant k \leqslant n. & \text{(5.94d)} \end{cases}$$

Denote

$$E^k = \|\delta_t u^{k+\frac{1}{2}}\|^2 + c^2\frac{|u^k|_1^2 + |u^{k+1}|_1^2}{2}.$$

(1) Take an inner product of (5.94a) with $\delta_t u^{\frac{1}{2}}$ and prove

$$E^0 \leqslant c^2|u^0|_1^2 + \frac{1}{4}\tau^2\|g^0\|^2.$$

(2) Take an inner product of (5.94b) with $\Delta_t u^k$ and prove

$$E^k \leqslant e^{\frac{3}{2}k\tau}\left(E^0 + \frac{3}{2}\tau\sum_{l=1}^{k}\|g^l\|^2\right), \quad 1 \leqslant k \leqslant n - 1$$

when $\tau \leqslant \frac{2}{3}$.

5.6 Let $\{u_{ij}^k\}$ be the solution of the difference scheme

$$
\begin{cases}
\dfrac{2}{\tau}AB\!\left(\delta_t u_{ij}^{\frac{1}{2}}\right) - \left(B\delta_x^2 + A\delta_y^2\right)u_{ij}^1 \\[2mm]
\qquad\qquad +\dfrac{\tau^2}{2}\delta_x^2\delta_y^2 u_{ij}^1 = g_{ij}^0, \quad (i,j)\in\omega, & (5.95\mathrm{b}) \\[4mm]
AB\delta_t^2 u_{ij}^k - \left(B\delta_x^2 + A\delta_y^2\right)u_{ij}^{\bar k} + \dfrac{\tau^2}{2}\delta_x^2\delta_y^2 u_{ij}^{\bar k} = g_{ij}^k, \\[2mm]
\qquad\qquad (i,j)\in\omega, \quad 1\leqslant k\leqslant n-1, & (5.95\mathrm{c}) \\[4mm]
u_{ij}^0 = \varphi(x_i, y_j), \quad (i,j)\in\omega, & (5.95\mathrm{d}) \\[2mm]
u_{ij}^k = 0, \quad (i,j)\in\gamma, \quad 0\leqslant k\leqslant n. & (5.95\mathrm{e})
\end{cases}
$$

$$\dfrac{2}{\tau}AB\!\left(\delta_t u_{ij}^{\frac{1}{2}}\right) - \left(B\delta_x^2 + A\delta_y^2\right)u_{ij}^1 \tag{5.95a}$$

(1) Take an inner product of (5.95b) with $\delta_t u^{\frac{1}{2}}$, and (5.95c) with $\Delta_t u^k$, respectively. Try to give a prior estimate of the difference solution.

(2) Take an inner product of (5.95b) with $-\Delta_h \delta_t u^{\frac{1}{2}}$, and (5.95c) with $-\Delta_h \Delta_t u^k$, respectively. Try to give an estimate of the difference solution.

5.7 Apply the ADI difference scheme (5.18) to calculate the problem

$$
\begin{cases}
\dfrac{\partial u}{\partial t} - \left(\dfrac{\partial^2 u}{\partial x^2} + \dfrac{\partial^2 u}{\partial y^2}\right) = (x^2 + y^2)t^2 \sin(xyt) \\[2mm]
\qquad\qquad +xy\cos(xyt),\ 0 < x, y < 1,\ 0 < t \leqslant 1, \\[2mm]
u(x, y, 0) = 0, \quad 0\leqslant x, y \leqslant 1, \\[2mm]
u(0, y, t) = 0, \quad u(1, y, t) = \sin(yt), \quad 0\leqslant y\leqslant 1, \quad 0 < t \leqslant 1, \\[2mm]
u(x, 0, t) = 0, \quad u(x, 1, t) = \sin(xt), \quad 0 < x < 1, \quad 0 < t \leqslant 1.
\end{cases}
$$

The exact solution of the problem is $u = \sin(xyt)$. Fill in Tables 5.9 and 5.10, respectively.

5.8 Apply the compact ADI difference scheme (5.41) to compute the problem given in Exercise 5.7. Fill in Tables 5.11 and 5.12, respectively.

5.9 Apply the ADI difference scheme (5.66) and (5.71) to compute the problem

$$
\begin{cases}
\dfrac{\partial^2 u}{\partial t^2} - \left(\dfrac{\partial^2 u}{\partial x^2} + \dfrac{\partial^2 u}{\partial y^2}\right) = -3e^{x+y}\sin t, \quad 0 < x, y < 1, \quad 0 < t \leqslant 1, \\[2mm]
u(x, y, 0) = 0, \quad \dfrac{\partial u(x,y,0)}{\partial t} = e^{x+y}, \quad 0\leqslant x, y \leqslant 1, \\[2mm]
u(0, y, t) = e^y \sin t, \quad u(1, y, t) = e^{1+y}\sin t, \quad 0\leqslant y\leqslant 1, \quad 0 < t \leqslant 1, \\[2mm]
u(x, 0, t) = e^x \sin t, \quad u(x, 1, t) = e^{1+x}\sin t, \quad 0 < x < 1, \quad 0 < t \leqslant 1.
\end{cases}
$$

Table 5.9 (Exercise 5.7)
Some numerical results
($h = 1/20$, $\tau = 1/20$)

(x, y, t)	NS	ES	$|ES-NS|$
$(0.5, 0.5, 0.1)$			
$(0.5, 0.5, 0.2)$			
$(0.5, 0.5, 0.3)$			
$(0.5, 0.5, 0.4)$			
$(0.5, 0.5, 0.5)$			
$(0.5, 0.5, 0.6)$			
$(0.5, 0.5, 0.7)$			
$(0.5, 0.5, 0.8)$			
$(0.5, 0.5, 0.9)$			
$(0.5, 0.5, 1.0)$			

Table 5.10 (Exercise 5.7)
Some numerical results
($h = 1/40$, $\tau = 1/40$)

(x, y, t)	NS	ES	$|ES-NS|$
$(0.5, 0.5, 0.1)$			
$(0.5, 0.5, 0.2)$			
$(0.5, 0.5, 0.3)$			
$(0.5, 0.5, 0.4)$			
$(0.5, 0.5, 0.5)$			
$(0.5, 0.5, 0.6)$			
$(0.5, 0.5, 0.7)$			
$(0.5, 0.5, 0.8)$			
$(0.5, 0.5, 0.9)$			
$(0.5, 0.5, 1.0)$			

Table 5.11 (Exercise 5.8)
Some numerical results
($h = 1/10$, $\tau = 1/100$)

(x, y, t)	NS	ES	$|ES-NS|$
$(0.5, 0.5, 0.1)$			
$(0.5, 0.5, 0.2)$			
$(0.5, 0.5, 0.3)$			
$(0.5, 0.5, 0.4)$			
$(0.5, 0.5, 0.5)$			
$(0.5, 0.5, 0.6)$			
$(0.5, 0.5, 0.7)$			
$(0.5, 0.5, 0.8)$			
$(0.5, 0.5, 0.9)$			
$(0.5, 0.5, 1.0)$			

The exact solution of the problem is $u = e^{x+y} \sin t$. Fill in Tables 5.13 and 5.14, respectively.

5.10 Apply the compact ADI difference scheme (5.84) and (5.89) to compute the problem given in Exercise 5.9. Fill in Tables 5.15 and 5.16, respectively.

Table 5.12 (Exercise 5.8)
Some numerical results
($h = 1/20$, $\tau = 1/400$)

(x, y, t)	NS	ES	|ES$-$NS|
$(0.5, 0.5, 0.1)$			
$(0.5, 0.5, 0.2)$			
$(0.5, 0.5, 0.3)$			
$(0.5, 0.5, 0.4)$			
$(0.5, 0.5, 0.5)$			
$(0.5, 0.5, 0.6)$			
$(0.5, 0.5, 0.7)$			
$(0.5, 0.5, 0.8)$			
$(0.5, 0.5, 0.9)$			
$(0.5, 0.5, 1.0)$			

Table 5.13 (Exercise 5.9)
Some numerical results
($h = 1/20$, $\tau = 1/20$)

(x, y, t)	NS	ES	|ES$-$NS|
$(0.5, 0.5, 0.1)$			
$(0.5, 0.5, 0.2)$			
$(0.5, 0.5, 0.3)$			
$(0.5, 0.5, 0.4)$			
$(0.5, 0.5, 0.5)$			
$(0.5, 0.5, 0.6)$			
$(0.5, 0.5, 0.7)$			
$(0.5, 0.5, 0.8)$			
$(0.5, 0.5, 0.9)$			
$(0.5, 0.5, 1.0)$			

Table 5.14 (Exercise 5.9)
Some numerical results
($h = 1/40$, $\tau = 1/40$)

(x, y, t)	NS	ES	|ES$-$NS|
$(0.5, 0.5, 0.1)$			
$(0.5, 0.5, 0.2)$			
$(0.5, 0.5, 0.3)$			
$(0.5, 0.5, 0.4)$			
$(0.5, 0.5, 0.5)$			
$(0.5, 0.5, 0.6)$			
$(0.5, 0.5, 0.7)$			
$(0.5, 0.5, 0.8)$			
$(0.5, 0.5, 0.9)$			
$(0.5, 0.5, 1.0)$			

Table 5.15 (Exercise 5.10)
Some numerical results
($h = 1/10, \ \tau = 1/100$)

(x, y, t)	NS	ES	|ES−NS|
$(0.5, 0.5, 0.1)$			
$(0.5, 0.5, 0.2)$			
$(0.5, 0.5, 0.3)$			
$(0.5, 0.5, 0.4)$			
$(0.5, 0.5, 0.5)$			
$(0.5, 0.5, 0.6)$			
$(0.5, 0.5, 0.7)$			
$(0.5, 0.5, 0.8)$			
$(0.5, 0.5, 0.9)$			
$(0.5, 0.5, 1.0)$			

Table 5.16 (Exercise 5.10)
Some numerical results
($h = 1/20, \ \tau = 1/400$)

(x, y, t)	NS	ES	|ES−NS|
$(0.5, 0.5, 0.1)$			
$(0.5, 0.5, 0.2)$			
$(0.5, 0.5, 0.3)$			
$(0.5, 0.5, 0.4)$			
$(0.5, 0.5, 0.5)$			
$(0.5, 0.5, 0.6)$			
$(0.5, 0.5, 0.7)$			
$(0.5, 0.5, 0.8)$			
$(0.5, 0.5, 0.9)$			
$(0.5, 0.5, 1.0)$			

References

1. D'Yakonov, E.G.: Difference schemes of second-order accuracy with a splitting operator for parabolic equations without mixed partial derivatives. Zh. Vychisl. Mat. Mat. Fiz. **4**, 935–941 (1964)
2. Ji, C.C., Du, R., Sun, Z.Z.: Stability and convergence of difference schemes for multidimensional parabolic equations with variable coefficients and mixed derivatives. Int. J. Comput. Math. **95**, 255–277 (2018)
3. Liao, H.L., Sun, Z.Z.: Maximum norm error bounds of ADI and compact ADI methods for solving parabolic equations. Numer. Methods Partial Differ. Equ. **26**, 37–60 (2010)
4. Liao, H.L., Sun, Z.Z.: Maximum norm error estimates of efficient difference schemes for second-order wave equations. J. Comput. Appl. Math. **235**, 2217–2233 (2011)
5. Liao, H.L., Sun, Z.Z.: A two-level compact ADI method for solving second-order wave equations. Int. J. Comput. Math. **90**, 1471–1488 (2013)
6. Peacemann, D.W., Rechford, H.H.: The numerical solution of parabolic and elliptic differential equations. J. Inst. Math. Appl. **15**, 239–248 (1955)
7. Sun, Z.Z.: Numerical Methods for Partial Differential Equations, 2nd edn. Science Press, Beijing (2012)
8. Sun, Z.Z., Li, X.L.: A compact alternate directional implicit difference method for the reaction diffusion equations. Math. Numer. Sin. **27**, 209–224 (2005)
9. Zhang, Q.F., Zhang, J.Y., Sun, Z.Z.: Optimal convergence rate of the explicit Euler method for convection-diffusion equations II: high dimensional cases. Numer. Methods Partial Differ. Equ. **39**(6), 4377–4402 (2023)

Chapter 6
Finite Difference Methods for Fractional Differential Equations

As a generalization of classical calculus, fractional calculus has evolved into an important branch of mathematics. It is widely believed that the concept originated from a letter by G. W. Leibniz (1646–1716) in 1695, in which the notion of a derivative of order one-half was discussed. Over the past three centuries, numerous mathematicians have made significant contributions to the development of this field.

Fractional differential equations (FDEs) have emerged as essential tools for modeling complex mechanical and physical phenomena. They have been found widespread applications in various fields, including anomalous diffusion, viscoelasticity, fluid dynamics, pipeline boundary layer effects, electromagnetism, signal processing and control, quantum economics, and fractal theory. However, obtaining analytical solutions to FDEs remains challenging, even for linear cases. As a result, developing effective numerical methods for simulating FDEs has become a critical focus of contemporary research.

In this chapter, several widely used definitions of fractional derivatives, along with some fundamental properties, will be presented. The L1 formula for the Caputo fractional derivative will then be derived. Subsequently, the numerical solutions to three types of initial-boundary value problems for time-fractional differential equations will be explored.

Z.-Z. Sun et al., *Numerical Solutions to Partial Differential Equations with Finite Difference Methods*, Springer Asia Pacific Mathematics Series 9, https://doi.org/10.1007/978-981-95-5563-5_6

319

6.1　Definitions and Properties of Fractional Derivatives

6.1.1　Fractional Integrals

Definition 6.1　Suppose α is a positive real number and the function $f(t)$ is defined on $[a, b]$. The α-th order fractional integral of the function $f(t)$ is defined by

$$_aD_t^{-\alpha} f(t) = \frac{1}{\Gamma(\alpha)} \int_a^t (t - \tau)^{\alpha-1} f(\tau)\mathrm{d}\tau, \quad t \in [a, b],$$

where $\Gamma(z)$ is the **Gamma** function in the form of

$$\Gamma(z) = \int_0^\infty \mathrm{e}^{-t} t^{z-1}\mathrm{d}t, \quad \mathrm{Re}(z) > 0.$$

Simple calculation produces

$$_aD_t^{-\alpha}(t - a)^p = \frac{\Gamma(1 + p)}{\Gamma(1 + p + \alpha)}(t - a)^{p+\alpha}, \quad p > -1.$$

6.1.2　Grünwald-Letnikov Fractional Derivatives

Definition 6.2　Suppose α is a positive real number and the function $f(t)$ is defined on $[a, b]$. Let n be a positive integer satisfying $n - 1 \leqslant \alpha < n$. The α-th order Grünwald-Letnikov (G-L) derivative of the function $f(t)$ is defined by

$$_aD_t^\alpha f(t) = \lim_{h \to 0} h^{-\alpha} \sum_{j=0}^{\lfloor (t-a)/h \rfloor} (-1)^j \binom{\alpha}{j} f(t - jh), \quad t \in [a, b],$$

where $t \in [a, b]$, $\lfloor z \rfloor$ is the maximum integer no more than z and $\binom{\alpha}{j}$ denotes the binomial coefficients in the form of

$$\binom{\alpha}{j} = \frac{\alpha(\alpha - 1) \cdots (\alpha - j + 1)}{j!}.$$

Suppose functions $f^{(k)}(t)$ $(k = 0, 1, 2, \ldots, n)$ are continuous on $[a, b]$ with n the minimum integer satisfying $\alpha < n$. It can be proved that

$$_aD_t^\alpha f(t) = \sum_{j=0}^{n-1} \frac{f^{(j)}(a)(t - a)^{j-\alpha}}{\Gamma(1 + j - \alpha)} + \frac{1}{\Gamma(n - \alpha)} \int_a^t \frac{f^{(n)}(\tau)\mathrm{d}\tau}{(t - \tau)^{\alpha-n+1}}.$$

6.1.3 Riemann-Liouville Fractional Derivatives

Definition 6.3 Suppose α is a positive real number and the function $f(t)$ is defined on $[a, b]$. Let n be a positive integer satisfying $n - 1 \leqslant \alpha < n$. The α-th order Riemann-Liouville (R-L) fractional derivative of the function $f(t)$ is defined by

$$_a\mathbf{D}_t^\alpha f(t) = \frac{\mathrm{d}^n}{\mathrm{d}t^n} \left(\frac{1}{\Gamma(n - \alpha)} \int_a^t \frac{f(\tau)\mathrm{d}\tau}{(t - \tau)^{\alpha - n + 1}} \right), \quad t \in [a, b].$$

It is easily known that

$$_a\mathbf{D}_t^\alpha f(t) = \frac{\mathrm{d}^n}{\mathrm{d}t^n} \left[_aD_t^{-(n-\alpha)} f(t) \right].$$

Some calculations yield

$$_a\mathbf{D}_t^\alpha (t - a)^p = \frac{\Gamma(1 + p)}{\Gamma(1 + p - \alpha)} (t - a)^{p - \alpha}, \quad p > -1.$$

It can be proved that

$$\frac{\mathrm{d}^m}{\mathrm{d}t^m} \left(_a\mathbf{D}_t^\alpha f(t) \right) = _a\mathbf{D}_t^{m+\alpha} f(t), \quad \alpha > 0, \ m \text{ is a positive integer.}$$

Note that there exists an equivalence relation between the R-L fractional derivative and the G-L fractional derivative. Specifically, for a positive real number α such that $n - 1 \leqslant \alpha < n$, if the function $f(t)$, defined on $[a, b]$, has continuous derivatives up to the $(n-1)$-th order and $f^{(n)}(t)$ is integrable on $[a, b]$, then the α-th order R-L fractional derivative of $f(t)$ is equivalent to its α-th order G-L fractional derivative.

6.1.4 Caputo Fractional Derivatives

Definition 6.4 Suppose α is a positive real number and the function $f(t)$ is defined on $[a, b]$. Let n be a positive integer satisfying $n - 1 < \alpha \leqslant n$. The α-th order Caputo fractional derivative of the function $f(t)$ is defined by

$$_a^C D_t^\alpha f(t) = \frac{1}{\Gamma(n - \alpha)} \int_a^t \frac{f^{(n)}(\tau)\mathrm{d}\tau}{(t - \tau)^{\alpha - n + 1}}, \quad t \in [a, b].$$

It is easily known that

$$_a^C D_t^\alpha f(t) = _aD_t^{-(n-\alpha)} \left[f^{(n)}(t) \right].$$

An elementary calculation shows

$$\,_{a}^{C}D_{t}^{\alpha}(t-a)^{p} = \frac{\Gamma(1+p)}{\Gamma(1+p-\alpha)}(t-a)^{p-\alpha}, \quad p > n-1 \geqslant 0.$$

If the function $f(t)$ has continuous derivatives up to the $(n+1)$-th order, then

$$\lim_{\alpha \to n-0} \,_{a}^{C}D_{t}^{\alpha}f(t)$$

$$= \lim_{\alpha \to n-0}\left[\frac{f^{(n)}(a)(t-a)^{n-\alpha}}{\Gamma(n-\alpha+1)} + \frac{1}{\Gamma(n-\alpha+1)}\int_{a}^{t}(t-\tau)^{n-\alpha}f^{(n+1)}(\tau)d\tau\right]$$

$$= f^{(n)}(a) + \int_{a}^{t} f^{(n+1)}(\tau)d\tau = f^{(n)}(t).$$

Note that there is also an equivalence relation between the Caputo fractional derivative and the R-L fractional derivative. Specifically, for a positive real number α and a positive integer n such that $0 \leqslant n-1 < \alpha < n$, if the function $f(t)$, defined on $[a, b]$, has continuous derivatives up to the $(n-1)$-th order and $f^{(n)}(t)$ is integrable on $[a, b]$, then

$$\,_{a}\mathbf{D}_{t}^{\alpha}f(t) = \,_{a}^{C}D_{t}^{\alpha}f(t) + \sum_{j=0}^{n-1}\frac{f^{(j)}(a)(t-a)^{j-\alpha}}{\Gamma(1+j-\alpha)}, \quad t \in [a, b].$$

In particular, when $\alpha \in (0, 1)$, it reads

$$\,_{a}\mathbf{D}_{t}^{\alpha}f(t) = \,_{a}^{C}D_{t}^{\alpha}f(t) + \frac{f(a)(t-a)^{-\alpha}}{\Gamma(1-\alpha)}.$$

It can be observed that if the function $f(t)$ satisfies

$$f^{(j)}(a) = 0, \quad j = 0, 1, \ldots, n-1,$$

then the α-th order Caputo fractional derivative and the α-th order R-L fractional derivative are equivalent.

From the definitions of the fractional derivatives above, it can be seen that the values of the fractional derivatives at a point t are related to the function values on the left-hand side of this point. Therefore, they are also termed the left G-L fractional derivative, the left R-L fractional derivative, and the left Caputo fractional derivative, respectively.

Similarly, the right G-L fractional derivative, the right R-L fractional derivative, and the right Caputo fractional derivative can be defined as follows:

$$
{}_tD_b^\alpha f(t) = \lim_{h\to 0} h^{-\alpha} \sum_{j=0}^{[(b-t)/h]} (-1)^j \binom{\alpha}{j} f(t+jh),
$$

$$
{}_tD_b^\alpha f(t) = (-1)^n \frac{\mathrm{d}^n}{\mathrm{d}t^n} \left(\frac{1}{\Gamma(n-\alpha)} \int_t^b \frac{f(\tau)\mathrm{d}\tau}{(\tau-t)^{\alpha-n+1}} \right),
$$

$$
{}_t^C D_b^\alpha f(t) = (-1)^n \frac{1}{\Gamma(n-\alpha)} \int_t^b \frac{f^{(n)}(\tau)\mathrm{d}\tau}{(\tau-t)^{\alpha-n+1}}.
$$

When $a = -\infty$, the left G-L fractional derivative reads

$$
{}_{-\infty}D_t^\alpha f(t) = \lim_{h\to 0} h^{-\alpha} \sum_{j=0}^{+\infty} (-1)^j \binom{\alpha}{j} f(t-jh).
$$

When $b = +\infty$, the right G-L fractional derivative reads

$$
{}_tD_{+\infty}^\alpha f(t) = \lim_{h\to 0} h^{-\alpha} \sum_{j=0}^{+\infty} (-1)^j \binom{\alpha}{j} f(t+jh).
$$

6.1.5 Riesz Fractional Derivatives

Definition 6.5 Suppose α is a positive real number and the function $f(t)$ is defined on $[a, b]$. Let n be a positive integer satisfying $n - 1 \leqslant \alpha < n$ and $\alpha \neq 2k+1$, $k = 0, 1, \ldots$. The α-th order Riesz fractional derivative of the function $f(t)$ is defined by

$$
\frac{\partial^\alpha f(x)}{\partial |x|^\alpha} = -\frac{1}{2\cos\left(\dfrac{\alpha\pi}{2}\right)} \left({}_a\mathbf{D}_x^\alpha f(x) + {}_x\mathbf{D}_b^\alpha f(x) \right), \quad x \in [a, b].
$$

It can be observed from the definition above that the Riesz fractional derivative can be interpreted as a weighted sum of the left R-L fractional derivative and the right R-L fractional derivative. Furthermore, the value of the Riesz fractional derivative at a point x depends on the function values on both sides of x.

6.2 Interpolation Approximations of Caputo Fractional Derivatives

6.2.1 Approximation to the Derivative of Order α When $0 < \alpha < 1$

For the Caputo fractional derivative of order α when $0 < \alpha < 1$, i.e.,

$$
{}_0^C D_t^\alpha f(t) = \frac{1}{\Gamma(1-\alpha)} \int_0^t \frac{f'(s)}{(t-s)^\alpha}\, ds,
$$

the simplest numerical method is the so-called L1 approximation, which is based on piecewise linear interpolation. This approach will be discussed in detail below.

Take a positive integer n. Denote $\tau = \frac{T}{n}$, $t_k = k\tau$, $0 \leqslant k \leqslant n$ and

$$
a_l^{(\alpha)} = (l+1)^{1-\alpha} - l^{1-\alpha}, \quad l \geqslant 0. \tag{6.1}
$$

Consequently,

$$
{}_0^C D_t^\alpha f(t)|_{t=t_k} = \frac{1}{\Gamma(1-\alpha)} \int_0^{t_k} \frac{f'(t)}{(t_k-t)^\alpha}\, dt = \frac{1}{\Gamma(1-\alpha)} \sum_{l=1}^{k} \int_{t_{l-1}}^{t_l} \frac{f'(t)}{(t_k-t)^\alpha}\, dt. \tag{6.2}
$$

On the interval $[t_{l-1}, t_l]$, making the linear interpolation of the function $f(t)$, we have

$$
L_{1,l}(t) = \frac{t_l - t}{\tau} f(t_{l-1}) + \frac{t - t_{l-1}}{\tau} f(t_l),
$$

$$
f(t) - L_{1,l}(t) = \frac{1}{2} f''(\xi_l)(t - t_{l-1})(t - t_l), \tag{6.3}
$$

where $\xi_l = \xi_l(t) \in (t_{l-1}, t_l)$.

Approximating the function $f(t)$ in (6.2) by $L_{1,l}(t)$ leads to

$$
{}_0^C D_t^\alpha f(t)|_{t=t_k}
$$

$$
\approx \frac{1}{\Gamma(1-\alpha)} \sum_{l=1}^{k} \int_{t_{l-1}}^{t_l} \frac{L_{1,l}'(t)}{(t_k-t)^\alpha}\, dt
$$

$$
= \frac{1}{\Gamma(1-\alpha)} \sum_{l=1}^{k} \frac{f(t_l) - f(t_{l-1})}{\tau} \cdot \int_{t_{l-1}}^{t_l} \frac{1}{(t_k-t)^\alpha}\, dt
$$

$$= \frac{1}{\Gamma(1-\alpha)} \sum_{l=1}^{k} \frac{f(t_l) - f(t_{l-1})}{\tau} \cdot \frac{1}{1-\alpha}\left[(t_k - t_{l-1})^{1-\alpha} - (t_k - t_l)^{1-\alpha}\right]$$

$$= \frac{\tau^{-\alpha}}{\Gamma(2-\alpha)} \sum_{l=1}^{k} \left[f(t_l) - f(t_{l-1})\right] \cdot \left[(k-l+1)^{1-\alpha} - (k-l)^{1-\alpha}\right]$$

$$= \frac{\tau^{-\alpha}}{\Gamma(2-\alpha)} \sum_{l=1}^{k} a_{k-l}^{(\alpha)}\left[f(t_l) - f(t_{l-1})\right]$$

$$= \frac{\tau^{-\alpha}}{\Gamma(2-\alpha)} \left[a_0^{(\alpha)} f(t_k) - \sum_{l=1}^{k-1} \left(a_{k-l-1}^{(\alpha)} - a_{k-l}^{(\alpha)}\right) f(t_l) - a_{k-1}^{(\alpha)} f(t_0)\right].$$

An approximate formula for ${}_0^C D_t^\alpha f(t)|_{t=t_k}$ is then derived as

$$D_t^\alpha f(t_k) \equiv \frac{\tau^{-\alpha}}{\Gamma(2-\alpha)} \left[a_0^{(\alpha)} f(t_k) - \sum_{l=1}^{k-1} \left(a_{k-l-1}^{(\alpha)} - a_{k-l}^{(\alpha)}\right) f(t_l) - a_{k-1}^{(\alpha)} f(t_0)\right], \quad (6.4)$$

which is usually called the **L1 formula**, or the **L1 approximation**.

Next the approximate error

$$R(f(t_k)) = {}_0^C D_t^\alpha f(t)|_{t=t_k} - D_t^\alpha f(t_k)$$

will be discussed.

Theorem 6.1 *If $f \in C^2[t_0, t_k]$, then*

$$|R(f(t_k))| \leqslant \frac{1}{2\Gamma(1-\alpha)} \left[\frac{1}{4} + \frac{\alpha}{(1-\alpha)(2-\alpha)}\right] \max_{t_0 \leqslant t \leqslant t_k} \left|f''(t)\right| \tau^{2-\alpha}.$$

Proof It follows from the definition of $R(f(t_k))$ that

$$R(f(t_k)) = \frac{1}{\Gamma(1-\alpha)} \sum_{l=1}^{k} \int_{t_{l-1}}^{t_l} \frac{f'(t)}{(t_k - t)^\alpha}\, dt - \frac{1}{\Gamma(1-\alpha)} \sum_{l=1}^{k} \int_{t_{l-1}}^{t_l} \frac{L_{1,l}'(t)}{(t_k - t)^\alpha}\, dt$$

$$= \frac{1}{\Gamma(1-\alpha)} \sum_{l=1}^{k} \int_{t_{l-1}}^{t_l} \left[f(t) - L_{1,l}(t)\right]' \frac{1}{(t_k - t)^\alpha}\, dt.$$

Applying the integration by parts and noticing (6.3) yield

$$R(f(t_k))$$

$$= -\frac{1}{\Gamma(1-\alpha)} \sum_{l=1}^{k} \int_{t_{l-1}}^{t_l} \left[f(t) - L_{1,l}(t) \right] \mathrm{d}\left(\frac{1}{(t_k - t)^\alpha} \right)$$

$$= -\frac{1}{\Gamma(1-\alpha)} \sum_{l=1}^{k} \int_{t_{l-1}}^{t_l} \left[f(t) - L_{1,l}(t) \right] \alpha (t_k - t)^{-\alpha-1} \mathrm{d}t$$

$$= \frac{1}{\Gamma(1-\alpha)} \sum_{l=1}^{k} \int_{t_{l-1}}^{t_l} \frac{1}{2} f''(\xi_l)(t - t_{l-1})(t_l - t)\alpha (t_k - t)^{-\alpha-1} \mathrm{d}t.$$

Hence,

$$\left| R(f(t_k)) \right| \leqslant \frac{1}{2\Gamma(1-\alpha)} \max_{t_0 \leqslant t \leqslant t_k} \left| f''(t) \right| \sum_{l=1}^{k} \int_{t_{l-1}}^{t_l} (t - t_{l-1})(t_l - t)\alpha (t_k - t)^{-\alpha-1} \mathrm{d}t.$$

$$(6.5)$$

Direct calculation gives

$$\sum_{l=1}^{k-1} \int_{t_{l-1}}^{t_l} (t - t_{l-1})(t_l - t)\alpha (t_k - t)^{-\alpha-1} \mathrm{d}t$$

$$\leqslant \frac{\tau^2}{4} \sum_{l=1}^{k-1} \int_{t_{l-1}}^{t_l} \alpha (t_k - t)^{-\alpha-1} \mathrm{d}t$$

$$= \frac{\tau^2}{4} \int_{t_0}^{t_{k-1}} \alpha (t_k - t)^{-\alpha-1} \mathrm{d}t$$

$$= \frac{\tau^2}{4} \left(\tau^{-\alpha} - t_k^{-\alpha} \right) \leqslant \frac{1}{4} \tau^{2-\alpha} \tag{6.6}$$

and

$$\int_{t_{k-1}}^{t_k} (t - t_{k-1})(t_k - t)\alpha (t_k - t)^{-\alpha-1} \mathrm{d}t$$

$$= \alpha \int_{t_{k-1}}^{t_k} (t - t_{k-1})(t_k - t)^{-\alpha} \mathrm{d}t$$

$$= \alpha \int_{0}^{\tau} (\tau - \xi)\xi^{-\alpha} \mathrm{d}\xi$$

$$= \frac{\alpha}{(1-\alpha)(2-\alpha)} \tau^{2-\alpha}. \tag{6.7}$$

Substituting (6.6) and (6.7) into (6.5) arrives at

$$\left| R(f(t_k)) \right| \leqslant \frac{1}{2\Gamma(1-\alpha)} \left[\frac{1}{4} + \frac{\alpha}{(1-\alpha)(2-\alpha)} \right] \max_{t_0 \leqslant t \leqslant t_k} \left| f''(t) \right| \tau^{2-\alpha}.$$

$\square$

The coefficient $\{a_l^{(\alpha)}\}$ satisfies the following properties:

Lemma 6.1 *Suppose* $\alpha \in (0, 1)$ *and* $\{a_l^{(\alpha)}\}$ *is defined by* (6.1), $l = 0, 1, 2, \ldots,$ *then:*

(I) $1 = a_0^{(\alpha)} > a_1^{(\alpha)} > a_2^{(\alpha)} > \cdots > a_l^{(\alpha)} > 0; \quad \lim_{l \to \infty} a_l^{(\alpha)} = 0.$

(II) $(1-\alpha)l^{-\alpha} < a_{l-1}^{(\alpha)} < (1-\alpha)(l-1)^{-\alpha}, \quad l \geqslant 1.$

6.2.2   Approximation to the Derivative of Order γ When $1 < \gamma < 2$

Now we consider the numerical approximation to the Caputo fractional derivative of order γ when $1 < \gamma < 2$:

$$_0^C D_t^\gamma f(t) = \frac{1}{\Gamma(2-\gamma)} \int_0^t \frac{f''(s)}{(t-s)^{\gamma-1}}\, ds.$$

Let

$$g(t) = f'(t), \quad \alpha = \gamma - 1.$$

Then

$$_0^C D_t^\gamma f(t) = \frac{1}{\Gamma(1-(\gamma-1))} \int_0^t \frac{g'(s)}{(t-s)^{\gamma-1}}\, ds = {}_0^C D_t^\alpha g(t),$$

which implies that the γ-th order derivative of the function $f(t)$ is precisely the α-th order derivative of the function $g(t)$.

By Theorem 6.1, we have

$$_0^C D_t^\alpha g(t)|_{t=t_k} = \frac{\tau^{-\alpha}}{\Gamma(2-\alpha)} \left[a_0^{(\alpha)} g(t_k) - \sum_{l=1}^{k-1} \left(a_{k-l-1}^{(\alpha)} - a_{k-l}^{(\alpha)} \right) g(t_l) - a_{k-1}^{(\alpha)} g(t_0) \right]$$

$$+ R(g(t_k)),$$

where

$$|R(g(t_k))| \leqslant \frac{1}{2\Gamma(1-\alpha)}\left[\frac{1}{4} + \frac{\alpha}{(1-\alpha)(2-\alpha)}\right] \cdot \max_{t_0 \leqslant t \leqslant t_k}\left|g''(t)\right| \tau^{2-\alpha}$$

$$= \frac{1}{2\Gamma(2-\gamma)}\left[\frac{1}{4} + \frac{\gamma-1}{(2-\gamma)(3-\gamma)}\right] \cdot \max_{t_0 \leqslant t \leqslant t_k}\left|f'''(t)\right| \tau^{3-\gamma}.$$

Denote

$$b_l^{(\gamma)} = a_l^{(\alpha)} = (l+1)^{1-\alpha} - l^{1-\alpha} = (l+1)^{2-\gamma} - l^{2-\gamma}, \quad l = 0, 1, 2, \ldots. \tag{6.8}$$

Then

$${}_0^C D_t^\gamma f(t)|_{t=t_k} = \frac{\tau^{1-\gamma}}{\Gamma(3-\gamma)}\left[b_0^{(\gamma)} g(t_k) - \sum_{l=1}^{k-1}\left(b_{k-l-1}^{(\gamma)} - b_{k-l}^{(\gamma)}\right)g(t_l) - b_{k-1}^{(\gamma)} g(t_0)\right]$$

$$+ R(g(t_k)). \tag{6.9}$$

Similarly, we have

$${}_0^C D_t^\gamma f(t)|_{t=t_{k-1}}$$

$$= \frac{\tau^{1-\gamma}}{\Gamma(3-\gamma)}\left[b_0^{(\gamma)} g(t_{k-1}) - \sum_{l=1}^{k-2}\left(b_{k-l-2}^{(\gamma)} - b_{k-l-1}^{(\gamma)}\right)g(t_l) - b_{k-2}^{(\gamma)} g(t_0)\right]$$

$$+ R(g(t_{k-1})). \tag{6.10}$$

Averaging (6.9) and (6.10) yields

$$\frac{1}{2}\left[{}_0^C D_t^\gamma f(t)|_{t=t_k} + {}_0^C D_t^\gamma f(t)|_{t=t_{k-1}}\right]$$

$$= \frac{\tau^{1-\gamma}}{\Gamma(3-\gamma)}\left[b_0^{(\gamma)} \frac{g(t_k) + g(t_{k-1})}{2} - \sum_{l=1}^{k-1}\left(b_{k-l-1}^{(\gamma)} - b_{k-l}^{(\gamma)}\right)\frac{g(t_l) + g(t_{l-1})}{2}\right.$$

$$\left. - b_{k-1}^{(\gamma)} g(t_0)\right] + \frac{1}{2}\left[R(g(t_k)) + R(g(t_{k-1}))\right]. \tag{6.11}$$

Noticing

$$\frac{g(t_l) + g(t_{l-1})}{2} = \frac{f'(t_l) + f'(t_{l-1})}{2}$$

$$= \frac{f(t_l) - f(t_{l-1})}{\tau} + \frac{\tau^2}{12} f'''(\eta_l), \quad \eta_l \in (t_{l-1}, t_l),$$

and denoting

$$\delta_t f^{l-\frac{1}{2}} = \frac{f(t_l) - f(t_{l-1})}{\tau},$$

it follows from (6.11) that

$$\frac{1}{2}\Big[{}_0^C D_t^\gamma f(t)|_{t=t_k} + {}_0^C D_t^\gamma f(t)|_{t=t_{k-1}}\Big]$$

$$= \frac{\tau^{1-\gamma}}{\Gamma(3-\gamma)}\Big[b_0^{(\gamma)}\delta_t f^{k-\frac{1}{2}} - \sum_{l=1}^{k-1}\big(b_{k-l-1}^{(\gamma)} - b_{k-l}^{(\gamma)}\big)\delta_t f^{l-\frac{1}{2}} - b_{k-1}^{(\gamma)} f'(t_0)\Big]$$

$$+ \frac{\tau^{1-\gamma}}{\Gamma(3-\gamma)}\Big[b_0^{(\gamma)}\frac{\tau^2}{12} f'''(\eta_k) - \sum_{l=1}^{k-1}\big(b_{k-l-1}^{(\gamma)} - b_{k-l}^{(\gamma)}\big)\frac{\tau^2}{12} f'''(\eta_l)\Big]$$

$$+ \frac{1}{2}[R(g(t_k)) + R(g(t_{k-1}))].$$

Let

$$\hat{R}^{k-\frac{1}{2}} = \frac{\tau^{1-\gamma}}{\Gamma(3-\gamma)}\Big[b_0^{(\gamma)}\frac{\tau^2}{12} f'''(\eta_k) - \sum_{l=1}^{k-1}\big(b_{k-l-1}^{(\gamma)} - b_{k-l}^{(\gamma)}\big)\frac{\tau^2}{12} f'''(\eta_l)\Big]$$

$$+ \frac{1}{2}[R(g(t_k)) + R(g(t_{k-1}))].$$

Then we have

$$\big|\hat{R}^{k-\frac{1}{2}}\big| \leqslant \left\{\frac{1}{6\Gamma(3-\gamma)} + \frac{1}{2\Gamma(2-\gamma)}\Big[\frac{1}{4} + \frac{\gamma-1}{(2-\gamma)(3-\gamma)}\Big]\right\} \max_{t_0 \leqslant t \leqslant t_k} \big|f'''(t)\big|\, \tau^{3-\gamma}.$$

$$(6.12)$$

Thus, the following theorem is obtained.

Theorem 6.2 *If $f \in C^3[t_0, t_k]$, then*

$$\frac{1}{2}\Big[{}_0^C D_t^\gamma f(t)|_{t=t_k} + {}_0^C D_t^\gamma f(t)|_{t=t_{k-1}}\Big]$$

$$= \frac{\tau^{1-\gamma}}{\Gamma(3-\gamma)}\Big[b_0^{(\gamma)}\delta_t f^{k-\frac{1}{2}} - \sum_{l=1}^{k-1}\big(b_{k-l-1}^{(\gamma)} - b_{k-l}^{(\gamma)}\big)\delta_t f^{l-\frac{1}{2}} - b_{k-1}^{(\gamma)} f'(t_0)\Big] + \hat{R}^{k-\frac{1}{2}},$$

$$(6.13)$$

where $\hat{R}^{k-\frac{1}{2}}$ satisfies (6.12).

6.3 Difference Methods for Time-Fractional Sub-Diffusion Equations

Consider the following initial-boundary value problem of the time-fractional sub-diffusion equation

$$
\begin{cases}
{}_{0}^{C}D_{t}^{\alpha}u(x,t) = u_{xx}(x,t) + f(x,t), & 0 < x < L, \quad 0 < t \leqslant T, & (6.14a) \\[2mm]
u(x,0) = \varphi(x), & 0 < x < L, & (6.14b) \\[2mm]
u(0,t) = \mu(t), \quad u(L,t) = v(t), & 0 \leqslant t \leqslant T, & (6.14c)
\end{cases}
$$

where $\alpha \in (0,1)$, $f(x,t)$, $\varphi(x)$, $\mu(t)$, and $v(t)$ are all known functions, and $\varphi(0) = \mu(0)$, $\varphi(L) = v(0)$.

Assume that the exact solution $u \in C^{(4,2)}([0,L] \times [0,T])$.

6.3.1 Derivation of the Difference Scheme

Take two positive integers m and n. Let $h = L/m$, $\tau = T/n$. Denote $x_i = ih$ ($0 \leqslant i \leqslant m$), $t_k = k\tau$ ($0 \leqslant k \leqslant n$), $\Omega_h = \{x_i \,|\, 0 \leqslant i \leqslant m\}$, $\Omega_\tau = \{t_k \,|\, 0 \leqslant k \leqslant n\}$, $\Omega_{h\tau} = \Omega_h \times \Omega_\tau$. In addition, denote

$$
s = \tau^{\alpha}\Gamma(2-\alpha), \quad \lambda = \frac{s}{h^2}.
$$

Define the following grid function spaces:

$$
\mathcal{U}_h = \{u \,|\, u = (u_0, u_1, \ldots, u_m)\}, \quad \overset{\circ}{\mathcal{U}}_h = \{u \,|\, u \in \mathcal{U}_h, u_0 = u_m = 0\}.
$$

For an arbitrary grid function $u \in \mathcal{U}_h$, introduce some notation as follows:

$$
\delta_x u_{i-\frac{1}{2}} = \frac{1}{h}(u_i - u_{i-1}), \quad \delta_x^2 u_i = \frac{1}{h^2}(u_{i+1} - 2u_i + u_{i-1}).
$$

Define the grid function

$$
U = \{U_i^k \,|\, 0 \leqslant i \leqslant m, 0 \leqslant k \leqslant n\}
$$

on $\Omega_{h\tau}$, where $U_i^k = u(x_i, t_k)$. In addition, denote $f_i^k = f(x_i, t_k)$.

Considering Eq. (6.14a) at the node point (x_i, t_k), we have

$$
{}_{0}^{C}D_{t}^{\alpha}u(x_i, t_k) = u_{xx}(x_i, t_k) + f_i^k, \quad 1 \leqslant i \leqslant m-1, \quad 1 \leqslant k \leqslant n.
$$

Applying the L1 formula (6.4) to approximate the time-fractional derivative and the second-order central difference quotient to approximate the second spatial derivative, it follows from Theorem 6.1 and Lemma 1.2 that

$$\frac{1}{s}\left[a_0^{(\alpha)}U_i^k - \sum_{l=1}^{k-1}(a_{k-l-1}^{(\alpha)} - a_{k-l}^{(\alpha)})U_i^l - a_{k-1}^{(\alpha)}U_i^0\right] = \delta_x^2 U_i^k + f_i^k + (R_1)_i^k,$$

$$1 \leqslant i \leqslant m-1, \quad 1 \leqslant k \leqslant n, \qquad (6.15)$$

where there is a positive constant c_1 such that

$$|(R_1)_i^k| \leqslant c_1(\tau^{2-\alpha} + h^2), \quad 1 \leqslant i \leqslant m-1, \quad 1 \leqslant k \leqslant n. \qquad (6.16)$$

Noticing the initial-boundary value conditions (6.14b)–(6.14c), we have

$$\begin{cases} U_i^0 = \varphi(x_i), & 1 \leqslant i \leqslant m-1, \\ U_0^k = \mu(t_k), & U_m^k = \nu(t_k), \quad 0 \leqslant k \leqslant n. \end{cases} \qquad (6.17)$$

Omitting the small term $(R_1)_i^k$ in (6.15) and replacing the exact solution U_i^k by its numerical one u_i^k, a difference scheme for solving the problem (6.14) is derived as

$$\begin{cases} \dfrac{1}{s}\left[a_0^{(\alpha)}u_i^k - \sum_{l=1}^{k-1}(a_{k-l-1}^{(\alpha)} - a_{k-l}^{(\alpha)})u_i^l - a_{k-1}^{(\alpha)}u_i^0\right] = \delta_x^2 u_i^k + f_i^k, \\[4mm] \qquad\qquad\qquad\qquad\qquad 1 \leqslant i \leqslant m-1,\ 1 \leqslant k \leqslant n, \quad (6.18a) \\[2mm] u_i^0 = \varphi(x_i), \quad 1 \leqslant i \leqslant m-1, \hfill (6.18b) \\[2mm] u_0^k = \mu(t_k), \quad u_m^k = \nu(t_k), \quad 0 \leqslant k \leqslant n. \hfill (6.18c) \end{cases}$$

In the following, the unique solvability, unconditional stability, and convergence of the difference scheme (6.18) will be examined.

6.3.2   Existence and Uniqueness of the Difference Scheme

Theorem 6.3 *The difference scheme* (6.18) *is uniquely solvable.*

Proof Denote

$$u^k = \left(u_0^k, u_1^k, \ldots, u_{m-1}^k, u_m^k\right).$$

From (6.18b) to (6.18c), the value of u^0 at the 0-th time level is known. Now, suppose the values of $u^0, u^1, \ldots, u^{k-1}$ at the first k-th time levels have been uniquely determined. The system of linear equations in u^k can then be derived from (6.18a) and (6.18c). To establish its unique solvability, it is sufficient to prove that the corresponding homogeneous system

$$
\begin{cases}
\dfrac{1}{s} u_i^k = \delta_x^2 u_i^k, & 1 \leqslant i \leqslant m-1, & \text{(6.19a)} \\[2mm]
u_0^k = u_m^k = 0 & & \text{(6.19b)}
\end{cases}
$$

admits only the trivial solution.

Suppose $\|u^k\|_\infty = |u_{i_k}^k|$, where $i_k \in \{1, 2, \ldots, m-1\}$. Reformulate (6.19a) as

$$
(1 + 2\lambda) u_i^k = \lambda \left(u_{i-1}^k + u_{i+1}^k \right), \quad 1 \leqslant i \leqslant m-1.
$$

Letting $i = i_k$ in the equality above and then taking the absolute value on both sides of the resultant equality, using the triangle inequality, it follows by noticing (6.19b) that

$$
(1 + 2\lambda) \|u^k\|_\infty \leqslant 2\lambda \|u^k\|_\infty,
$$

which implies that $\|u^k\|_\infty = 0$. Hence, $u^k = 0$.

By induction, the conclusion is true. $\qquad\qquad\qquad\qquad\qquad\qquad\qquad\qquad\qquad$ $\square$

6.3.3   Stability of the Difference Solution

Theorem 6.4 *Suppose $\{v_i^k \mid 0 \leqslant i \leqslant m, 0 \leqslant k \leqslant n\}$ is the solution of the difference scheme*

$$
\begin{cases}
\dfrac{1}{s} \left[a_0^{(\alpha)} v_i^k - \displaystyle\sum_{l=1}^{k-1} (a_{k-l-1}^{(\alpha)} - a_{k-l}^{(\alpha)}) v_i^l - a_{k-1}^{(\alpha)} v_i^0 \right] = \delta_x^2 v_i^k + f_i^k, \\[4mm]
\qquad\qquad\qquad\qquad\qquad 1 \leqslant i \leqslant m-1, \quad 1 \leqslant k \leqslant n, & \text{(6.20a)} \\[2mm]
v_i^0 = \varphi(x_i), \quad 1 \leqslant i \leqslant m-1, & \text{(6.20b)} \\[2mm]
v_0^k = 0, \quad v_m^k = 0, \quad 0 \leqslant k \leqslant n. & \text{(6.20c)}
\end{cases}
$$

Then it holds that

$$
\|v^k\|_\infty \leqslant \|v^0\|_\infty + \Gamma(1-\alpha) \max_{1 \leqslant l \leqslant k} \{t_l^\alpha \|f^l\|_\infty\}, \quad 1 \leqslant k \leqslant n,
$$

where

$$\|f^l\|_\infty = \max_{1 \leqslant i \leqslant m-1} |f_i^l|.$$

Proof Rewrite Eq. (6.20a) as

$$a_0^{(\alpha)} v_i^k = \sum_{l=1}^{k-1} (a_{k-l-1}^{(\alpha)} - a_{k-l}^{(\alpha)}) v_i^l + a_{k-1}^{(\alpha)} v_i^0$$

$$+ \lambda(v_{i-1}^k - 2v_i^k + v_{i+1}^k) + s f_i^k, \quad 1 \leqslant i \leqslant m-1, \quad 1 \leqslant k \leqslant n.$$

In other words,

$$(a_0^{(\alpha)} + 2\lambda) v_i^k = \sum_{l=1}^{k-1} (a_{k-l-1}^{(\alpha)} - a_{k-l}^{(\alpha)}) v_i^l + a_{k-1}^{(\alpha)} v_i^0$$

$$+ \lambda(v_{i-1}^k + v_{i+1}^k) + s f_i^k, \quad 1 \leqslant i \leqslant m-1, \quad 1 \leqslant k \leqslant n.$$

Suppose $\|v^k\|_\infty = |v_{i_k}^k|$, where $i_k \in \{1, 2, \ldots, m-1\}$. By setting $i = i_k$ in the equality above and then taking the absolute values of both sides, and applying the triangle inequality, it follows, upon noticing (6.20c), that

$$(a_0^{(\alpha)} + 2\lambda)\|v^k\|_\infty \leqslant \sum_{l=1}^{k-1} \left(a_{k-l-1}^{(\alpha)} - a_{k-l}^{(\alpha)}\right) \|v^l\|_\infty + a_{k-1}^{(\alpha)} \|v^0\|_\infty$$

$$+ 2\lambda\|v^k\|_\infty + s\|f^k\|_\infty, \quad 1 \leqslant k \leqslant n.$$

Thus,

$$a_0^{(\alpha)} \|v^k\|_\infty \leqslant \sum_{l=1}^{k-1} \left(a_{k-l-1}^{(\alpha)} - a_{k-l}^{(\alpha)}\right) \|v^l\|_\infty$$

$$+ a_{k-1}^{(\alpha)} \left(\|v^0\|_\infty + \frac{s}{a_{k-1}^{(\alpha)}} \|f^k\|_\infty\right), \quad 1 \leqslant k \leqslant n.$$

It follows from Lemma 6.1 that

$$\frac{s}{a_{k-1}^{(\alpha)}} \leqslant \frac{\tau^\alpha \Gamma(2-\alpha)}{(1-\alpha)k^{-\alpha}} = t_k^\alpha \Gamma(1-\alpha).$$

Hence,

$$\|v^k\|_\infty \leqslant \sum_{l=1}^{k-1} \left(a_{k-l-1}^{(\alpha)} - a_{k-l}^{(\alpha)}\right) \|v^l\|_\infty$$

$$+a_{k-1}^{(\alpha)} \left(\|v^0\|_\infty + t_k^\alpha \Gamma(1-\alpha)\|f^k\|_\infty\right), \quad 1 \leqslant k \leqslant n. \quad (6.21)$$

The application of induction into (6.21) can easily yield

$$\|v^k\|_\infty \leqslant \|v^0\|_\infty + \Gamma(1-\alpha)\max_{1\leqslant l\leqslant k}\left\{t_l^\alpha\|f^l\|_\infty\right\}, \quad 1 \leqslant k \leqslant n.$$

$\square$

6.3.4　Convergence of the Difference Solution

Theorem 6.5 *Suppose* $\{U_i^k \mid 0 \leqslant i \leqslant m, 0 \leqslant k \leqslant n\}$ *and* $\{u_i^k \mid 0 \leqslant i \leqslant m, 0 \leqslant k \leqslant n\}$ *are solutions of the problem* (6.14) *and the difference scheme* (6.18), *respectively. Denote*

$$e_i^k = U_i^k - u_i^k, \quad 0 \leqslant i \leqslant m, \quad 0 \leqslant k \leqslant n.$$

Then it holds that

$$\|e^k\|_\infty \leqslant c_1 T^\alpha \Gamma(1-\alpha)(\tau^{2-\alpha}+h^2), \quad 1 \leqslant k \leqslant n.$$

Proof Subtracting (6.18) from (6.15) and (6.17), the system of error equations reads

$$\begin{cases} \dfrac{1}{s}\left[a_0^{(\alpha)}e_i^k - \sum_{l=1}^{k-1}(a_{k-l-1}^{(\alpha)} - a_{k-l}^{(\alpha)})e_i^l - a_{k-1}^{(\alpha)}e_i^0\right] = \delta_x^2 e_i^k + (R_1)_i^k, \\[2mm] \qquad\qquad\qquad\qquad 1 \leqslant i \leqslant m-1, \quad 1 \leqslant k \leqslant n, \\[2mm] e_i^0 = 0, \quad 1 \leqslant i \leqslant m-1, \\[2mm] e_0^k = 0, \quad e_m^k = 0, \quad 0 \leqslant k \leqslant n. \end{cases}$$

Noticing (6.16), the application of Theorem 6.4 gives

$$\|e^k\|_\infty \leqslant \|e^0\|_\infty + t_k^\alpha \Gamma(1-\alpha)\max_{1\leqslant l\leqslant k}\|(R_1)^l\|_\infty$$

$$\leqslant t_k^\alpha \Gamma(1-\alpha)c_1(\tau^{2-\alpha}+h^2)$$

$$\leqslant c_1 T^\alpha \Gamma(1-\alpha)(\tau^{2-\alpha}+h^2), \quad 1 \leqslant k \leqslant n.$$

$\square$

6.3.5 *Numerical Examples*

Example 6.1 Use the L1 scheme (6.18) to compute the problem

$$
\begin{cases}
{}^{C}_{0}D^{\alpha}_{t}u(x,t) = u_{xx}(x,t) + e^{x}t^{4}\left[\dfrac{\Gamma(5+\alpha)}{24} - t^{\alpha}\right], & 0 < x < 1, \quad 0 < t \leqslant 1, \\[2mm]
u(x,0) = 0, \quad 0 \leqslant x \leqslant 1, \\[2mm]
u(0,t) = t^{4+\alpha}, \quad u(1,t) = e \cdot t^{4+\alpha}, \quad 0 < t \leqslant 1.
\end{cases}
$$

The exact solution of this problem is $u(x,t) = e^{x}t^{4+\alpha}$.

For different values of $\alpha = 0.1,\ 0.5,\ 0.9$, Table 6.1 records the maximum error

$$
E_{\infty}(h,\tau) = \max_{0\leqslant i\leqslant m,\,0\leqslant k\leqslant n}|u(x_{i},t_{k}) - u^{k}_{i}|
$$

of numerical solutions with the fixed sufficient small spatial step size $h = 1/20000$ and varying temporal step sizes.

It can be observed from Table 6.1 that the numerical accuracy achieves an order of $2 - \alpha$ in time when the spatial step size is sufficiently small. Additionally, for $\alpha = 0.9$, the error surfaces of the numerical solutions with different step sizes are presented in Fig. 6.1.

Table 6.1 (Example 6.1)
The maximum errors
$E_{\infty}(h,\tau)$ and numerical
accuracy of the solution in
time with different τ

α	τ	$E_{\infty}(h,\tau)$	$\log_{2}\dfrac{E_{\infty}(h,2\tau)}{E_{\infty}(h,\tau)}$
0.9	1/640	1.123e−3	
	1/1280	5.243e−4	1.0985
	1/2560	2.447e−4	1.0992
	1/5120	1.142e−4	1.0996
	1/10240	5.328e−5	1.0999
0.5	1/640	4.018e−5	
	1/1280	1.431e−5	1.4894
	1/2560	5.085e−6	1.4926
	1/5120	1.804e−6	1.4949
	1/10240	6.395e−7	1.4963
0.1	1/640	6.443e−7	
	1/1280	1.824e−7	1.8209
	1/2560	5.134e−8	1.8288
	1/5120	1.441e−8	1.8330
	1/10240	3.991e−9	1.8522

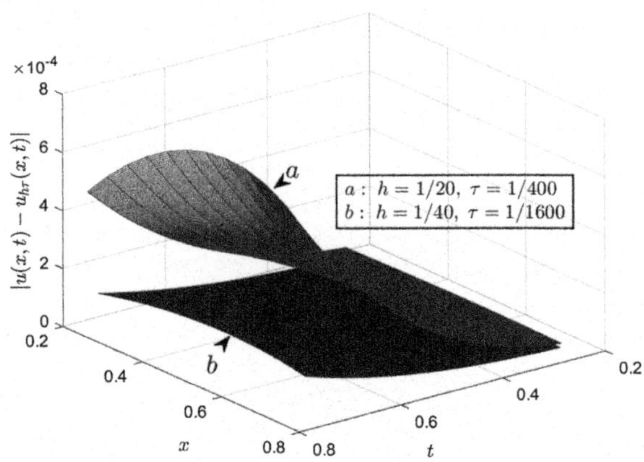

Fig. 6.1 (Example 6.1) The numerical error surfaces with different step sizes ($\alpha = 0.9$)

6.4   Difference Methods for Time-Fractional Wave Equations

Consider the following initial-boundary value problem of the time-fractional wave equation:

$$
\begin{cases}
{}_0^C D_t^\gamma u(x,t) = u_{xx}(x,t) + f(x,t), & 0 < x < L, \quad 0 < t \leqslant T, & \text{(6.22a)} \\[2mm]
u(x,0) = \varphi(x), \quad u_t(x,0) = \psi(x), & 0 < x < L, & \text{(6.22b)} \\[2mm]
u(0,t) = \mu(t), \quad u(L,t) = v(t), & 0 \leqslant t \leqslant T, & \text{(6.22c)}
\end{cases}
$$

where $\gamma \in (1,2)$, $f(x,t), \varphi(x), \psi(x), \mu(t)$ and $v(t)$ are given functions, and $\varphi(0) = \mu(0)$, $\varphi(L) = v(0)$, $\psi(0) = \mu'(0)$, $\psi(L) = v'(0)$. Assume that the exact solution $u \in C^{(4,3)}([0,L] \times [0,T])$.

For any grid function $v = \{v_i^k \mid 0 \leqslant i \leqslant m, \ 0 \leqslant k \leqslant n\}$ defined on $\Omega_{h\tau}$, define

$$
v_i^{k-\frac{1}{2}} = \frac{1}{2}\left(v_i^k + v_i^{k-1}\right), \qquad \delta_t v_i^{k-\frac{1}{2}} = \frac{1}{\tau}\left(v_i^k - v_i^{k-1}\right).
$$

6.4.1   Derivation of the Difference Scheme

Define the grid function

$$
U = \{U_i^k \mid 0 \leqslant i \leqslant m, \ 0 \leqslant k \leqslant n\}
$$

on $\Omega_{h\tau}$ with $U_i^k = u(x_i, t_k)$. In addition, denote

$$
f_i^k = f(x_i, t_k), \qquad f_i^{k-\frac{1}{2}} = \frac{1}{2}\left(f_i^k + f_i^{k-1}\right), \qquad \psi_i = \psi(x_i), \qquad \eta = \tau^{\gamma-1}\Gamma(3-\gamma).
$$

Considering Eq. (6.22a) at the node point (x_i, t_k), we have

$$_0^C D_t^\gamma u(x_i, t_k) = u_{xx}(x_i, t_k) + f_i^k, \quad 1 \leqslant i \leqslant m - 1, \quad 0 \leqslant k \leqslant n.$$

Taking an average at two adjacent time levels arrives at

$$\frac{1}{2}\left[{}_0^C D_t^\gamma u(x_i, t_k) + {}_0^C D_t^\gamma u(x_i, t_{k-1})\right] = \frac{1}{2}\left[u_{xx}(x_i, t_k) + u_{xx}(x_i, t_{k-1})\right] + f_i^{k-\frac{1}{2}},$$

$$1 \leqslant i \leqslant m - 1, \quad 1 \leqslant k \leqslant n.$$

By applying the L1 formula (6.13) to approximate the time-fractional derivative and the second-order central difference quotient for the second spatial derivative, it follows from Theorem 6.2 and Lemma 1.2 that

$$\frac{1}{\eta}\left[b_0^{(\gamma)}\delta_t U_i^{k-\frac{1}{2}} - \sum_{l=1}^{k-1}\left(b_{k-l-1}^{(\gamma)} - b_{k-l}^{(\gamma)}\right)\delta_t U_i^{l-\frac{1}{2}} - b_{k-1}^{(\gamma)}\psi_i\right]$$

$$= \delta_x^2 U_i^{k-\frac{1}{2}} + f_i^{k-\frac{1}{2}} + (R_2)_i^k, \quad 1 \leqslant i \leqslant m - 1, \quad 1 \leqslant k \leqslant n, \quad (6.23)$$

where there is a positive constant c_2 such that

$$|(R_2)_i^k| \leqslant c_2(\tau^{3-\gamma} + h^2), \quad 1 \leqslant i \leqslant m - 1, \quad 1 \leqslant k \leqslant n, \quad (6.24)$$

with $\{b_l^{(\gamma)}\}$ defined by (6.8).

Noticing the initial-boundary value conditions (6.22b)–(6.22c), we obtain

$$\begin{cases} U_i^0 = \varphi(x_i), & 1 \leqslant i \leqslant m - 1, \\ U_0^k = \mu(t_k), \quad U_m^k = v(t_k), & 0 \leqslant k \leqslant n. \end{cases} \quad (6.25)$$

Neglecting the small term $(R_2)_i^k$ in (6.23) and replacing the exact solution U_i^k by its numerical one u_i^k, a difference scheme for solving the problem (6.22) is developed as

$$\begin{cases} \dfrac{1}{\eta}\left[b_0^{(\gamma)}\delta_t u_i^{k-\frac{1}{2}} - \displaystyle\sum_{l=1}^{k-1}(b_{k-l-1}^{(\gamma)} - b_{k-l}^{(\gamma)})\delta_t u_i^{l-\frac{1}{2}} - b_{k-1}^{(\gamma)}\psi_i\right] = \delta_x^2 u_i^{k-\frac{1}{2}} + f_i^{k-\frac{1}{2}}, \\ \qquad\qquad\qquad\qquad 1 \leqslant i \leqslant m - 1, \quad 1 \leqslant k \leqslant n, \qquad\qquad (6.26a) \\ u_i^0 = \varphi(x_i), \quad 1 \leqslant i \leqslant m - 1, \qquad\qquad\qquad\qquad\qquad\qquad (6.26b) \\ u_0^k = \mu(t_k), \quad u_m^k = v(t_k), \quad 0 \leqslant k \leqslant n. \qquad\qquad\qquad\quad (6.26c) \end{cases}$$

6.4.2   Existence and Uniqueness of the Difference Scheme

Theorem 6.6 *The difference scheme* (6.26) *is uniquely solvable.*

Proof Denote

$$u^k = \left(u_0^k, u_1^k, \ldots, u_{m-1}^k, u_m^k \right).$$

From (6.26b) to (6.26c), the value of u^0 at the 0-th time level is given. Now, suppose the values of $u^0, u^1, \ldots, u^{k-1}$ at the first k-th time levels have been uniquely determined. The system of linear equations in u^k can then be derived from (6.26a) and (6.26c). To establish its unique solvability, it suffices to prove that the corresponding homogeneous system

$$\begin{cases} \dfrac{1}{\eta\tau} u_i^k = \dfrac{1}{2}\delta_x^2 u_i^k, & 1 \leqslant i \leqslant m-1, & (6.27a) \\[2ex] u_0^k = u_m^k = 0 & & (6.27b) \end{cases}$$

has only the trivial solution.

Taking the inner product on both sides of (6.27a) with u^k and using (6.27b), we obtain

$$\frac{1}{\eta\tau} \|u^k\|^2 = \frac{1}{2}\left(\delta_x^2 u^k, u^k \right) = -\frac{1}{2}\|\delta_x u^k\|^2 \leqslant 0,$$

which implies that $\|u^k\| = 0$. Consequently, $u^k = 0$ follows from the combination with (6.27b).

By induction, the difference scheme (6.26) is uniquely solvable. $\square$

6.4.3   Stability of the Difference Solution

Theorem 6.7 *Suppose* $\{v_i^k \,|\, 0 \leqslant i \leqslant m, 0 \leqslant k \leqslant n\}$ *is the solution of the difference scheme*

$$\begin{cases} \dfrac{1}{\eta}\left[b_0^{(\gamma)}\delta_t v_i^{k-\frac{1}{2}} - \sum_{l=1}^{k-1} (b_{k-l-1}^{(\gamma)} - b_{k-l}^{(\gamma)})\delta_t v_i^{l-\frac{1}{2}} - b_{k-1}^{(\gamma)}\psi_i \right] = \delta_x^2 v_i^{k-\frac{1}{2}} + f_i^{k-\frac{1}{2}}, \\[1ex] \hspace{5cm} 1 \leqslant i \leqslant m-1, \quad 1 \leqslant k \leqslant n, \quad (6.28a) \\[1ex] v_i^0 = \varphi(x_i), \hspace{1.5cm} 1 \leqslant i \leqslant m-1, \hspace{3cm} (6.28b) \\[1ex] v_0^k = 0, \quad v_m^k = 0, \quad 0 \leqslant k \leqslant n. \hspace{3.2cm} (6.28c) \end{cases}$$

Then it holds that

$$\|\delta_x v^k\|^2 \leqslant \|\delta_x v^0\|^2 + \frac{t_k^{2-\gamma}}{\Gamma(3-\gamma)}\|\psi\|^2 + \Gamma(2-\gamma)t_k^{\gamma-1}\cdot\tau\sum_{l=1}^{k}\|f^{l-\frac{1}{2}}\|^2, \quad 1\leqslant k\leqslant n,$$

$$(6.29)$$

where

$$\|\psi\|^2 = h\sum_{i=1}^{m-1}\psi_i^2, \quad \|f^{l-\frac{1}{2}}\|^2 = h\sum_{i=1}^{m-1}\left(f_i^{l-\frac{1}{2}}\right)^2.$$

Proof Taking the inner product on both sides of (6.28a) with $\eta\delta_t v^{k-\frac{1}{2}}$ yields

$$b_0^{(\gamma)}\|\delta_t v^{k-\frac{1}{2}}\|^2 = \sum_{l=1}^{k-1}\left(b_{k-l-1}^{(\gamma)} - b_{k-l}^{(\gamma)}\right)\left(\delta_t v^{l-\frac{1}{2}}, \delta_t v^{k-\frac{1}{2}}\right)$$

$$+ b_{k-1}^{(\gamma)}(\psi, \delta_t v^{k-\frac{1}{2}}) + \eta\left(\delta_x^2 v^{k-\frac{1}{2}}, \delta_t v^{k-\frac{1}{2}}\right)$$

$$+ \eta\left(f^{k-\frac{1}{2}}, \delta_t v^{k-\frac{1}{2}}\right), \quad 1\leqslant k\leqslant n. \tag{6.30}$$

Noticing (6.28c), it follows from the summation by parts that

$$\left(\delta_x^2 v^{k-\frac{1}{2}}, \delta_t v^{k-\frac{1}{2}}\right) = -\left(\delta_x v^{k-\frac{1}{2}}, \delta_x\delta_t v^{k-\frac{1}{2}}\right) = -\frac{1}{2\tau}\left(\|\delta_x v^k\|^2 - \|\delta_x v^{k-1}\|^2\right).$$

$$(6.31)$$

Substituting (6.31) into (6.30) and using the Cauchy-Schwarz inequality arrive at

$$b_0^{(\gamma)}\|\delta_t v^{k-\frac{1}{2}}\|^2 + \frac{\eta}{2\tau}\left(\|\delta_x v^k\|^2 - \|\delta_x v^{k-1}\|^2\right)$$

$$= \sum_{l=1}^{k-1}(b_{k-l-1}^{(\gamma)} - b_{k-l}^{(\gamma)})(\delta_t v^{l-\frac{1}{2}}, \delta_t v^{k-\frac{1}{2}}) + b_{k-1}^{(\gamma)}(\psi, \delta_t v^{k-\frac{1}{2}}) + \eta\left(f^{k-\frac{1}{2}}, \delta_t v^{k-\frac{1}{2}}\right)$$

$$\leqslant \frac{1}{2}\sum_{l=1}^{k-1}(b_{k-l-1}^{(\gamma)} - b_{k-l}^{(\gamma)})\left(\|\delta_t v^{l-\frac{1}{2}}\|^2 + \|\delta_t v^{k-\frac{1}{2}}\|^2\right)$$

$$+ \frac{1}{2}b_{k-1}^{(\gamma)}\left(\|\psi\|^2 + \|\delta_t v^{k-\frac{1}{2}}\|^2\right) + \eta\left(f^{k-\frac{1}{2}}, \delta_t v^{k-\frac{1}{2}}\right),$$

namely,

$$b_0^{(\gamma)}\|\delta_t v^{k-\frac{1}{2}}\|^2 + \frac{\eta}{\tau}(\|\delta_x v^k\|^2 - \|\delta_x v^{k-1}\|^2)$$

$$\leqslant \sum_{l=1}^{k-1}(b_{k-l-1}^{(\gamma)} - b_{k-l}^{(\gamma)})\|\delta_t v^{l-\frac{1}{2}}\|^2 + b_{k-1}^{(\gamma)}\|\psi\|^2 + 2\eta\big(f^{k-\frac{1}{2}}, \delta_t v^{k-\frac{1}{2}}\big), \quad 1 \leqslant k \leqslant n.$$

The inequality above can be further reformulated as

$$\|\delta_x v^k\|^2 + \frac{\tau}{\eta}\sum_{l=1}^{k} b_{k-l}^{(\gamma)}\|\delta_t v^{l-\frac{1}{2}}\|^2$$

$$\leqslant \|\delta_x v^{k-1}\|^2 + \frac{\tau}{\eta}\sum_{l=1}^{k-1} b_{k-l-1}^{(\gamma)}\|\delta_t v^{l-\frac{1}{2}}\|^2$$

$$+ \frac{\tau}{\eta}b_{k-1}^{(\gamma)}\|\psi\|^2 + 2\tau\big(f^{k-\frac{1}{2}}, \delta_t v^{k-\frac{1}{2}}\big), \quad 1 \leqslant k \leqslant n.$$

Let

$$F^0 = \|\delta_x v^0\|^2, \quad F^k = \|\delta_x v^k\|^2 + \frac{\tau}{\eta}\sum_{l=1}^{k} b_{k-l}^{(\gamma)}\|\delta_t v^{l-\frac{1}{2}}\|^2, \quad k \geqslant 1.$$

Then

$$F^k \leqslant F^{k-1} + \frac{\tau}{\eta}b_{k-1}^{(\gamma)}\|\psi\|^2 + 2\tau\big(f^{k-\frac{1}{2}}, \delta_t v^{k-\frac{1}{2}}\big), \quad 1 \leqslant k \leqslant n.$$

In the inequality above, replacing k by l and summing over l from 1 to k give

$$F^k \leqslant F^0 + \frac{\tau}{\eta}\sum_{l=1}^{k} b_{l-1}^{(\gamma)}\|\psi\|^2 + 2\tau\sum_{l=1}^{k}\big(f^{l-\frac{1}{2}}, \delta_t v^{l-\frac{1}{2}}\big)$$

$$\leqslant F^0 + \frac{\tau}{\eta}\sum_{l=0}^{k-1} b_l^{(\gamma)}\|\psi\|^2 + \tau\sum_{l=1}^{k}\left(\frac{\eta}{b_{k-l}^{(\gamma)}}\|f^{l-\frac{1}{2}}\|^2 + \frac{b_{k-l}^{(\gamma)}}{\eta}\|\delta_t v^{l-\frac{1}{2}}\|^2\right), \quad 1 \leqslant k \leqslant n.$$

Thus,

$$\|\delta_x v^k\|^2 \leqslant \|\delta_x v^0\|^2 + \frac{\tau}{\eta}\sum_{l=0}^{k-1} b_l^{(\gamma)}\|\psi\|^2 + \tau\sum_{l=1}^{k}\frac{\eta}{b_{k-l}^{(\gamma)}}\|f^{l-\frac{1}{2}}\|^2, \quad 1 \leqslant k \leqslant n.$$

$$(6.32)$$

It follows from the definition of $\{b_l^{(\gamma)}\}$ and Lemma 6.1 that

$$(2-\gamma)(l+1)^{1-\gamma} < b_l^{(\gamma)} < (2-\gamma)l^{1-\gamma},$$

and then

$$b_{k-l}^{(\gamma)} > (2-\gamma)(k-l+1)^{1-\gamma} \geqslant (2-\gamma)k^{1-\gamma}, \quad 1 \leqslant l \leqslant k.$$

Moreover, we have

$$\frac{\eta}{b_{k-l}^{(\gamma)}} \leqslant \frac{\tau^{\gamma-1}\Gamma(3-\gamma)}{(2-\gamma)k^{1-\gamma}} = \Gamma(2-\gamma)(k\tau)^{\gamma-1} = \Gamma(2-\gamma)t_k^{\gamma-1},$$

which leads to

$$\tau \sum_{l=1}^{k} \frac{\eta}{b_{k-l}^{(\gamma)}} \| f^{l-\frac{1}{2}} \|^2 \leqslant \Gamma(2-\gamma)t_k^{\gamma-1} \cdot \tau \sum_{l=1}^{k} \| f^{l-\frac{1}{2}} \|^2. \tag{6.33}$$

In addition, it follows from the definition of $b_l^{(\gamma)}$ that

$$\frac{\tau}{\eta} \sum_{l=0}^{k-1} b_l^{(\gamma)} = \frac{\tau}{\tau^{\gamma-1}\Gamma(3-\gamma)} \sum_{l=0}^{k-1} [(l+1)^{2-\gamma} - l^{2-\gamma}]$$

$$= \frac{\tau^{2-\gamma}}{\Gamma(3-\gamma)} k^{2-\gamma} = \frac{t_k^{2-\gamma}}{\Gamma(3-\gamma)}. \tag{6.34}$$

Substituting (6.33) and (6.34) into (6.32) yields (6.29). $\square$

6.4.4 Convergence of the Difference Solution

Theorem 6.8 *Suppose $\{U_i^k \mid 0 \leqslant i \leqslant m, 0 \leqslant k \leqslant n\}$ and $\{u_i^k \mid 0 \leqslant i \leqslant m, 0 \leqslant k \leqslant n\}$ are solutions of the problem (6.22) and the difference scheme (6.26), respectively. Denote*

$$e_i^k = U_i^k - u_i^k, \quad 0 \leqslant i \leqslant m, \quad 0 \leqslant k \leqslant n.$$

Then it holds that

$$\| e^k \|_\infty \leqslant \frac{c_2 L}{2} \sqrt{T^\gamma \Gamma(2-\gamma)}(\tau^{3-\gamma} + h^2), \quad 1 \leqslant k \leqslant n.$$

Proof Subtracting (6.26) from (6.23) and (6.25), the system of error equations reads

$$\begin{cases} \dfrac{1}{\eta}\left[b_0^{(\gamma)}\delta_t e_i^{k-\frac{1}{2}} - \displaystyle\sum_{l=1}^{k-1}(b_{k-l-1}^{(\gamma)} - b_{k-l}^{(\gamma)})\delta_t e_i^{l-\frac{1}{2}} - b_{k-1}^{(\gamma)}\cdot 0\right] = \delta_x^2 e_i^{k-\frac{1}{2}} + (R_2)_i^k, \\[2mm] \qquad\qquad\qquad\qquad\qquad\qquad 1\leqslant i\leqslant m-1, \quad 1\leqslant k\leqslant n, \\[2mm] e_i^0 = 0, \quad 1\leqslant i\leqslant m-1, \\[2mm] e_0^k = 0, \quad e_m^k = 0, \quad 0\leqslant k\leqslant n. \end{cases}$$

Noticing (6.24), it follows by using Theorem 6.7 that

$$\|\delta_x e^k\|^2 \leqslant t_k^{\gamma-1}\Gamma(2-\gamma)\tau\sum_{l=1}^{k}\|(R_2)^l\|^2$$

$$\leqslant T^\gamma\Gamma(2-\gamma)Lc_2^2(\tau^{3-\gamma}+h^2)^2, \quad 1\leqslant k\leqslant n.$$

Taking the square root on both sides of the inequality above and noticing Lemma 1.4 produce

$$\|e^k\|_\infty \leqslant \frac{\sqrt{L}}{2}\|\delta_x e^k\| \leqslant \frac{c_2 L}{2}\sqrt{T^\gamma\Gamma(2-\gamma)}(\tau^{3-\gamma}+h^2), \quad 1\leqslant k\leqslant n.$$

$$\square$$

6.4.5 Numerical Examples

Example 6.2 Use the difference scheme (6.26) to compute the problem

$$\begin{cases} {}_0^C D_t^\gamma u(x,t) = u_{xx}(x,t) + e^x t^4\left[\dfrac{\Gamma(5+\gamma)}{24} - t^\gamma\right], \quad 0 < x < 1, \quad 0 < t \leqslant 1, \\[2mm] u(x,0) = 0, \quad u_t(x,0) = 0, \quad 0\leqslant x\leqslant 1, \\[2mm] u(0,t) = t^{4+\gamma}, \quad u(1,t) = e\cdot t^{4+\gamma}, \quad 0 < t \leqslant 1. \end{cases}$$

The exact solution of this problem is $u(x,t) = e^x t^{4+\gamma}$.

For different values of $\gamma = 1.1,\ 1.5,\ 1.9$ and the fixed sufficient small spatial step size $h = 1/20000$, Table 6.2 lists the maximum error

$$E_\infty(h,\tau) = \max_{0\leqslant i\leqslant m, 0\leqslant k\leqslant n}|u(x_i,t_k) - u_i^k|$$

Table 6.2 (Example 6.2) The maximum errors $E_\infty(h, \tau)$ and numerical accuracy of the solution in time with different τ

γ	τ	$E_\infty(h, \tau)$	$\log_2 \frac{E_\infty(h, 2\tau)}{E_\infty(h, \tau)}$
1.9	1/640	3.846e−3	
	1/1280	1.797e−3	1.0980
	1/2560	8.390e−4	1.0989
	1/5120	3.916e−4	1.0994
	1/10240	1.827e−4	1.0997
1.5	1/640	1.730e−4	
	1/1280	6.181e−5	1.4847
	1/2560	2.202e−5	1.4894
	1/5120	7.823e−6	1.4927
	1/10240	2.775e−6	1.4951
1.1	1/640	2.617e−6	
	1/1280	7.481e−7	1.8065
	1/2560	2.107e−7	1.8279
	1/5120	5.712e−8	1.8832
	1/10240	1.395e−8	2.0336

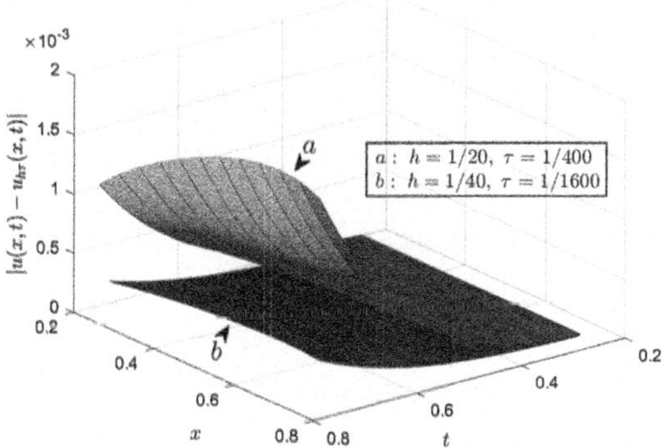

Fig. 6.2 (Example 6.2) The numerical error surfaces with different step sizes ($\gamma = 1.9$)

of numerical solutions with the varying temporal step sizes. It can be seen from Table 6.2 that the $(3 - \gamma)$-th order accuracy can be obtained in time. For $\gamma = 1.9$, Fig. 6.2 illustrates the error surfaces of numerical solutions with different grid step sizes.

6.5 Difference Methods for Time-Fractional Mixed Diffusion and Wave Equations

Consider the difference method for solving the following initial-boundary problem of the time-fractional mixed diffusion and wave equation

$$
\begin{cases}
{}_0^C D_t^\gamma u(x,t) + {}_0^C D_t^\alpha u(x,t) = u_{xx}(x,t) + f(x,t), \\
\qquad\qquad\qquad\qquad\qquad\qquad 0 < x < L,\ 0 < t \leqslant T, & \text{(6.35a)} \\
u(x,0) = \varphi(x), \quad u_t(x,0) = \psi(x), \quad 0 < x < L, & \text{(6.35b)} \\
u(0,t) = \mu(t), \quad u(L,t) = v(t), \quad 0 \leqslant t \leqslant T, & \text{(6.35c)}
\end{cases}
$$

where $f(x,t), \varphi(x), \psi(x), \mu(t), v(t)$ are all given functions, and $\varphi(0) = \mu(0)$, $\varphi(L) = v(0)$, $\psi(0) = \mu'(0)$, $\psi(L) = v'(0)$, ${}_0^C D_t^\alpha u(x,t)$ and ${}_0^C D_t^\gamma u(x,t)$ are Caputo fractional derivatives of $u(x,t)$ with respect to t:

$$
{}_0^C D_t^\alpha u(x,t) = \frac{1}{\Gamma(1-\alpha)} \int_0^t \frac{\partial u(x,s)}{\partial s}\, \frac{\mathrm{d}s}{(t-s)^\alpha}, \quad 0 < \alpha < 1,
$$

$$
{}_0^C D_t^\gamma u(x,t) = \frac{1}{\Gamma(2-\gamma)} \int_0^t \frac{\partial^2 u(x,s)}{\partial s^2}\, \frac{\mathrm{d}s}{(t-s)^{\gamma-1}}, \quad 1 < \gamma < 2.
$$

Assume that the exact solution $u \in C^{(4,3)}([0,L] \times [0,T])$.

6.5.1 *Derivation of the Difference Scheme*

Define the grid function

$$
U = \{U_i^k \,|\, 0 \leqslant i \leqslant m,\ 0 \leqslant k \leqslant n\}
$$

on $\Omega_{h\tau}$, where $U_i^k = u(x_i, t_k)$. In addition, denote

$$
f_i^k = f(x_i, t_k), \quad f_i^{k-\frac{1}{2}} = \frac{1}{2}(f_i^k + f_i^{k-1}), \quad \psi_i = \psi(x_i),
$$

$$
s_\gamma = \tau^{\gamma-1}\Gamma(3-\gamma), \quad s_\alpha = \tau^{\alpha-1}\Gamma(2-\alpha).
$$

Considering Eq. (6.35a) at the node points (x_i, t_k) and (x_i, t_{k-1}), respectively, and taking the average give

$$\frac{1}{2}\left[{}_{0}^{C}D_t^{\gamma}u(x_i, t_k) + {}_{0}^{C}D_t^{\gamma}u(x_i, t_{k-1})\right] + \frac{1}{2}\left[{}_{0}^{C}D_t^{\alpha}u(x_i, t_k) + {}_{0}^{C}D_t^{\alpha}u(x_i, t_{k-1})\right]$$

$$= \frac{1}{2}\left[u_{xx}(x_i, t_k) + u_{xx}(x_i, t_{k-1})\right] + f_i^{k-\frac{1}{2}}, \quad 1 \leqslant i \leqslant m-1, \quad 1 \leqslant k \leqslant n.$$

$$(6.36)$$

Applying Theorems 6.1 and 6.2, we have

$${}_{0}^{C}D_t^{\alpha}u(x_i, t_k)$$

$$= \frac{\tau^{-\alpha}}{\Gamma(2-\alpha)}\left[a_0^{(\alpha)}U_i^k - \sum_{l=1}^{k-1}(a_{k-l-1}^{(\alpha)} - a_{k-l}^{(\alpha)})U_i^l - a_{k-1}^{(\alpha)}U_i^0\right] + O(\tau^{2-\alpha})$$

$$= \frac{\tau^{1-\alpha}}{\Gamma(2-\alpha)}\sum_{l=1}^{k}a_{k-l}^{(\alpha)}\delta_t U_i^{l-\frac{1}{2}} + O(\tau^{2-\alpha}), \quad 1 \leqslant i \leqslant m-1, \quad 0 \leqslant k \leqslant n$$

$$(6.37)$$

and

$$\frac{1}{2}\left[{}_{0}^{C}D_t^{\gamma}u(x_i, t_k) + {}_{0}^{C}D_t^{\gamma}u(x_i, t_{k-1})\right]$$

$$= \frac{\tau^{1-\gamma}}{\Gamma(3-\gamma)}\left[b_0^{(\gamma)}\delta_t U_i^{k-\frac{1}{2}} - \sum_{l=1}^{k-1}(b_{k-l-1}^{(\gamma)} - b_{k-l}^{(\gamma)})\delta_t U_i^{l-\frac{1}{2}} - b_{k-1}^{(\gamma)}u_t(x_i, t_0)\right]$$

$$+ O(\tau^{3-\gamma}), \quad 1 \leqslant i \leqslant m-1, \quad 1 \leqslant k \leqslant n. \tag{6.38}$$

Substituting (6.37) and (6.38) into (6.36) yields

$$\frac{\tau^{1-\gamma}}{\Gamma(3-\gamma)}\left[b_0^{(\gamma)}\delta_t U_i^{k-\frac{1}{2}} - \sum_{l=1}^{k-1}(b_{k-l-1}^{(\gamma)} - b_{k-l}^{(\gamma)})\delta_t U_i^{l-\frac{1}{2}} - b_{k-1}^{(\gamma)}\psi_i\right]$$

$$+ \frac{\tau^{1-\alpha}}{\Gamma(2-\alpha)}\left[\frac{a_0^{(\alpha)}}{2}\delta_t U_i^{k-\frac{1}{2}} + \sum_{l=1}^{k-1}\frac{a_{k-l}^{(\alpha)} + a_{k-l-1}^{(\alpha)}}{2}\delta_t U_i^{l-\frac{1}{2}}\right]$$

$$= \delta_x^2 U_i^{k-\frac{1}{2}} + f_i^{k-\frac{1}{2}} + (R_3)_i^k, \quad 1 \leqslant i \leqslant m-1, \quad 1 \leqslant k \leqslant n, \tag{6.39}$$

where there exists a positive constant c_3 such that

$$\left|(R_3)_i^k\right| \leqslant c_3(\tau^{\min\{2-\alpha, 3-\gamma\}} + h^2), \quad 1 \leqslant i \leqslant m-1, \quad 1 \leqslant k \leqslant n. \tag{6.40}$$

Noticing the initial-boundary value conditions

$$
\begin{cases}
U_i^0 = \varphi(x_i), & 1 \leqslant i \leqslant m-1, \\
U_0^k = \mu(t_k), & U_m^k = v(t_k), \quad 0 \leqslant k \leqslant n,
\end{cases}
\tag{6.41}
$$

omitting the small term $(R_3)_i^k$ in (6.39), and replacing the exact solution U_i^k with its numerical one u_i^k, it arrives at a difference scheme for solving the problem (6.35) as follows:

$$
\begin{cases}
\dfrac{1}{s_\gamma}\left[b_0^{(\gamma)} \delta_t u_i^{k-\frac{1}{2}} - \displaystyle\sum_{l=1}^{k-1} (b_{k-l-1}^{(\gamma)} - b_{k-l}^{(\gamma)})\delta_t u_i^{l-\frac{1}{2}} - b_{k-1}^{(\gamma)} \psi_i \right] \\[2ex]
\quad + \dfrac{1}{s_\alpha}\left[\dfrac{a_0^{(\alpha)}}{2} \delta_t u_i^{k-\frac{1}{2}} + \displaystyle\sum_{l=1}^{k-1} \dfrac{a_{k-l}^{(\alpha)} + a_{k-1-l}^{(\alpha)}}{2} \delta_t u_i^{l-\frac{1}{2}} \right] = \delta_x^2 u_i^{k-\frac{1}{2}} + f_i^{k-\frac{1}{2}}, \\[2ex]
\hspace{6cm} 1 \leqslant i \leqslant m-1, \quad 1 \leqslant k \leqslant n, \hfill \text{(6.42a)} \\[1ex]
u_i^0 = \varphi(x_i), \quad 1 \leqslant i \leqslant m-1, \hfill \text{(6.42b)} \\[1ex]
u_0^k = \mu(t_k), \quad u_m^k = v(t_k), \quad 0 \leqslant k \leqslant n. \hfill \text{(6.42c)}
\end{cases}
$$

6.5.2 Existence and Uniqueness of the Difference Scheme

Theorem 6.9 *The difference scheme* (6.42) *is uniquely solvable.*

Proof Denote

$$
u^k = \left(u_0^k, u_1^k, \ldots, u_{m-1}^k, u_m^k \right).
$$

The value of u^0 at the 0-th time level is determined by (6.42b)–(6.42c).

Now suppose the values of $u^0, u^1, \ldots, u^{k-1}$ at the first k-th time levels have been uniquely determined, and then from (6.42a) and (6.42c), we can get the system of linear equations in u^k. To verify its unique solvability, it suffices to show that the corresponding homogeneous one

$$
\begin{cases}
\left(\dfrac{b_0^{(\gamma)}}{s_\gamma} + \dfrac{a_0^{(\alpha)}}{2s_\alpha} \right) \dfrac{1}{\tau} u_i^k - \dfrac{1}{2}\delta_x^2 u_i^k = 0, & 1 \leqslant i \leqslant m-1, \hfill \text{(6.43a)} \\[2ex]
u_0^k = 0, \quad u_m^k = 0 \hfill \text{(6.43b)}
\end{cases}
$$

has only the trivial solution.

Taking the inner product on both sides of (6.43a) with u^k and using the summation by parts, it follows by noticing (6.43b) that

$$\left(\frac{b_0^{(\gamma)}}{s_\gamma} + \frac{a_0^{(\alpha)}}{2s_\alpha}\right)\frac{1}{\tau}\|u^k\|^2 + \frac{1}{2}\|\delta_x u^k\|^2 = 0.$$

Thus, $u^k = 0$.

By induction, the difference scheme (6.42) has only one solution. $\square$

6.5.3   Stability of the Difference Solution

For the analysis of the stability and convergence of the difference scheme (6.42), three lemmas are provided in advance.

Lemma 6.2 *Suppose that $\{b_l^{(\gamma)}\}$ is defined by (6.8). Then for any positive integer p and any grid functions $\psi, V_1, V_2, \ldots, V_p \in \overset{\circ}{\mathcal{U}}_h$, it holds that*

$$\sum_{k=1}^{p}\left(b_0^{(\gamma)}V^k - \sum_{l=1}^{k-1}(b_{k-l-1}^{(\gamma)} - b_{k-l}^{(\gamma)})V^l - b_{k-1}^{(\gamma)}\psi, V^k\right)$$

$$\geq \frac{1}{2}\left(\sum_{l=1}^{p}b_{p-l}^{(\gamma)}\|V^l\|^2 - \sum_{l=1}^{p}b_{l-1}^{(\gamma)}\|\psi\|^2\right).$$

Proof Some direct calculations show that

$$\sum_{k=1}^{p}\left(b_0^{(\gamma)}V^k - \sum_{l=1}^{k-1}(b_{k-l-1}^{(\gamma)} - b_{k-l}^{(\gamma)})V^l - b_{k-1}^{(\gamma)}\psi, V^k\right)$$

$$= \sum_{k=1}^{p}\left[b_0^{(\gamma)}\|V^k\|^2 - \sum_{l=1}^{k-1}(b_{k-l-1}^{(\gamma)} - b_{k-l}^{(\gamma)})(V^l, V^k) - b_{k-1}^{(\gamma)}(\psi, V^k)\right]$$

$$\geq \sum_{k=1}^{p}\left[b_0^{(\gamma)}\|V^k\|^2 - \frac{1}{2}\sum_{l=1}^{k-1}(b_{k-l-1}^{(\gamma)} - b_{k-l}^{(\gamma)})(\|V^l\|^2 + \|V^k\|^2)\right.$$

$$\left. - \frac{1}{2}b_{k-1}^{(\gamma)}(\|\psi\|^2 + \|V^k\|^2)\right]$$

$$= \frac{1}{2} \sum_{k=1}^{p} \left(\sum_{l=1}^{k} b_{k-l}^{(\gamma)} \|V^l\|^2 - \sum_{l=1}^{k-1} b_{k-l-1}^{(\gamma)} \|V^l\|^2 - b_{k-1}^{(\gamma)} \|\psi\|^2 \right)$$

$$= \frac{1}{2} \left(\sum_{l=1}^{p} b_{p-l}^{(\gamma)} \|V^l\|^2 - \sum_{l=1}^{p} b_{l-1}^{(\gamma)} \|\psi\|^2 \right).$$

$\square$

Lemma 6.3 ([5]) *Suppose that* $\{g_0, g_1, \ldots, g_l, \ldots\}$ *is a real sequence satisfying*

$$g_l \geqslant 0, \quad g_l - g_{l-1} \leqslant 0, \quad g_{l+1} - 2g_l + g_{l-1} \geqslant 0,$$

and then for any positive integer p *and any grid functions* $V_1, V_2, \ldots, V_p \in \overset{\circ}{\mathcal{U}}_h$*, it holds that*

$$\sum_{k=1}^{p} \left(\sum_{q=0}^{k-1} g_q V_{k-q}, V_k \right) \geqslant 0.$$

Lemma 6.4 *Suppose that* $\{a_l^{(\alpha)}\}$ *is defined by* (6.1). *Then for any positive integer* p *and any grid functions* $V_1, V_2, \ldots, V_p \in \overset{\circ}{\mathcal{U}}_h$*, it holds that*

$$\sum_{k=1}^{p} \left(\frac{a_0^{(\alpha)}}{2} V^k + \sum_{l=1}^{k-1} \frac{a_{k-l}^{(\alpha)} + a_{k-1-l}^{(\alpha)}}{2} V^l, V^k \right) \geqslant -a_0^{(\alpha)} \sum_{k=1}^{p} \|V^k\|^2.$$

Proof Rewrite the left-hand side of the inequality above as

$$\sum_{k=1}^{p} \left(\frac{a_0^{(\alpha)}}{2} V^k + \sum_{l=1}^{k-1} \frac{a_{k-l}^{(\alpha)} + a_{k-1-l}^{(\alpha)}}{2} V^l, V^k \right)$$

$$= \sum_{k=1}^{p} \left(\frac{3}{2} a_0^{(\alpha)} V^k + \sum_{l=1}^{k-1} \frac{a_{k-l}^{(\alpha)} + a_{k-l-1}^{(\alpha)}}{2} V^l, V^k \right) - \sum_{k=1}^{p} a_0^{(\alpha)} \|V^k\|^2.$$

It is straightforward to verify that the coefficient of the first term on the right-hand side of the above equality satisfies the conditions of Lemma 6.3 and is therefore nonnegative. Consequently, the result follows. $\square$

For the solution of the difference scheme (6.42), we have the following estimate.

Theorem 6.10 *Suppose that* $\{u_i^k \mid 0 \leqslant i \leqslant m, \ 0 \leqslant k \leqslant n\}$ *is the solution of the difference scheme* (6.42) *with* $\mu(t) \equiv 0$ *and* $v(t) \equiv 0$. *Denote* $\tau_0 =$

$\left(\frac{T^{1-\gamma}\Gamma(2-\alpha)}{4\Gamma(2-\gamma)}\right)^{1/(1-\alpha)}$. *If $\tau \leqslant \tau_0$, then it holds that*

$$\|\delta_x u^k\|^2 \leqslant \|\delta_x u^0\|^2 + \frac{T^{2-\gamma}}{\Gamma(3-\gamma)}\|\psi\|^2 + 2\Gamma(2-\gamma)T^{\gamma-1}\tau\sum_{l=1}^{k}\|f^{l-\frac{1}{2}}\|^2, \quad 1 \leqslant k \leqslant n.$$

Proof Taking the inner product on both sides of (6.42a) with $\delta_t u^{k-\frac{1}{2}}$ and summing over k from 1 to p yield

$$\frac{1}{s_\gamma}\sum_{k=1}^{p}\left[b_0^{(\gamma)}\|\delta_t u^{k-\frac{1}{2}}\|^2 - \sum_{l=1}^{k-1}(b_{k-l-1}^{(\gamma)} - b_{k-l}^{(\gamma)})(\delta_t u^{l-\frac{1}{2}}, \delta_t u^{k-\frac{1}{2}})\right.$$

$$\left. - b_{k-1}^{(\gamma)}(\psi, \delta_t u^{k-\frac{1}{2}})\right]$$

$$+ \frac{1}{s_\alpha}\sum_{k=1}^{p}\left(\frac{a_0^{(\alpha)}}{2}\delta_t u^{k-\frac{1}{2}} + \sum_{l=1}^{k-1}\frac{a_{k-l}^{(\alpha)} + a_{k-l-1}^{(\alpha)}}{2}\delta_t u^{l-\frac{1}{2}}, \delta_t u^{k-\frac{1}{2}}\right)$$

$$+ \frac{1}{2\tau}\left(\|\delta_x u^p\|^2 - \|\delta_x u^0\|^2\right)$$

$$= \sum_{k=1}^{p}\left(f^{k-\frac{1}{2}}, \delta_t u^{k-\frac{1}{2}}\right), \quad 1 \leqslant p \leqslant n.$$

It follows by using Lemmas 6.2 and 6.4 that

$$\frac{1}{2s_\gamma}\left(\sum_{l=1}^{p}b_{p-l}^{(\gamma)}\|\delta_t u^{l-\frac{1}{2}}\|^2 - \sum_{k=1}^{p}b_{k-1}^{(\gamma)}\|\psi\|^2\right) - \frac{1}{s_\alpha}\sum_{k=1}^{p}\|\delta_t u^{k-\frac{1}{2}}\|^2$$

$$+ \frac{1}{2\tau}\left(\|\delta_x u^p\|^2 - \|\delta_x u^0\|^2\right)$$

$$\leqslant \sum_{k=1}^{p}\left(f^{k-\frac{1}{2}}, \delta_t u^{k-\frac{1}{2}}\right), \quad 1 \leqslant p \leqslant n.$$

Moreover, we have

$$\frac{1}{4s_\gamma}\sum_{l=1}^{p}b_{p-l}^{(\gamma)}\|\delta_t u^{l-\frac{1}{2}}\|^2 + \left(\frac{1}{4s_\gamma}\sum_{l=1}^{p}b_{p-l}^{(\gamma)}\|\delta_t u^{l-\frac{1}{2}}\|^2 - \frac{1}{s_\alpha}\sum_{l=1}^{p}\|\delta_t u^{l-\frac{1}{2}}\|^2\right)$$

$$+ \frac{1}{2\tau}\left(\|\delta_x u^p\|^2 - \|\delta_x u^0\|^2\right)$$

$$\leqslant \frac{1}{2s_\gamma}\sum_{k=1}^{p}b_{k-1}^{(\gamma)}\|\psi\|^2 + \sum_{k=1}^{p}\left(f^{k-\frac{1}{2}}, \delta_t u^{k-\frac{1}{2}}\right), \quad 1 \leqslant p \leqslant n. \tag{6.44}$$

For the second term on the left-hand side of (6.44), by noticing

$$b_l^{(\gamma)} > (2-\gamma)(l+1)^{1-\gamma} \geqslant (2-\gamma)n^{1-\gamma}, \quad 0 \leqslant l \leqslant n-1,$$

we get

$$\frac{1}{4s_\gamma} \sum_{l=1}^{p} b_{p-l}^{(\gamma)} \|\delta_t u^{l-\frac{1}{2}}\|^2 - \frac{1}{s_\alpha} \sum_{l=1}^{p} \|\delta_t u^{l-\frac{1}{2}}\|^2$$

$$\geqslant \frac{T^{1-\gamma}}{4\Gamma(2-\gamma)} \sum_{k=1}^{p} \|\delta_t u^{k-\frac{1}{2}}\|^2 - \frac{1}{s_\alpha} \sum_{l=1}^{p} \|\delta_t u^{l-\frac{1}{2}}\|^2$$

$$= \left(\frac{T^{1-\gamma}}{4\Gamma(2-\gamma)} - \frac{\tau^{1-\alpha}}{\Gamma(2-\alpha)}\right) \sum_{l=1}^{p} \|\delta_t u^{l-\frac{1}{2}}\|^2, \quad 1 \leqslant p \leqslant n. \tag{6.45}$$

Hence, when $\tau \leqslant \tau_0$, the right-hand side of (6.45) is nonnegative. It follows from (6.44) that

$$\frac{T^{1-\gamma}}{4\Gamma(2-\gamma)} \sum_{l=1}^{p} \|\delta_t u^{l-\frac{1}{2}}\|^2 + \frac{1}{2\tau} \left(\|\delta_x u^p\|^2 - \|\delta_x u^0\|^2\right)$$

$$\leqslant \frac{1}{2s_\gamma} \sum_{k=1}^{p} b_{k-1}^{(\gamma)} \|\psi\|^2 + \sum_{k=1}^{p} \left(f^{k-\frac{1}{2}}, \delta_t u^{k-\frac{1}{2}}\right)$$

$$= \frac{1}{2s_\gamma} p^{2-\gamma} \|\psi\|^2 + \sum_{k=1}^{p} \left(f^{k-\frac{1}{2}}, \delta_t u^{k-\frac{1}{2}}\right)$$

$$\leqslant \frac{1}{2s_\gamma} p^{2-\gamma} \|\psi\|^2 + \sum_{k=1}^{p} \left(\frac{T^{1-\gamma}}{4\Gamma(2-\gamma)} \|\delta_t u^{k-\frac{1}{2}}\|^2 + \Gamma(2-\gamma)T^{\gamma-1}\|f^{k-\frac{1}{2}}\|^2\right),$$

$$1 \leqslant p \leqslant n.$$

Therefore, it is easy to obtain

$$\|\delta_x u^p\|^2$$

$$\leqslant \|\delta_x u^0\|^2 + \frac{\tau^{2-\gamma}}{\Gamma(3-\gamma)} p^{2-\gamma} \|\psi\|^2 + 2\Gamma(2-\gamma)T^{\gamma-1}\tau \sum_{k=1}^{p} \|f^{k-\frac{1}{2}}\|^2$$

$$\leqslant \|\delta_x u^0\|^2 + \frac{T^{2-\gamma}}{\Gamma(3-\gamma)} \|\psi\|^2 + 2\Gamma(2-\gamma)T^{\gamma-1}\tau \sum_{k=1}^{p} \|f^{k-\frac{1}{2}}\|^2, \quad 1 \leqslant p \leqslant n.$$

$$\square$$

6.5.4 Convergence of the Difference Solution

From Theorem 6.10, it is easy to get the following convergence result.

Theorem 6.11 *Suppose $\{U_i^k \mid 0 \leqslant i \leqslant m, 0 \leqslant k \leqslant n\}$ and $\{u_i^k \mid 0 \leqslant i \leqslant m,\ 0 \leqslant k \leqslant n\}$ are solutions of the problem (6.35) and the difference scheme (6.42), respectively. Denote*

$$e_i^k = U_i^k - u_i^k, \quad 0 \leqslant i \leqslant m, \quad 0 \leqslant k \leqslant n.$$

Then when $\tau \leqslant \tau_0$, it holds that

$$\|e^k\|_\infty \leqslant \frac{L}{2}\sqrt{2\Gamma(2-\gamma)T^\gamma L}\, c_3\big(\tau^{\min\{2-\alpha,3-\gamma\}} + h^2\big), \quad 1 \leqslant k \leqslant n,$$

where c_3 and τ_0 are defined by (6.40) and Theorem 6.10, respectively.

Proof Subtracting (6.42) from (6.39) and (6.41), the system of error equations is produced as

$$
\begin{cases}
\dfrac{1}{s_\gamma}\Big(b_0^{(\gamma)}\delta_t e_i^{k-\frac{1}{2}} - \displaystyle\sum_{l=1}^{k-1}(b_{k-l-1}^{(\gamma)} - b_{k-l}^{(\gamma)})\delta_t e_i^{l-\frac{1}{2}} - b_{k-1}^{(\gamma)}\cdot 0\Big) \\[2mm]
\quad +\dfrac{1}{s_\alpha}\Big(\dfrac{a_0^{(\alpha)}}{2}\delta_t e_i^{k-\frac{1}{2}} + \displaystyle\sum_{l=1}^{k-1}\dfrac{a_{k-l}^{(\alpha)} + a_{k-1-l}^{(\alpha)}}{2}\delta_t e_i^{l-\frac{1}{2}}\Big) = \delta_x^2 e_i^{k-\frac{1}{2}} + (R_3)_i^k, \\[2mm]
\hspace{5cm} 1 \leqslant i \leqslant m-1, \quad 1 \leqslant k \leqslant n, \\[2mm]
e_i^0 = 0, \quad 1 \leqslant i \leqslant m-1, \\[2mm]
e_0^k = 0, \quad e_m^k = 0, \quad 0 \leqslant k \leqslant n.
\end{cases}
$$

By Theorem 6.10, it follows by noticing (6.40) that

$$
\begin{aligned}
\|\delta_x e^k\|^2 &\leqslant 2\Gamma(2-\gamma)T^{\gamma-1}\tau\sum_{l=1}^{k}\|(R_3)^l\|^2 \\[2mm]
&\leqslant 2\Gamma(2-\gamma)T^{\gamma-1}k\tau L\Big[c_3\big(\tau^{\min\{2-\alpha,3-\gamma\}} + h^2\big)\Big]^2 \\[2mm]
&\leqslant 2\Gamma(2-\gamma)T^\gamma L c_3^2\big(\tau^{\min\{2-\alpha,3-\gamma\}} + h^2\big)^2, \quad 1 \leqslant k \leqslant n.
\end{aligned}
$$

The desired result follows by taking the square root of both sides of the inequality above and applying Lemma 1.4. $\qquad\square$

6.5.5   Numerical Examples

Example 6.3   Use the difference scheme (6.42) to compute the problem

$$
\begin{cases}
{}_0^C D_t^\gamma u(x,t) + {}_0^C D_t^\alpha u(x,t) = u_{xx}(x,t) + e^x t^4 \Bigg[\dfrac{\Gamma(5+\gamma)}{24} \\
\qquad + \dfrac{\Gamma(5+\alpha)}{24} + \dfrac{\Gamma(5+\gamma)}{\Gamma(5+\gamma-\alpha)} t^{\gamma-\alpha} + \dfrac{\Gamma(5+\alpha)}{\Gamma(5+\alpha-\gamma)} t^{\alpha-\gamma} - t^\gamma - t^\alpha \Bigg], \\
\qquad\qquad\qquad\qquad 0 < x < 1, \quad 0 < t \leqslant 1, \\
u(x,0) = 0, \quad u_t(x,0) = 0, \quad 0 \leqslant x \leqslant 1, \\
u(0,t) = t^{4+\gamma} + t^{4+\alpha}, \quad u(1,t) = e \cdot (t^{4+\gamma} + t^{4+\alpha}), \quad 0 < t \leqslant 1.
\end{cases}
$$

The exact solution of this problem is $u(x,t) = e^x(t^{4+\gamma} + t^{4+\alpha})$.

For different values of (γ, α) and temporal step size, when $h = 1/10000$ is fixed, Table 6.3 records the maximum errors of solutions defined by

$$
E_\infty(h,\tau) = \max_{0 \leqslant i \leqslant m, 0 \leqslant k \leqslant n} |u(x_i, t_k) - u_i^k|.
$$

It can be observed that the temporal order $\min\{3 - \gamma, 2 - \alpha\}$ is achievable when the spatial step size is sufficiently small. Figure 6.3 illustrates the error surfaces of the numerical solutions for $\gamma = 1.9$ and $\alpha = 0.9$ with different step sizes.

6.6   Summary and Extension

In [3, 4, 9], various techniques have been employed to demonstrate the accuracy of order $2 - \alpha$ for the L1 formula. The proof presented in Sect. 6.2 is based on [3, 4]. The contents of Sects. 6.3 and 6.4 are derived from [9], while the material in Sect. 6.5 is taken from [8]. For the case where the fractional derivative is approximated using the L1 formula and the spatial derivative is approximated using the compact difference scheme, the relevant work can be found in [2]. For the initial-boundary value problems of time-fractional sub-diffusion and wave equations, the L2-1_σ method is also applicable, as described in [1, 6]. Readers interested in difference methods for fractional differential equations are encouraged to refer to [7].

For the problem (6.14), if the solution $u \in C^{(2,1)}([0, L] \times [0, T])$, then we have $\lim\limits_{t \to 0^+} {}_0^C D_t^\alpha u(x,t) = 0$. Let $t \to 0^+$ on both sides of (6.14a). The following necessary condition between $\varphi(x)$ and $f(x,0)$ should be satisfied:

$$
\begin{cases}
-\varphi''(x) = f(x,0), \quad 0 < x < L, & \text{(6.46a)} \\
\varphi(0) = \mu(0), \quad \varphi(L) = \nu(0). & \text{(6.46b)}
\end{cases}
$$

Table 6.3 (Example 6.3) The maximum errors $E_\infty(h, \tau)$ and numerical accuracy of the solution in time with different τ

(γ, α)	τ	$E_\infty(h, \tau)$	$\log_2 \frac{E_\infty(h, 2\tau)}{E_\infty(h, \tau)}$
$(1.9, 0.1)$	1/2560	1.204e$-$3	
	1/5120	5.620e$-$4	1.0995
	1/10240	2.622e$-$4	1.0997
$(1.9, 0.5)$	1/2560	1.252e$-$3	
	1/5120	5.835e$-$4	1.1011
	1/10240	2.720e$-$4	1.1010
$(1.9, 0.9)$	1/2560	1.508e$-$3	
	1/5120	7.037e$-$4	1.0994
	1/10240	3.284e$-$4	1.0997
$(1.5, 0.1)$	1/2560	2.913e$-$5	
	1/5120	1.035e$-$5	1.4925
	1/10240	3.670e$-$6	1.4962
$(1.5, 0.5)$	1/2560	3.864e$-$5	
	1/5120	1.374e$-$5	1.4918
	1/10240	4.875e$-$6	1.4946
$(1.5, 0.9)$	1/2560	3.752e$-$4	
	1/5120	1.716e$-$4	1.1282
	1/10240	7.886e$-$5	1.1220
$(1.1, 0.1)$	1/2560	2.934e$-$7	
	1/5120	8.128e$-$8	1.8519
	1/10240	1.919e$-$8	2.0827
$(1.1, 0.5)$	1/2560	9.072e$-$6	
	1/5120	3.211e$-$6	1.4985
	1/10240	1.134e$-$6	1.5018
$(1.1, 0.9)$	1/2560	3.848e$-$4	
	1/5120	1.796e$-$4	1.0996
	1/10240	8.377e$-$5	1.0999

Fig. 6.3 (Example 6.3) The numerical error surfaces with different step sizes ($\gamma = 1.9$, $\alpha = 0.9$)

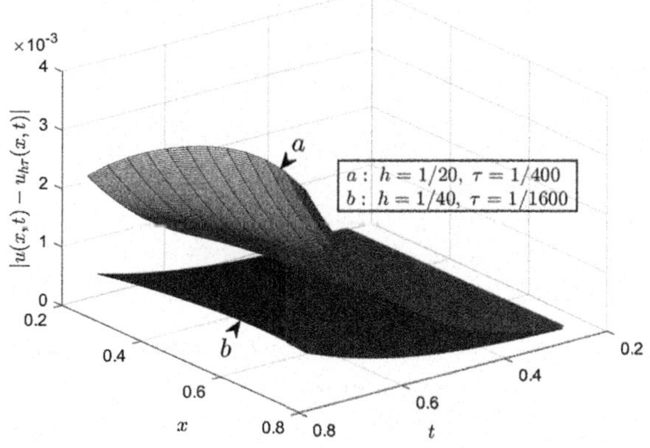

Hence, for a given $f(x, 0)$, the value of $\varphi(x)$ is uniquely determined. Conversely, the value of $f(x, 0)$ is also determined by $\varphi(x)$. That is, $\varphi(x)$ and $f(x, 0)$ are not independent.

For the problem (6.22), if the solution $u \in C^{(2,2)}([0, L] \times [0, T])$, then we have $\lim\limits_{t \to 0^+} {}_0^C D_t^\gamma u(x, t) = 0$. Let $t \to 0^+$ in (6.22a). The necessary condition (6.46) between $\varphi(x)$ and $f(x, 0)$ should be satisfied. In addition, if $u \in C^{(2,3)}([0, L] \times [0, T])$, then $u_{tt}(x, 0) = 0$, and the following constraint condition should be satisfied:

$$\begin{cases} -\psi''(x) = f_t(x, 0), & 0 < x < L, & \text{(6.47a)} \\ \psi(0) = \mu_t(0), & \psi(L) = \nu_t(0). & \text{(6.47b)} \end{cases}$$

For the problem (6.35), if the solution $u \in C^{(2,2)}([0, L] \times [0, T])$, then we have $\lim\limits_{t \to 0^+} {}_0^C D_t^\alpha u(x, t) = 0$, $\lim\limits_{t \to 0^+} {}_0^C D_t^\gamma u(x, t) = 0$. Let $t \to 0^+$ in (6.35a). The same necessary condition (6.46) should be satisfied. In addition, if $u \in C^{(2,3)}([0, L] \times [0, T])$, then $u_{tt}(x, 0) = 0$ and (6.47) should be satisfied.

6.7   Exercise

6.1   Compute ${}_a\mathbf{D}_t^\alpha (t - a)^p$ and ${}_a^C D_t^\alpha (t - a)^p$ by definition.

6.2   Consider the numerical approximation of

$$\left. {}_0^C D_t^\gamma f(t) \right|_{t=t_{k-\frac{1}{2}}} = \frac{1}{\Gamma(2 - \gamma)} \int_{t_0}^{t_{k-\frac{1}{2}}} f''(t)(t_{k-\frac{1}{2}} - t)^{1-\gamma} dt, \quad 1 \leq k \leq n.$$

Rewrite it as the sum on some small intervals:

$$\left. {}_0^C D_t^\gamma f(t) \right|_{t=t_{k-\frac{1}{2}}} = \frac{1}{\Gamma(2 - \gamma)} \Bigg[\int_{t_0}^{t_{\frac{1}{2}}} f''(t)(t_{k-\frac{1}{2}} - t)^{1-\gamma} dt$$

$$+ \sum_{l=1}^{k-1} \int_{t_{l-\frac{1}{2}}}^{t_{l+\frac{1}{2}}} f''(t)(t_{k-\frac{1}{2}} - t)^{1-\gamma} dt \Bigg].$$

Make the quadratic Hermite interpolation polynomial of the function $f(t)$ based on the points $(t_0, f(t_0))$, $(t_0, f'(t_0))$, $(t_1, f(t_1))$, one has

$$H_{2,0}(t) = f(t_0) + f'(t_0)(t - t_0) + \frac{1}{\tau}\left(\delta_t f^{\frac{1}{2}} - f'(t_0)\right)(t - t_0)^2.$$

It is easily known that

$$H_{2,0}''(t) = \frac{2}{\tau}\left(\delta_t f^{\frac{1}{2}} - f'(t_0)\right). \tag{6.48}$$

Make the quadratic Newton interpolation polynomial of the function $f(t)$ based on three points $(t_{k-1}, f(t_{k-1}))$, $(t_k, f(t_k))$, $(t_{k+1}, f(t_{k+1}))$, and one has

$$N_{2,k}(t) = f(t_{k-1}) + \left(\delta_t f^{k-\frac{1}{2}}\right)(t - t_{k-1}) + \frac{1}{2}\left(\delta_t^2 f^k\right)(t - t_{k-1})(t - t_k),$$

where

$$\delta_t^2 f^k = \frac{1}{\tau}\left(\delta_t f^{k+\frac{1}{2}} - \delta_t f^{k-\frac{1}{2}}\right).$$

It is easily known that

$$N_{2,k}''(t) = \delta_t^2 f^k. \tag{6.49}$$

It follows from (6.48) and (6.49) that

$$\begin{aligned}
&{}^C_0 D_t^\gamma f(t)|_{t=t_{k-\frac{1}{2}}} \\
&\approx \frac{1}{\Gamma(2-\gamma)}\left[\int_{t_0}^{t_{\frac{1}{2}}} H_{2,0}''(t)(t_{k-\frac{1}{2}} - t)^{1-\gamma}\,\mathrm{d}t\right. \\
&\qquad\left. + \sum_{l=1}^{k-1}\int_{t_{l-\frac{1}{2}}}^{t_{l+\frac{1}{2}}} N_{2,l}''(t)(t_{k-\frac{1}{2}} - t)^{1-\gamma}\,\mathrm{d}t\right] \\
&= \frac{1}{\Gamma(2-\gamma)}\left[\int_{t_0}^{t_{\frac{1}{2}}} \frac{2}{\tau}\left(\delta_t f^{\frac{1}{2}} - f'(t_0)\right)(t_{k-\frac{1}{2}} - t)^{1-\gamma}\,\mathrm{d}t\right. \\
&\qquad\left. + \sum_{l=1}^{k-1}\int_{t_{l-\frac{1}{2}}}^{t_{l+\frac{1}{2}}} \left(\delta_t^2 f^l\right)(t_{k-\frac{1}{2}} - t)^{1-\gamma}\,\mathrm{d}t\right] \\
&= \frac{1}{\Gamma(2-\gamma)}\left[\hat{b}_0^{(k,\gamma)}\delta_t f^{k-\frac{1}{2}} - \sum_{l=1}^{k-1}\left(\hat{b}_{k-l-1}^{(k,\gamma)} - \hat{b}_{k-l}^{(k,\gamma)}\right)\delta_t f^{l-\frac{1}{2}} - \hat{b}_{k-1}^{(k,\gamma)} f'(t_0)\right].
\end{aligned} \tag{6.50}$$

(1) Show the expression of $\hat{b}_l^{(k,\gamma)}$ $(0 \leqslant l \leqslant k-1)$.
(2) Show that the coefficients in (6.50) satisfy

$$\hat{b}_0^{(k,\gamma)} > \hat{b}_1^{(k,\gamma)} > \hat{b}_2^{(k,\gamma)} > \cdots > \hat{b}_{k-1}^{(k,\gamma)} > 0.$$

6.3 For the initial-boundary value problem (6.14) of the time-fractional sub-diffusion equation, consider the spatially compact difference scheme as follows:

$$
\begin{cases}
\dfrac{\tau^{-\alpha}}{\Gamma(2-\alpha)} \mathcal{A}\left[a_0^{(\alpha)} u_i^k - \sum_{l=1}^{k-1} \left(a_{k-l-1}^{(\alpha)} - a_{k-l}^{(\alpha)} \right) u_i^l - a_{k-1}^{(\alpha)} u_i^0 \right] = \delta_x^2 u_i^k + \mathcal{A} f_i^k, \\[2ex]
\qquad\qquad\qquad\qquad\qquad\qquad\qquad 1 \leqslant i \leqslant m-1, \quad 1 \leqslant k \leqslant n, \\[2ex]
u_i^0 = \varphi(x_i), \quad 1 \leqslant i \leqslant m-1, \\[1ex]
u_0^k = \mu(t_k), \quad u_m^k = v(t_k), \quad 0 \leqslant k \leqslant n.
\end{cases}
$$

(1) Show the expression of local truncation errors of the difference scheme.
(2) Show the unique solvability of the difference scheme.
(3) Show the stability of the difference solution with respect to the initial value and the function on the right-hand side.
(4) Show the convergence of the difference solution.

6.4 For the initial-boundary value problem (6.22) of the time-fractional wave equation, consider the spatially compact difference scheme as follows:

$$
\begin{cases}
\dfrac{\tau^{1-\gamma}}{\Gamma(3-\gamma)} \mathcal{A}\left[b_0^{(\gamma)} \delta_t u_i^{k-\frac{1}{2}} - \sum_{l=1}^{k-1} \left(b_{k-l-1}^{(\gamma)} - b_{k-l}^{(\gamma)} \right) \delta_t u_i^{l-\frac{1}{2}} - b_{k-1}^{(\gamma)} \psi_i \right] \\[2ex]
\qquad\qquad\qquad = \delta_x^2 u_i^{k-\frac{1}{2}} + \mathcal{A} f_i^{k-\frac{1}{2}}, \quad 1 \leqslant i \leqslant m-1, \quad 1 \leqslant k \leqslant n, \\[2ex]
u_i^0 = \varphi(x_i), \quad 1 \leqslant i \leqslant m-1, \\[1ex]
u_0^k = \mu(t_k), \quad u_m^k = v(t_k), \quad 0 \leqslant k \leqslant n.
\end{cases}
$$

(1) Show the expression of local truncation errors of the difference scheme.
(2) Show the unique solvability of the difference scheme.
(3) Show the stability of the difference solution with respect to the initial value and the function on the right-hand side.
(4) Show the convergence of the difference solution.

References

1. Alikhanov, A.A.: A new difference scheme for the time fractional diffusion equation. J. Comput. Phys. **280**, 424–438 (2015)
2. Gao, G.H., Sun, Z.Z.: A compact finite difference scheme for the fractional sub-diffusion equations. J. Comput. Phys. **230**(3), 586–595 (2015)
3. Gao, G.H., Sun, Z.Z., Zhang, H.W.: A new fractional numerical differentiation formula to approximate the Caputo fractional derivative and its applications. J. Comput. Phys. **259**, 33–50 (2014)
4. Li, M., Xiong, X.T., Wang, Y.J.: A numerical evaluation and regularization of Caputo fractional derivatives. J. Phys.: Conf. Ser. **290**, 012011 (2011)
5. Lopez-Marcos, J.: A difference scheme for a nonlinear partial integrodifferential equation. SIAM J. Numer. Anal. **27**, 20–31 (1990)
6. Sun, H., Sun, Z.Z., Gao, G.H.: Some temporal second order difference schemes for fractional wave equations. Numer. Methods Partial Differ. Equations **32**, 970–1001 (2016)
7. Sun, Z.Z., Gao, G.H.: Fractional Differential Equations: Finite Difference Methods. De Gruyter, Berlin (2023)
8. Sun, Z.Z., Ji, C.C., Du, R.L.: A new analytical technique of the L-type difference schemes for time fractional mixed sub-diffusion and diffusion-wave equations. Appl. Math. Lett. **102**, 106115 (2020)
9. Sun, Z.Z., Wu, X.: A fully discrete difference scheme for a diffusion-wave system. Appl. Numer. Math. **56**, 193–209 (2006)

Chapter 7
Finite Difference Methods for the Schrödinger Equation

The Schrödinger equation forms the foundation of modern quantum mechanics, revealing the fundamental laws governing the motion of matter at the microscopic scale. Its significance in quantum mechanics is comparable to that of Newton's three laws in classical mechanics and Maxwell's equations in electromagnetism. The Schrödinger equation also plays a crucial role in various important applications, including plasma physics, nonlinear photonics, water waves, and bimolecular dynamics, among others.

7.1　The Schrödinger Equation

Consider the following initial-boundary value problem of the Schrödinger equation:

$$
\begin{cases}
\mathrm{i}u_t + u_{xx} + q|u|^2 u = 0, & 0 < x < L,\, 0 < t \leqslant T, & (7.1\text{a}) \\[2mm]
u(x,0) = \varphi(x), & 0 < x < L, & (7.1\text{b}) \\[2mm]
u(0,t) = 0, \quad u(L,t) = 0, & 0 \leqslant t \leqslant T, & (7.1\text{c})
\end{cases}
$$

where q is a real constant, $\mathrm{i} = \sqrt{-1}$ is the imaginary unit, $\varphi(x)$ is a complex function, and $u(x,t)$ is the unknown complex function. The solution to the problem (7.1) satisfies the following two conservation laws.

Theorem 7.1 *Suppose $u(x,t)$ is the solution of the problem (7.1). Denote*

$$
Q(t) = \int_0^L |u(x,t)|^2 \mathrm{d}x, \quad E(t) = \int_0^L \left(|u_x(x,t)|^2 - \frac{q}{2}|u(x,t)|^4 \right) \mathrm{d}x.
$$

© Science Press 2026
Z.-Z. Sun et al., *Numerical Solutions to Partial Differential Equations
with Finite Difference Methods*, Springer Asia Pacific Mathematics Series 9,
https://doi.org/10.1007/978-981-95-5563-5_7

Then we have

$$Q(t) = Q(0), \quad 0 < t \leqslant T, \tag{7.2}$$

$$E(t) = E(0), \quad 0 < t \leqslant T. \tag{7.3}$$

Proof

(I) Multiplying Eq. (7.1a) on both sides by $\bar{u}(x, t)$ and integrating the result with respect to x from 0 to L, we have

$$i \int_0^L u_t(x, t)\bar{u}(x, t)\mathrm{d}x + \int_0^L u_{xx}(x, t)\bar{u}(x, t)\mathrm{d}x + q \int_0^L |u(x, t)|^4 \mathrm{d}x = 0. \tag{7.4}$$

Using the boundary value condition (7.1c), we have

$$\int_0^L u_{xx}(x, t)\bar{u}(x, t)\mathrm{d}x = \int_0^L \left[(u_x(x, t)\bar{u}(x, t))_x - u_x(x, t)\bar{u}_x(x, t) \right]\mathrm{d}x$$

$$= - \int_0^L |u_x(x, t)|^2 \mathrm{d}x.$$

Taking the imaginary parts on both sides of (7.4), it follows by noticing

$$\mathrm{Re}\{u_t(x, t)\bar{u}(x, t)\} = \left(\frac{1}{2}|u(x, t)|^2 \right)_t,$$

that

$$\frac{\mathrm{d}}{\mathrm{d}t} \int_0^L \frac{1}{2}|u(x, t)|^2 \mathrm{d}x = 0.$$

Therefore, (7.2) is valid.

(II) Multiplying Eq. (7.1a) on both sides by $-\bar{u}_t(x, t)$ and integrating the result with respect to x from 0 to L, we have

$$-i \int_0^L |u_t(x, t)|^2 \mathrm{d}x - \int_0^L u_{xx}(x, t)\bar{u}_t(x, t)\mathrm{d}x$$

$$- q \int_0^L |u(x, t)|^2 u(x, t)\bar{u}_t(x, t)\mathrm{d}x = 0. \tag{7.5}$$

It follows from the summation by parts and (7.1c) that

$$
-\int_0^L u_{xx}(x,t)\bar{u}_t(x,t)\mathrm{d}x = -\int_0^L [(u_x(x,t)\bar{u}_t(x,t))_x - u_x(x,t)\bar{u}_{xt}(x,t)]\mathrm{d}x
$$

$$
= \int_0^L u_x(x,t)\bar{u}_{xt}(x,t)\mathrm{d}x.
$$

Taking the real parts on both sides of the above equality gives

$$
-\mathrm{Re}\left\{ \int_0^L u_{xx}(x,t)\bar{u}_t(x,t)\mathrm{d}x \right\}
$$

$$
= \int_0^L \frac{\partial}{\partial t}\left(\frac{1}{2}|u_x(x,t)|^2 \right)\mathrm{d}x = \frac{1}{2}\frac{\mathrm{d}}{\mathrm{d}t}\int_0^L |u_x(x,t)|^2\mathrm{d}x.
$$

Taking the real parts on both sides of (7.5) and using the above result, it yields

$$
\frac{1}{2}\frac{\mathrm{d}}{\mathrm{d}t}\int_0^L |u_x(x,t)|^2\mathrm{d}x - \frac{q}{4}\frac{\mathrm{d}}{\mathrm{d}t}\int_0^L |u(x,t)|^4\mathrm{d}x = 0.
$$

Consequently, (7.3) holds.

$\square$

When $q \leqslant 0$, it follows from (7.3) that

$$
|u(\cdot,t)|_1^2 \leqslant E(0).
$$

Now we consider $q > 0$. In combination of (7.2) and Lemma 1.1, we have

$$
\int_0^L |u(x,t)|^4\mathrm{d}x \leqslant \|u(\cdot,t)\|_\infty^2 \|u(\cdot,t)\|^2
$$

$$
\leqslant \left(\varepsilon|u(\cdot,t)|_1^2 + \frac{1}{4\varepsilon}\|u(\cdot,t)\|^2 \right)\|u(\cdot,t)\|^2
$$

$$
= \left[\varepsilon|u(\cdot,t)|_1^2 + \frac{1}{4\varepsilon}Q(0) \right]Q(0).
$$

According to (7.3), we have

$$
|u(\cdot,t)|_1^2 = \frac{q}{2}\int_0^L |u(x,t)|^4\mathrm{d}x + E(0) \leqslant \frac{q}{2}\left[\varepsilon|u(\cdot,t)|_1^2 + \frac{1}{4\varepsilon}Q(0) \right]Q(0) + E(0).
$$

When $Q(0) = 0$, we have

$$
|u(\cdot,t)|_1^2 = 0.
$$

When $Q(0) \neq 0$, taking $\varepsilon = \dfrac{1}{q\,Q(0)}$, one gives

$$|u(\cdot,t)|_1^2 \leqslant \frac{q}{4\varepsilon} Q^2(0) + 2E(0) = \frac{1}{4} q^2 Q^3(0) + 2E(0). \tag{7.6}$$

Using Lemma 1.1 again, it follows from (7.6) that

$$\|u(\cdot,t)\|_\infty^2 \leqslant \frac{L}{4} |u(\cdot,t)|_1^2 \leqslant \frac{L}{4}\left(\frac{1}{4} q^2 Q^3(0) + 2E(0)\right).$$

In the present chapter, denote

$$c_0 = \max_{0 \leqslant t \leqslant T} \|u(\cdot,t)\|_\infty. \tag{7.7}$$

7.2 Two-Level Nonlinear Difference Scheme

Divide the spatial domain $[0, L]$ uniformly into m subintervals and denote $h = L/m$, $x_j = jh, 0 \leqslant j \leqslant m$, $\Omega_h = \{x_j \mid 0 \leqslant j \leqslant m\}$. Similarly, divide the temporal domain $[0, T]$ into n equal subintervals and denote $\tau = T/n$, $t_k = k\tau, 0 \leqslant k \leqslant n$, $\Omega_\tau = \{t_k \mid 0 \leqslant k \leqslant n\}$.

Define

$$\mathcal{W}_h = \big\{v \mid v = \{v_j \mid 0 \leqslant j \leqslant m\} \text{ is a grid function on } \Omega_h \,; v_j \in \mathcal{C}, 0 \leqslant j \leqslant m\big\},$$

$$\mathring{\mathcal{W}}_h = \big\{v \mid v = \{v_j \mid 0 \leqslant j \leqslant m\} \in \mathcal{W}_h \text{ and } v_0 = v_m = 0\big\}.$$

For any grid functions $u,\ v \in \mathcal{W}_h$, define the following notation:

$$(u, v) = h\left(\frac{1}{2} u_0 \bar{v}_0 + \sum_{j=1}^{m-1} u_j \bar{v}_j + \frac{1}{2} u_m \bar{v}_m\right), \quad \|u\| = \sqrt{(u, u)},$$

$$(\delta_x u, \delta_x v) = h \sum_{j=1}^{m} (\delta_x u_{j-\frac{1}{2}})(\delta_x \bar{v}_{j-\frac{1}{2}}), \quad \|\delta_x u\| = \sqrt{(\delta_x u, \delta_x u)},$$

$$\|u\|_p = \sqrt[p]{h\left(\frac{1}{2}|u_0|^p + \sum_{j=1}^{m-1} |u_j|^p + \frac{1}{2}|u_m|^p\right)},$$

$$\|u\|_\infty = \max_{0 \leqslant j \leqslant m} |u_j|, \quad |u|_1 = \|\delta_x u\|.$$

7.2.1 *Derivation of the Difference Scheme*

Considering Eq. (7.1a) at the point $(x_j, t_{k+\frac{1}{2}})$, we have

$$iu_t(x_j, t_{k+\frac{1}{2}}) + u_{xx}(x_j, t_{k+\frac{1}{2}}) + q|u(x_j, t_{k+\frac{1}{2}})|^2 u(x_j, t_{k+\frac{1}{2}}) = 0,$$

$$1 \leqslant j \leqslant m - 1, \quad 0 \leqslant k \leqslant n - 1.$$

Using the Taylor expansion and the numerical differential formula, we arrive at

$$i\delta_t U_j^{k+\frac{1}{2}} + \delta_x^2 U_j^{k+\frac{1}{2}} + \frac{q}{2}\left(|U_j^k|^2 + |U_j^{k+1}|^2\right) U_j^{k+\frac{1}{2}} = (R_1)_j^k,$$

$$1 \leqslant j \leqslant m - 1, \quad 0 \leqslant k \leqslant n - 1, \tag{7.8}$$

where there is a positive constant c_1 such that

$$\begin{cases} |(R_1)_j^k| \leqslant c_1(\tau^2 + h^2), & 1 \leqslant j \leqslant m - 1, \quad 0 \leqslant k \leqslant n - 1, \tag{7.9a} \\[2mm] |\delta_t (R_1)_j^{k+\frac{1}{2}}| \leqslant c_1(\tau^2 + h^2), & 1 \leqslant j \leqslant m - 1, \quad 0 \leqslant k \leqslant n - 2. \tag{7.9b} \end{cases}$$

The Taylor expansion with the integral remainder is necessary to obtain (7.9b).
  Noticing the initial-boundary value conditions (7.1b)–(7.1c), we have

$$\begin{cases} U_j^0 = \varphi(x_j), & 1 \leqslant j \leqslant m - 1, \\ U_0^k = 0, \quad U_m^k = 0, & 0 \leqslant k \leqslant n. \end{cases} \tag{7.10}$$

Omitting the small term in (7.8) and replacing U_j^k with u_j^k, a difference scheme
reads

$$\begin{cases} i\delta_t u_j^{k+\frac{1}{2}} + \delta_x^2 u_j^{k+\frac{1}{2}} + \frac{q}{2}(|u_j^k|^2 + |u_j^{k+1}|^2)u_j^{k+\frac{1}{2}} = 0, \\[2mm] \qquad\qquad\qquad 1 \leqslant j \leqslant m - 1, \quad 0 \leqslant k \leqslant n - 1, \tag{7.11a} \\[2mm] u_j^0 = \varphi(x_j), \quad 1 \leqslant j \leqslant m - 1, \tag{7.11b} \\[2mm] u_0^k = 0, \quad u_m^k = 0, \quad 0 \leqslant k \leqslant n. \tag{7.11c} \end{cases}$$

7.2.2   Conservation and Boundedness of the Difference Solution

Denote

$$Q^k = h \sum_{j=1}^{m-1} |u_j^k|^2, \quad E^k = h \sum_{j=0}^{m-1} |\delta_x u_{j+\frac{1}{2}}^k|^2 - \frac{q}{2} h \sum_{j=1}^{m-1} |u_j^k|^4.$$

Theorem 7.2 *Suppose $\{u_j^k \mid 0 \leqslant j \leqslant m, 0 \leqslant k \leqslant n\}$ is the solution of the difference scheme (7.11), then we have*

$$Q^k = Q^0, \quad 1 \leqslant k \leqslant n, \tag{7.12}$$

$$E^k = E^0, \quad 1 \leqslant k \leqslant n. \tag{7.13}$$

Proof

(I) Taking the inner product on both sides of (7.11a) with $u^{k+\frac{1}{2}}$ gives

$$i(\delta_t u^{k+\frac{1}{2}}, u^{k+\frac{1}{2}}) + \left(\delta_x^2 u^{k+\frac{1}{2}}, u^{k+\frac{1}{2}}\right) + \frac{q}{2} h \sum_{j=1}^{m-1} (|u_j^k|^2 + |u_j^{k+1}|^2)|u_j^{k+\frac{1}{2}}|^2 = 0,$$

$$0 \leqslant k \leqslant n - 1. \tag{7.14}$$

It follows by noticing $u_0^{k+\frac{1}{2}} = 0$ and $u_m^{k+\frac{1}{2}} = 0$ that

$$\left(\delta_x^2 u^{k+\frac{1}{2}}, u^{k+\frac{1}{2}}\right) = -\|\delta_x u^{k+\frac{1}{2}}\|^2.$$

Noticing

$$\mathrm{Re}\left\{(\delta_t u^{k+\frac{1}{2}}, u^{k+\frac{1}{2}})\right\} = \frac{1}{2\tau}(\|u^{k+1}\|^2 - \|u^k\|^2),$$

and taking the imaginary parts on both sides of (7.14), we have

$$\frac{1}{2\tau}(\|u^{k+1}\|^2 - \|u^k\|^2) = 0, \quad 0 \leqslant k \leqslant n - 1,$$

which implies

$$\|u^{k+1}\|^2 = \|u^k\|^2, \quad 0 \leqslant k \leqslant n - 1.$$

Therefore, (7.12) is valid.

(II) Taking the inner product on both sides of (7.11a) with $-\delta_t u^{k+\frac{1}{2}}$ yields

$$-\mathrm{i}\|\delta_t u^{k+\frac{1}{2}}\|^2 - \left(\delta_x^2 u^{k+\frac{1}{2}}, \delta_t u^{k+\frac{1}{2}}\right) - \frac{q}{2} h \sum_{j=1}^{m-1} \left(|u_j^k|^2 + |u_j^{k+1}|^2\right) u_j^{k+\frac{1}{2}} \delta_t \bar{u}_j^{k+\frac{1}{2}} = 0.$$

$$(7.15)$$

It follows by noticing $\delta_t u_0^{k+\frac{1}{2}} = 0$ and $\delta_t u_m^{k+\frac{1}{2}} = 0$ that

$$\mathrm{Re}\left\{-(\delta_x^2 u^{k+\frac{1}{2}}, \delta_t u^{k+\frac{1}{2}})\right\} = \mathrm{Re}\left\{h \sum_{j=1}^{m} \left(\delta_x u_{j-\frac{1}{2}}^{k+\frac{1}{2}}\right)\left(\delta_x \delta_t \bar{u}_{j-\frac{1}{2}}^{k+\frac{1}{2}}\right)\right\}$$

$$= \frac{1}{2\tau}(|u^{k+1}|_1^2 - |u^k|_1^2).$$

Taking the real parts on both sides of (7.15), we have

$$\frac{1}{2\tau}(|u^{k+1}|_1^2 - |u^k|_1^2) - \frac{q}{2} \cdot \frac{1}{2\tau}\left(h \sum_{j=1}^{m-1} |u_j^{k+1}|^4 - h \sum_{j=1}^{m-1} |u_j^k|^4\right) = 0, \quad 0 \leqslant k \leqslant n-1,$$

which implies

$$|u^{k+1}|_1^2 - \frac{q}{2} h \sum_{j=1}^{m-1} |u_j^{k+1}|^4 = |u^k|_1^2 - \frac{q}{2} h \sum_{j=1}^{m-1} |u_j^k|^4, \quad 0 \leqslant k \leqslant n-1.$$

Hence, (7.13) is valid.

$$\square$$

Theorem 7.3 *Suppose $\{u_j^k \mid 0 \leqslant j \leqslant m, 0 \leqslant k \leqslant n\}$ is the solution of the difference scheme (7.11). Then we have*

$$\|u^k\|_\infty^2 \leqslant \frac{L}{2}\left(\frac{1}{8} q^2 \|u^0\|^6 + |u^0|_1^2 - \frac{q}{2}\|u^0\|_4^4\right), \quad 1 \leqslant k \leqslant n. \qquad (7.16)$$

Proof By Theorem 7.2, we have

$$\|u^k\|^2 = \|u^0\|^2, \quad 1 \leqslant k \leqslant n, \qquad (7.17)$$

$$|u^k|_1^2 - \frac{q}{2}\|u^k\|_4^4 = |u^0|_1^2 - \frac{q}{2}\|u^0\|_4^4, \quad 1 \leqslant k \leqslant n. \qquad (7.18)$$

When $q \leqslant 0$, it follows from (7.18) that

$$|u^k|_1^2 \leqslant |u^k|_1^2 - \frac{q}{2}\|u^k\|_4^4 = |u^0|_1^2 - \frac{q}{2}\|u^0\|_4^4.$$

Therefore,

$$\|u^k\|_\infty^2 \leqslant \frac{L}{4}|u^k|_1^2 \leqslant \frac{L}{4}\left(|u^0|_1^2 - \frac{q}{2}\|u^0\|_4^4\right).$$

Hence, (7.16) is valid.

When $q > 0$, it follows from (7.18) that

$$|u^k|_1^2 = |u^0|_1^2 - \frac{q}{2}\|u^0\|_4^4 + \frac{q}{2}\|u^k\|_4^4, \quad 1 \leqslant k \leqslant n. \tag{7.19}$$

It is easy to know that

$$\|u^k\|_4^4 = h\sum_{j=1}^{m-1}|u_j^k|^4 \leqslant \|u^k\|_\infty^2\|u^k\|^2.$$

Thus,

$$\|u^k\|_4^4 \leqslant \left(\varepsilon|u^k|_1^2 + \frac{1}{4\varepsilon}\|u^k\|^2\right)\|u^k\|^2. \tag{7.20}$$

Using (7.20) on the right-hand side of (7.19) and noticing (7.17), we have

$$|u^k|_1^2 \leqslant |u^0|_1^2 - \frac{q}{2}\|u^0\|_4^4 + \frac{q}{2}\left(\varepsilon|u^k|_1^2 + \frac{1}{4\varepsilon}\|u^k\|^2\right)\|u^k\|^2$$

$$= |u^0|_1^2 - \frac{q}{2}\|u^0\|_4^4 + \frac{q}{2}\left(\varepsilon|u^k|_1^2 + \frac{1}{4\varepsilon}\|u^0\|^2\right)\|u^0\|^2.$$

When $\|u^0\|^2 = 0$, we have

$$|u^k|_1^2 = 0.$$

When $\|u^0\|^2 \neq 0$, it follows by taking $\varepsilon = 1/(q\|u^0\|^2)$ that

$$|u^k|_1^2 \leqslant 2\left(\frac{1}{8}q^2\|u^0\|^6 + |u^0|_1^2 - \frac{q}{2}\|u^0\|_4^4\right).$$

Therefore,

$$\|u^k\|_\infty^2 \leqslant \frac{L}{4}|u^k|_1^2 \leqslant \frac{L}{2}\left(\frac{1}{8}q^2\|u^0\|^6 + |u^0|_1^2 - \frac{q}{2}\|u^0\|_4^4\right), \quad 1 \leqslant k \leqslant n.$$

$\square$

Colollary 7.1 *Suppose $\{u_j^k \mid 0 \leqslant j \leqslant m, 0 \leqslant k \leqslant n\}$ is the solution of the difference scheme (7.11), then there is a positive constant c_2 such that*

$$\|u^k\|_\infty \leqslant c_2, \quad 1 \leqslant k \leqslant n. \tag{7.21}$$

7.2.3 *Existence and Uniqueness of the Difference Solution*

The existence of the difference solution will be demonstrated with the aid of the following Browder theorem.

Theorem 7.4 (Browder Theorem [1, 2]) *Let $(H, (\cdot, \cdot))$ be a finite-dimensional inner product space with the associated norm $\|\cdot\|$, and $\Pi : H \to H$ be a continuous operator. Assume moreover that there exists an $\alpha > 0$, for any $z \in H$ and $\|z\| = \alpha$, it holds that*

$$\mathrm{Re}(\Pi(z), z) \geqslant 0.$$

Then there is a $z^* \in H$ satisfying $\|z^*\| \leqslant \alpha$ such that $\Pi(z^*) = 0$.

Theorem 7.5

 (I) *The difference scheme (7.11) has a solution.*
(II) *When $\tau < 1/(3c_2^2|q|)$, the solution of the difference scheme (7.11) is unique.*

Proof When $q = 0$, the difference scheme (7.11) is linear, and its unique solvability can be easily established. In the remainder of the discussion, we focus on the case where $q \neq 0$.

 (I) Suppose $\{u_j^k \mid 0 \leqslant j \leqslant m\}$ has been obtained. Rewrite (7.11) as

$$\begin{cases} \mathrm{i}\dfrac{2}{\tau}(u_j^{k+\frac{1}{2}} - u_j^k) + \delta_x^2 u_j^{k+\frac{1}{2}} + \dfrac{q}{2}(|u_j^k|^2 + |2u_j^{k+\frac{1}{2}} - u_j^k|^2)u_j^{k+\frac{1}{2}} = 0, \\ \qquad\qquad\qquad\qquad 1 \leqslant j \leqslant m - 1, \\ u_0^{k+\frac{1}{2}} = 0, \quad u_m^{k+\frac{1}{2}} = 0. \end{cases} \tag{7.22}$$

Once $\{u_j^{k+\frac{1}{2}} \mid 0 \leqslant j \leqslant m\}$ is obtained, then we have

$$u_j^{k+1} = 2u_j^{k+\frac{1}{2}} - u_j^k, \quad 0 \leqslant j \leqslant m.$$

Let

$$w_j = u_j^{k+\frac{1}{2}}, \quad 0 \leqslant j \leqslant m.$$

Then it follows from (7.22) that the system in $\{w_j \mid 0 \leqslant j \leqslant m\}$ reads

$$
\begin{cases}
\mathrm{i}\dfrac{2}{\tau}(w_j - u_j^k) + \delta_x^2 w_j + \dfrac{q}{2}(|u_j^k|^2 \\
\qquad\qquad\qquad + |2w_j - u_j^k|^2)w_j = 0, \ 1 \leqslant j \leqslant m-1, & (7.23\mathrm{a}) \\
w_0 = 0, \quad w_m = 0. & (7.23\mathrm{b})
\end{cases}
$$

Rewrite (7.23a) as

$$
\frac{2}{\tau}(w_j - u_j^k) - \mathrm{i}\delta_x^2 w_j - \mathrm{i}\frac{q}{2}(|u_j^k|^2 + |2w_j - u_j^k|^2)w_j = 0, \quad 1 \leqslant j \leqslant m-1.
$$

Define the operator $\Pi : \overset{\circ}{\mathcal{W}}_h \to \overset{\circ}{\mathcal{W}}_h$ by

$$
\Pi(w)_j =
\begin{cases}
\dfrac{2}{\tau}(w_j - u_j^k) - \mathrm{i}\delta_x^2 w_j - \mathrm{i}\dfrac{q}{2}(|u_j^k|^2 + |2w_j - u_j^k|^2)w_j, & 1 \leqslant j \leqslant m-1, \\
0, & j = 0, \ m.
\end{cases}
$$

Taking the inner product of $\Pi(w)$ with w, we have

$$
\big(\Pi(w), w\big) = \frac{2}{\tau}\big((w, w) - (u^k, w)\big) - \mathrm{i}(\delta_x^2 w, w)
$$

$$
- \mathrm{i}\frac{q}{2}h \sum_{j=1}^{m-1}(|u_j^k|^2 + |2w_j - u_j^k|^2)|w_j|^2
$$

$$
= \frac{2}{\tau}\big(\|w\|^2 - (u^k, w)\big) + \mathrm{i}|w|_1^2
$$

$$
- \mathrm{i}\frac{q}{2}h \sum_{j=1}^{m-1}(|u_j^k|^2 + |2w_j - u_j^k|^2)|w_j|^2.
$$

Therefore,

$$
\mathrm{Re}\{(\Pi(w), w)\} = \frac{2}{\tau}\big(\|w\|^2 - \mathrm{Re}(u^k, w)\big) \geqslant \frac{2}{\tau}\|w\|\big(\|w\| - \|u^k\|\big).
$$

When $\|w\| = \|u^k\|$, it follows that $\mathrm{Re}\{(\Pi(w), w)\} \geqslant 0$.

By the Browder theorem (Theorem 7.4), there is a $w^* \in \overset{\circ}{\mathcal{W}}_h$ satisfying $\|w^*\| \leqslant \|u^k\|$ such that $\Pi(w^*) = 0$. Therefore there exists a solution to the difference scheme (7.11).

(II) Suppose (7.23) has another solution $v = \{v_j \mid 0 \leqslant j \leqslant m\} \in \overset{\circ}{\mathcal{W}}_h$ satisfying

$$
\begin{cases}
\mathrm{i}\dfrac{2}{\tau}(v_j - u_j^k) + \delta_x^2 v_j + \dfrac{q}{2}(|u_j^k|^2 + |2v_j - u_j^k|^2)v_j = 0, & 1 \leqslant j \leqslant m - 1, \\[2mm]
v_0 = 0, \quad v_m = 0.
\end{cases}
$$

$$(7.24)$$

Denote

$$
\theta_j = w_j - v_j, \quad 0 \leqslant j \leqslant m.
$$

Subtracting (7.24) from (7.23) arrives at

$$
\begin{cases}
\mathrm{i}\dfrac{2}{\tau}\theta_j + \delta_x^2 \theta_j + \dfrac{q}{2}(|u_j^k|^2\theta_j + |2w_j - u_j^k|^2 w_j - |2v_j - u_j^k|^2 v_j) = 0, \\[2mm]
\hspace{6cm} 1 \leqslant j \leqslant m - 1, & (7.25\mathrm{a}) \\[2mm]
\theta_0 = 0, \quad \theta_m = 0. & (7.25\mathrm{b})
\end{cases}
$$

Taking the inner product on both sides of (7.25a) with $\mathrm{i}\theta$ gives

$$
\frac{2}{\tau}\|\theta\|^2 - \mathrm{i}(\delta_x^2\theta, \theta) - \mathrm{i}\cdot\frac{q}{2}h\sum_{j=1}^{m-1}|u_j^k|^2 \cdot |\theta_j|^2
$$

$$
- \mathrm{i}\frac{q}{2}h\sum_{j=1}^{m-1}(|2w_j - u_j^k|^2 w_j - |2v_j - u_j^k|^2 v_j)\bar{\theta}_j = 0. \qquad (7.26)
$$

Noticing

$$
|2w_j - u_j^k|^2 w_j - |2v_j - u_j^k|^2 v_j
$$

$$
= |2w_j - u_j^k|^2(w_j - v_j) + \left(|2w_j - u_j^k|^2 - |2v_j - u_j^k|^2\right)v_j
$$

$$
= |2w_j - u_j^k|^2\theta_j + \left[\left(2w_j - u_j^k\right)\overline{2w_j - u_j^k} - \left(2v_j - u_j^k\right)\overline{2v_j - u_j^k}\right]v_j
$$

$$
= |2w_j - u_j^k|^2\theta_j + \left[2(w_j - v_j)\overline{2w_j - u_j^k} + (2v_j - u_j^k)\overline{2(w_j - v_j)}\right]v_j
$$

$$
= |2w_j - u_j^k|^2\theta_j + 2\left[\theta_j\overline{2w_j - u_j^k} + (2v_j - u_j^k)\bar{\theta}_j\right]v_j,
$$

and then taking the real parts on both sides of (7.26), we get

$$
\frac{2}{\tau}\|\theta\|^2 \leqslant |q|h\sum_{j=1}^{m-1}\left(|\theta_j|^2|2w_j - u_j^k| + |2v_j - u_j^k|\cdot|\theta_j|^2\right)|v_j| \leqslant 6c_2^2|q|\cdot\|\theta\|^2.
$$

When $\tau < 1/(3c_2^2|q|)$, it follows that $\|\theta\|^2 = 0$, which implies $\theta_j = 0, 0 \leqslant j \leqslant m$. Therefore, the difference scheme (7.11) is uniquely solvable.

$\square$

7.2.4   *Convergence of the Difference Solution*

Theorem 7.6 *Suppose* $\{U_j^k \mid 0 \leqslant j \leqslant m, 0 \leqslant k \leqslant n\}$ *is the solution of the problem* (7.1) *and* $\{u_j^k \mid 0 \leqslant j \leqslant m, 0 \leqslant k \leqslant n\}$ *is the solution of the difference scheme* (7.11). *Denote*

$$e_j^k = U_j^k - u_j^k, \quad 0 \leqslant j \leqslant m, \quad 0 \leqslant k \leqslant n.$$

There is a positive constant c_3 *such that*

$$\|e^k\| \leqslant c_3(\tau^2 + h^2), \quad 1 \leqslant k \leqslant n.$$

Proof Subtracting (7.11) from (7.8) and (7.10), one obtains the system of error equations as

$$
\begin{cases}
\mathrm{i}\delta_t e_j^{k+\frac{1}{2}} + \delta_x^2 e_j^{k+\frac{1}{2}} + \dfrac{q}{2}\left[(|U_j^k|^2 + |U_j^{k+1}|^2)U_j^{k+\frac{1}{2}} - (|u_j^k|^2 + |u_j^{k+1}|^2)u_j^{k+\frac{1}{2}}\right] \\
\qquad = (R_1)_j^k, \quad 1 \leqslant j \leqslant m-1, \quad 0 \leqslant k \leqslant n-1, & \text{(7.27a)} \\[2mm]
e_j^0 = 0, \quad 1 \leqslant j \leqslant m-1, & \text{(7.27b)} \\[2mm]
e_0^k = 0, \quad e_m^k = 0, \quad 0 \leqslant k \leqslant n. & \text{(7.27c)}
\end{cases}
$$

Denote

$$G(U)_j^{k+\frac{1}{2}} = (|U_j^k|^2 + |U_j^{k+1}|^2)U_j^{k+\frac{1}{2}}.$$

Taking the inner product on both sides of (7.27a) with $e^{k+\frac{1}{2}}$ produces

$$\mathrm{i}(\delta_t e^{k+\frac{1}{2}}, e^{k+\frac{1}{2}}) + (\delta_x^2 e^{k+\frac{1}{2}}, e^{k+\frac{1}{2}}) + \frac{q}{2}h\sum_{j=1}^{m-1}\left[G(U)_j^{k+\frac{1}{2}} - G(u)_j^{k+\frac{1}{2}}\right]\bar{e}_j^{k+\frac{1}{2}}$$

$$= ((R_1)^k, e^{k+\frac{1}{2}}), \quad 0 \leqslant k \leqslant n-1. \tag{7.28}$$

Noticing

$$\left(\delta_x^2 e^{k+\frac{1}{2}}, e^{k+\frac{1}{2}}\right) = -\|\delta_x e^{k+\frac{1}{2}}\|^2,$$

$$G(U)_j^{k+\frac{1}{2}} - G(u)_j^{k+\frac{1}{2}}$$

$$= \left(|U_j^k|^2 + |U_j^{k+1}|^2\right)\left(U_j^{k+\frac{1}{2}} - u_j^{k+\frac{1}{2}}\right)$$

$$+ \left(|U_j^k|^2 + |U_j^{k+1}|^2 - |u_j^k|^2 - |u_j^{k+1}|^2\right) u_j^{k+\frac{1}{2}}$$

$$= \left(|U_j^k|^2 + |U_j^{k+1}|^2\right) e_j^{k+\frac{1}{2}} + \left(e_j^k \bar{U}_j^k + u_j^k \bar{e}_j^k + e_j^{k+1} \bar{U}_j^{k+1} + u_j^{k+1} \bar{e}_j^{k+1}\right) u_j^{k+\frac{1}{2}}$$

$$\tag{7.29}$$

and then taking the imaginary parts on both sides of (7.28), we obtain

$$\frac{1}{2\tau}\left(\|e^{k+1}\|^2 - \|e^k\|^2\right)$$

$$= \mathrm{Im}\left\{-\frac{q}{2}h\sum_{j=1}^{m-1}\left(e_j^k \bar{U}_j^k + u_j^k \bar{e}_j^k + e_j^{k+1} \bar{U}_j^{k+1} + u_j^{k+1} \bar{e}_j^{k+1}\right) u_j^{k+\frac{1}{2}} \bar{e}_j^{k+\frac{1}{2}} + \left((R_1)^k, e^{k+\frac{1}{2}}\right)\right\}$$

$$\leqslant \frac{|q|}{2}h\sum_{j=1}^{m-1}\left(c_0|e_j^k| + c_2|e_j^k| + c_0|e_j^{k+1}| + c_2|e_j^{k+1}|\right) c_2 |\bar{e}_j^{k+\frac{1}{2}}| + \|(R_1)^k\| \cdot \|e^{k+\frac{1}{2}}\|$$

$$\leqslant \frac{|q|(c_0 + c_2)c_2}{2}\left(\|e^k\| + \|e^{k+1}\|\right)\|e^{k+\frac{1}{2}}\| + \|(R_1)^k\| \cdot \|e^{k+\frac{1}{2}}\|$$

$$\leqslant \frac{|q|(c_0 + c_2)c_2}{2}\left(\|e^k\| + \|e^{k+1}\|\right) \cdot \frac{\|e^k\| + \|e^{k+1}\|}{2} + \|(R_1)^k\| \cdot \frac{\|e^k\| + \|e^{k+1}\|}{2},$$

$$0 \leqslant k \leqslant n - 1.$$

When $\frac{1}{2}(\|e^{k+1}\| + \|e^k\|) \neq 0$, we get from division by $\frac{1}{2}(\|e^{k+1}\| + \|e^k\|)$ that

$$\frac{1}{\tau}\left(\|e^{k+1}\| - \|e^k\|\right) \leqslant \frac{|q|}{2}(c_0 + c_2)c_2\left(\|e^k\| + \|e^{k+1}\|\right) + \|(R_1)^k\|.$$

When $\frac{1}{2}(\|e^{k+1}\| + \|e^k\|) = 0$, the above result also holds. Therefore, we have

$$\left[1 - \frac{|q|}{2}(c_0 + c_2)c_2\tau\right]\|e^{k+1}\| \leqslant \left[1 + \frac{|q|}{2}(c_0 + c_2)c_2\tau\right]\|e^k\| + \tau\|(R_1)^k\|,$$

$$0 \leqslant k \leqslant n - 1. \tag{7.30}$$

When $\frac{|q|}{2}(c_0 + c_2)c_2\tau \leqslant \frac{1}{3}$, it follows by noticing (7.9a) that

$$
\begin{aligned}
\|e^{k+1}\| &\leqslant \left[1 + \frac{3|q|}{2}(c_0 + c_2)c_2\tau\right]\|e^k\| + \frac{3}{2}\tau\|(R_1)^k\| \\
&\leqslant \left[1 + \frac{3|q|}{2}(c_0 + c_2)c_2\tau\right]\|e^k\| + \frac{3}{2}\sqrt{L}c_1\tau(\tau^2 + h^2), \quad 0 \leqslant k \leqslant n-1.
\end{aligned}
$$

When $q = 0$, it follows from (7.30) and (7.9a) that

$$
\|e^k\| \leqslant \|e^0\| + \sqrt{L}c_1\, k\tau(\tau^2 + h^2) \leqslant \sqrt{L}c_1 T(\tau^2 + h^2), \quad 1 \leqslant k \leqslant n
$$

by recursion of

$$
\|e^{k+1}\| \leqslant \|e^k\| + \sqrt{L}c_1\tau(\tau^2 + h^2), \quad 0 \leqslant k \leqslant n-1.
$$

When $q \neq 0$, it follows from Lemma 3.3 (Gronwall inequality) that

$$
\|e^k\| \leqslant e^{\frac{3|q|}{2}(c_0 + c_2)c_2 T} \frac{\sqrt{L}c_1}{|q|(c_0 + c_2)c_2}(\tau^2 + h^2), \quad 1 \leqslant k \leqslant n.
$$

$\square$

Theorem 7.7 *Suppose $\{U_j^k \mid 0 \leqslant j \leqslant m, 0 \leqslant k \leqslant n\}$ is the solution of the problem (7.1) and $\{u_j^k \mid 0 \leqslant j \leqslant m, 0 \leqslant k \leqslant n\}$ is the solution of the difference scheme (7.11). Denote*

$$
e_j^k = U_j^k - u_j^k, \quad 0 \leqslant j \leqslant m, \ 0 \leqslant k \leqslant n.
$$

Then there is a constant c_4 such that

$$
\|e^k\|_\infty \leqslant c_4(\tau^2 + h^2), \quad 1 \leqslant k \leqslant n. \tag{7.31}
$$

Proof Taking the inner product on both sides of (7.27a) with $-\delta_t e^{k+\frac{1}{2}}$ gives

$$
-\mathrm{i}\|\delta_t e^{k+\frac{1}{2}}\|^2 - (\delta_x^2 e^{k+\frac{1}{2}}, \delta_t e^{k+\frac{1}{2}}) - \frac{q}{2}h\sum_{j=1}^{m-1}[G(U)_j^{k+\frac{1}{2}} - G(u)_j^{k+\frac{1}{2}}]\delta_t \bar{e}_j^{k+\frac{1}{2}}
$$

$$
= -\left((R_1)^k, \delta_t e^{k+\frac{1}{2}}\right), \quad 0 \leqslant k \leqslant n-1.
$$

Taking the real parts on both sides of the above equality, we have

$$\frac{1}{2\tau}\left(|e^{k+1}|_1^2 - |e^k|_1^2\right) = \mathrm{Re}\left\{\frac{q}{2}h\sum_{j=1}^{m-1}[G(U)_j^{k+\frac{1}{2}} - G(u)_j^{k+\frac{1}{2}}]\delta_t\bar{e}_j^{k+\frac{1}{2}}\right\}$$

$$+\mathrm{Re}\left\{-\left((R_1)^k, \delta_t e^{k+\frac{1}{2}}\right)\right\}, \quad 0 \leqslant k \leqslant n-1. \quad (7.32)$$

It follows from (7.27a) that

$$\delta_t e_j^{k+\frac{1}{2}} = i\delta_x^2 e_j^{k+\frac{1}{2}} + i\frac{q}{2}[G(U)_j^{k+\frac{1}{2}} - G(u)_j^{k+\frac{1}{2}}] - i(R_1)_j^k.$$

Then the first term on the right-hand side of (7.32) becomes

$$\mathrm{Re}\left\{\frac{q}{2}h\sum_{j=1}^{m-1}[G(U)_j^{k+\frac{1}{2}} - G(u)_j^{k+\frac{1}{2}}]\delta_t\bar{e}_j^{k+\frac{1}{2}}\right\}$$

$$= \frac{q}{2}\mathrm{Re}\left\{h\sum_{j=1}^{m-1}\overline{G(U)_j^{k+\frac{1}{2}} - G(u)_j^{k+\frac{1}{2}}}\,\delta_t e_j^{k+\frac{1}{2}}\right\}$$

$$= \frac{q}{2}\mathrm{Re}\left\{h\sum_{j=1}^{m-1}\overline{G(U)_j^{k+\frac{1}{2}} - G(u)_j^{k+\frac{1}{2}}}\left[i\delta_x^2 e_j^{k+\frac{1}{2}} + i\frac{q}{2}[G(U)_j^{k+\frac{1}{2}} - G(u)_j^{k+\frac{1}{2}}]\right.\right.$$

$$\left.\left.-i(R_1)_j^k\right]\right\}$$

$$= \frac{q}{2}\mathrm{Re}\left\{h\sum_{j=1}^{m-1}\overline{G(U)_j^{k+\frac{1}{2}} - G(u)_j^{k+\frac{1}{2}}}\left[i\delta_x^2 e_j^{k+\frac{1}{2}} - i(R_1)_j^k\right]\right\}$$

$$= \frac{q}{2}\mathrm{Re}\left\{ih\sum_{j=1}^{m-1}\overline{G(U)_j^{k+\frac{1}{2}} - G(u)_j^{k+\frac{1}{2}}}\,\delta_x^2 e_j^{k+\frac{1}{2}}\right.$$

$$\left.-ih\sum_{j=1}^{m-1}\overline{G(U)_j^{k+\frac{1}{2}} - G(u)_j^{k+\frac{1}{2}}}\,(R_1)_j^k\right\}$$

$$= \frac{q}{2}\mathrm{Re}\left\{-ih\sum_{j=1}^{m}\overline{\delta_x\left(G(U^{k+\frac{1}{2}}) - G(u^{k+\frac{1}{2}})\right)_{j-\frac{1}{2}}}\,\delta_x e_{j-\frac{1}{2}}^{k+\frac{1}{2}}\right.$$

$$-ih \sum_{j=1}^{m-1} \overline{G(U)_j^{k+\frac{1}{2}} - G(u)_j^{k+\frac{1}{2}}} \, (R_1)_j^k \Bigg\}$$

$$\leqslant \frac{|q|}{2} \Big(|G(U^{k+\frac{1}{2}}) - G(u^{k+\frac{1}{2}})|_1 \cdot |e^{k+\frac{1}{2}}|_1 + \|G(U^{k+\frac{1}{2}}) - G(u^{k+\frac{1}{2}})\| \cdot \|(R_1)^k\| \Big).$$

$$(7.33)$$

With the help of (7.29), (7.7), and (7.21), there are two constants c_5 and c_6 satisfying

$$\|G(U^{k+\frac{1}{2}}) - G(u^{k+\frac{1}{2}})\| \leqslant c_5(\|e^k\| + \|e^{k+1}\|), \tag{7.34}$$

$$|G(U^{k+\frac{1}{2}}) - G(u^{k+\frac{1}{2}})|_1 \leqslant c_6(\|e^k\| + \|e^{k+1}\| + |e^k|_1 + |e^{k+1}|_1). \tag{7.35}$$

Inserting (7.34) and (7.35) into (7.33) leads to

$$\mathrm{Re}\Big\{ \frac{q}{2}h \sum_{j=1}^{m-1} \big[G(U)_j^{k+\frac{1}{2}} - G(u)_j^{k+\frac{1}{2}} \big] \delta_t \bar{e}_j^{k+\frac{1}{2}} \Big\}$$

$$\leqslant \frac{|q|}{2} \cdot \Big[c_6(\|e^k\| + \|e^{k+1}\| + |e^k|_1 + |e^{k+1}|_1) \cdot \frac{1}{2}(|e^k|_1 + |e^{k+1}|_1)$$

$$+ c_5(\|e^k\| + \|e^{k+1}\|)\|(R_1)^k\| \Big]. \tag{7.36}$$

Substituting (7.36) into (7.32), replacing k by l, and summing over l from 0 to k, we have

$$\frac{1}{2\tau}|e^{k+1}|_1^2 \leqslant \sum_{l=0}^{k} \frac{|q|}{4} \Big[c_6(\|e^l\| + \|e^{l+1}\| + |e^l|_1 + |e^{l+1}|_1)(|e^l|_1 + |e^{l+1}|_1)$$

$$+ 2c_5(\|e^l\| + \|e^{l+1}\|)\|(R_1)^l\| \Big] + \mathrm{Re}\Big\{ -h \sum_{j=1}^{m-1}\sum_{l=0}^{k} (R_1)_j^l \delta_t \bar{e}_j^{l+\frac{1}{2}} \Big\},$$

$$0 \leqslant k \leqslant n-1. \tag{7.37}$$

Now we analyze the last term in the above inequality. Noticing

$$\sum_{l=0}^{k} (R_1)_j^l \delta_t \bar{e}_j^{l+\frac{1}{2}}$$

$$= \sum_{l=0}^{k} (R_1)_j^l \frac{\bar{e}_j^{l+1} - \bar{e}_j^l}{\tau}$$

$$= \frac{1}{\tau} \left(\sum_{l=0}^{k} (R_1)_j^l \bar{e}_j^{l+1} - \sum_{l=-1}^{k-1} (R_1)_j^{l+1} \bar{e}_j^{l+1} \right)$$

$$= \frac{1}{\tau} \left[(R_1)_j^k \bar{e}_j^{k+1} - \sum_{l=0}^{k-1} \left((R_1)_j^{l+1} - (R_1)_j^l \right) \bar{e}_j^{l+1} - (R_1)_j^0 \bar{e}_j^0 \right]$$

$$= \frac{1}{\tau} (R_1)_j^k \bar{e}_j^{k+1} - \sum_{l=0}^{k-1} \left[\delta_t (R_1)_j^{l+\frac{1}{2}} \right] \bar{e}_j^{l+1},$$

we have

$$\mathrm{Re} \left\{ -h \sum_{j=1}^{m-1} \sum_{l=0}^{k} (R_1)_j^l \delta_t \bar{e}_j^{l+\frac{1}{2}} \right\}$$

$$= \mathrm{Re} \left\{ -h \sum_{j=1}^{m-1} \left[\frac{1}{\tau} (R_1)_j^k \bar{e}_j^{k+1} - \sum_{l=0}^{k-1} \left(\delta_t (R_1)_j^{l+\frac{1}{2}} \right) \bar{e}_j^{l+1} \right] \right\}$$

$$\leqslant \frac{1}{\tau} \| (R_1)^k \| \cdot \| e^{k+1} \| + \sum_{l=0}^{k-1} \| \delta_t (R_1)^{l+\frac{1}{2}} \| \cdot \| e^{l+1} \|. \tag{7.38}$$

Substituting (7.38) into (7.37) and multiplying both sides of the resultant inequality by 2τ, we have

$$|e^{k+1}|_1^2 \leqslant 2\tau \sum_{l=0}^{k} \frac{|q|}{4} \Big[c_6 (\|e^l\| + \|e^{l+1}\| + |e^l|_1 + |e^{l+1}|_1)(|e^l|_1 + |e^{l+1}|_1)$$

$$+ 2c_5 (\|e^l\| + \|e^{l+1}\|) \| (R_1)^l \| \Big] + 2 \| (R_1)^k \| \cdot \| e^{k+1} \|$$

$$+ 2\tau \sum_{l=0}^{k-1} \| \delta_t (R_1)^{l+\frac{1}{2}} \| \cdot \| e^{l+1} \|.$$

With the application of (7.9), Lemma 1.4, and Theorem 7.6, we have

$$|e^{k+1}|_1^2 \leqslant 2\tau \sum_{l=0}^{k} \frac{|q|}{4} \left[c_6 \left(1 + \frac{L}{\sqrt{6}} \right) (|e^l|_1 + |e^{l+1}|_1)^2 + 4\sqrt{L} c_1 c_3 c_5 (\tau^2 + h^2)^2 \right]$$

$$+ 2\sqrt{L} c_1 (\tau^2 + h^2) c_3 (\tau^2 + h^2) + 2\tau \sum_{l=0}^{k-1} \sqrt{L} c_1 (\tau^2 + h^2) \cdot c_3 (\tau^2 + h^2),$$

$$0 \leqslant k \leqslant n - 1.$$

Hence, there is a constant c_7 satisfying

$$|e^{k+1}|_1^2 \leqslant c_7\tau \sum_{l=1}^{k} |e^l|_1^2 + c_7(\tau^2 + h^2)^2, \quad 0 \leqslant k \leqslant n-1.$$

It reveals that (7.31) is valid from Lemma 5.1 (Gronwall inequality) and Lemma 1.4.

$$\square$$

7.2.5   Numerical Examples

Example 7.1 Use the difference scheme (7.11) to compute the following initial-boundary value problem

$$\begin{cases} iu_t + u_{xx} + 2|u|^2u = 0, & 0 < x < 80, \quad 0 < t \leqslant 1, \\ u(x,0) = \text{sech}(x-40)\exp(2i(x-40)), & 0 \leqslant x \leqslant 80, \\ u(0,t) = 0, \quad u(80,t) = 0, & 0 < t \leqslant 1. \end{cases} \tag{7.39}$$

The exact solution of this problem is $u(x,t) = \text{sech}(x - 4t - 40)\exp(2i(x - 40) - 3it)$.

From the proof of Theorem 7.5, the system (7.23), which is nonlinear, need be solved. To address this, we employ a simple iteration method

$$\begin{cases} i\dfrac{2}{\tau}\left(w_j^{(l+1)} - u_j^k\right) + \delta_x^2 w_j^{(l+1)} + \dfrac{q}{2}\left(|u_j^k|^2 + |2w_j^{(l)} - u_j^k|^2\right)w_j^{(l+1)} = 0, & 1 \leqslant j \leqslant m-1, \\ w_0^{(l+1)} = 0, \quad w_m^{(l+1)} = 0, \end{cases}$$

until $\|w^{(l+1)} - w^{(l)}\|_\infty \leqslant 1e-8$, where l denotes the iterative index. Take an initial iteration value as

$$w_j^{(0)} = u_j^k, \quad 0 \leqslant j \leqslant m.$$

When $\{w_j \mid 1 \leqslant j \leqslant m-1\}$ is obtained, take

$$u_j^{k+1} = 2w_j - u_j^k, \quad 1 \leqslant j \leqslant m-1.$$

For different values of the step sizes, Tables 7.1 and 7.2 present the maximum norm errors of the numerical solutions obtained using the difference scheme (7.11), defined by

$$E_\infty(h,\tau) = \max_{0 \leqslant k \leqslant n, 0 \leqslant j \leqslant m} |u(x_j, t_k) - u_j^k|.$$

Table 7.1 (Example 7.1) The maximum norm errors of numerical solutions with different spatial step sizes ($\tau = 1/1600$)

h	$E_\infty(h, \tau)$	$E_\infty(2h, \tau)/E_\infty(h, \tau)$
1/5	1.394626e−01	
1/10	3.350696e−02	4.1622
1/20	8.292061e−03	4.0408
1/40	2.070994e−03	4.0039

Table 7.2 (Example 7.1) The maximum norm errors of numerical solutions with different temporal step sizes ($h = 1/400$)

τ	$E_\infty(h, \tau)$	$E_\infty(h, 2\tau)/E_\infty(h, \tau)$
1/20	3.465874e−02	
1/40	8.494186e−03	4.0803
1/80	2.124567e−03	3.9981
1/160	5.428168e−04	3.9140

Fig. 7.1 (Example 7.1) The curves of energy E^k for the difference scheme (7.11)

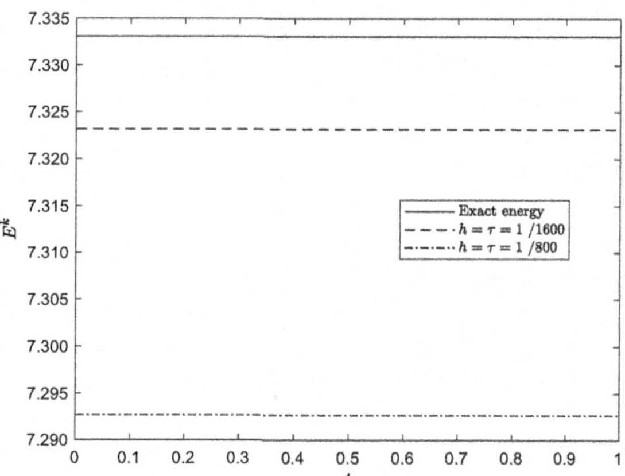

From Table 7.1, it can be observed that when the spatial step size h is halved, the maximum norm error is reduced to one-quarter of its original value. A similar trend is evident in Table 7.2: When the temporal step size τ is halved, the maximum norm error is also reduced to one-quarter of the original. The curves of the invariant E^k defined in Theorem 7.2 are presented in Fig. 7.1.

7.3   Three-Level Linearized Difference Scheme

7.3.1   Derivation of the Difference Scheme

Considering Eq. (7.1a) at the point $(x_j, t_{\frac{1}{2}})$, we have

$$\mathrm{i}u_t(x_j, t_{\frac{1}{2}}) + u_{xx}(x_j, t_{\frac{1}{2}}) + q|u(x_j, t_{\frac{1}{2}})|^2 u(x_j, t_{\frac{1}{2}}) = 0, \quad 1 \leqslant j \leqslant m - 1.$$

It follows from the numerical differentiation formula that

$$i\delta_t U_j^{\frac{1}{2}} + \delta_x^2 U_j^{\frac{1}{2}} + q|\hat{u}_j|^2 U_j^{\frac{1}{2}} = (R_2)_j^0, \quad 1 \leqslant j \leqslant m-1, \tag{7.40}$$

where

$$\hat{u}_j = u(x_j, 0) + \frac{\tau}{2} u_t(x_j, 0), \quad 1 \leqslant j \leqslant m-1,$$

and there is a constant c_8 such that

$$|(R_2)_j^0| \leqslant c_8(\tau^2 + h^2), \quad 1 \leqslant j \leqslant m-1. \tag{7.41}$$

Considering Eq. (7.1a) at the node point (x_j, t_k), we have

$$iu_t(x_j, t_k) + u_{xx}(x_j, t_k) + q|u(x_j, t_k)|^2 u(x_j, t_k) = 0, \ 1 \leqslant j \leqslant m-1, \ 1 \leqslant k \leqslant n-1.$$

It follows from the numerical differentiation formula that

$$i\Delta_t U_j^k + \delta_x^2 U_j^{\bar{k}} + q|U_j^k|^2 U_j^{\bar{k}} = (R_2)_j^k, \quad 1 \leqslant j \leqslant m-1, \quad 1 \leqslant k \leqslant n-1, \tag{7.42}$$

and there is a constant c_9 such that

$$\begin{cases} |(R_2)_j^k| \leqslant c_9(\tau^2 + h^2), & 1 \leqslant j \leqslant m-1, \quad 1 \leqslant k \leqslant n-1, \\ |\Delta_t (R_2)_j^k| \leqslant c_9(\tau^2 + h^2), & 1 \leqslant j \leqslant m-1, \quad 2 \leqslant k \leqslant n-2. \end{cases} \tag{7.43}$$

Noticing the initial-boundary value conditions (7.1b)–(7.1c), we have

$$\begin{cases} U_j^0 = \varphi(x_j), & 1 \leqslant j \leqslant m-1, \\ U_0^k = 0, \quad U_m^k = 0, & 0 \leqslant k \leqslant n. \end{cases} \tag{7.44}$$

Omitting the small terms $(R_2)_j^0$ and $(R_2)_j^k$ in (7.40) and (7.42), respectively, a linearized difference scheme for the problem (7.1) reads

$$\begin{cases} i\delta_t u_j^{\frac{1}{2}} + \delta_x^2 u_j^{\frac{1}{2}} + q|\hat{u}_j|^2 u_j^{\frac{1}{2}} = 0, & 1 \leqslant j \leqslant m-1, & \text{(7.45a)} \\ i\Delta_t u_j^k + \delta_x^2 u_j^{\bar{k}} + q|u_j^k|^2 u_j^{\bar{k}} = 0, & 1 \leqslant j \leqslant m-1, \quad 1 \leqslant k \leqslant n-1, & \text{(7.45b)} \\ u_j^0 = \varphi(x_j), & 1 \leqslant j \leqslant m-1, & \text{(7.45c)} \\ u_0^k = 0, \quad u_m^k = 0, & 0 \leqslant k \leqslant n. & \text{(7.45d)} \end{cases}$$

7.3.2 *Conservation and Boundedness of the Difference Solution*

Theorem 7.8 *Suppose* $\{u_j^k \mid 0 \leqslant j \leqslant m, 0 \leqslant k \leqslant n\}$ *is the solution of the difference scheme (7.45). Denote*

$$E^k = \frac{1}{2}(|u^{k+1}|_1^2 + |u^k|_1^2) - \frac{q}{2}h \sum_{j=1}^{m-1} |u_j^k|^2 \cdot |u_j^{k+1}|^2, \quad 0 \leqslant k \leqslant n-1.$$

Then we have

$$\|u^k\|^2 = \|u^0\|^2, \quad 1 \leqslant k \leqslant n, \tag{7.46}$$

$$\frac{1}{2}(|u^1|_1^2 + |u^0|_1^2) - \frac{q}{2}h \sum_{j=1}^{m-1} |\hat{u}_j|^2 |u_j^1|^2 = |u^0|_1^2 - \frac{q}{2}h \sum_{j=1}^{m-1} |\hat{u}_j|^2 |u_j^0|^2, \tag{7.47}$$

$$E^k = E^0, \quad 1 \leqslant k \leqslant n-1. \tag{7.48}$$

Proof

(I) Taking the inner product on both sides of (7.45a) with $u^{\frac{1}{2}}$ gives

$$\mathrm{i}(\delta_t u^{\frac{1}{2}}, u^{\frac{1}{2}}) + (\delta_x^2 u^{\frac{1}{2}}, u^{\frac{1}{2}}) + qh \sum_{j=1}^{m-1} |\hat{u}_j|^2 |u_j^{\frac{1}{2}}|^2 = 0. \tag{7.49}$$

Noticing $u_0^{\frac{1}{2}} = 0$ and $u_m^{\frac{1}{2}} = 0$, we have

$$(\delta_x^2 u^{\frac{1}{2}}, u^{\frac{1}{2}}) = -|u^{\frac{1}{2}}|_1^2.$$

Taking the imaginary parts on both sides of (7.49), we have

$$\frac{1}{2\tau}(\|u^1\|^2 - \|u^0\|^2) = 0,$$

which implies

$$\|u^1\|^2 = \|u^0\|^2. \tag{7.50}$$

Taking the inner product on both sides of (7.45b) with $u^{\bar{k}}$ gives

$$\mathrm{i}(\Delta_t u^k, u^{\bar{k}}) + (\delta_x^2 u^{\bar{k}}, u^{\bar{k}}) + qh \sum_{j=1}^{m-1} |u_j^k|^2 |u_j^{\bar{k}}|^2 = 0. \tag{7.51}$$

Noticing $u_0^{\bar{k}} = 0$ and $u_m^{\bar{k}} = 0$, we have

$$(\delta_x^2 u^{\bar{k}}, u^{\bar{k}}) = -|u^{\bar{k}}|_1^2.$$

Taking the imaginary parts on both sides of (7.51), we have

$$\frac{1}{4\tau}(\|u^{k+1}\|^2 - \|u^{k-1}\|^2) = 0, \quad 1 \leqslant k \leqslant n-1,$$

which implies

$$\|u^{k+1}\|^2 = \|u^{k-1}\|^2, \quad 1 \leqslant k \leqslant n-1. \tag{7.52}$$

Combining (7.50) with (7.52) leads to (7.46).

(II) Taking the inner product on both sides of (7.45a) with $-\delta_t u^{\frac{1}{2}}$ gives

$$-\,\mathrm{i}\|\delta_t u^{\frac{1}{2}}\|^2 - (\delta_x^2 u^{\frac{1}{2}}, \delta_t u^{\frac{1}{2}}) - qh\sum_{j=1}^{m-1} |\hat{u}_j|^2 u_j^{\frac{1}{2}}\delta_t \bar{u}_j^{\frac{1}{2}} = 0. \tag{7.53}$$

Noticing $\delta_t u_0^{\frac{1}{2}} = 0$ and $\delta_t u_m^{\frac{1}{2}} = 0$, we have

$$-(\delta_x^2 u^{\frac{1}{2}}, \delta_t u^{\frac{1}{2}}) = h\sum_{j=0}^{m-1} \left(\delta_x u^{\frac{1}{2}}_{j+\frac{1}{2}}\right)\left(\delta_t \delta_x \bar{u}^{\frac{1}{2}}_{j+\frac{1}{2}}\right).$$

Taking the real parts on both sides of (7.53), we have

$$\frac{1}{2\tau}\left(|u^1|_1^2 - |u^0|_1^2\right) - qh\sum_{j=1}^{m-1} |\hat{u}_j|^2 \cdot \frac{1}{2\tau}\left(|u_j^1|^2 - |u_j^0|^2\right) = 0,$$

that is

$$\frac{1}{2}(|u^1|_1^2 + |u^0|_1^2) - \frac{q}{2}h\sum_{j=1}^{m-1} |\hat{u}_j|^2|u_j^1|^2 = |u^0|_1^2 - \frac{q}{2}h\sum_{j=1}^{m-1} |\hat{u}_j|^2|u_j^0|^2,$$

which is precisely (7.47).

Taking the inner product on both sides of (7.45b) with $-\Delta_t u^k$ gives

$$-\,\mathrm{i}\,\|\Delta_t u^k\|^2 - (\delta_x^2 u^{\bar{k}}, \Delta_t u^k) - qh\sum_{j=1}^{m-1} |u_j^k|^2 u_j^{\bar{k}}\Delta_t \bar{u}_j^k = 0.$$

Taking the real parts on both sides of the above equality, we get

$$\frac{1}{4\tau}\left(|u^{k+1}|_1^2 - |u^{k-1}|_1^2\right) - qh\sum_{j=1}^{m-1}|u_j^k|^2 \cdot \frac{1}{4\tau}\left(|u_j^{k+1}|^2 - |u_j^{k-1}|^2\right) = 0.$$

Thus,

$$\frac{1}{2}\left(|u^{k+1}|_1^2 + |u^k|_1^2\right) - \frac{q}{2}h\sum_{j=1}^{m-1}|u_j^k|^2 \cdot |u_j^{k+1}|^2$$

$$=\frac{1}{2}\left(|u^k|_1^2 + |u^{k-1}|_1^2\right) - \frac{q}{2}h\sum_{j=1}^{m-1}|u_j^{k-1}|^2 \cdot |u_j^k|^2, \quad 1 \leqslant k \leqslant n-1,$$

which implies that (7.48) is true.

$\square$

Colollary 7.2 *Suppose* $\{u_j^k \mid 0 \leqslant j \leqslant m, 0 \leqslant k \leqslant n\}$ *is the solution of the difference scheme* (7.45), *and then there is a constant* c_{10} *satisfying*

$$\|u^k\|_\infty \leqslant c_{10}, \quad 1 \leqslant k \leqslant n. \tag{7.54}$$

7.3.3 Existence and Uniqueness of the Difference Solution

Theorem 7.9 *The difference scheme* (7.45) *is uniquely solvable.*

Proof Denote

$$u^k = \left(u_0^k, u_1^k, \ldots, u_{m-1}^k, u_m^k\right).$$

(I) From (7.45c) and (7.45d), the value of u^0 has been determined.

(II) The difference Eqs. (7.45a) and (7.45d) form a system of linear equations in u^1. Consider its homogeneous one

$$\begin{cases} \mathrm{i}\dfrac{1}{\tau}u_j^1 + \dfrac{1}{2}\delta_x^2 u_j^1 + \dfrac{1}{2}q|\hat{u}_j|^2 u_j^1 = 0, & 1 \leqslant j \leqslant m-1, \quad (7.55\mathrm{a}) \\[3mm] u_0^1 = 0, \quad u_m^1 = 0. & (7.55\mathrm{b}) \end{cases}$$

Taking the inner product on both sides of (7.55a) with u^1, with the help of the summation by parts, it follows by noticing (7.55b) and taking the imaginary parts that

$$\|u^1\|^2 = 0.$$

This means that

$$u_j^1 = 0, \quad 0 \leqslant j \leqslant m.$$

Hence the value of u^1 at the first time level is uniquely determined by the difference scheme.

(III) Suppose that the values of u^{k-1} and u^k have been obtained, then a system of linear equations in u^{k+1} can be determined from (7.45b) and (7.45d). Consider its homogeneous one

$$
\begin{cases}
\mathrm{i} \cdot \dfrac{1}{2\tau} u_j^{k+1} + \dfrac{1}{2}\delta_x^2 u_j^{k+1} + \dfrac{q}{2}|u_j^k|^2 u_j^{k+1} = 0, & 1 \leqslant j \leqslant m-1, \quad \text{(7.56a)} \\[2mm]
u_0^{k+1} = 0, \quad u_m^{k+1} = 0. & \text{(7.56b)}
\end{cases}
$$

Taking the inner product on both sides of (7.56a) with u^{k+1} and using the summation by parts, noticing (7.56b), it follows by taking the imaginary parts that

$$\frac{1}{2\tau}\|u^{k+1}\|^2 = 0,$$

which implies

$$u_j^{k+1} = 0, \quad 0 \leqslant j \leqslant m.$$

Therefore, the difference Eqs. (7.45b) and (7.45d) admit a unique solution u^{k+1}.

By induction, the conclusion is true. $\qquad\qquad\qquad\qquad\qquad\qquad\qquad\square$

7.3.4　Convergence of the Difference Solution

Theorem 7.10 *Suppose $\{U_j^k \,|\, 0 \leqslant j \leqslant m, 0 \leqslant k \leqslant n\}$ is the solution of the problem (7.1) and $\{u_j^k \,|\, 0 \leqslant j \leqslant m, 0 \leqslant k \leqslant n\}$ is the solution of the difference scheme (7.45). Denote*

$$e_j^k = U_j^k - u_j^k, \quad 0 \leqslant j \leqslant m, \quad 0 \leqslant k \leqslant n.$$

Then there is a constant c_{11} such that

$$\|e^k\| \leqslant c_{11}(\tau^2 + h^2), \quad 0 \leqslant k \leqslant n. \tag{7.57}$$

Proof Subtracting (7.45) from (7.40), (7.42), and (7.44), we have the system of error equations:

$$\begin{cases} i\delta_t e_j^{\frac{1}{2}} + \delta_x^2 e_j^{\frac{1}{2}} + q|\hat{u}_j|^2 e_j^{\frac{1}{2}} = (R_2)_j^0, \quad 1 \leqslant j \leqslant m-1, & (7.58a) \\[2mm] i\Delta_t e_j^k + \delta_x^2 e_j^{\bar{k}} + q(|U_j^k|^2 U_j^{\bar{k}} - |u_j^k|^2 u_j^{\bar{k}}) = (R_2)_j^k, \\[1mm] \qquad\qquad\qquad 1 \leqslant j \leqslant m-1, \quad 1 \leqslant k \leqslant n-1, & (7.58b) \\[2mm] e_j^0 = 0, \quad 1 \leqslant j \leqslant m-1, & (7.58c) \\[2mm] e_0^k = 0, \quad e_m^k = 0, \quad 0 \leqslant k \leqslant n. & (7.58d) \end{cases}$$

It follows from (7.58d) that

$$\begin{cases} e_0^{\frac{1}{2}} = 0, \quad e_m^{\frac{1}{2}} = 0, \\[2mm] e_0^{\bar{k}} = 0, \quad e_m^{\bar{k}} = 0, \quad 1 \leqslant k \leqslant n-1. \end{cases}$$

(I) Taking the inner product on both sides of (7.58a) with $e^{\frac{1}{2}}$ gives

$$i(\delta_t e^{\frac{1}{2}}, e^{\frac{1}{2}}) + (\delta_x^2 e^{\frac{1}{2}}, e^{\frac{1}{2}}) + qh \sum_{j=1}^{m-1} |\hat{u}_j|^2 |e_j^{\frac{1}{2}}|^2 = ((R_2)^0, e^{\frac{1}{2}}). \tag{7.59}$$

Noticing

$$(\delta_x^2 e^{\frac{1}{2}}, e^{\frac{1}{2}}) = -\|\delta_x e^{\frac{1}{2}}\|^2$$

and taking the imaginary parts on both sides of (7.59), we have

$$\frac{1}{2\tau}(\|e^1\|^2 - \|e^0\|^2) = \mathrm{Im}\left\{((R_2)^0, e^{\frac{1}{2}})\right\} \leqslant \|(R_2)^0\| \cdot \|e^{\frac{1}{2}}\|.$$

Noticing (7.58c) and (7.58d), it reduces to

$$\frac{1}{2\tau}\|e^1\|^2 \leqslant \|(R_2)^0\| \cdot \frac{1}{2}\|e^1\|,$$

which implies by noticing (7.41) that

$$\|e^1\| \leqslant \tau\|(R_2)^0\| \leqslant \tau c_8 \sqrt{L}(\tau^2 + h^2) \leqslant c_8 \sqrt{L}(\tau^2 + h^2). \tag{7.60}$$

(II) Taking the inner product on both sides of (7.58b) with $e^{\bar{k}}$ gives

$$\mathrm{i}(\Delta_t e^k, e^{\bar{k}}) + (\delta_x^2 e^{\bar{k}}, e^{\bar{k}}) + qh \sum_{j=1}^{m-1} (|U_j^k|^2 U_j^{\bar{k}} - |u_j^k|^2 u_j^{\bar{k}}) \bar{e}_j^{\bar{k}} = ((R_2)^k, e^{\bar{k}}).$$

(7.61)

Noticing

$$(\delta_x^2 e^{\bar{k}}, e^{\bar{k}}) = -\|\delta_x e^{\bar{k}}\|^2,$$

$$(|U_j^k|^2 U_j^{\bar{k}} - |u_j^k|^2 u_j^{\bar{k}}) \bar{e}_j^{\bar{k}} = \left[|U_j^k|^2 e_j^{\bar{k}} + (|U_j^k|^2 - |u_j^k|^2) u_j^{\bar{k}} \right] \bar{e}_j^{\bar{k}}$$

$$= |U_j^k|^2 |e_j^{\bar{k}}|^2 + (e_j^k \bar{U}_j^k + u_j^k \bar{e}_j^k) u_j^{\bar{k}} \bar{e}_j^{\bar{k}},$$

(7.62)

and taking the imaginary parts on both sides of (7.61), with the help of (7.7) and (7.54), we get

$$\frac{1}{4\tau} (\|e^{k+1}\|^2 - \|e^{k-1}\|^2)$$

$$= -q \,\mathrm{Im} \left\{ h \sum_{j=1}^{m-1} (e_j^k \bar{U}_j^k + u_j^k \bar{e}_j^k) u_j^{\bar{k}} \bar{e}_j^{\bar{k}} \right\} + \mathrm{Im} \left\{ h \sum_{j=1}^{m-1} (R_2)_j^k \bar{e}_j^{\bar{k}} \right\}$$

$$\leqslant |q|(c_0 + c_{10})c_{10}\|e^k\| \cdot \|e^{\bar{k}}\| + \|(R_2)^k\| \cdot \|e^{\bar{k}}\|$$

$$\leqslant \left(|q|(c_0 + c_{10})c_{10}\|e^k\| + \|(R_2)^k\| \right) \frac{\|e^{k+1}\| + \|e^{k-1}\|}{2}, \quad 1 \leqslant k \leqslant n-1.$$

Noticing (7.43), it further implies that

$$\frac{1}{2\tau} (\|e^{k+1}\| - \|e^{k-1}\|)$$

$$\leqslant |q|(c_0 + c_{10})c_{10}\|e^k\| + \|(R_2)^k\|$$

$$\leqslant |q|(c_0 + c_{10})c_{10}\|e^k\| + \sqrt{L}c_9(\tau^2 + h^2), \quad 1 \leqslant k \leqslant n-1,$$

or

$$\|e^{k+1}\| \leqslant \|e^{k-1}\| + 2|q|(c_0 + c_{10})c_{10}\tau\|e^k\| + 2\sqrt{L}c_9\tau(\tau^2 + h^2), \quad 1 \leqslant k \leqslant n-1.$$

It follows from the above inequality that

$$\max\{\|e^{k+1}\|, \|e^k\|\} \leqslant \left[1 + 2|q|(c_0 + c_{10})c_{10}\tau \right] \max\{\|e^k\|, \|e^{k-1}\|\}$$

$$+ 2\sqrt{L}c_9\tau(\tau^2 + h^2), \quad 1 \leqslant k \leqslant n-1.$$

When $q = 0$, we have

$$\max\{\|e^{k+1}\|, \|e^k\|\} \leqslant \max\{\|e^1\|, \|e^0\|\} + 2\sqrt{L}c_9 k\tau(\tau^2 + h^2)$$

$$\leqslant (c_8 + 2c_9 T)\sqrt{L}(\tau^2 + h^2), \quad 0 \leqslant k \leqslant n - 1.$$

When $q \neq 0$, it follows from Lemma 3.3 (Gronwall inequality) that

$$\max\{\|e^{k+1}\|, \|e^k\|\}$$

$$\leqslant e^{2|q|(c_0+c_{10})c_{10}T} \left[\max\{\|e^1\|, \|e^0\|\} + \frac{\sqrt{L}c_9}{|q|(c_0 + c_{10})c_{10}}(\tau^2 + h^2) \right]$$

$$\leqslant e^{2|q|(c_0+c_{10})c_{10}T} \left[c_8 + \frac{c_9}{|q|(c_0 + c_{10})c_{10}} \right] \sqrt{L}(\tau^2 + h^2), \quad 0 \leqslant k \leqslant n - 1.$$

$$\square$$

Theorem 7.11 *Suppose $\{U_j^k \,|\, 0 \leqslant j \leqslant m, 0 \leqslant k \leqslant n\}$ is the solution of the problem (7.1) and $\{u_j^k \,|\, 0 \leqslant j \leqslant m, 0 \leqslant k \leqslant n\}$ is the solution of the difference scheme (7.45). Denote*

$$e_j^k = U_j^k - u_j^k, \quad 0 \leqslant j \leqslant m, \ 0 \leqslant k \leqslant n.$$

Then there is a constant c_{12} satisfying

$$\|e^k\|_\infty \leqslant c_{12}(\tau^2 + h^2), \quad 0 \leqslant k \leqslant n.$$

Proof

(I) Taking the inner product on both sides of (7.58a) with $-\delta_t e^{\frac{1}{2}}$ gives

$$-\mathrm{i}\|\delta_t e^{\frac{1}{2}}\|^2 - (\delta_x^2 e^{\frac{1}{2}}, \delta_t e^{\frac{1}{2}}) - qh \sum_{j=1}^{m-1} |\hat{u}_j|^2 e_j^{\frac{1}{2}} \delta_t \bar{e}_j^{\frac{1}{2}} = -((R_2)^0, \delta_t e^{\frac{1}{2}}). \quad (7.63)$$

Noticing

$$-\left(\delta_x^2 e^{\frac{1}{2}}, \delta_t e^{\frac{1}{2}}\right) = h \sum_{j=1}^{m} \left(\delta_x e_{j-\frac{1}{2}}^{\frac{1}{2}}\right)\left(\delta_t \delta_x \bar{e}_{j-\frac{1}{2}}^{\frac{1}{2}}\right)$$

and taking the real parts on both sides of (7.63), we get

$$\frac{1}{2\tau}(|e^1|_1^2 - |e^0|_1^2) - qh\sum_{j=0}^{m-1}|\hat{u}_j|^2\frac{|e_j^1|^2 - |e_j^0|^2}{2\tau} = \mathrm{Re}\left\{-((R_2)^0, \delta_t e^{\frac{1}{2}})\right\}$$

$$\leqslant \|(R_2)^0\| \cdot \|\delta_t e^{\frac{1}{2}}\|,$$

which follows from

$$e_j^0 = 0, \quad 0 \leqslant j \leqslant m$$

that

$$\frac{1}{2\tau}|e^1|_1^2 - q \cdot \frac{1}{2\tau}h\sum_{j=0}^{m-1}|\hat{u}_j|^2|e_j^1|^2 \leqslant \frac{1}{\tau}\|(R_2)^0\| \cdot \|e^1\|.$$

Multiplying 2τ on both sides of the above inequality produces

$$|e^1|_1^2 \leqslant |q|h\sum_{j=0}^{m-1}|\hat{u}_j|^2|e_j^1|^2 + 2\|(R_2)^0\| \cdot \|e^1\|. \tag{7.64}$$

Denote

$$c_{13} = \max_{0\leqslant x\leqslant L}|u_t(x,0)|.$$

Then we have

$$|\hat{u}_j| \leqslant c_0 + \frac{\tau}{2}c_{13} \leqslant c_0 + c_{13}, \quad 1 \leqslant j \leqslant m-1.$$

It follows from (7.64), (7.41) and (7.60) that

$$|e^1|_1^2 \leqslant |q|(c_0 + c_{13})^2\|e^1\|^2 + 2\|(R_2)^0\| \cdot \|e^1\|$$

$$\leqslant |q|(c_0 + c_{13})^2[c_8\sqrt{L}(\tau^2 + h^2)]^2 + 2\sqrt{L}c_8(\tau^2 + h^2)c_8\sqrt{L}(\tau^2 + h^2)$$

$$= [|q|(c_0 + c_{13})^2 + 2][c_8\sqrt{L}(\tau^2 + h^2)]^2,$$

which simplifies further to

$$|e^1|_1 \leqslant \sqrt{|q|(c_0 + c_{13})^2 + 2}\, c_8\sqrt{L}(\tau^2 + h^2). \tag{7.65}$$

(II) Taking the inner product on both sides of (7.58b) with $-\Delta_t e^k$ gives

$$-\mathrm{i}\|\Delta_t e^k\|^2 - \left(\delta_x^2 e^{\bar{k}}, \Delta_t e^k\right)$$

$$= qh\sum_{j=1}^{m-1}(|U_j^k|^2 U_j^{\bar{k}} - |u_j^k|^2 u_j^{\bar{k}})\Delta_t \bar{e}_j^k - h\sum_{j=1}^{m-1}(R_2)_j^k\Delta_t \bar{e}_j^k, \quad 1\leqslant k\leqslant n-1.$$

$$(7.66)$$

For the second term on the left-hand side of (7.66), we have

$$-\left(\delta_x^2 e^{\bar{k}}, \Delta_t e^k\right) = h\sum_{j=0}^{m-1}(\delta_x e_{j+\frac{1}{2}}^{\bar{k}})\Delta_t \delta_x \bar{e}_{j+\frac{1}{2}}^k.$$

Then taking the real parts of the above equality results in

$$\mathrm{Re}\left\{-h\sum_{j=1}^{m-1}(\delta_x^2 e_j^{\bar{k}})\Delta_t \bar{e}_j^k\right\} = \frac{1}{4\tau}(|e^{k+1}|_1^2 - |e^{k-1}|_1^2).$$

$$(7.67)$$

Now we analyze the first term on the right-hand side of (7.66).
It follows from (7.58b) that

$$\Delta_t e_j^k = \mathrm{i}\delta_x^2 e_j^{\bar{k}} + \mathrm{i}q\left(|U_j^k|^2 U_j^{\bar{k}} - |u_j^k|^2 u_j^{\bar{k}}\right) - \mathrm{i}(R_2)_j^k.$$

Thus, we have

$$\mathrm{Re}\left\{qh\sum_{j=1}^{m-1}\left(|U_j^k|^2 U_j^{\bar{k}} - |u_j^k|^2 u_j^{\bar{k}}\right)\Delta_t \bar{e}_j^k\right\}$$

$$=\mathrm{Re}\left\{qh\sum_{j=1}^{m-1}\overline{|U_j^k|^2 U_j^{\bar{k}} - |u_j^k|^2 u_j^{\bar{k}}}\,\Delta_t e_j^k\right\}$$

$$=\mathrm{Re}\left\{qh\sum_{j=1}^{m-1}\overline{|U_j^k|^2 U_j^{\bar{k}} - |u_j^k|^2 u_j^{\bar{k}}}\left[\mathrm{i}\delta_x^2 e_j^{\bar{k}} + \mathrm{i}q\left(|U_j^k|^2 U_j^{\bar{k}} - |u_j^k|^2 u_j^{\bar{k}}\right) - \mathrm{i}(R_2)_j^k\right]\right\}$$

$$=\mathrm{Re}\left\{qh\sum_{j=1}^{m-1}\overline{|U_j^k|^2 U_j^{\bar{k}} - |u_j^k|^2 u_j^{\bar{k}}}\left(\mathrm{i}\delta_x^2 e_j^{\bar{k}} - \mathrm{i}(R_2)_j^k\right)\right\}.$$

Noticing (7.62), (7.7), and (7.54), there is a constant c_{14} such that

$$
\mathrm{Re}\left\{ qh \sum_{j=1}^{m-1} \left(|U_j^k|^2 U_j^{\bar{k}} - |u_j^k|^2 u_j^{\bar{k}} \right) \Delta_t \bar{e}_j^k \right\}
$$

$$
\leqslant c_{14}\left(\|e^{k-1}\|^2 + \|e^k\|^2 + \|e^{k+1}\|^2 + |e^{k-1}|_1^2 + |e^k|_1^2 + |e^{k+1}|_1^2 + \|(R_2)^k\|^2 \right).
$$

$$(7.68)$$

Taking the real parts on both sides of (7.66) and combining (7.67) with (7.68) arrive at

$$
\frac{1}{4\tau}\left(|e^{k+1}|_1^2 - |e^{k-1}|_1^2 \right)
$$

$$
\leqslant c_{14}\left(\|e^{k-1}\|^2 + \|e^k\|^2 + \|e^{k+1}\|^2 + |e^{k-1}|_1^2 + |e^k|_1^2 + |e^{k+1}|_1^2 + \|(R_2)^k\|^2 \right)
$$

$$
+ \mathrm{Re}\left\{ -h \sum_{j=1}^{m-1} (R_2)_j^k \Delta_t \bar{e}_j^k \right\}, \quad 1 \leqslant k \leqslant n-1.
$$

Replacing k with l and summing over l from 1 to k yield

$$
\frac{1}{4\tau}\left(|e^{k+1}|_1^2 + |e^k|_1^2 - |e^1|_1^2 - |e^0|_1^2 \right)
$$

$$
\leqslant c_{14} \sum_{l=1}^{k} \left(\|e^{l-1}\|^2 + \|e^l\|^2 + \|e^{l+1}\|^2 + |e^{l-1}|_1^2 + |e^l|_1^2 + |e^{l+1}|_1^2 + \|(R_2)^l\|^2 \right)
$$

$$
+ \mathrm{Re}\left\{ -h \sum_{j=1}^{m-1} \sum_{l=1}^{k} (R_2)_j^l \Delta_t \bar{e}_j^l \right\}, \quad 1 \leqslant k \leqslant n-1. \qquad (7.69)
$$

Noticing

$$
\sum_{l=1}^{k} (R_2)_j^l \Delta_t \bar{e}_j^l
$$

$$
= \frac{1}{2\tau} \sum_{l=1}^{k} (R_2)_j^l \left(\bar{e}_j^{l+1} - \bar{e}_j^{l-1} \right)
$$

$$
= \frac{1}{2\tau} \left[\sum_{l=2}^{k+1} (R_2)_j^{l-1} \bar{e}_j^l - \sum_{l=0}^{k-1} (R_2)_j^{l+1} \bar{e}_j^l \right]
$$

$$
=\frac{1}{2\tau}\left[(R_2)_j^k \bar{e}_j^{k+1} + (R_2)_j^{k-1}\bar{e}_j^k - \sum_{l=2}^{k-1}\left((R_2)_j^{l+1} - (R_2)_j^{l-1}\right)\bar{e}_j^l - (R_2)_j^2\bar{e}_j^1\right]
$$

$$
=\frac{1}{2\tau}\left[(R_2)_j^k \bar{e}_j^{k+1} + (R_2)_j^{k-1}\bar{e}_j^k - 2\tau\sum_{l=2}^{k-1}\left(\Delta_t (R_2)_j^l\right)\bar{e}_j^l - (R_2)_j^2\bar{e}_j^1\right],
$$

we have

$$
\left|h\sum_{j=1}^{m-1}\sum_{l=1}^{k-1}(R_2)_j^l \Delta_t \bar{e}_j^l\right|
$$

$$
=\left|\frac{1}{2\tau}h\sum_{j=1}^{m-1}(R_2)_j^k \bar{e}_j^{k+1} + \frac{1}{2\tau}h\sum_{j=1}^{m-1}(R_2)_j^{k-1}\bar{e}_j^k\right.
$$

$$
\left.-\sum_{l=2}^{k-1}h\sum_{j=1}^{m-1}\left(\Delta_t (R_2)_j^l\right)\bar{e}_j^l - \frac{1}{2\tau}h\sum_{j=1}^{m-1}(R_2)_j^2\bar{e}_j^1\right|
$$

$$
\leqslant \frac{1}{2\tau}\|(R_2)^k\|\cdot\|e^{k+1}\| + \frac{1}{2\tau}\|(R_2)^{k-1}\|\cdot\|e^k\|
$$

$$
+\sum_{l=2}^{k-1}\|\Delta_t (R_2)^l\|\cdot\|e^l\| + \frac{1}{2\tau}\|(R_2)^2\|\cdot\|e^1\|. \tag{7.70}
$$

Substituting (7.70) into (7.69) and multiplying 4τ on both sides lead to

$$
|e^{k+1}|_1^2 + |e^k|_1^2 - |e^1|_1^2 - |e^0|_1^2
$$

$$
\leqslant 4c_{14}\tau\sum_{l=1}^{k}(\|e^{l-1}\|^2 + \|e^l\|^2 + \|e^{l+1}\|^2 + |e^{l-1}|_1^2 + |e^l|_1^2 + |e^{l+1}|_1^2 + \|(R_2)^l\|^2)
$$

$$
+ 2\|(R_2)^k\|\cdot\|e^{k+1}\| + 2\|(R_2)^{k-1}\|\cdot\|e^k\| + 4\tau\sum_{l=2}^{k-1}\|\Delta_t (R_2)^l\|\cdot\|e^l\|
$$

$$
+ 2\|(R_2)^2\|\cdot\|e^1\|, \quad 1\leqslant k\leqslant n-1.
$$

Noticing

$$
2\|(R_2)^k\|\cdot\|e^{k+1}\| \leqslant \frac{3}{L^2}\|e^{k+1}\|^2 + \frac{L^2}{3}\|(R_2)^k\|^2 \leqslant \frac{1}{2}|e^{k+1}|_1^2 + \frac{L^2}{3}\|(R_2)^k\|^2,
$$

$$
2\|(R_2)^{k-1}\|\cdot\|e^k\| \leqslant \frac{3}{L^2}\|e^k\|^2 + \frac{L^2}{3}\|(R_2)^{k-1}\|^2 \leqslant \frac{1}{2}|e^k|_1^2 + \frac{L^2}{3}\|(R_2)^{k-1}\|^2,
$$

$$
2\|(R_2)^2\|\cdot\|e^1\| \leqslant \|(R_2)^2\|^2 + \|e^1\|^2,
$$

and in combination of (7.57), (7.65), and (7.43), there is a constant c_{15} such that

$$|e^{k+1}|_1^2 + |e^k|_1^2$$

$$\leqslant c_{15}\tau \sum_{l=1}^{k}(|e^{l+1}|_1^2 + 2|e^l|_1^2 + |e^{l-1}|_1^2) + c_{15}(\tau^2 + h^2)^2, \quad 1 \leqslant k \leqslant n-1,$$

which can be rearranged as

$$(1 - c_{15}\tau)(|e^{k+1}|_1^2 + |e^k|_1^2)$$

$$\leqslant 2c_{15}\tau \sum_{l=0}^{k-1}(|e^{l+1}|_1^2 + |e^l|_1^2) + c_{15}(\tau^2 + h^2)^2, \quad 1 \leqslant k \leqslant n-1.$$

When $c_{15}\tau \leqslant \dfrac{1}{3}$, we have

$$|e^{k+1}|_1^2 + |e^k|_1^2$$

$$\leqslant 3c_{15}\tau \sum_{l=0}^{k-1}(|e^{l+1}|_1^2 + |e^l|_1^2) + \frac{3}{2}c_{15}(\tau^2 + h^2)^2, \quad 1 \leqslant k \leqslant n-1.$$

It follows by using Lemma 5.1 (Gronwall inequality) and noticing (7.65) that

$$|e^{k+1}|_1^2 + |e^k|_1^2$$

$$\leqslant e^{3c_{15}k\tau}\left[|e^1|_1^2 + |e^0|_1^2 + \frac{3}{2}c_{15}(\tau^2 + h^2)^2\right]$$

$$\leqslant e^{3c_{15}T}\left[(|q|(c_0 + c_{13})^2 + 2)c_8^2 L + \frac{3}{2}c_{15}\right](\tau^2 + h^2)^2, \quad 1 \leqslant k \leqslant n-1.$$

Noticing Lemma 1.4, we obtain

$$\|e^k\|_\infty \leqslant \frac{\sqrt{L}}{2}|e^k|_1 \leqslant \frac{\sqrt{L}}{2} \cdot e^{\frac{3}{2}c_{15}T}\sqrt{(|q|(c_0 + c_{13})^2 + 2)c_8^2 L + \frac{3}{2}c_{15}(\tau^2 + h^2)},$$

$$1 \leqslant k \leqslant n.$$

$\square$

Table 7.3 (Example 7.2) The maximum norm errors of numerical solutions with different spatial step sizes ($\tau = 1/1600$)

h	$E_\infty(h, \tau)$	$E_\infty(2h, \tau)/E_\infty(h, \tau)$
1/5	1.394862e−01	
1/10	3.352911e−02	4.1602
1/20	8.313797e−03	4.0329
1/40	2.092628e−03	3.9729

Table 7.4 (Example 7.2) The maximum norm errors of numerical solutions with different temporal step sizes ($h = 1/400$)

τ	$E_\infty(h, \tau)$	$E_\infty(h, 2\tau)/E_\infty(h, \tau)$
1/20	2.041883e−01	
1/40	4.420347e−02	4.6193
1/80	1.083048e−02	4.0814
1/160	2.711934e−03	3.9936

Fig. 7.2 (Example 7.2) The curves of energy E^k for the difference scheme (7.45)

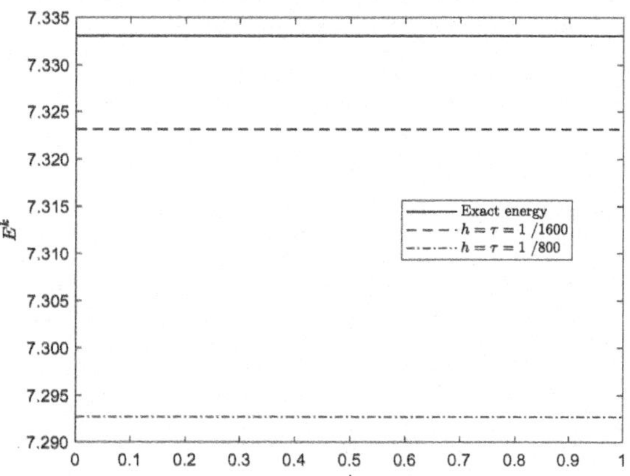

7.3.5   *Numerical Examples*

Example 7.2 Use the difference scheme (7.45) to compute the initial-boundary value problem (7.39) in Example 7.1.

The Thomas algorithm is borrowed to solve the difference scheme (7.45).

For different values of step sizes, Tables 7.3 and 7.4 collect the maximum errors of numerical solutions defined by

$$E_\infty(h, \tau) = \max_{0 \leqslant k \leqslant n, 0 \leqslant j \leqslant m} \left| u(x_j, t_k) - u_j^k \right|.$$

From Table 7.3, it can be observed that when the spatial step size h is halved, the maximum norm error is reduced to one-quarter of its original value. Similarly, from Table 7.4, it is evident that when the temporal step size τ is halved, the maximum norm error is also reduced to one-quarter of the original. The curves of the invariant E^k defined in Theorem 7.8 are presented in Fig. 7.2.

7.4   Summary and Extension

In this chapter, two difference schemes for the initial-boundary value problem of the one-dimensional Schrödinger equation are developed and analyzed theoretically.

The first difference scheme is a two-time-level nonlinear scheme. It is demonstrated that the solution of the difference scheme satisfies two energy conservation laws, and an error estimate for the solution in the maximum norm is provided. Browder's theorem is applied to establish the existence of the solution. Additionally, it is proved that the solution of the difference scheme is unique and exhibits second-order unconditional convergence in both time and space in the maximum norm. This work is based on [1, 6].

The second difference scheme is three-time-level linearized. It is demonstrated that the solution of the difference scheme satisfies two energy conservation laws, and an error estimate for the difference solution in the maximum norm is provided. The unique solvability and convergence of the difference scheme are also established.

In [8], Wang and Guo developed a two-level nonlinear compact difference scheme and a three-level linearized compact difference scheme for the Schrödinger equation. In [3, 10], the authors derived a two-level nonlinear difference scheme of order two in time and four in space using a five-point spatial stencil. Building upon the work in [3, 10], a three-level linearized difference scheme of order two in time and four in space was proposed on a five-point spatial stencil in [5]. In [4, 6, 7], difference methods for solving the coupled system of Schrödinger equations were considered. Additionally, in [9], the periodic boundary value problem for the two-dimensional Schrödinger equation was investigated.

7.5   Exercise

7.1   Replace $\hat{u}_j$ with u_j^0 in (7.45a) and establish a difference scheme for the problem (7.1) as follows:

$$\begin{cases} \mathrm{i}\delta_t u_j^{\frac{1}{2}} + \delta_x^2 u_j^{\frac{1}{2}} + q|u_j^0|^2 u_j^{\frac{1}{2}} = 0, & 1 \leqslant j \leqslant m-1, \\ \mathrm{i}\Delta_t u_j^k + \delta_x^2 u_j^{\bar{k}} + q|u_j^k|^2 u_j^{\bar{k}} = 0, & 1 \leqslant j \leqslant m-1, \quad 1 \leqslant k \leqslant n-1, \\ u_j^0 = \varphi(x_j), & 1 \leqslant j \leqslant m-1, \\ u_0^k = 0, \quad u_m^k = 0, & 0 \leqslant k \leqslant n. \end{cases} \tag{7.71}$$

(1)   Suppose $\{u_j^k \mid 0 \leqslant j \leqslant m, \ 0 \leqslant k \leqslant n\}$ is the solution of (7.71). Denote

$$E^k = \frac{1}{2}(|u^{k+1}|_1^2 + |u^k|_1^2) - \frac{q}{2}h\sum_{j=1}^{m-1}|u_j^k|^2 \cdot |u_j^{k+1}|^2, \quad 0 \leqslant k \leqslant n-1.$$

Try to prove that

$$E^k = |u^0|_1^2 - \frac{q}{2}h\sum_{j=1}^{m-1}|u_j^0|^4, \quad 0 \leqslant k \leqslant n-1.$$

(2) Denote

$$e_j^k = u(x_j, t_k) - u_j^k, \quad 0 \leqslant j \leqslant m, \quad 0 \leqslant k \leqslant n.$$

Try to show that there is a constant c_{16} such that

$$\|e^k\| \leqslant c_{16}(\tau^2 + h^2), \quad 1 \leqslant k \leqslant n.$$

7.2 Suppose $u(x,t)$ is the solution of the initial-boundary value problem of Kuramoto-Tsuzuki equation in the form of

$$\begin{cases} u_t = (1 + ic_1)u_{xx} + u - (1 + ic_2)|u|^2u, & 0 < x < L, \quad 0 < t \leqslant T, \\ u(x, 0) = \varphi(x), & 0 \leqslant x \leqslant L, \\ u_x(0, t) = 0, \quad u_x(L, t) = 0, & 0 < t \leqslant T, \end{cases}$$

$$(7.72)$$

where c_1 and c_2 are real constants, functions $\varphi(x)$ and $u(x, t)$ are complex, and $\varphi_x(0) = \varphi_x(L) = 0$. Try to prove that

$$\int_0^L |u(x, t)|^2 dx \leqslant e^{2t}\|\varphi\|^2, \quad 0 < t \leqslant T.$$

7.3 For the problem (7.72), a two-level nonlinear difference scheme is derived as follows:

$$\begin{cases} \delta_t u_0^{k+\frac{1}{2}} = (1 + ic_1)\frac{2}{h}\delta_x u_{\frac{1}{2}}^{k+\frac{1}{2}} + u_0^{k+\frac{1}{2}} - (1 + ic_2)\left|u_0^{k+\frac{1}{2}}\right|^2 u_0^{k+\frac{1}{2}}, \\ \qquad\qquad 0 \leqslant k \leqslant n-1, \\ \delta_t u_j^{k+\frac{1}{2}} = (1 + ic_1)\delta_x^2 u_j^{k+\frac{1}{2}} + u_j^{k+\frac{1}{2}} - (1 + ic_2)\left|u_j^{k+\frac{1}{2}}\right|^2 u_j^{k+\frac{1}{2}}, \\ \qquad\qquad 1 \leqslant j \leqslant m-1, \quad 0 \leqslant k \leqslant n-1, \\ \delta_t u_m^{k+\frac{1}{2}} = (1 + ic_1)\left(-\frac{2}{h}\delta_x u_{m-\frac{1}{2}}^{k+\frac{1}{2}}\right) + u_m^{k+\frac{1}{2}} - (1 + ic_2)\left|u_m^{k+\frac{1}{2}}\right|^2 u_m^{k+\frac{1}{2}}, \\ \qquad\qquad 0 \leqslant k \leqslant n-1, \\ u_j^0 = \varphi(x_j), \quad 0 \leqslant j \leqslant m. \end{cases}$$

$$(7.73)$$

(1) Show the local truncation error of the difference scheme (7.73).
(2) Show that there is a solution of the difference scheme (7.73).
(3) Show that the solution of the difference scheme (7.73) satisfies

$$\|u^k\| \leqslant e^{\frac{3}{2}T}\|u^0\|, \quad 1 \leqslant k \leqslant n,$$

where

$$\|u^k\| = \sqrt{h\left(\frac{1}{2}|u_0^k|^2 + \sum_{j=1}^{m-1}|u_j^k|^2 + \frac{1}{2}|u_m|^2\right)}.$$

7.4   For the problem (7.72), the following three-level linearized difference scheme is proposed:

$$\begin{cases}
\Delta_t u_0^k = (1+ic_1)\dfrac{2}{h}\delta_x u_{\frac{1}{2}}^{\bar{k}} + u_0^{\bar{k}} - (1+ic_2)|u_0^k|^2 u_0^{\bar{k}}, & 1 \leqslant k \leqslant n-1, \\[2mm]
\Delta_t u_j^k = (1+ic_1)\delta_x^2 u_j^{\bar{k}} + u_j^{\bar{k}} - (1+ic_2)|u_j^k|^2 u_j^{\bar{k}}, & 1 \leqslant j \leqslant m-1, \quad 1 \leqslant k \leqslant n-1, \\[2mm]
\Delta_t u_m^k = (1+ic_1)\left(-\dfrac{2}{h}\delta_x u_{m-\frac{1}{2}}^{\bar{k}}\right) + u_m^{\bar{k}} - (1+ic_2)|u_m^k|^2 u_m^{\bar{k}}, & 1 \leqslant k \leqslant n-1, \\[2mm]
\delta_t u_0^{\frac{1}{2}} = (1+ic_1)\dfrac{2}{h}\delta_x u_{\frac{1}{2}}^{\frac{1}{2}} + u_0^{\frac{1}{2}} - (1+ic_2)|\hat{u}_0|^2 u_0^{\frac{1}{2}}, \\[2mm]
\delta_t u_j^{\frac{1}{2}} = (1+ic_1)\delta_x^2 u_j^{\frac{1}{2}} + u_j^{\frac{1}{2}} - (1+ic_2)|\hat{u}_j|^2 u_j^{\frac{1}{2}}, & 1 \leqslant j \leqslant m-1, \\[2mm]
\delta_t u_m^{\frac{1}{2}} = (1+ic_1)\left(-\dfrac{2}{h}\delta_x u_{m-\frac{1}{2}}^{\frac{1}{2}}\right) + u_m^{\frac{1}{2}} - (1+ic_2)|\hat{u}_m|^2 u_m^{\frac{1}{2}}, \\[2mm]
u_j^0 = \varphi(x_j), & 0 \leqslant j \leqslant m,
\end{cases}$$

$$(7.74)$$

where

$$\hat{u}_j = \varphi(x_j) + \frac{\tau}{2}u_t(x_j, 0), \quad 0 \leqslant j \leqslant m.$$

(1) Show the local truncation error of the difference scheme.
(2) Show that the solution of the difference scheme (7.74) satisfies

$$\begin{cases}
\|u^1\| \leqslant \frac{1+\tau}{1-\tau}\|u^0\|; \\[2mm]
\|u^{k+1}\| \leqslant \frac{1+\tau}{1-\tau}\|u^{k-1}\|, & 1 \leqslant k \leqslant n-1.
\end{cases}$$

(3) Show the unique solvability of the difference scheme (7.74).

References

1. Akrivis, G.D.: Finite difference discretization of the cubic Schrödinger equation. IMA J. Numer. Anal. **13**, 115–124 (1993)
2. Browder, F.E.: Existence and uniqueness theorems for solutions of nonlinear boundary value problems. Proc. Sympos. Appl. Math., Amer. Math. Soc., Providence, R. I. **17**, 24–49 (1965)
3. Cui, J., Sun, Z.Z., Wu, H.W.: A high accurate and conservative difference scheme for the solution of nonlinear Schrödinger equation. Numer. Math. A J. Chinese Univ. **37**(1), 31–52 (2015)
4. Speúlveda, M., Vera, O.: Numerical methods for a coupled nonlinear Schrödinger system. Bol. Soc. Esp. Mat. Apl. **43**, 95–102 (2008)
5. Sun, Z.Z., Zhang, Q.F., Gao, G.H.: Finite Difference Methods For Nonlinear Evolution Equations. De Gruyter; Science Press, Berlin (2023)
6. Sun, Z.Z., Zhao, D.D.: On the L_∞ convergence of a difference scheme for coupled nonlinear Schrödinger equations. Comput. Math. Appl. **59**, 3286–3300 (2010)
7. Wang, T.C.: Maximum norm error bound of a linearized difference scheme for a coupled nonlinear Schrödinger equations. J. Comput. Appl. Math. **235**, 4237–4250 (2011)
8. Wang, T.C., Guo, B.L.: Unconditional convergence of two conservative compact difference schemes for non-linear Schrödinger equation in one dimension. Sci. Sin. Math. **41**(3), 207–233 (2011)
9. Wang, T.C., Guo, B.L., Xu, Q.: Fourth-order compact and energy conservative difference schemes for the nonlinear Schrödinger equation in two dimensions. J. Comput. Phys. **243**, 382–399 (2013)
10. Zhang, R.P., Cao, S.S.: A high accurate and conservative numerical scheme for nonlinear Schrödinger equation. Numer. Math. A J. Chinese Univ. **29**(3), 226–235 (2007)

Chapter 8
Finite Difference Methods for the Burgers' Equation

The Burgers' equation models a variety of physical phenomena, including fluid dynamics, nonlinear acoustics, gas dynamics, and traffic flow. It can be considered as a simplified version of the Navier-Stokes equation in the context of fluid dynamics. In recent years, there has been growing interest among researchers in developing numerical methods for solving the Burgers' equation.

8.1 The Burgers' Equation

Consider the initial-boundary value problem of the one-dimensional nonlinear Burgers' equation:

$$
\begin{cases}
u_t + u u_x = v u_{xx}, & 0 < x < L, \quad 0 < t \leqslant T, & (8.1a) \\[2mm]
u(x, 0) = \varphi(x), & 0 < x < L, & (8.1b) \\[2mm]
u(0, t) = 0, \quad u(L, t) = 0, & 0 \leqslant t \leqslant T, & (8.1c)
\end{cases}
$$

where $v > 0$ is the dynamic viscosity coefficient, and $\varphi(x)$ is a given function satisfying $\varphi(0) = \varphi(L) = 0$.

Before introducing a numerical difference scheme, we first apply the energy method to derive a priori estimate for the solution of problem (8.1).

Theorem 8.1 *Let $u(x, t)$ be the solution of the problem (8.1). Denote*

$$
E(t) = \int_0^L u^2(x, t) \mathrm{d}x + 2v \int_0^t \left[\int_0^L u_x^2(x, s) \mathrm{d}x \right] \mathrm{d}s.
$$

© Science Press 2026

Z.-Z. Sun et al., *Numerical Solutions to Partial Differential Equations with Finite Difference Methods*, Springer Asia Pacific Mathematics Series 9, https://doi.org/10.1007/978-981-95-5563-5_8

Then

$$E(t) = E(0), \quad 0 < t \leqslant T.$$

Proof   Multiplying (8.1a) by $u(x, t)$ on both sides, we have

$$\left(\frac{1}{2}u^2\right)_t + \left(\frac{1}{3}u^3\right)_x = \nu\left[(uu_x)_x - u_x^2\right].$$

Integrating the above equality on both sides with respect to x on the interval $[0, L]$ and noticing (8.1c), we arrive at

$$\frac{1}{2}\frac{\mathrm{d}}{\mathrm{d}t}\int_0^L u^2(x, t)\mathrm{d}x + \nu \int_0^L u_x^2(x, t)\mathrm{d}x = 0,$$

which can be rewritten as

$$\frac{1}{2}\frac{\mathrm{d}}{\mathrm{d}t}\left\{\int_0^L u^2(x, t)\mathrm{d}x + 2\nu \int_0^t \left[\int_0^L u_x^2(x, s)\mathrm{d}x\right]\mathrm{d}s\right\} = 0,$$

that is to say

$$\frac{\mathrm{d}E(t)}{\mathrm{d}t} = 0, \quad 0 < t \leqslant T.$$

Consequently,

$$E(t) = E(0), \quad 0 < t \leqslant T.$$

$\square$

The following corollary can be directly derived from Theorem 8.1.

Colollary 8.1 *Let $u(x, t)$ be the solution of the problem (8.1). Then we have*

$$\|u(\cdot, t)\|^2 \leqslant E(t) = E(0) = \|\varphi\|^2.$$

Theorem 8.2 *Let $u(x, t)$ be the solution of the problem (8.1). Then we have*

$$|u(\cdot, t)|_1 \leqslant \|\varphi'\|\mathrm{e}^{\frac{c^2}{2\nu}\left(\frac{1}{L}+\frac{c^2}{\nu^2}\right)t}, \quad 0 < t \leqslant T,$$

where $c = \|\varphi\|$.

Proof Taking the inner product on both sides of (8.1a) with $-u_{xx}$ gives

$$\frac{1}{2}\frac{\mathrm{d}}{\mathrm{d}t}\|u_x(\cdot,t)\|^2 + \nu\|u_{xx}(\cdot,t)\|^2$$

$$= \int_0^L u(x,t)u_x(x,t)u_{xx}(x,t)\mathrm{d}x$$

$$\leqslant \|u_x(\cdot,t)\|_\infty \int_0^L \left|u(x,t)u_{xx}(x,t)\right|\mathrm{d}x$$

$$\leqslant \|u_x(\cdot,t)\|_\infty \cdot \|u(\cdot,t)\| \cdot \|u_{xx}(\cdot,t)\|$$

$$\leqslant c\|u_x(\cdot,t)\|_\infty \cdot \|u_{xx}(\cdot,t)\|$$

$$\leqslant \frac{\nu}{2}\|u_{xx}(\cdot,t)\|^2 + \frac{c^2}{2\nu}\|u_x(\cdot,t)\|_\infty^2$$

$$\leqslant \frac{\nu}{2}\|u_{xx}(\cdot,t)\|^2 + \frac{c^2}{2\nu}\left[\varepsilon\|u_{xx}(\cdot,t)\|^2 + \left(\frac{1}{\varepsilon}+\frac{1}{L}\right)\|u_x(\cdot,t)\|^2\right],$$

where Corollary 8.1 and Lemma 1.1 are used in the third and last inequality, respectively. Taking $\varepsilon = \frac{\nu^2}{c^2}$, we have

$$\frac{\mathrm{d}}{\mathrm{d}t}\|u_x(\cdot,t)\|^2 \leqslant \frac{c^2}{\nu}\left(\frac{1}{L}+\frac{c^2}{\nu^2}\right)\|u_x(\cdot,t)\|^2, \quad 0 < t \leqslant T.$$

Using the Gronwall inequality, we have

$$|u(\cdot,t)|_1^2 = \|u_x(\cdot,t)\|^2 \leqslant \|u_x(\cdot,0)\|^2 \mathrm{e}^{\frac{c^2}{\nu}\left(\frac{1}{L}+\frac{c^2}{\nu^2}\right)t} = \|\varphi'\|^2 \mathrm{e}^{\frac{c^2}{\nu}\left(\frac{1}{L}+\frac{c^2}{\nu^2}\right)t}, \quad 0 < t \leqslant T.$$

$$\square$$

According to Theorem 8.2 and Lemma 1.1, the following corollary can be obtained.

Colollary 8.2 *Let $u(x,t)$ be the solution of the problem (8.1). Then we have*

$$\|u(\cdot,t)\|_\infty \leqslant \frac{\sqrt{L}}{2}\|\varphi'\|\mathrm{e}^{\frac{c^2}{2\nu}\left(\frac{1}{L}+\frac{c^2}{\nu^2}\right)t}, \quad 0 < t \leqslant T.$$

8.2   Two-Level Nonlinear Difference Scheme

In order to derive the difference scheme, introduce the same notation, grid function spaces, inner products, and norms as those in Chaps. 1 and 3. In addition, for $u \in \mathcal{U}_h$, denote

$$\Delta_x u_i = \frac{1}{2h}(u_{i+1} - u_{i-1}).$$

It is easy to know that

$$\Delta_x u_i = \frac{1}{2}(\delta_x u_{i-\frac{1}{2}} + \delta_x u_{i+\frac{1}{2}}).$$

8.2.1   Derivation of the Difference Scheme

Define the grid function $U = \{U_i^k \mid 0 \leqslant i \leqslant m, 0 \leqslant k \leqslant n\}$ on $\Omega_{h\tau}$, where

$$U_i^k = u(x_i, t_k), \quad 0 \leqslant i \leqslant m, \quad 0 \leqslant k \leqslant n.$$

Considering Eq. (8.1a) at the point $(x_i, t_{k+\frac{1}{2}})$, we have

$$u_t(x_i, t_{k+\frac{1}{2}}) + u(x_i, t_{k+\frac{1}{2}})u_x(x_i, t_{k+\frac{1}{2}}) = \nu u_{xx}(x_i, t_{k+\frac{1}{2}}),$$
$$1 \leqslant i \leqslant m - 1, \quad 0 \leqslant k \leqslant n - 1. \tag{8.2}$$

With the help of Lemma 1.2, we obtain

$$u_t(x_i, t_{k+\frac{1}{2}}) = \delta_t U_i^{k+\frac{1}{2}} + O(\tau^2), \tag{8.3}$$

$$u(x_i, t_{k+\frac{1}{2}}) = \frac{1}{3}[u(x_{i-1}, t_{k+\frac{1}{2}}) + u(x_i, t_{k+\frac{1}{2}}) + u(x_{i+1}, t_{k+\frac{1}{2}})] + O(h^2)$$
$$= \frac{1}{3}(U_{i-1}^{k+\frac{1}{2}} + U_i^{k+\frac{1}{2}} + U_{i+1}^{k+\frac{1}{2}}) + O(\tau^2 + h^2), \tag{8.4}$$

$$u_x(x_i, t_{k+\frac{1}{2}}) = \frac{1}{2}[u_x(x_i, t_k) + u_x(x_i, t_{k+1})] + O(\tau^2)$$
$$= \frac{1}{2}(\Delta_x U_i^k + \Delta_x U_i^{k+1}) + O(\tau^2 + h^2)$$
$$= \Delta_x U_i^{k+\frac{1}{2}} + O(\tau^2 + h^2), \tag{8.5}$$

$$u_{xx}(x_i, t_{k+\frac{1}{2}}) = \frac{1}{2}[u_{xx}(x_i, t_k) + u_{xx}(x_i, t_{k+1})] + O(\tau^2)$$

$$= \frac{1}{2}\left(\delta_x^2 U_i^k + \delta_x^2 U_i^{k+1}\right) + O(\tau^2 + h^2)$$

$$= \delta_x^2 U_i^{k+\frac{1}{2}} + O(\tau^2 + h^2). \tag{8.6}$$

Substituting (8.3)–(8.6) into (8.2) leads to

$$\delta_t U_i^{k+\frac{1}{2}} + \frac{1}{3}\left(U_{i-1}^{k+\frac{1}{2}} + U_i^{k+\frac{1}{2}} + U_{i+1}^{k+\frac{1}{2}}\right)\Delta_x U_i^{k+\frac{1}{2}} = \nu \delta_x^2 U_i^{k+\frac{1}{2}} + (R_1)_i^k,$$

$$1 \leqslant i \leqslant m-1, \quad 0 \leqslant k \leqslant n-1, \tag{8.7}$$

and there is a positive constant c_1 such that

$$|(R_1)_i^k| \leqslant c_1(\tau^2 + h^2), \quad 1 \leqslant i \leqslant m-1, \quad 0 \leqslant k \leqslant n-1. \tag{8.8}$$

Noticing the initial-boundary value conditions (8.1b)–(8.1c), we have

$$\begin{cases} U_i^0 = \varphi(x_i), & 1 \leqslant i \leqslant m-1, \\ U_0^k = 0, \quad U_m^k = 0, & 0 \leqslant k \leqslant n. \end{cases} \tag{8.9}$$

Omitting the small term $(R_1)_i^k$ in (8.7) and replacing U_i^k by u_i^k, a difference scheme for solving the problem (8.1) reads

$$\begin{cases} \delta_t u_i^{k+\frac{1}{2}} + \dfrac{1}{3}(u_{i-1}^{k+\frac{1}{2}} + u_i^{k+\frac{1}{2}} + u_{i+1}^{k+\frac{1}{2}})\Delta_x u_i^{k+\frac{1}{2}} = \nu \delta_x^2 u_i^{k+\frac{1}{2}}, \\ \qquad\qquad 1 \leqslant i \leqslant m-1, \quad 0 \leqslant k \leqslant n-1, \qquad\qquad (8.10a) \\ u_i^0 = \varphi(x_i), \quad 1 \leqslant i \leqslant m-1, \qquad\qquad\qquad\qquad\qquad (8.10b) \\ u_0^k = 0, \quad u_m^k = 0, \quad 0 \leqslant k \leqslant n. \qquad\qquad\qquad\quad\, (8.10c) \end{cases}$$

The difference scheme (8.10) is a two-time-level nonlinear one.

8.2.2   Conservation and Boundedness of the Difference Solution

The nonlinear term in (8.10a) can be rewritten as follows:

$$\frac{1}{3}\left(u_{i-1}^{k+\frac{1}{2}} + u_i^{k+\frac{1}{2}} + u_{i+1}^{k+\frac{1}{2}}\right)\Delta_x u_i^{k+\frac{1}{2}}$$

$$= \frac{1}{3} \left[u_i^{k+\frac{1}{2}} \Delta_x u_i^{k+\frac{1}{2}} + \left(u_{i+1}^{k+\frac{1}{2}} + u_{i-1}^{k+\frac{1}{2}} \right) \Delta_x u_i^{k+\frac{1}{2}} \right]$$

$$= \frac{1}{3} \left[u_i^{k+\frac{1}{2}} \Delta_x u_i^{k+\frac{1}{2}} + \Delta_x \left(u_i^{k+\frac{1}{2}} u_i^{k+\frac{1}{2}} \right) \right].$$

For any $v, w \in \mathcal{U}_h$, define

$$\psi(v, w)_i = \frac{1}{3} [v_i \Delta_x w_i + \Delta_x(vw)_i], \quad 1 \leqslant i \leqslant m - 1.$$

Then it is easy to check that

$$\frac{1}{3} \left(u_{i-1}^{k+\frac{1}{2}} + u_i^{k+\frac{1}{2}} + u_{i+1}^{k+\frac{1}{2}} \right) \Delta_x u_i^{k+\frac{1}{2}} = \psi \left(u^{k+\frac{1}{2}}, u^{k+\frac{1}{2}} \right)_i,$$

$$\frac{1}{3} \left(U_{i-1}^{k+\frac{1}{2}} + U_i^{k+\frac{1}{2}} + U_{i+1}^{k+\frac{1}{2}} \right) \Delta_x U_i^{k+\frac{1}{2}} = \psi \left(U^{k+\frac{1}{2}}, U^{k+\frac{1}{2}} \right)_i.$$

And (8.10a) can be rewritten as

$$\delta_t u_i^{k+\frac{1}{2}} + \psi(u^{k+\frac{1}{2}}, u^{k+\frac{1}{2}})_i = v \delta_x^2 u_i^{k+\frac{1}{2}}, \quad 1 \leqslant i \leqslant m - 1, \quad 0 \leqslant k \leqslant n - 1.$$

Reformulate uu_x as $\frac{1}{3}[uu_x + (u^2)_x]$. Then $\psi(U^{k+\frac{1}{2}}, U^{k+\frac{1}{2}})_i$ can be viewed as the discretization of $\frac{1}{3}[uu_x + (u^2)_x]$ at the point $(x_i, t_{k+\frac{1}{2}})$.

The operator ψ satisfies the following property.

Lemma 8.1 *Suppose $v \in \mathcal{U}_h$, $w \in \overset{\circ}{\mathcal{U}}_h$, then*

$$(\psi(v, w), w) = 0.$$

Proof

$$(\psi(v, w), w)$$

$$= \frac{1}{3}(v \Delta_x w + \Delta_x(vw), w)$$

$$= \frac{1}{3} \left[(v \Delta_x w, w) + (\Delta_x(vw), w) \right]$$

$$= \frac{1}{3} \left[(\Delta_x w, vw) + (\Delta_x(vw), w) \right]$$

$$= 0.$$

$\square$

Theorem 8.3 *Let $\{u_i^k \mid 0 \leqslant i \leqslant m, 0 \leqslant k \leqslant n\}$ be the solution of the difference scheme (8.10). Denote*

$$E^k = \|u^k\|^2 + 2v\tau \sum_{l=0}^{k-1} |u^{l+\frac{1}{2}}|_1^2, \quad 0 \leqslant k \leqslant n,$$

then

$$E^k = E^0, \quad 1 \leqslant k \leqslant n. \tag{8.11}$$

Proof   Notice that the difference equation (8.10a) can be written as

$$\delta_t u_i^{k+\frac{1}{2}} + \psi(u^{k+\frac{1}{2}}, u^{k+\frac{1}{2}})_i - v\delta_x^2 u_i^{k+\frac{1}{2}} = 0, \quad 1 \leqslant i \leqslant m-1, \quad 0 \leqslant k \leqslant n-1.$$

Taking the inner product on both sides of the above equality with $u^{k+\frac{1}{2}}$ gives

$$(\delta_t u^{k+\frac{1}{2}}, u^{k+\frac{1}{2}}) + (\psi(u^{k+\frac{1}{2}}, u^{k+\frac{1}{2}}), u^{k+\frac{1}{2}}) - v(\delta_x^2 u^{k+\frac{1}{2}}, u^{k+\frac{1}{2}}) = 0.$$

Noticing $u^{k+\frac{1}{2}} \in \overset{\circ}{\mathcal{U}}_h$, we have

$$(\delta_t u^{k+\frac{1}{2}}, u^{k+\frac{1}{2}}) = \frac{1}{2\tau}(\|u^{k+1}\|^2 - \|u^k\|^2),$$

$$(\psi(u^{k+\frac{1}{2}}, u^{k+\frac{1}{2}}), u^{k+\frac{1}{2}}) = 0,$$

$$-(\delta_x^2 u^{k+\frac{1}{2}}, u^{k+\frac{1}{2}}) = |u^{k+\frac{1}{2}}|_1^2.$$

Hence,

$$\frac{1}{2\tau}(\|u^{k+1}\|^2 - \|u^k\|^2) + v|u^{k+\frac{1}{2}}|_1^2 = 0, \quad 0 \leqslant k \leqslant n-1.$$

Replacing k by l in the above equality and summing over l from 0 to $k-1$, we have

$$\frac{1}{2\tau}(\|u^k\|^2 - \|u^0\|^2) + v\sum_{l=0}^{k-1} |u^{l+\frac{1}{2}}|_1^2 = 0, \quad 1 \leqslant k \leqslant n,$$

which yields (8.11) after the reformulation. $\square$

From Theorem 8.3, we can get the following corollary.

Colollary 8.3 *Let $\{u_i^k \mid 0 \leqslant i \leqslant m, 0 \leqslant k \leqslant n\}$ be the solution of the difference scheme (8.10). Then we have*

$$\|u^k\| \leqslant \|u^0\|, \quad 1 \leqslant k \leqslant n.$$

Denote $c_2 = \|u^0\|$.

8.2.3　Existence and Uniqueness of the Difference Solution

We will prove the existence of the difference solution by means of the Browder theorem (Theorem 7.4).

Theorem 8.4 *The difference scheme (8.10) has a solution.*

Proof From (8.10b)–(8.10c), the value of u^0 has been determined. Suppose that the numerical solution u^k has been determined. Let

$$w_i = u_i^{k+\frac{1}{2}}, \quad 0 \leqslant i \leqslant m,$$

we then have

$$u_i^{k+1} = 2w_i - u_i^k, \quad 0 \leqslant i \leqslant m.$$

Combining (8.10a) with (8.10c), we have the system of nonlinear equations in $w = (w_0, w_1, \ldots, w_m)$:

$$\begin{cases} \dfrac{2}{\tau}(w_i - u_i^k) + \psi(w, w)_i - \nu \delta_x^2 w_i = 0, & 1 \leqslant i \leqslant m - 1, & (8.12a) \\[2mm] w_0 = 0, \quad w_m = 0. & & (8.12b) \end{cases}$$

For any $u, v \in \overset{\circ}{\mathcal{U}}_h$, define the inner product

$$(u, v) = h \sum_{i=1}^{m-1} u_i v_i.$$

Then $\overset{\circ}{\mathcal{U}}_h$ is an inner product space with the induced norm $\|u\| = \sqrt{(u, u)}$.

Define the operator $\Pi : \overset{\circ}{\mathcal{U}}_h \to \overset{\circ}{\mathcal{U}}_h$ by

$$\Pi(w)_i = \begin{cases} \dfrac{2}{\tau}(w_i - u_i^k) + \psi(w, w)_i - \nu \delta_x^2 w_i, & 1 \leqslant i \leqslant m - 1, \\[2mm] 0, & i = 0, m. \end{cases}$$

Then $\Pi(w)$ is a continuous function in $\overset{\circ}{\mathcal{U}}_h$. Applying Lemma 8.1, we have

$$
\begin{aligned}
(\Pi(w), w) &= \frac{2}{\tau}\big[(w, w) - (u^k, w)\big] + \big(\psi(w, w), w\big) - v\big(\delta_x^2 w, w\big) \\
&= \frac{2}{\tau}\big[\|w\|^2 - (u^k, w)\big] + v|w|_1^2 \\
&\geqslant \frac{2}{\tau}\big(\|w\|^2 - \|u^k\| \cdot \|w\|\big) \\
&= \frac{2}{\tau}\big(\|w\| - \|u^k\|\big) \cdot \|w\|.
\end{aligned}
$$

Therefore, $(\Pi(w), w) \geqslant 0$ when $\|w\| = \|u^k\|$. By Theorem 7.4, there is a $w^* \in \overset{\circ}{\mathcal{U}}_h$ satisfying $\|w^*\| \leqslant \|u^k\|$ such that $\Pi(w^*) = 0$. This implies that the system (8.12) has a solution w^*.

By induction, the conclusion is true. $\square$

Theorem 8.5 *When* $\tau < \frac{4v^3}{c_2^4}$, *the solution of the difference scheme (8.10) is unique.*

Proof From the proof for Theorem 8.4, it suffices to show that the solution of (8.12) is unique.

Suppose (8.12) has two solutions $X, Y \in \overset{\circ}{\mathcal{U}}_h$, which means that X, Y satisfy

$$
\begin{cases}
\dfrac{2}{\tau}(X_i - u_i^k) + \psi(X, X)_i - v\delta_x^2 X_i = 0, & 1 \leqslant i \leqslant m - 1, \\
X_0 = 0, \quad X_m = 0;
\end{cases}
\tag{8.13}
$$

$$
\begin{cases}
\dfrac{2}{\tau}(Y_i - u_i^k) + \psi(Y, Y)_i - v\delta_x^2 Y_i = 0, & 1 \leqslant i \leqslant m - 1, \\
Y_0 = 0, \quad Y_m = 0.
\end{cases}
\tag{8.14}
$$

Let

$$
z = X - Y.
$$

Subtracting (8.14) from (8.13) produces

$$
\begin{cases}
\dfrac{2}{\tau}z_i + \psi(X, X)_i - \psi(Y, Y)_i - v\delta_x^2 z_i = 0, & 1 \leqslant i \leqslant m - 1, & \text{(8.15a)} \\
z_0 = 0, \quad z_m = 0. & & \text{(8.15b)}
\end{cases}
$$

According to Corollary 8.3, we have

$$
\|X\| \leqslant c_2, \quad \|Y\| \leqslant c_2.
$$

Taking the inner product of (8.15a) on both sides with z and using the summation by parts, it follows by noticing (8.15b) that

$$\frac{2}{\tau}\|z\|^2 + (\psi(X, X) - \psi(Y, Y), z) + v|z|_1^2 = 0. \tag{8.16}$$

Noticing

$$\psi(X, X) - \psi(Y, Y) = \psi(X, X) - \psi(X - z, X - z) = \psi(z, X) + \psi(X, z) - \psi(z, z)$$

and $z \in \overset{\circ}{\mathcal{U}}_h$, by means of Lemma 8.1, we have

$$(\psi(X, X) - \psi(Y, Y), z) = (\psi(z, X), z).$$

Therefore,

$$- (\psi(X, X) - \psi(Y, Y), z)$$

$$= -\frac{1}{3}h \sum_{i=1}^{m-1} [z_i \Delta_x X_i + \Delta_x(zX)_i] z_i$$

$$= \frac{1}{3}h \sum_{i=1}^{m-1} [X_i \Delta_x(z_i^2) + (zX)_i \Delta_x z_i]$$

$$\leqslant \frac{1}{3}(2\|z\|_\infty \cdot \|X\| \cdot |z|_1 + \|z\|_\infty \cdot \|X\| \cdot |z|_1)$$

$$= \|X\| \cdot \|z\|_\infty \cdot |z|_1$$

$$\leqslant c_2 \|z\|_\infty \cdot |z|_1.$$

It follows from (8.16) that

$$\frac{2}{\tau}\|z\|^2 + v|z|_1^2 \leqslant c_2 \|z\|_\infty \cdot |z|_1.$$

By Lemma 1.4, for any $\varepsilon > 0$, we have

$$\|z\|_\infty \leqslant \varepsilon |z|_1 + \frac{1}{2\varepsilon}\|z\|.$$

Consequently,

$$\frac{2}{\tau}\|z\|^2 + v|z|_1^2 \leqslant c_2 \left(\varepsilon |z|_1 + \frac{1}{2\varepsilon}\|z\|\right) |z|_1$$

$$= c_2 \varepsilon |z|_1^2 + \frac{c_2}{2\varepsilon}\|z\| \cdot |z|_1$$

$$\leqslant c_2\varepsilon |z|_1^2 + c_2\varepsilon |z|_1^2 + \frac{1}{4c_2\varepsilon}\left(\frac{c_2}{2\varepsilon}\right)^2 \|z\|^2$$

$$= 2c_2\varepsilon |z|_1^2 + \frac{c_2}{16\varepsilon^3}\|z\|^2,$$

in which taking $\varepsilon = \frac{\nu}{2c_2}$ gives

$$\frac{2}{\tau}\|z\|^2 \leqslant \frac{c_2^4}{2\nu^3}\|z\|^2.$$

When $\tau < \frac{4\nu^3}{c_2^4}$, the above inequality implies $\|z\| = 0$, which reveals that the solution of (8.12) is unique. $\square$

Remark 8.1 The constant c_2 (equals to $\|u^0\|$) is an approximate value of $\|\varphi\| \equiv \sqrt{\int_0^L \varphi^2(x)\mathrm{d}x}$ for which the composite Trapezoid formula is used.

8.2.4 Convergence of the Difference Solution

Denote

$$c_3 = \max_{0\leqslant x\leqslant L,0\leqslant t\leqslant T} |u_x(x,t)|. \tag{8.17}$$

Theorem 8.6 *Let* $\{U_i^k \mid 0 \leqslant i \leqslant m, 0 \leqslant k \leqslant n\}$ *be the solution of the problem (8.1) and* $\{u_i^k \mid 0 \leqslant i \leqslant m, 0 \leqslant k \leqslant n\}$ *be the solution of the difference scheme (8.10). Denote*

$$e_i^k = U_i^k - u_i^k, \quad 0 \leqslant i \leqslant m, \quad 0 \leqslant k \leqslant n.$$

Then there is a constant c_4 *such that*

$$\|e^k\| \leqslant c_4(\tau^2 + h^2), \quad 0 \leqslant k \leqslant n.$$

Proof   Subtracting (8.10) from (8.7) and (8.9), the system of error equations reads

$$\begin{cases} \delta_t e_i^{k+\frac{1}{2}} + \psi(U^{k+\frac{1}{2}}, U^{k+\frac{1}{2}})_i - \psi(u^{k+\frac{1}{2}}, u^{k+\frac{1}{2}})_i = \nu\delta_x^2 e_i^{k+\frac{1}{2}} + (R_1)_i^k, \\ \qquad\qquad\qquad\qquad\qquad 1 \leqslant i \leqslant m-1, \quad 0 \leqslant k \leqslant n-1, \quad (8.18a) \\ e_i^0 = 0, \quad 1 \leqslant i \leqslant m-1, \qquad\qquad\qquad\qquad\qquad (8.18b) \\ e_0^k = 0, \quad e_m^k = 0, \quad 0 \leqslant k \leqslant n. \qquad\qquad\qquad\quad (8.18c) \end{cases}$$

Taking the inner product on both sides of (8.18a) with $e^{k+\frac{1}{2}}$ produces

$$(\delta_t e^{k+\frac{1}{2}}, e^{k+\frac{1}{2}}) + (\psi(U^{k+\frac{1}{2}}, U^{k+\frac{1}{2}}) - \psi(u^{k+\frac{1}{2}}, u^{k+\frac{1}{2}}), e^{k+\frac{1}{2}}) + \nu|e^{k+\frac{1}{2}}|_1^2$$

$$= ((R_1)^k, e^{k+\frac{1}{2}}), \quad 0 \leqslant k \leqslant n-1. \tag{8.19}$$

It is easy to know that

$$(\delta_t e^{k+\frac{1}{2}}, e^{k+\frac{1}{2}}) = \frac{1}{2\tau}(\|e^{k+1}\|^2 - \|e^k\|^2). \tag{8.20}$$

Noticing

$$\psi(U^{k+\frac{1}{2}}, U^{k+\frac{1}{2}}) - \psi(u^{k+\frac{1}{2}}, u^{k+\frac{1}{2}})$$

$$= \psi(U^{k+\frac{1}{2}}, U^{k+\frac{1}{2}}) - \psi(U^{k+\frac{1}{2}} - e^{k+\frac{1}{2}}, U^{k+\frac{1}{2}} - e^{k+\frac{1}{2}})$$

$$= \psi(e^{k+\frac{1}{2}}, U^{k+\frac{1}{2}}) + \psi(U^{k+\frac{1}{2}}, e^{k+\frac{1}{2}}) - \psi(e^{k+\frac{1}{2}}, e^{k+\frac{1}{2}}),$$

and using Lemma 8.1 again, we have

$$-(\psi(U^{k+\frac{1}{2}}, U^{k+\frac{1}{2}}) - \psi(u^{k+\frac{1}{2}}, u^{k+\frac{1}{2}}), e^{k+\frac{1}{2}})$$

$$= -(\psi(e^{k+\frac{1}{2}}, U^{k+\frac{1}{2}}), e^{k+\frac{1}{2}})$$

$$= -\frac{1}{3}h \sum_{i=1}^{m-1} \left[e_i^{k+\frac{1}{2}} \Delta_x U_i^{k+\frac{1}{2}} + \Delta_x\left(e^{k+\frac{1}{2}}U^{k+\frac{1}{2}}\right)_i \right] e_i^{k+\frac{1}{2}}$$

$$= -\frac{1}{3}h \sum_{i=1}^{m-1} \left(e_i^{k+\frac{1}{2}}\right)^2 \Delta_x U_i^{k+\frac{1}{2}} + \frac{1}{3}h \sum_{i=1}^{m-1} e_i^{k+\frac{1}{2}} U_i^{k+\frac{1}{2}} \Delta_x e_i^{k+\frac{1}{2}}$$

$$= -\frac{1}{3}h \sum_{i=1}^{m-1} \left(e_i^{k+\frac{1}{2}}\right)^2 \Delta_x U_i^{k+\frac{1}{2}} - \frac{1}{6}h \sum_{i=0}^{m-1} \frac{U_{i+1}^{k+\frac{1}{2}} - U_i^{k+\frac{1}{2}}}{h} e_i^{k+\frac{1}{2}} e_{i+1}^{k+\frac{1}{2}}$$

$$\leqslant \frac{1}{2}c_3 \|e^{k+\frac{1}{2}}\|^2. \tag{8.21}$$

Substituting (8.20) and (8.21) into (8.19) arrives at

$$\frac{1}{2\tau}(\|e^{k+1}\|^2 - \|e^k\|^2)$$

$$\leqslant \frac{1}{2}c_3 \|e^{k+\frac{1}{2}}\|^2 + \|(R_1)^k\| \cdot \|e^{k+\frac{1}{2}}\|$$

$$\leqslant \frac{1}{2}c_3 \left(\frac{\|e^k\| + \|e^{k+1}\|}{2} \right)^2 + \|(R_1)^k\| \cdot \frac{\|e^k\| + \|e^{k+1}\|}{2}, \quad 0 \leqslant k \leqslant n-1.$$

It follows after eliminating $\dfrac{\|e^{k+1}\| + \|e^k\|}{2}$ on both sides of the above inequality that

$$\frac{1}{\tau}(\|e^{k+1}\| - \|e^k\|) \leqslant \frac{c_3}{4}(\|e^k\| + \|e^{k+1}\|) + \|(R_1)^k\|, \quad 0 \leqslant k \leqslant n-1,$$

i.e.,

$$\left(1 - \frac{c_3}{4}\tau\right) \|e^{k+1}\| \leqslant \left(1 + \frac{c_3\tau}{4}\right) \|e^k\| + \tau\|(R_1)^k\|, \quad 0 \leqslant k \leqslant n-1.$$

When $\dfrac{c_3}{4}\tau \leqslant \dfrac{1}{3}$, in view of (8.8), we have

$$\|e^{k+1}\| \leqslant \left(1 + \frac{3c_3\tau}{4}\right) \|e^k\| + \frac{3}{2}\tau\|(R_1)^k\|$$

$$\leqslant \left(1 + \frac{3c_3}{4}\tau\right) \|e^k\| + \frac{3}{2}c_1\sqrt{L}\tau(\tau^2 + h^2), \quad 0 \leqslant k \leqslant n-1.$$

Using Lemma 3.3 (Gronwall inequality) and noticing $\|e^0\| = 0$, we obtain

$$\|e^{k+1}\| \leqslant e^{\frac{3c_3}{4}k\tau} \cdot \frac{2c_1\sqrt{L}}{c_3}(\tau^2 + h^2) \leqslant c_4(\tau^2 + h^2), \quad 0 \leqslant k \leqslant n-1,$$

where $c_4 = e^{\frac{3c_3}{4}T} \cdot \dfrac{2c_1\sqrt{L}}{c_3}$. $\qquad\qquad\qquad\qquad\qquad\qquad\qquad\qquad\qquad\qquad\square$

Similar to the proof for Theorem 8.2, the maximum error estimate can be obtained. Denote

$$c_0 = \max_{0 \leqslant x \leqslant L, 0 \leqslant t \leqslant T} |u(x, t)|.$$

Theorem 8.7 *Let* $\{U_i^k \mid 0 \leqslant i \leqslant m, 0 \leqslant k \leqslant n\}$ *be the solution of the problem (8.1) and* $\{u_i^k \mid 0 \leqslant i \leqslant m, 0 \leqslant k \leqslant n\}$ *be the solution of the difference scheme (8.10). Denote*

$$e_i^k = U_i^k - u_i^k, \quad 0 \leqslant i \leqslant m, \quad 0 \leqslant k \leqslant n$$

and

$$c_5 = \frac{125\left(c_0\sqrt{L} + c_2\right)^4}{16\nu^3} + \frac{5\left(c_0\sqrt{L} + c_2\right)^2}{4\nu L} + \frac{5(c_0 + c_3)^2}{4\nu}\left(1 + \frac{L^2}{6}\right).$$

Then we have

$$\|e^k\|_\infty \leqslant \frac{1}{4}Lc_1 e^{\frac{3}{2}c_5 T}\sqrt{\frac{5}{\nu c_5}}(\tau^2 + h^2), \quad 0 \leqslant k \leqslant n.$$

Proof Subtracting (8.10) from (8.7) and (8.9), the system of error equations reads

$$\begin{cases} \delta_t e_i^{k+\frac{1}{2}} + \psi(U^{k+\frac{1}{2}}, U^{k+\frac{1}{2}})_i - \psi(u^{k+\frac{1}{2}}, u^{k+\frac{1}{2}})_i = \nu\delta_x^2 e_i^{k+\frac{1}{2}} + (R_1)_i^k, \\ \qquad\qquad\qquad\qquad\qquad 1 \leqslant i \leqslant m-1, \quad 0 \leqslant k \leqslant n-1, \quad (8.22a) \\ e_i^0 = 0, \quad 1 \leqslant i \leqslant m-1, \qquad\qquad\qquad\qquad\qquad\qquad\qquad (8.22b) \\ e_0^k = 0, \quad e_m^k = 0, \quad 0 \leqslant k \leqslant n. \qquad\qquad\qquad\qquad\qquad (8.22c) \end{cases}$$

It follows from Corollary 8.3 that

$$\|e^k\| = \|U^k - u^k\| \leqslant \|U^k\| + \|u^k\| \leqslant c_0\sqrt{L} + c_2, \quad 0 \leqslant k \leqslant n. \qquad (8.23)$$

In view of (8.22c), we have

$$e_0^{k+\frac{1}{2}} = 0, \quad e_m^{k+\frac{1}{2}} = 0, \quad \delta_t e_0^{k+\frac{1}{2}} = 0, \quad \delta_t e_m^{k+\frac{1}{2}} = 0, \quad 0 \leqslant k \leqslant n-1. \qquad (8.24)$$

Taking the inner product on both sides of (8.22a) with $-\delta_x^2 e^{k+\frac{1}{2}}$ gives

$$\left(\delta_t e^{k+\frac{1}{2}}, -\delta_x^2 e^{k+\frac{1}{2}}\right) + \nu\|\delta_x^2 e^{k+\frac{1}{2}}\|^2$$

$$= \left(\psi(U^{k+\frac{1}{2}}, U^{k+\frac{1}{2}}) - \psi(u^{k+\frac{1}{2}}, u^{k+\frac{1}{2}}), \delta_x^2 e^{k+\frac{1}{2}}\right) - \left((R_1)^k, \delta_x^2 e^{k+\frac{1}{2}}\right)$$

$$= \left(\psi(U^{k+\frac{1}{2}}, e^{k+\frac{1}{2}}) + \psi(e^{k+\frac{1}{2}}, U^{k+\frac{1}{2}}), \delta_x^2 e^{k+\frac{1}{2}}\right) - \left(\psi(e^{k+\frac{1}{2}}, e^{k+\frac{1}{2}}), \delta_x^2 e^{k+\frac{1}{2}}\right)$$

$$- \left((R_1)^k, \delta_x^2 e^{k+\frac{1}{2}}\right), \quad 0 \leqslant k \leqslant n-1. \qquad (8.25)$$

For the first term on the left-hand side of (8.25), we apply the summation by parts and notice (8.24) to obtain

$$-\left(\delta_t e^{k+\frac{1}{2}}, \delta_x^2 e^{k+\frac{1}{2}}\right) = \frac{1}{2\tau}\left(|e^{k+1}|_1^2 - |e^k|_1^2\right). \qquad (8.26)$$

For the first term on the right-hand side of (8.25), using the Cauchy-Schwarz inequality and Lemma 8.1, we have

$$\left(\psi(U^{k+\frac{1}{2}}, e^{k+\frac{1}{2}}) + \psi(e^{k+\frac{1}{2}}, U^{k+\frac{1}{2}}), \delta_x^2 e^{k+\frac{1}{2}}\right)$$

$$= \frac{h}{3} \sum_{i=1}^{m-1} \left[U_i^{k+\frac{1}{2}} \Delta_x e_i^{k+\frac{1}{2}} + e_i^{k+\frac{1}{2}} \Delta_x U_i^{k+\frac{1}{2}} + 2\Delta_x (Ue)_i^{k+\frac{1}{2}}\right] \delta_x^2 e_i^{k+\frac{1}{2}}$$

$$= \frac{h}{3} \sum_{i=1}^{m-1} \left(U_i^{k+\frac{1}{2}} \Delta_x e_i^{k+\frac{1}{2}} + 3e_i^{k+\frac{1}{2}} \Delta_x U_i^{k+\frac{1}{2}} + U_{i+1}^{k+\frac{1}{2}} \delta_x e_{i+\frac{1}{2}}^{k+\frac{1}{2}} + U_{i-1}^{k+\frac{1}{2}} \delta_x e_{i-\frac{1}{2}}^{k+\frac{1}{2}}\right)$$

$$\times \delta_x^2 e_i^{k+\frac{1}{2}}$$

$$\leqslant (c_0 + c_3) \left(|e^{k+\frac{1}{2}}|_1 + \|e^{k+\frac{1}{2}}\|\right) \|\delta_x^2 e^{k+\frac{1}{2}}\|$$

$$= (c_0 + c_3)|e^{k+\frac{1}{2}}|_1 \|\delta_x^2 e^{k+\frac{1}{2}}\| + (c_0 + c_3)\|e^{k+\frac{1}{2}}\| \cdot \|\delta_x^2 e^{k+\frac{1}{2}}\|$$

$$\leqslant \left(\frac{\nu}{5} \|\delta_x^2 e^{k+\frac{1}{2}}\|^2 + \frac{5(c_0 + c_3)^2}{4\nu} |e^{k+\frac{1}{2}}|_1^2\right)$$

$$+ \left(\frac{\nu}{5} \|\delta_x^2 e^{k+\frac{1}{2}}\|^2 + \frac{5(c_0 + c_3)^2}{4\nu} \|e^{k+\frac{1}{2}}\|^2\right)$$

$$\leqslant \frac{2\nu}{5} \|\delta_x^2 e^{k+\frac{1}{2}}\|^2 + \frac{5(c_0 + c_3)^2}{4\nu} \left(1 + \frac{L^2}{6}\right) |e^{k+\frac{1}{2}}|_1^2. \tag{8.27}$$

For the second term on the right-hand side of (8.25), we obtain

$$- \left(\psi(e^{k+\frac{1}{2}}, e^{k+\frac{1}{2}}), \delta_x^2 e^{k+\frac{1}{2}}\right)$$

$$= -\frac{h}{3} \sum_{i=1}^{m-1} \left(e_{i+1}^{k+\frac{1}{2}} + e_i^{k+\frac{1}{2}} + e_{i-1}^{k+\frac{1}{2}}\right) \Delta_x e_i^{k+\frac{1}{2}} \cdot \delta_x^2 e_i^{k+\frac{1}{2}}$$

$$\leqslant \frac{h}{3} \sum_{i=1}^{m-1} \left(|e_{i+1}^{k+\frac{1}{2}}| + |e_i^{k+\frac{1}{2}}| + |e_{i-1}^{k+\frac{1}{2}}|\right) |\Delta_x e_i^{k+\frac{1}{2}}| \cdot |\delta_x^2 e_i^{k+\frac{1}{2}}|$$

$$\leqslant \|\delta_x e^{k+\frac{1}{2}}\|_\infty \cdot \frac{h}{3} \sum_{i=1}^{m-1} \left(|e_{i+1}^{k+\frac{1}{2}}| + |e_i^{k+\frac{1}{2}}| + |e_{i-1}^{k+\frac{1}{2}}|\right) \cdot |\delta_x^2 e_i^{k+\frac{1}{2}}|$$

$$\leqslant \|\delta_x e^{k+\frac{1}{2}}\|_\infty \|e^{k+\frac{1}{2}}\| \cdot \|\delta_x^2 e^{k+\frac{1}{2}}\|$$

$$\leqslant \left(c_0\sqrt{L} + c_2\right) \|\delta_x e^{k+\frac{1}{2}}\|_\infty \|\delta_x^2 e^{k+\frac{1}{2}}\|$$

$$\leqslant \frac{\nu}{5} \|\delta_x^2 e^{k+\frac{1}{2}}\|^2 + \frac{5}{4\nu} \left(c_0\sqrt{L} + c_2\right)^2 \|\delta_x e^{k+\frac{1}{2}}\|_\infty^2$$

$$\leqslant \frac{\nu}{5}\|\delta_x^2 e^{k+\frac{1}{2}}\|^2 + \frac{5}{4\nu}\left(c_0\sqrt{L}+c_2\right)^2\left[\varepsilon\|\delta_x^2 e^{k+\frac{1}{2}}\|^2 + \left(\frac{1}{\varepsilon}+\frac{1}{L}\right)|e^{k+\frac{1}{2}}|_1^2\right]$$

$$= \frac{2\nu}{5}\|\delta_x^2 e^{k+\frac{1}{2}}\|^2 + \frac{5}{4\nu}\left(c_0\sqrt{L}+c_2\right)^2\left(\frac{25\left(c_0\sqrt{L}+c_2\right)^2}{4\nu^2}+\frac{1}{L}\right)|e^{k+\frac{1}{2}}|_1^2,$$

$$(8.28)$$

where in the fourth inequality, we have used (8.23); In the last inequality, we have used Exercise 1.6; In the last equality, we have taken $\varepsilon = \frac{4}{25}\nu^2\left(c_0\sqrt{L}+c_2\right)^{-2}$.

For the third term on the right-hand side of (8.25), using the Cauchy-Schwarz inequality and noticing (8.8), we have

$$-\left((R_1)^k, \delta_x^2 e^{k+\frac{1}{2}}\right) \leqslant \frac{\nu}{5}\|\delta_x^2 e^{k+\frac{1}{2}}\|^2 + \frac{5}{4\nu}\|(R_1)^k\|^2$$

$$\leqslant \frac{\nu}{5}\|\delta_x^2 e^{k+\frac{1}{2}}\|^2 + \frac{5}{4\nu}Lc_1^2(\tau^2+h^2)^2. \qquad (8.29)$$

Inserting (8.26)–(8.29) into (8.25) yields

$$\frac{1}{2\tau}\left(|e^{k+1}|_1^2 - |e^k|_1^2\right)$$

$$\leqslant \left[\frac{125\left(c_0\sqrt{L}+c_2\right)^4}{16\nu^3} + \frac{5\left(c_0\sqrt{L}+c_2\right)^2}{4\nu L} + \frac{5(c_0+c_3)^2}{4\nu}\left(1+\frac{L^2}{6}\right)\right]|e^{k+\frac{1}{2}}|_1^2$$

$$+ \frac{5}{4\nu}Lc_1^2(\tau^2+h^2)^2, \quad 0\leqslant k\leqslant n-1.$$

It is easy to know that

$$(1-c_5\tau)|e^{k+1}|_1^2 \leqslant (1+c_5\tau)|e^k|_1^2 + \frac{5Lc_1^2}{2\nu}\tau(\tau^2+h^2)^2, \quad 0\leqslant k\leqslant n-1.$$

When $c_5\tau \leqslant \frac{1}{3}$, we have

$$|e^{k+1}|_1^2 \leqslant (1+3c_5\tau)|e^k|_1^2 + \frac{15Lc_1^2}{4\nu}\tau(\tau^2+h^2)^2, \quad 0\leqslant k\leqslant n-1.$$

Using Lemma 3.3 (Gronwall inequality), we get

$$|e^k|_1^2 \leqslant e^{3c_5k\tau}\frac{5Lc_1^2}{4\nu c_5}(\tau^2+h^2)^2 \leqslant e^{3c_5T}\frac{5Lc_1^2}{4\nu c_5}(\tau^2+h^2)^2, \quad 0\leqslant k\leqslant n.$$

Taking the square root on both sides of the above inequality and using Lemma 1.4 (imbedding inequality), the conclusion is obtained. $\qquad\square$

8.2.5 Numerical Examples

Example 8.1 Use the difference scheme (8.10) to compute the initial-boundary value problem [1]

$$\begin{cases} u_t + uu_x = vu_{xx}, & 0 \leqslant x \leqslant 1, \quad 0 < t \leqslant 1, \\ u(x,0) = 2\pi\dfrac{\sin(\pi x)}{100 + \cos(\pi x)}, & 0 < x < 1, \\ u(0,t) = 0, \quad u(1,t) = 0, & 0 \leqslant t \leqslant 1. \end{cases} \tag{8.30}$$

The exact solution of this problem is

$$u(x,t) = 2\pi\frac{\sin(\pi x)\exp(-\pi^2 vt)}{100 + \cos(\pi x)\exp(-\pi^2 vt)}.$$

From the proof of Theorem 8.4, it is necessary to solve the nonlinear system (8.12) with respect to $\{w_i \mid 1 \leqslant i \leqslant m - 1\}$. The Newton iteration method is employed to solve this system. Once the value of $\{w_i \mid 1 \leqslant i \leqslant m - 1\}$ is obtained, the numerical solution at the next time level can be computed by

$$u_i^{k+1} = 2w_i - u_i^k, \quad 1 \leqslant i \leqslant m - 1.$$

Let $v = 1$. For different values of step sizes, the maximum norm errors of the numerical solutions obtained by the difference scheme (8.10) defined by

$$E_\infty(h, \tau) = \max_{0 \leqslant k \leqslant n, 0 \leqslant j \leqslant m} \left|u(x_j, t_k) - u_j^k\right|$$

are reported in Tables 8.1 and 8.2

It can be observed from Table 8.1 that when the spatial step size h is halved, the maximum norm error is reduced to one-fourth of the original, indicating the second-order spatial accuracy. Similarly, Table 8.2 shows that halving the temporal

Table 8.1 (Example 8.1) The maximum norm errors of numerical solutions with different spatial step sizes ($\tau = 1/1600$)

h	$E_\infty(h, \tau)$	$E_\infty(2h, \tau)/E_\infty(h, \tau)$
1/10	1.902120e−04	
1/20	4.746880e−05	4.0071
1/40	1.181041e−05	4.0192
1/80	2.897518e−06	4.0760
1/160	6.694009e−07	4.3285

Table 8.2 (Example 8.1)
The maximum norm errors of
numerical solutions with
different temporal step sizes
($h = 1/1600$)

τ	$E_\infty(h, \tau)$	$E_\infty(h, 2\tau)/E_\infty(h, \tau)$
1/10	2.110609e−03	
1/20	4.821319e−04	4.3777
1/40	1.180817e−04	4.0830
1/80	2.936737e−05	4.0208
1/160	7.327130e−06	4.0080

Fig. 8.1 (Example 8.1) The
curves of energy E^k for the
difference scheme (8.10)

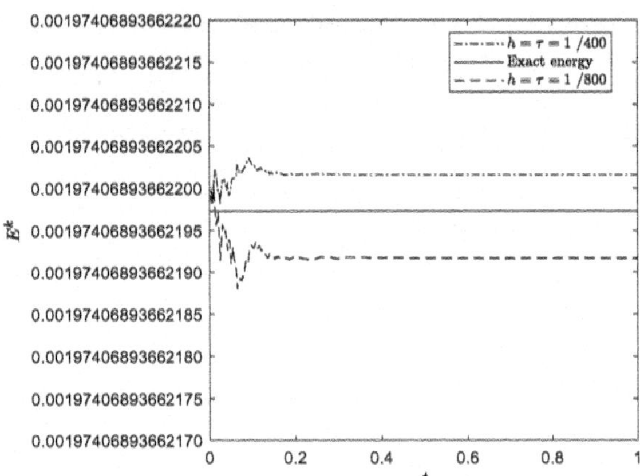

step size τ also leads to a reduction of the maximum norm error by a factor of
four, confirming the second-order temporal accuracy. The curves of the invariant
E^k, defined in Theorem 8.3, are presented in Fig. 8.1.

8.3 Three-Level Linearized Difference Scheme

8.3.1 Derivation of the Difference Scheme

Considering Eq. (8.1a) at the node point (x_i, t_0) and noticing (8.1b), we have

$$u_t(x_i, 0) = \nu\varphi''(x_i) - \varphi(x_i)\varphi'(x_i), \quad 1 \leqslant i \leqslant m - 1.$$

Denote

$$\hat{u}_i = \varphi(x_i) + \frac{\tau}{2}[\nu\varphi''(x_i) - \varphi(x_i)\varphi'(x_i)], \quad 0 \leqslant i \leqslant m.$$

Considering Eq. (8.1a) at the point $(x_i, t_{\frac{1}{2}})$ and using the Taylor expansion, we get

$$\delta_t U_i^{\frac{1}{2}} + \psi(\hat{u}, U^{\frac{1}{2}})_i = \nu \delta_x^2 U_i^{\frac{1}{2}} + (R_2)_i^0, \quad 1 \leqslant i \leqslant m - 1, \tag{8.31}$$

where there is a constant c_6 satisfying

$$|(R_2)_i^0| \leqslant c_6(\tau^2 + h^2), \quad 1 \leqslant i \leqslant m - 1. \tag{8.32}$$

Considering Eq. (8.1a) at the node point (x_i, t_k) and using the Taylor expansion, we get

$$\Delta_t U_i^k + \psi(U^k, U^{\bar{k}})_i = \nu \delta_x^2 U_i^{\bar{k}} + (R_2)_i^k, \quad 1 \leqslant i \leqslant m - 1, \quad 1 \leqslant k \leqslant n - 1, \tag{8.33}$$

where there is a constant c_7 satisfying

$$|(R_2)_i^k| \leqslant c_7(\tau^2 + h^2), \quad 1 \leqslant i \leqslant m - 1, \quad 1 \leqslant k \leqslant n - 1. \tag{8.34}$$

Noticing the initial-boundary value conditions (8.1b)–(8.1c), we have

$$\begin{cases} U_i^0 = \varphi(x_i), & 1 \leqslant i \leqslant m - 1, \\ U_0^k = 0, \quad U_m^k = 0, & 0 \leqslant k \leqslant n. \end{cases} \tag{8.35}$$

Omitting the small terms $(R_2)_i^0$ and $(R_2)_i^k$ in (8.31) and (8.33), respectively, a difference scheme for solving the problem (8.1) is derived in the form of

$$\begin{cases} \delta_t u_i^{\frac{1}{2}} + \psi(\hat{u}, u^{\frac{1}{2}})_i = \nu \delta_x^2 u_i^{\frac{1}{2}}, & 1 \leqslant i \leqslant m - 1, & \text{(8.36a)} \\ \Delta_t u_i^k + \psi(u^k, u^{\bar{k}})_i = \nu \delta_x^2 u_i^{\bar{k}}, & 1 \leqslant i \leqslant m - 1, \quad 1 \leqslant k \leqslant n - 1, & \text{(8.36b)} \\ u_i^0 = \varphi(x_i), & 1 \leqslant i \leqslant m - 1, & \text{(8.36c)} \\ u_0^k = 0, \quad u_m^k = 0, & 0 \leqslant k \leqslant n. & \text{(8.36d)} \end{cases}$$

8.3.2   *Conservation and Boundedness of the Difference Solution*

Theorem 8.8 *Let $\{u_i^k \mid 0 \leqslant i \leqslant m, 0 \leqslant k \leqslant n\}$ be the solution of (8.36). Then we have*

$$\begin{cases} \dfrac{1}{2}(\|u^1\|^2 + \|u^0\|^2) + \nu\tau |u^{\frac{1}{2}}|_1^2 = \|u^0\|^2, & \text{(8.37a)} \\ E^k = E^0, \quad k = 1, 2, 3, \ldots, n - 1, & \text{(8.37b)} \end{cases}$$

where

$$E^k = \frac{1}{2}(\|u^{k+1}\|^2 + \|u^k\|^2) + 2v\tau \sum_{l=1}^{k} |u^{\bar{l}}|_1^2, \quad k = 0, 1, \ldots, n-1.$$

Proof (I) Taking the inner product on both sides of (8.36a) with $u^{\frac{1}{2}}$, we have

$$(\delta_t u^{\frac{1}{2}}, u^{\frac{1}{2}}) + (\psi(\hat{u}, u^{\frac{1}{2}}), u^{\frac{1}{2}}) = v(\delta_x^2 u^{\frac{1}{2}}, u^{\frac{1}{2}}).$$

In view of

$$(\delta_t u^{\frac{1}{2}}, u^{\frac{1}{2}}) = \frac{1}{2\tau}(\|u^1\|^2 - \|u^0\|^2),$$

$$(\psi(\hat{u}, u^{\frac{1}{2}}), u^{\frac{1}{2}}) = 0,$$

$$(\delta_x^2 u^{\frac{1}{2}}, u^{\frac{1}{2}}) = -|u^{\frac{1}{2}}|_1^2,$$

it yields

$$\frac{1}{2\tau}(\|u^1\|^2 - \|u^0\|^2) + v|u^{\frac{1}{2}}|_1^2 = 0. \tag{8.38}$$

Consequently,

$$\frac{1}{2}(\|u^1\|^2 + \|u^0\|^2) + v\tau|u^{\frac{1}{2}}|_1^2 = \|u^0\|^2.$$

(II) Taking the inner product on both sides of (8.36b) with $u^{\bar{k}}$, we have

$$(\Delta_t u^k, u^{\bar{k}}) + (\psi(u^k, u^{\bar{k}}), u^{\bar{k}}) = v(\delta_x^2 u^{\bar{k}}, u^{\bar{k}}), \quad 1 \leqslant k \leqslant n-1.$$

It is easy to get

$$\frac{1}{4\tau}(\|u^{k+1}\|^2 - \|u^{k-1}\|^2) + v|u^{\bar{k}}|_1^2 = 0, \quad 1 \leqslant k \leqslant n-1, \tag{8.39}$$

or

$$\frac{1}{2\tau}\left(\frac{\|u^{k+1}\|^2 + \|u^k\|^2}{2} - \frac{\|u^k\|^2 + \|u^{k-1}\|^2}{2}\right) + v|u^{\bar{k}}|_1^2 = 0, \quad 1 \leqslant k \leqslant n-1,$$

which can be rewritten as

$$\frac{1}{2\tau}\left(E^k - E^{k-1}\right) = 0, \quad 1 \leqslant k \leqslant n-1.$$

Hence,

$$E^k = E^0, \quad 1 \leqslant k \leqslant n - 1.$$

$\square$

Remark 8.2 It is worth mentioning that (8.37a) and (8.37b) can be unified as

$$\frac{1}{2}(\|u^{k+1}\|^2 + \|u^k\|^2) + \nu\tau|u^{\frac{1}{2}}|_1^2 + 2\nu\tau\sum_{l=1}^{k}|u^{\bar{l}}|_1^2 = \|u^0\|^2, \quad k = 0, 1, \ldots, n - 1.$$

Remark 8.3 It follows from (8.38) and (8.39) that

$$\|u^k\| \leqslant \|u^0\|, \quad 1 \leqslant k \leqslant n.$$

8.3.3 Existence and Uniqueness of the Difference Solution

Theorem 8.9 *The difference scheme (8.36) is uniquely solvable.*

Proof From (8.36c) and (8.36d), the value of u^0 at the initial time level is given. Based on (8.36a) and (8.36d), the corresponding system of linear equations in u^1 can be formulated. Consider the associated homogeneous system

$$\begin{cases} \dfrac{1}{\tau}u_i^1 + \dfrac{1}{2}\psi(\hat{u}, u^1)_i = \dfrac{1}{2}\nu\delta_x^2 u_i^1, & 1 \leqslant i \leqslant m - 1, & (8.40a) \\[2mm] u_0^1 = 0, \quad u_m^1 = 0. & & (8.40b) \end{cases}$$

Taking the inner product on both sides of (8.40a) with u^1 yields

$$\frac{1}{\tau}\|u^1\|^2 + \frac{1}{2}(\psi(\hat{u}, u^1), u^1) = \frac{1}{2}\nu\left(\delta_x^2 u^1, u^1\right).$$

In view of $(\psi(\hat{u}, u^1), u^1) = 0$ and $(\delta_x^2 u^1, u^1) = -|u^1|_1^2$, we have

$$\frac{1}{\tau}\|u^1\|^2 + \frac{1}{2}\nu|u^1|_1^2 = 0.$$

Hence, $\|u^1\| = 0$, which implies that the system (8.40) admits only the trivial solution. Thus, the value of u^1 is uniquely determined by (8.36a) and (8.36d).

Now, suppose that the numerical solutions u^{k-1} at the $(k-1)$-th time level and u^k at the k-th time level have been determined. In this case, a system of linear equations in u^{k+1} can be obtained from (8.36b) and (8.36d). To investigate its solvability, we consider the associated homogeneous system

$$\begin{cases} \dfrac{1}{2\tau} u_i^{k+1} + \dfrac{1}{2} \psi(u^k, u^{k+1})_i = \dfrac{1}{2} \nu\, \delta_x^2 u_i^{k+1}, & 1 \leqslant i \leqslant m-1, & (8.41\mathrm{a}) \\[2mm] u_0^{k+1} = 0, \quad u_m^{k+1} = 0. & & (8.41\mathrm{b}) \end{cases}$$

Taking the inner product on both sides of (8.41a) with u^{k+1} produces

$$\frac{1}{2\tau} \|u^{k+1}\|^2 + \frac{1}{2}(\psi(u^k, u^{k+1}), u^{k+1}) = \frac{1}{2}\nu(\delta_x^2 u^{k+1}, u^{k+1}).$$

In view of $(\psi(u^k, u^{k+1}), u^{k+1}) = 0$ and $(\delta_x^2 u^{k+1}, u^{k+1}) = -|u^{k+1}|_1^2$, we obtain

$$\frac{1}{2\tau}\|u^{k+1}\|^2 + \frac{1}{2}\nu|u^{k+1}|_1^2 = 0.$$

Hence, $\|u^{k+1}\| = 0$, which implies that the system (8.41) has only the trivial solution. Thus, (8.36b) and (8.36d) determine u^{k+1} uniquely.

     By induction, the theorem is true. $\square$

8.3.4   Convergence of the Difference Solution

Theorem 8.10   *Let $\{U_i^k \mid 0 \leqslant i \leqslant m, 0 \leqslant k \leqslant n\}$ be the solution of the problem (8.1) and $\{u_i^k \mid 0 \leqslant i \leqslant m, 0 \leqslant k \leqslant n\}$ be the solution of the difference scheme (8.36). Denote*

$$e_i^k = U_i^k - u_i^k, \quad 0 \leqslant i \leqslant m,\ 0 \leqslant k \leqslant n.$$

Suppose $\dfrac{\tau}{2\nu} \leqslant 1$ and $\dfrac{L(\sqrt{L}c_3 + 1)^2}{2\nu}\tau \leqslant \frac{1}{3}$. Then there is a constant c_8, when $\tau^2 + h^2 \leqslant \dfrac{1}{c_8}$, such that

$$|e^k|_1 \leqslant c_8(\tau^2 + h^2), \quad 0 \leqslant k \leqslant n, \tag{8.42}$$

$$\|e^k\|_\infty \leqslant \frac{\sqrt{L}}{2} c_8(\tau^2 + h^2), \quad 0 \leqslant k \leqslant n. \tag{8.43}$$

Proof   If (8.42) holds, then by Lemma 1.4 (imbedding theorem), we have

$$\|e^k\|_\infty \leqslant \frac{\sqrt{L}}{2}|e^k|_1 \leqslant \frac{\sqrt{L}}{2} c_8(\tau^2 + h^2), \quad 0 \leqslant k \leqslant n,$$

which says that (8.43) is true, so that we only need to prove (8.42).

Subtracting (8.36) from (8.31), (8.33), and (8.35), we get the system of error equations:

$$
\begin{cases}
\delta_t e_i^{\frac{1}{2}} + \psi(\hat{u}, e^{\frac{1}{2}})_i = v\delta_x^2 e_i^{\frac{1}{2}} + (R_2)_i^0, \quad 1 \leqslant i \leqslant m-1, & (8.44a) \\[2mm]
\Delta_t e_i^k + \psi(U^k, U^{\bar{k}})_i - \psi(u^k, u^{\bar{k}})_i = v\delta_x^2 e_i^{\bar{k}} + (R_2)_i^k, & \\[1mm]
\qquad\qquad 1 \leqslant i \leqslant m-1, \quad 1 \leqslant k \leqslant n-1, & (8.44b) \\[2mm]
e_i^0 = 0, \quad 1 \leqslant i \leqslant m-1, & (8.44c) \\[2mm]
e_0^k = 0, \quad e_m^k = 0, \quad 0 \leqslant k \leqslant n. & (8.44d)
\end{cases}
$$

We will use the method of induction to prove the desired result.
From (8.44c)–(8.44d), we have

$$
|e^0|_1 = 0. \tag{8.45}
$$

Hence, (8.42) is true for $k = 0$.

(I) Taking the inner product on both sides of (8.44a) with $\delta_t e^{\frac{1}{2}}$ gives

$$
\|\delta_t e^{\frac{1}{2}}\|^2 + (\psi(\hat{u}, e^{\frac{1}{2}}), \delta_t e^{\frac{1}{2}}) = v(\delta_x^2 e^{\frac{1}{2}}, \delta_t e^{\frac{1}{2}}) + ((R_2)^0, \delta_t e^{\frac{1}{2}}),
$$

Noticing

$$
e_i^0 = 0, \quad 0 \leqslant i \leqslant m,
$$

we have

$$
\frac{1}{\tau^2} \|e^1\|^2 + \frac{1}{2\tau}(\psi(\hat{u}, e^1), e^1) = -\frac{v}{2\tau}|e^1|_1^2 + \frac{1}{\tau}((R_2)^0, e^1).
$$

With the help of

$$
(\psi(\hat{u}, e^1), e^1) = 0,
$$

it yields

$$
\frac{1}{\tau^2}\|e^1\|^2 + \frac{v}{2\tau}|e^1|_1^2 = \frac{1}{\tau}((R_2)^0, e^1) \leqslant \frac{1}{\tau^2}\|e^1\|^2 + \frac{1}{4}\|(R_2)^0\|^2.
$$

Combining with (8.32) yields

$$
|e^1|_1^2 \leqslant \frac{2\tau}{v} \cdot \frac{1}{4}\|(R_2)^0\|^2 \leqslant \frac{\tau}{2v} L c_6^2 (\tau^2 + h^2)^2.
$$

When $\tau \leqslant 2\nu$, we have

$$|e^1|_1^2 \leqslant Lc_6^2(\tau^2 + h^2)^2,$$

or

$$|e^1|_1 \leqslant \sqrt{L}c_6(\tau^2 + h^2). \tag{8.46}$$

(II) Taking the inner product on both sides of (8.44b) with $\Delta_t e^k$ gives

$$\|\Delta_t e^k\|^2 + \big(\psi(U^k, U^{\bar{k}}) - \psi(u^k, u^{\bar{k}}), \Delta_t e^k\big) = \nu\big(\delta_x^2 e^{\bar{k}}, \Delta_t e^k\big) + \big((R_2)^k, \Delta_t e^k\big),$$
$$1 \leqslant k \leqslant n - 1,$$

or

$$\|\Delta_t e^k\|^2 + \frac{\nu}{4\tau}\big(|e^{k+1}|_1^2 - |e^{k-1}|_1^2\big)$$
$$= -(\psi(U^k, U^{\bar{k}}) - \psi(u^k, u^{\bar{k}}), \Delta_t e^k) + \big((R_2)^k, \Delta_t e^k\big), \quad 1 \leqslant k \leqslant n - 1. \tag{8.47}$$

It follows from (8.17) that

$$|U^k|_1 \leqslant \sqrt{L}c_3, \quad \|U^k\|_\infty \leqslant \frac{\sqrt{L}}{2}|U^k|_1 \leqslant \frac{L}{2}c_3, \quad 0 \leqslant k \leqslant n. \tag{8.48}$$

Suppose now (8.42) is true for $k = 1, 2, \ldots, l$, then when $c_8(\tau^2 + h^2) \leqslant 1$, we have

$$\begin{cases} |u^k|_1 \leqslant |U^k|_1 + |e^k|_1 \leqslant \sqrt{L}c_3 + 1, & 1 \leqslant k \leqslant l, \\ \|u^k\|_\infty \leqslant \dfrac{\sqrt{L}}{2}|u^k|_1 \leqslant \dfrac{\sqrt{L}}{2}(\sqrt{L}c_3 + 1), & 1 \leqslant k \leqslant l. \end{cases} \tag{8.49}$$

Noticing

$$\psi(U^k, U^{\bar{k}})_i - \psi(u^k, u^{\bar{k}})_i$$
$$= \psi(e^k, U^{\bar{k}})_i + \psi(u^k, e^{\bar{k}})_i$$
$$= \frac{1}{3}\Big[e_i^k \Delta_x U_i^{\bar{k}} + \Delta_x(e^k U^{\bar{k}})_i\Big] + \frac{1}{3}\Big[u_i^k \Delta_x e_i^{\bar{k}} + \Delta_x(u^k e^{\bar{k}})_i\Big]$$
$$= \frac{1}{3}\Big[e_i^k \Delta_x U_i^{\bar{k}} + \frac{1}{2}\Big(\delta_x e_{i+\frac{1}{2}}^k\Big)U_{i+1}^{\bar{k}} + e_i^k \Delta_x U_i^{\bar{k}} + \frac{1}{2}\Big(\delta_x e_{i-\frac{1}{2}}^k\Big)U_{i-1}^{\bar{k}}\Big]$$
$$+ \frac{1}{3}\Big[u_i^k \Delta_x e_i^{\bar{k}} + \frac{1}{2}\Big(\delta_x u_{i+\frac{1}{2}}^k\Big)e_{i+1}^{\bar{k}} + u_i^k \Delta_x e_i^{\bar{k}} + \frac{1}{2}(\delta_x u_{i-\frac{1}{2}}^k)e_{i-1}^{\bar{k}}\Big]$$

and (8.48)–(8.49), we get

$$- (\psi(U^k, U^{\bar{k}}) - \psi(u^k, u^{\bar{k}}), \Delta_t e^k)$$

$$\leqslant \frac{1}{3}(\|e^k\|_\infty |U^{\bar{k}}|_1 + \|U^{\bar{k}}\|_\infty |e^k|_1 + \|e^k\|_\infty |U^{\bar{k}}|_1)\|\Delta_t e^k\|$$

$$+ \frac{1}{3}(\|u^k\|_\infty |e^{\bar{k}}|_1 + \|e^{\bar{k}}\|_\infty |u^k|_1 + \|u^k\|_\infty |e^{\bar{k}}|_1)\|\Delta_t e^k\|$$

$$\leqslant \frac{1}{3}\left(2\sqrt{L}c_3\|e^k\|_\infty + \frac{L}{2}c_3|e^k|_1\right)\|\Delta_t e^k\|$$

$$+ \frac{1}{3}\left(2 \cdot \frac{\sqrt{L}}{2}(\sqrt{L}c_3 + 1)|e^{\bar{k}}|_1 + (\sqrt{L}c_3 + 1)\|e^{\bar{k}}\|_\infty\right)\|\Delta_t e^k\|$$

$$\leqslant \frac{1}{3}\left(2\sqrt{L}c_3\frac{\sqrt{L}}{2}|e^k|_1 + \frac{L}{2}c_3|e^k|_1\right)\|\Delta_t e^k\|$$

$$+ \frac{1}{3}\left[\sqrt{L}(\sqrt{L}c_3 + 1)|e^{\bar{k}}|_1 + (\sqrt{L}c_3 + 1)\frac{\sqrt{L}}{2}|e^{\bar{k}}|_1\right]\|\Delta_t e^k\|$$

$$= \frac{1}{2}Lc_3|e^k|_1 \cdot \|\Delta_t e^k\| + \frac{1}{2}\sqrt{L}(\sqrt{L}c_3 + 1)|e^{\bar{k}}|_1 \cdot \|\Delta_t e^k\|$$

$$\leqslant \frac{1}{4}\|\Delta_t e^k\|^2 + \frac{L^2 c_3^2}{4}|e^k|_1^2 + \frac{1}{4}\|\Delta_t e^k\|^2 + \frac{L(\sqrt{L}c_3 + 1)^2}{4}|e^{\bar{k}}|_1^2, \quad 1 \leqslant k \leqslant l.$$

Moreover, we have

$$((R_2)^k, \Delta_t e^k) \leqslant \frac{1}{2}\|\Delta_t e^k\|^2 + \frac{1}{2}\|(R_2)^k\|^2.$$

Inserting the above two inequalities into (8.47) and using (8.34), we arrive at

$$\frac{v}{4\tau}(|e^{k+1}|_1^2 - |e^{k-1}|_1^2)$$

$$\leqslant \frac{L^2 c_3^2}{4}|e^k|_1^2 + \frac{L(\sqrt{L}c_3 + 1)^2}{4}|e^{\bar{k}}|_1^2 + \frac{1}{2}\|(R_2)^k\|^2$$

$$\leqslant \frac{L^2 c_3^2}{4}|e^k|_1^2 + \frac{L(\sqrt{L}c_3 + 1)^2}{4} \cdot \frac{|e^{k+1}|_1^2 + |e^{k-1}|_1^2}{2} + \frac{1}{2}Lc_7^2(\tau^2 + h^2)^2, \quad 1 \leqslant k \leqslant l.$$

Multiplying both sides of the above inequality by $\frac{4\tau}{\nu}$ and rearranging the result lead to

$$|e^{k+1}|_1^2 \leqslant |e^{k-1}|_1^2 + \frac{L^2 c_3^2}{\nu}\tau |e^k|_1^2 + \frac{1}{2\nu}L(\sqrt{L}c_3 + 1)^2\tau\left(|e^{k+1}|_1^2 + |e^{k-1}|_1^2\right)$$

$$+ \frac{2}{\nu}Lc_7^2\tau(\tau^2 + h^2)^2, \quad 1 \leqslant k \leqslant l,$$

that is,

$$\left[1 - \frac{L(\sqrt{L}c_3 + 1)^2}{2\nu}\tau\right]|e^{k+1}|_1^2$$

$$\leqslant \frac{L^2 c_3^2}{\nu}\tau |e^k|_1^2 + \left[1 + \frac{L(\sqrt{L}c_3 + 1)^2}{2\nu}\tau\right]|e^{k-1}|_1^2 + \frac{2}{\nu}Lc_7^2\tau(\tau^2 + h^2)^2, \ 1 \leqslant k \leqslant l.$$

When $\frac{L(\sqrt{L}c_3 + 1)^2}{2\nu}\tau \leqslant \frac{1}{3}$, it yields

$$|e^{k+1}|_1^2 \leqslant \frac{3L^2 c_3^2}{2\nu}\tau |e^k|_1^2 + \left[1 + \frac{3L(\sqrt{L}c_3 + 1)^2}{2\nu}\tau\right]|e^{k-1}|_1^2 + \frac{3}{\nu}Lc_7^2\tau(\tau^2 + h^2)^2,$$

$$1 \leqslant k \leqslant l.$$

It is easy to know that

$$\max\left\{|e^k|_1^2, |e^{k+1}|_1^2\right\}$$

$$\leqslant \left[1 + \frac{3L^2 c_3^2 + 3L(\sqrt{L}c_3 + 1)^2}{2\nu}\tau\right]\max\{|e^{k-1}|_1^2, |e^k|_1^2\} + \frac{3}{\nu}Lc_7^2\tau(\tau^2 + h^2)^2,$$

$$1 \leqslant k \leqslant l.$$

Applying Lemma 3.3 (Gronwall inequality) produces

$$\max\{|e^l|_1^2, |e^{l+1}|_1^2\}$$

$$\leqslant \exp\left\{\frac{3L^2 c_3^2 + 3L(\sqrt{L}c_3 + 1)^2}{2\nu}T\right\}$$

$$\cdot \left(\max\{|e^0|_1^2, |e^1|_1^2\} + \frac{2Lc_7^2}{L^2 c_3^2 + L(\sqrt{L}c_3 + 1)^2}(\tau^2 + h^2)^2\right).$$

It follows by noticing (8.45) and (8.46) that

$$|e^{l+1}|_1^2 \leqslant \exp\left\{\frac{3L^2c_3^2 + 3L(\sqrt{L}c_3 + 1)^2}{2\nu}T\right\}$$
$$\cdot\left(Lc_6^2 + \frac{2c_7^2}{Lc_3^2 + (\sqrt{L}c_3 + 1)^2}\right)(\tau^2 + h^2)^2$$
$$\equiv c_8^2(\tau^2 + h^2)^2,$$

where

$$c_8 = \exp\left\{\frac{3L^2c_3^2 + 3L(\sqrt{L}c_3 + 1)^2}{4\nu}T\right\}\cdot\left(Lc_6^2 + \frac{2c_7^2}{Lc_3^2 + (\sqrt{L}c_3 + 1)^2}\right)^{\frac{1}{2}}.$$

Thus, (8.42) also holds for $k = l+1$. By induction, it is concluded that (8.42) holds for $k = 0, 1, 2, \ldots, n$. $\square$

8.3.5   Numerical Examples

Example 8.2 Apply the difference scheme (8.36) to compute the initial-boundary value problem (8.30).

The Thomas algorithm is used to solve the difference scheme (8.36).

Take $\nu = 1$. For different values of step sizes, Tables 8.3 and 8.4 record the maximum errors of numerical solutions using the difference scheme (8.36) defined by

$$E_\infty(h, \tau) = \max_{0 \leqslant k \leqslant n, 0 \leqslant j \leqslant m} \left|u(x_j, t_k) - u_j^k\right|.$$

It can be seen from Table 8.3 that when the step size h decreases by half, the maximum norm error is reduced to a quarter of the original. From Table 8.4, we can find that when the step size τ decreases by half, the maximum norm error is also reduced to a quarter of the original. Figure 8.2 displays the curves of invariant E^k defined in Theorem 8.8.

Table 8.3 (Example 8.2) The maximum errors of numerical solutions with different spatial step sizes ($\tau = 1/1600$)

h	$E_\infty(h, \tau)$	$E_\infty(2h, \tau)/E_\infty(h, \tau)$
1/10	1.899969e−04	
1/20	4.725070e−05	4.0071
1/40	1.159189e−05	4.0192
1/80	2.678947e−06	4.0760
1/160	4.508333e−07	4.3285

Table 8.4 (Example 8.2) The maximum errors of numerical solutions with different temporal step sizes ($h = 1/1600$)

τ	$E_\infty(h, \tau)$	$E_\infty(h, 2\tau)/E_\infty(h, \tau)$
1/10	8.320394e−03	
1/20	2.111490e−03	4.3777
1/40	4.821587e−04	4.0830
1/80	1.180908e−04	4.0208
1/160	2.936970e−05	4.0080

Fig. 8.2 (Example 8.2) The curves of energy E^k for the difference scheme (8.36)

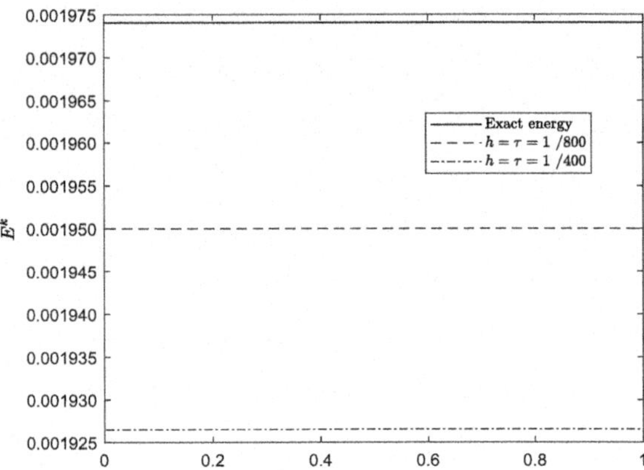

8.4   Summary and Extension

In this chapter, we discuss finite difference methods for solving the Burgers' equation. First, it is demonstrated that the solution to the problem (8.1) satisfies the conservation of energy, and an estimate of the solution in the maximum norm is provided. Then, in Sects. 8.2 and 8.3, a two-time-level nonlinear difference scheme and a three-time-level linearized difference scheme are introduced, respectively. The existence, uniqueness, boundedness, and convergence of the solutions to the difference schemes are established. The content of this chapter is primarily based on [2, 6].

For the problem (8.1), another two-level linearized difference scheme can be derived:

$$\begin{cases} \delta_t u_i^{k+\frac{1}{2}} + \dfrac{1}{2}(u_i^k \Delta_x u_i^{k+1} + u_i^{k+1} \Delta_x u_i^k) = \nu \delta_x^2 u_i^{k+\frac{1}{2}}, \\ \qquad\qquad 1 \leqslant i \leqslant m-1, \quad 0 \leqslant k \leqslant n-1, \\ u_i^0 = \varphi(x_i), \quad 1 \leqslant i \leqslant m-1, \\ u_0^k = 0, \quad u_m^k = 0, \quad 0 \leqslant k \leqslant n. \end{cases} \tag{8.50}$$

It can also be demonstrated that the difference scheme (8.50) is uniquely solvable and convergent, with the convergence order of two in both time and space in the maximum norm [2].

In [3], a compact difference method for solving the Burgers' equation was studied. The generalized Burgers' equation was addressed in [5]. Additionally, in [4], a difference method for the two-dimensional Burgers' equation was considered.

We prove the existence of the solution to the system of nonlinear equations (8.12) by means of the Browder theorem [7]. In fact, there is another related theorem, the Leray-Schauder theorem [7], which can be applied in conjunction with the Browder theorem.

Theorem 8.11 (Leray-Schauder Theorem [7]) *Let H be a finite dimensional inner product space with the associated norm $\| \cdot \|$. Consider the operator $T_\lambda(w)$: $H \to H$, where $\lambda \in [0, 1]$ is a parameter. If $T_\lambda(w)$ satisfies the following conditions:*

(I) $T_\lambda(w)$ is a continuous operator in H;
(II) $T_0(w) = 0$ has a unique solution;
(III) All possible solutions to $T_\lambda(w) = 0$ are uniformly bounded.

Then for any $\lambda \in [0, 1]$, there exists a solution satisfying $T_\lambda(w) = 0$. Especially, there is a solution to $T_1(w) = 0$.

Now we shall use the above theorem to prove Theorem 8.4, that is to show that there is a solution to (8.12).

Let $H = \overset{\circ}{\mathcal{U}}_h$. For any $w \in \overset{\circ}{\mathcal{U}}_h$, define

$$\begin{cases} T_\lambda(w)_i = \dfrac{2}{\tau}(w_i - u_i^k) + \lambda \psi(w, w)_i - v\delta_x^2 w_i, & 1 \leqslant i \leqslant m - 1, \\ T_\lambda(w)_0 = 0, \quad T_\lambda(w)_m = 0. \end{cases}$$

It is easy to know that (I) $T_\lambda(w)$ is continuous; (II) $T_0(w)_i = 0$, $i = 0, 1, \ldots, m$ is a strictly diagonally dominant tridiagonal system of linear equations; thus, there is a unique solution. Now we check (III). Suppose w is a possible solution to $T_\lambda(w) = 0$. Taking the inner product of $T_\lambda(w) = 0$ with w, we have

$$\frac{2}{\tau}\big((w, w) - (u^k, w)\big) + \lambda\big(\psi(w, w), w\big) - v\big(\delta_x^2 w, w\big) = 0.$$

With the aid of

$$(\psi(w, w), w) = 0, \qquad -(\delta_x^2 w, w) = |w|_1^2,$$

we have

$$\frac{2}{\tau}\big(\|w\|^2 - (u^k, w)\big) + v|w|_1^2 = 0.$$

Consequently,

$$\|w\|^2 \leqslant (u^k, w) \leqslant \|u^k\| \cdot \|w\|.$$

It is easy to know that

$$\|w\| \leqslant \|u^k\|.$$

Therefore, the condition (III) is satisfied. With the help of the Leray-Schauder theorem, there is a solution to (8.12).

8.5　Exercise

8.1　Consider the initial-boundary value problem of the regularized long-wave equation as follows:

$$\begin{cases} u_t - \mu u_{xxt} + \gamma u u_x + u_x = 0, & 0 < x < L, \quad 0 < t \leqslant T, \\ u(x, 0) = \varphi(x), & 0 < x < L, \\ u(0, t) = 0, \quad u(L, t) = 0, & 0 \leqslant t \leqslant T, \end{cases} \tag{8.51}$$

where μ, γ are two positive constants and $\varphi(0) = \varphi(L) = 0$. For the problem (8.51), a difference scheme is proposed as

$$\begin{cases} \delta_t u_i^{k+\frac{1}{2}} - \mu \delta_t \delta_x^2 u_i^{k+\frac{1}{2}} + \gamma \psi(u^{k+\frac{1}{2}}, u^{k+\frac{1}{2}})_i + \Delta_x u_i^{k+\frac{1}{2}} = 0, \\ \qquad\qquad\qquad 1 \leqslant i \leqslant m-1, \quad 0 \leqslant k \leqslant n-1, \\ u_i^0 = \varphi(x_i), \quad 1 \leqslant i \leqslant m-1, \\ u_0^k = 0, \quad u_m^k = 0, \quad 0 \leqslant k \leqslant n. \end{cases}$$

(1) Analyze the local truncation error of the difference scheme.
(2) Analyze the conservation of the difference scheme.
(3) Show the existence of the difference solution.
(4) Prove the convergence of the difference solution.

8.2 For the problem (8.51), the following difference scheme is derived:

$$
\begin{cases}
\delta_t u_i^{\frac{1}{2}} - \mu \delta_t \delta_x^2 u_i^{\frac{1}{2}} + \gamma \psi(u^0, u^{\frac{1}{2}})_i + \Delta_x u_i^{\frac{1}{2}} = 0, \\
\qquad\qquad\qquad\qquad 1 \leqslant i \leqslant m - 1, \\
\Delta_t u_i^k - \mu \Delta_t \delta_x^2 u_i^k + \gamma \psi(u^k, u^{\bar{k}})_i + \Delta_x u_i^{\bar{k}} = 0, \\
\qquad\qquad\qquad 1 \leqslant i \leqslant m - 1, \quad 1 \leqslant k \leqslant n - 1, \\
u_i^0 = \varphi(x_i), \quad 1 \leqslant i \leqslant m - 1, \\
u_0^k = 0, \quad u_m^k = 0, \quad 0 \leqslant k \leqslant n.
\end{cases}
$$

(1) Analyze the local truncation error of the difference scheme.
(2) Analyze the conservation of the difference scheme.
(3) Show the existence of the difference solution.
(4) Prove the convergence of the difference solution.

References

1. Cole, J.D.: On a quasi-linear parabolic equation occurring in aerodynamics. Q. Appl. Math. **9**, 225–236 (1951)
2. Sun, H., Sun, Z.Z.: On two linearized difference schemes for the Burgers' equation. Int. J. Comput. Math. **92**(6), 1160–1179 (2015)
3. Wang, X.P., Zhang, Q.F., Sun, Z.Z.: The pointwise error estimates of two energy-preserving fourth-order compact schemes for viscous Burgers' equation. Adv. Comput. Math. **47**, 23 (2021)
4. Xu, P.P., Sun, Z.Z.: A second order accurate difference scheme for the two-dimensional Burgers system. Numer. Methods Partial Differ. Equations **25**(1), 172–194 (2009)
5. Zhang, Q.F., Qin, Y.F., Wang, X.P., Sun, Z.Z.: The study of exact and numerical solutions of the generalized viscous Burgers' equation. Appl. Math. Lett. **112**, 106719 (2021)
6. Zhang, Q.F., Wang, X.P., Sun, Z.Z.: The pointwise estimates of a conservative difference scheme for Burgers' equation. Numer. Methods Partial Differ. Equations **36**, 1611–1628 (2020)
7. Zhou, Y.L.: Applications of Discrete Functional Analysis to the Finite Difference Method. International Academic Publishers, Beijing (1991)

Chapter 9
Finite Difference Methods for the Korteweg-de Vries Equation

The Korteweg-de Vries (KdV) equation is a prominent example of nonlinear dispersion equations. Due to its infinite conservation laws, its widespread applications has been found in various fields, including solid, liquid, gas, and plasma physics. In 1895, Dutch mathematicians Diederik Korteweg and Gustav de Vries discovered the KdV equation while studying the long and medium amplitude motion of shallow water waves. The KdV equation represents a class of partial differential equations governing unidirectionally moving shallow water waves. Although Boussinesq introduced the KdV equation in 1877, it was later recognized that the equation can also describe a variety of physical phenomena, such as magnetic current waves, ionic sound waves in plasma, and pressure waves in liquid-gas mixtures.

9.1   The Korteweg-de Vries Equation

In this chapter, we study the difference methods for solving the initial-boundary value problem of the KdV equation

$$
\begin{cases}
u_t + \gamma u u_x + u_{xxx} = 0, & 0 < x < L, \quad 0 < t \leqslant T, & (9.1a) \\[2mm]
u(x,0) = \varphi(x), & 0 < x < L, & (9.1b) \\[2mm]
u(0,t) = 0, \quad u(L,t) = 0, \quad u_x(L,t) = 0, & 0 \leqslant t \leqslant T, & (9.1c)
\end{cases}
$$

where γ is a constant and $\varphi(0) = \varphi(L) = \varphi'(L) = 0$.

Before introducing the difference method, we first employ the energy method to provide a priori estimate of the solution to the problem (9.1).

© Science Press 2026

Z.-Z. Sun et al., *Numerical Solutions to Partial Differential Equations
with Finite Difference Methods*, Springer Asia Pacific Mathematics Series 9,
https://doi.org/10.1007/978-981-95-5563-5_9

Theorem 9.1 *Let $u(x, t)$ be the solution of the problem (9.1). Denote*

$$E(t) = \int_0^L u^2(x, t)\,dx + \int_0^t u_x^2(0, s)\,ds,$$

Then

$$E(t) = E(0), \quad 0 < t \leqslant T. \tag{9.2}$$

Proof Taking the inner product on both sides of (9.1a) with u gives

$$\int_0^L u(x, t)u_t(x, t)\,dx + \gamma \int_0^L u^2(x, t)u_x(x, t)\,dx + \int_0^L u(x, t)u_{xxx}(x, t)\,dx = 0. \tag{9.3}$$

Now each term on the left-hand side of the above equality is analyzed.

For the first term,

$$\int_0^L u(x, t)u_t(x, t)\,dx = \frac{1}{2}\frac{d}{dt}\int_0^L u^2(x, t)\,dx. \tag{9.4}$$

For the second one, in view of the boundary value condition (9.1c), we have

$$\int_0^L u^2(x, t)u_x(x, t)\,dx = \frac{1}{3}u^3(x, t)\Big|_{x=0}^L = 0. \tag{9.5}$$

For the third one, in view of the boundary value condition (9.1c), we have

$$\int_0^L u(x, t)u_{xxx}(x, t)\,dx = \left[u(x, t)u_{xx}(x, t) - \frac{1}{2}u_x^2(x, t)\right]\Big|_{x=0}^L = \frac{1}{2}u_x^2(0, t). \tag{9.6}$$

Substituting (9.4)–(9.6) into (9.3) arrives at

$$\frac{1}{2}\frac{d}{dt}\int_0^L u^2(x, t)\,dx + \frac{1}{2}u_x^2(0, t) = 0,$$

that is,

$$\frac{d}{dt}\left[\int_0^L u^2(x, t)\,dx + \int_0^t u_x^2(0, s)\,ds\right] = 0.$$

Hence,

$$E(t) = E(0), \quad 0 < t \leqslant T.$$

$\square$

The equality (9.2) is called *the conservation law of energy.*

9.2   First-Order Difference Scheme in Space

Assume that there is a solution $u \in C_{x,t}^{4,3}([0, L] \times [0, T])$ to the problem (9.1).
Similar to that in Chap. 8, for any $v, w \in \mathcal{U}_h$, denote

$$\psi(v, w)_i = \frac{1}{3}\left[v_i \Delta_x w_i + \Delta_x(vw)_i\right], \quad 1 \leqslant i \leqslant m - 1.$$

9.2.1   Derivation of the Difference Scheme

Taking $x = L$ in Eq. (9.1a) and noticing $u(L, t) = 0$, we have

$$u_{xxx}(L, t) = 0.$$

Considering Eq. (9.1a) at the point (x_i, t) gives

$$u_t(x_i, t) + \gamma u(x_i, t)u_x(x_i, t) + u_{xxx}(x_i, t) = 0, \quad 1 \leqslant i \leqslant m - 1. \tag{9.7}$$

Now the numerical approximation of $u_{xxx}(x_i, t)$ is first addressed.
From the numerical differentiation formula (Lemma 1.2), we obtain

$$
\begin{aligned}
u_{xxx}&(x_i, t)\\
&= \frac{1}{h}\left(u_{xx}(x_{i+1}, t) - u_{xx}(x_i, t)\right) + O(h)\\
&= \frac{1}{h}\left[\left(\frac{u(x_{i+2}, t) - 2u(x_{i+1}, t) + u(x_i, t)}{h^2} + O(h^2)\right)\right.\\
&\qquad \left. - \left(\frac{u(x_{i+1}, t) - 2u(x_i, t) + u(x_{i-1}, t)}{h^2} + O(h^2)\right)\right] + O(h)\\
&= \frac{1}{h^3}\left[u(x_{i+2}, t) - 3u(x_{i+1}, t) + 3u(x_i, t) - u(x_{i-1}, t)\right]\\
&\quad + O(h), \quad 1 \leqslant i \leqslant m - 2
\end{aligned}
\tag{9.8}
$$

and

$$u_{xxx}(x_{m-1}, t)$$

$$= \frac{1}{h}\Big(u_{xx}(x_m, t) - u_{xx}(x_{m-1}, t)\Big) + O(h)$$

$$= \frac{1}{h}\bigg[\frac{2}{h}\Big(u_x(x_m, t) - \frac{u(x_m, t) - u(x_{m-1}, t)}{h}\Big) + \frac{h}{3}u_{xxx}(x_m, t) + O(h^2)$$

$$- \Big(\frac{u(x_m, t) - 2u(x_{m-1}, t) + u(x_{m-2}, t)}{h^2} + O(h^2)\Big)\bigg] + O(h)$$

$$= \frac{1}{h}\bigg[-\frac{2}{h}\cdot\frac{u(x_m, t) - u(x_{m-1}, t)}{h} - \frac{u(x_m, t) - 2u(x_{m-1}, t) + u(x_{m-2}, t)}{h^2}\bigg]$$

$$+ O(h). \tag{9.9}$$

Define the grid function

$$U_i^k = u(x_i, t_k), \quad 0 \leqslant i \leqslant m, \quad 0 \leqslant k \leqslant n.$$

Averaging (9.7) on two adjacent time levels $t = t_k$ and $t = t_{k+1}$ and combining (9.8) with (9.9), we get

$$\begin{cases} \delta_t U_i^{k+\frac{1}{2}} + \gamma \psi(U^{k+\frac{1}{2}}, U^{k+\frac{1}{2}})_i + \delta_x^2(\delta_x U_{i+\frac{1}{2}}^{k+\frac{1}{2}}) = (R_1)_i^k, \\[2mm] \qquad\qquad 1 \leqslant i \leqslant m - 2, \quad 0 \leqslant k \leqslant n - 1, \\[2mm] \delta_t U_{m-1}^{k+\frac{1}{2}} + \gamma \psi(U^{k+\frac{1}{2}}, U^{k+\frac{1}{2}})_{m-1} + \frac{1}{h}\Big(-\frac{2}{h}\delta_x U_{m-\frac{1}{2}}^{k+\frac{1}{2}} - \delta_x^2 U_{m-1}^{k+\frac{1}{2}}\Big) = (R_1)_{m-1}^k, \\[2mm] \qquad\qquad 0 \leqslant k \leqslant n - 1 \end{cases} \tag{9.10}$$

and there is a constant c_1 satisfying

$$|(R_1)_i^k| \leqslant c_1(\tau^2 + h), \quad 1 \leqslant i \leqslant m - 1, \quad 0 \leqslant k \leqslant n - 1. \tag{9.11}$$

Noticing the initial-boundary value conditions (9.1b)–(9.1c), we have

$$\begin{cases} U_i^0 = \varphi(x_i), \quad 1 \leqslant i \leqslant m - 1, \\[2mm] U_0^k = 0, \quad U_m^k = 0, \quad 0 \leqslant k \leqslant n. \end{cases} \tag{9.12}$$

Omitting the small term $(R_1)_i^k$ in (9.10) and noticing (9.12), a difference scheme for solving (9.1) reads

$$
\begin{cases}
\delta_t u_i^{k+\frac{1}{2}} + \gamma \psi(u^{k+\frac{1}{2}}, u^{k+\frac{1}{2}})_i + \delta_x^2(\delta_x u_{i+\frac{1}{2}}^{k+\frac{1}{2}}) = 0, \\[2mm]
\qquad\qquad 1 \leqslant i \leqslant m-2, \quad 0 \leqslant k \leqslant n-1, & (9.13a) \\[3mm]
\delta_t u_{m-1}^{k+\frac{1}{2}} + \gamma \psi(u^{k+\frac{1}{2}}, u^{k+\frac{1}{2}})_{m-1} + \frac{1}{h}\left(-\frac{2}{h}\delta_x u_{m-\frac{1}{2}}^{k+\frac{1}{2}} - \delta_x^2 u_{m-1}^{k+\frac{1}{2}}\right) = 0, \\[2mm]
\qquad\qquad 0 \leqslant k \leqslant n-1, & (9.13b) \\[3mm]
u_i^0 = \varphi(x_i), \quad 1 \leqslant i \leqslant m-1, & (9.13c) \\[3mm]
u_0^k = 0, \quad u_m^k = 0, \quad 0 \leqslant k \leqslant n. & (9.13d)
\end{cases}
$$

9.2.2   Existence of the Difference Solution

Lemma 9.1 *For an arbitrary $w \in \overset{\circ}{\mathcal{U}}_h$, it holds*

$$
h \sum_{i=1}^{m-2}(\delta_x^2 \delta_x w_{i+\frac{1}{2}})w_i + \left(-\frac{2}{h}\delta_x w_{m-\frac{1}{2}} - \delta_x^2 w_{m-1}\right)w_{m-1}
$$

$$
= \frac{1}{2}h|w|_2^2 + \frac{1}{2}(\delta_x w_{\frac{1}{2}})^2 + \frac{3}{2}(\delta_x w_{m-\frac{1}{2}})^2,
$$

where

$$
|w|_2^2 = h \sum_{i=1}^{m-1}(\delta_x^2 w_i)^2.
$$

Proof Some direct calculations give

$$
h \sum_{l=1}^{m-2}(\delta_x^2 \delta_x w_{i+\frac{1}{2}})w_i + \left(-\frac{2}{h}\delta_x w_{m-\frac{1}{2}} - \delta_x^2 w_{m-1}\right)w_{m-1}
$$

$$
= \sum_{i=1}^{m-2}(\delta_x^2 w_{i+1} - \delta_x^2 w_i)w_i + \left(2\delta_x w_{m-\frac{1}{2}} + h\delta_x^2 w_{m-1}\right)\delta_x w_{m-\frac{1}{2}}
$$

$$
= \sum_{i=2}^{m-1}(\delta_x^2 w_i)w_{i-1} - \sum_{i=1}^{m-2}(\delta_x^2 w_i)w_i + \left(3\delta_x w_{m-\frac{1}{2}} - \delta_x w_{m-\frac{3}{2}}\right)(\delta_x w_{m-\frac{1}{2}})
$$

$$= -h \sum_{i=1}^{m-1} (\delta_x^2 w_i)(\delta_x w_{i-\frac{1}{2}}) + (\delta_x^2 w_{m-1}) w_{m-1} + 3(\delta_x w_{m-\frac{1}{2}})^2 - (\delta_x w_{m-\frac{3}{2}})(\delta_x w_{m-\frac{1}{2}})$$

$$= \sum_{i=1}^{m-1} (\delta_x w_{i-\frac{1}{2}})^2 - \sum_{i=1}^{m-1} (\delta_x w_{i+\frac{1}{2}})(\delta_x w_{i-\frac{1}{2}}) + 2(\delta_x w_{m-\frac{1}{2}})^2$$

$$= \frac{1}{2} \sum_{i=0}^{m-2} (\delta_x w_{i+\frac{1}{2}})^2 + \frac{1}{2} \sum_{i=1}^{m-1} (\delta_x w_{i-\frac{1}{2}})^2 - \sum_{i=1}^{m-1} (\delta_x w_{i+\frac{1}{2}})(\delta_x w_{i-\frac{1}{2}}) + 2(\delta_x w_{m-\frac{1}{2}})^2$$

$$= \frac{1}{2} \sum_{i=1}^{m-1} (\delta_x w_{i+\frac{1}{2}} - \delta_x w_{i-\frac{1}{2}})^2 + \frac{1}{2}(\delta_x w_{\frac{1}{2}})^2 + \frac{3}{2}(\delta_x w_{m-\frac{1}{2}})^2.$$

$\square$

Theorem 9.2 *The difference scheme (9.13) admits a solution.*

Proof From (9.13c)–(9.13d), the value of u^0 has been uniquely determined. Suppose the value of u^k has been determined. Let

$$w_i = u_i^{k+\frac{1}{2}}, \quad 0 \leqslant i \leqslant m.$$

Then we have the system of nonlinear equations in w as

$$\begin{cases} \frac{2}{\tau}(w_i - u_i^k) + \gamma \psi(w, w)_i + \delta_x^2(\delta_x w_{i+\frac{1}{2}}) = 0, & 1 \leqslant i \leqslant m - 2, \\ \frac{2}{\tau}(w_{m-1} - u_{m-1}^k) + \gamma \psi(w, w)_{m-1} + \frac{1}{h}\left(-\frac{2}{h}\delta_x w_{m-\frac{1}{2}} - \delta_x^2 w_{m-1} \right) = 0, \\ w_0 = 0, \quad w_m = 0. \end{cases}$$

$$(9.14)$$

For any $w \in \mathring{\mathcal{U}}_h$, define the operator $\Pi : \mathring{\mathcal{U}}_h \to \mathring{\mathcal{U}}_h$ by

$$\Pi(w)_i = \begin{cases} \dfrac{2}{\tau}(w_i - u_i^k) + \gamma \psi(w, w)_i + \delta_x^2(\delta_x w_{i+\frac{1}{2}}), & 1 \leqslant i \leqslant m - 2, \\[2ex] \dfrac{2}{\tau}(w_{m-1} - u_{m-1}^k) + \gamma \psi(w, w)_{m-1} \\[1ex] \quad + \dfrac{1}{h}\left(-\dfrac{2}{h}\delta_x w_{m-\frac{1}{2}} - \delta_x^2 w_{m-1} \right), & i = m - 1, \\[2ex] 0, & i = 0, m. \end{cases}$$

Direct calculations produce

$$(\Pi(w), w) = \frac{2}{\tau}\big[\|w\|^2 - (u^k, w)\big] + \gamma(\psi(w, w), w)$$

$$+ h \sum_{i=1}^{m-2} (\delta_x^2 \delta_x w_{i+\frac{1}{2}}) w_i + \Big(-\frac{2}{h}\delta_x w_{m-\frac{1}{2}} - \delta_x^2 w_{m-1}\Big) w_{m-1}.$$

It follows from Lemma 9.1 and $(\psi(w, w), w) = 0$ that

$$(\Pi(w), w) \geqslant \frac{2}{\tau}\big[\|w\|^2 - (u^k, w)\big] \geqslant \frac{2}{\tau}\|w\|(\|w\| - \|u^k\|).$$

When $\|w\| = \|u^k\|$, it follows $(\Pi(w), w) \geqslant 0$.

By Theorem 7.4 (Browder theorem), there is a $w^* \in \overset{\circ}{\mathcal{U}}_h$ satisfying $\|w^*\| \leqslant \|u^k\|$ such that

$$\Pi(w^*) = 0.$$

By induction, the conclusion is true. $\square$

9.2.3 *Conservation and Boundedness of the Difference Solution*

Theorem 9.3 *Let $\{u_i^k \mid 0 \leqslant i \leqslant m, 0 \leqslant k \leqslant n\}$ be the solution of the difference scheme (9.13). Denote*

$$E^k = \|u^{k+1}\|^2 + \tau \sum_{l=0}^{k}\Big[(\delta_x u_{\frac{1}{2}}^{l+\frac{1}{2}})^2 + 3(\delta_x u_{m-\frac{1}{2}}^{l+\frac{1}{2}})^2 + h|u^{l+\frac{1}{2}}|_2^2\Big].$$

Then

$$E^k = \|u^0\|^2, \quad 0 \leqslant k \leqslant n - 1.$$

Proof Multiplying (9.13a) by $hu_i^{k+\frac{1}{2}}$ and (9.13b) by $hu_{m-1}^{k+\frac{1}{2}}$ and adding the results together yield

$$\frac{1}{2\tau}(\|u^{k+1}\|^2 - \|u^k\|^2) + \gamma(\psi(u^{k+\frac{1}{2}}, u^{k+\frac{1}{2}}), u^{k+\frac{1}{2}})$$

$$+ h \sum_{i=1}^{m-2}(\delta_x^2 \delta_x u_{i+\frac{1}{2}}^{k+\frac{1}{2}}) u_i^{k+\frac{1}{2}} + \Big(-\frac{2}{h}\delta_x u_{m-\frac{1}{2}}^{k+\frac{1}{2}} - \delta_x^2 u_{m-1}^{k+\frac{1}{2}}\Big) u_{m-1}^{k+\frac{1}{2}} = 0. \quad (9.15)$$

It follows from Lemma 9.1 that

$$h \sum_{i=1}^{m-2} \left(\delta_x^2 \delta_x u_{i+\frac{1}{2}}^{k+\frac{1}{2}} \right) u_i^{k+\frac{1}{2}} + \left(-\frac{2}{h} \delta_x u_{m-\frac{1}{2}}^{k+\frac{1}{2}} - \delta_x^2 u_{m-1}^{k+\frac{1}{2}} \right) u_{m-1}^{k+\frac{1}{2}}$$

$$= \frac{1}{2} h |u^{k+\frac{1}{2}}|_2^2 + \frac{1}{2} \left(\delta_x u_{\frac{1}{2}}^{k+\frac{1}{2}} \right)^2 + \frac{3}{2} \left(\delta_x u_{m-\frac{1}{2}}^{k+\frac{1}{2}} \right)^2 .$$

Inserting the above equality into (9.15) and noticing $\left(\psi(u^{k+\frac{1}{2}}, u^{k+\frac{1}{2}}), u^{k+\frac{1}{2}} \right) = 0$, we obtain

$$\frac{1}{2\tau} \left(\|u^{k+1}\|^2 - \|u^k\|^2 \right) + \frac{1}{2} h |u^{k+\frac{1}{2}}|_2^2 + \frac{1}{2} \left(\delta_x u_{\frac{1}{2}}^{k+\frac{1}{2}} \right)^2 + \frac{3}{2} \left(\delta_x u_{m-\frac{1}{2}}^{k+\frac{1}{2}} \right)^2 = 0,$$

$$0 \leqslant k \leqslant n - 1.$$

Replacing k in the above equality by l and summing over l from 0 to k, we have

$$\|u^{k+1}\|^2 + \tau \sum_{l=0}^{k} \left[\left(\delta_x u_{\frac{1}{2}}^{l+\frac{1}{2}} \right)^2 + 3 \left(\delta_x u_{m-\frac{1}{2}}^{l+\frac{1}{2}} \right)^2 + h |u^{l+\frac{1}{2}}|_2^2 \right] = \|u^0\|^2,$$

$$0 \leqslant k \leqslant n - 1,$$

that is,

$$E^k = \|u^0\|^2, \quad 0 \leqslant k \leqslant n - 1.$$

$\square$

9.2.4   Convergence of the Difference Solution

Theorem 9.4 *Suppose $\{U_i^k \mid 0 \leqslant i \leqslant m, 0 \leqslant k \leqslant n\}$ is the solution of the problem (9.1) and $\{u_i^k \mid 0 \leqslant i \leqslant m, 0 \leqslant k \leqslant n\}$ is the solution of the difference scheme (9.13). Denote*

$$e_i^k = U_i^k - u_i^k, \quad 0 \leqslant i \leqslant m, \quad 0 \leqslant k \leqslant n.$$

Then there is a constant c_2 such that

$$\|e^k\| \leqslant c_2(\tau^2 + h), \quad 0 \leqslant k \leqslant n.$$

Proof Subtracting (9.13) from (9.10) and (9.12), the system of error equations is produced as

$$
\begin{cases}
\delta_t e_i^{k+\frac{1}{2}} + \gamma\big[\psi(U^{k+\frac{1}{2}}, U^{k+\frac{1}{2}})_i - \psi(u^{k+\frac{1}{2}}, u^{k+\frac{1}{2}})_i\big] + \delta_x^2 \delta_x e_{i+\frac{1}{2}}^{k+\frac{1}{2}} = (R_1)_i^k, \\
\qquad\qquad\qquad\qquad 1 \leqslant i \leqslant m-2, \quad 0 \leqslant k \leqslant n-1, & (9.16a) \\[4pt]
\delta_t e_{m-1}^{k+\frac{1}{2}} + \gamma\big[\psi(U^{k+\frac{1}{2}}, U^{k+\frac{1}{2}})_{m-1} - \psi(u^{k+\frac{1}{2}}, u^{k+\frac{1}{2}})_{m-1}\big] \\[2pt]
+ \dfrac{1}{h}\Big(-\dfrac{2}{h}\delta_x e_{m-\frac{1}{2}}^{k+\frac{1}{2}} - \delta_x^2 e_{m-1}^{k+\frac{1}{2}} \Big) = (R_1)_{m-1}^k, \quad 0 \leqslant k \leqslant n-1, & (9.16b) \\[6pt]
e_i^0 = 0, \quad 1 \leqslant i \leqslant m-1, & (9.16c) \\[4pt]
e_0^k = 0, \quad e_m^k = 0, \quad 0 \leqslant k \leqslant n. & (9.16d)
\end{cases}
$$

Multiplying (9.16a) by $h e_i^{k+\frac{1}{2}}$ and (9.16b) by $h e_{m-1}^{k+\frac{1}{2}}$ and adding the results together give

$$
\frac{1}{2\tau}\big(\|e^{k+1}\|^2 - \|e^k\|^2\big) + \gamma\Big(\psi(U^{k+\frac{1}{2}}, U^{k+\frac{1}{2}}) - \psi(u^{k+\frac{1}{2}}, u^{k+\frac{1}{2}}), e^{k+\frac{1}{2}}\Big)
$$

$$
+ h \sum_{i=1}^{m-2} \big(\delta_x^2 \delta_x e_{i+\frac{1}{2}}^{k+\frac{1}{2}}\big) e_i^{k+\frac{1}{2}} + \Big(-\frac{2}{h}\delta_x e_{m-\frac{1}{2}}^{k+\frac{1}{2}} - \delta_x^2 e_{m-1}^{k+\frac{1}{2}} \Big) e_{m-1}^{k+\frac{1}{2}}
$$

$$
= \big((R_1)^k, e^{k+\frac{1}{2}}\big), \quad 0 \leqslant k \leqslant n-1. \tag{9.17}
$$

It follows from Lemma 9.1 that

$$
h \sum_{i=1}^{m-2} \big(\delta_x^2 \delta_x e_{i+\frac{1}{2}}^{k+\frac{1}{2}}\big) e_i^{k+\frac{1}{2}} + \Big(-\frac{2}{h}\delta_x e_{m-\frac{1}{2}}^{k+\frac{1}{2}} - \delta_x^2 e_{m-1}^{k+\frac{1}{2}} \Big) e_{m-1}^{k+\frac{1}{2}}
$$

$$
= \frac{1}{2}h^2 \sum_{i=1}^{m-1} \big(\delta_x^2 e_i^{k+\frac{1}{2}}\big)^2 + \frac{1}{2}\big(\delta_x e_{\frac{1}{2}}^{k+\frac{1}{2}}\big)^2 + \frac{3}{2}\big(\delta_x e_{m-\frac{1}{2}}^{k+\frac{1}{2}}\big)^2. \tag{9.18}
$$

Next, we will analyze the second term on the left-hand side of (9.17). It follows from Lemma 8.1 that

$$
\big(\psi(U^{k+\frac{1}{2}}, U^{k+\frac{1}{2}}) - \psi(u^{k+\frac{1}{2}}, u^{k+\frac{1}{2}}), e^{k+\frac{1}{2}}\big)
$$

$$
= \big(\psi(U^{k+\frac{1}{2}}, U^{k+\frac{1}{2}}) - \psi(U^{k+\frac{1}{2}} - e^{k+\frac{1}{2}}, U^{k+\frac{1}{2}} - e^{k+\frac{1}{2}}), e^{k+\frac{1}{2}}\big)
$$

$$
= \big(\psi(e^{k+\frac{1}{2}}, U^{k+\frac{1}{2}}) + \psi(U^{k+\frac{1}{2}}, e^{k+\frac{1}{2}}) - \psi(e^{k+\frac{1}{2}}, e^{k+\frac{1}{2}}), e^{k+\frac{1}{2}}\big)
$$

$$
= \big(\psi(e^{k+\frac{1}{2}}, U^{k+\frac{1}{2}}), e^{k+\frac{1}{2}}\big)
$$

$$= \frac{1}{3} h \sum_{i=1}^{m-1} \left[e_i^{k+\frac{1}{2}} \Delta_x U_i^{k+\frac{1}{2}} + \Delta_x (eU)_i^{k+\frac{1}{2}} \right] e_i^{k+\frac{1}{2}}$$

$$= \frac{1}{3} \left[h \sum_{i=1}^{m-1} (\Delta_x U_i^{k+\frac{1}{2}})(e_i^{k+\frac{1}{2}})^2 + \frac{1}{2} \sum_{i=1}^{m-1} (e_{i+1}^{k+\frac{1}{2}} U_{i+1}^{k+\frac{1}{2}} - e_{i-1}^{k+\frac{1}{2}} U_{i-1}^{k+\frac{1}{2}}) e_i^{k+\frac{1}{2}} \right]$$

$$= \frac{1}{3} \left[h \sum_{i=1}^{m-1} (\Delta_x U_i^{k+\frac{1}{2}})(e_i^{k+\frac{1}{2}})^2 + \frac{1}{2} h \sum_{i=1}^{m-2} e_{i+1}^{k+\frac{1}{2}} e_i^{k+\frac{1}{2}} \delta_x U_{i+\frac{1}{2}}^{k+\frac{1}{2}} \right].$$

Denote

$$\hat{c}_1 = \max_{0 \leqslant x \leqslant L, 0 \leqslant t \leqslant T} |u_x(x,t)|.$$

Then

$$-(\psi(U^{k+\frac{1}{2}}, U^{k+\frac{1}{2}}) - \psi(u^{k+\frac{1}{2}}, u^{k+\frac{1}{2}}), e^{k+\frac{1}{2}})$$

$$\leqslant \frac{1}{3} \hat{c}_1 \left[h \sum_{i=1}^{m-1} (e_i^{k+\frac{1}{2}})^2 + \frac{1}{2} h \sum_{i=1}^{m-2} |e_{i+1}^{k+\frac{1}{2}} e_i^{k+\frac{1}{2}}| \right]$$

$$\leqslant \frac{1}{2} \hat{c}_1 \| e^{k+\frac{1}{2}} \|^2. \tag{9.19}$$

Substituting (9.18) and (9.19) into (9.17) arrives at

$$\frac{1}{2\tau} (\|e^{k+1}\|^2 - \|e^k\|^2) \leqslant \frac{1}{2} |\gamma| \hat{c}_1 \|e^{k+\frac{1}{2}}\|^2 + \|(R_1)^k\| \cdot \|e^{k+\frac{1}{2}}\|$$

$$\leqslant \frac{1}{2} |\gamma| \hat{c}_1 \left(\frac{\|e^{k+1}\| + \|e^k\|}{2} \right)^2 + \|(R_1)^k\| \cdot \frac{\|e^{k+1}\| + \|e^k\|}{2},$$

$$0 \leqslant k \leqslant n-1.$$

Noticing (9.11), it follows after eliminating $\frac{1}{2}(\|e^{k+1}\| + \|e^k\|)$ on both sides that

$$\frac{1}{\tau} (\|e^{k+1}\| - \|e^k\|) \leqslant \frac{1}{2} |\gamma| \hat{c}_1 \frac{\|e^{k+1}\| + \|e^k\|}{2} + \|(R_1)^k\|$$

$$\leqslant \frac{1}{2} |\gamma| \hat{c}_1 \frac{\|e^{k+1}\| + \|e^k\|}{2} + \sqrt{L} c_1 (\tau^2 + h), \quad 0 \leqslant k \leqslant n-1.$$

$$\tag{9.20}$$

(I) $\gamma = 0$: It follows from (9.20) that

$$\|e^{k+1}\| \leqslant \|e^0\| + (k+1)\sqrt{L}c_1\tau(\tau^2 + h) \leqslant T\sqrt{L}c_1(\tau^2 + h), \quad 0 \leqslant k \leqslant n-1.$$

(II) $\gamma \neq 0$: When $\frac{|\gamma|\hat{c}_1}{4}\tau \leqslant \frac{1}{3}$, it follows from (9.20) that

$$\|e^{k+1}\| \leqslant \left(1 + \frac{3|\gamma|\hat{c}_1}{4}\tau\right)\|e^k\| + \frac{3}{2}\sqrt{L}c_1\tau(\tau^2 + h), \quad 0 \leqslant k \leqslant n-1.$$

Applying Lemma 3.3 (Gronwall inequality) leads to

$$\|e^{k+1}\| \leqslant e^{\frac{3|\gamma|\hat{c}_1}{4}T}\frac{2\sqrt{L}c_1}{|\gamma|\hat{c}_1}(\tau^2 + h) \equiv c_2(\tau^2 + h), \quad 0 \leqslant k \leqslant n-1.$$

$\square$

9.2.5   Numerical Examples

Example 9.1   Use the difference scheme (9.13) to compute the initial-boundary value problem

$$\begin{cases} u_t - 6uu_x + u_{xxx} = 0, & 0 < x < 1, \quad 0 < t \leqslant 1, \\ u(x, 0) = x(x-1)^2(x^3 - 2x^2 + 2), & 0 \leqslant x \leqslant 1, \\ u(0, t) = 0, \quad u(1, t) = 0, \quad u_x(1, t) = 0, & 0 < t \leqslant 1. \end{cases} \tag{9.21}$$

From the proof of Theorem 9.2, the system of nonlinear equations (9.14) in the unknown $\{w_i \mid 1 \leqslant i \leqslant m-1\}$ needs to be solved. To solve this system, we apply the Newton iteration method. Once the value of $\{w_i \mid 1 \leqslant i \leqslant m-1\}$ is obtained, the value of u^{k+1} is computed using the formula

$$u_i^{k+1} = 2w_i - u_i^k, \quad 1 \leqslant i \leqslant m-1.$$

For different values of step sizes, Tables 9.1 and 9.2 list the posterior errors of numerical solutions using the difference scheme (9.13) defined by

$$F(h, \tau) = \max_{0 \leqslant k \leqslant n} \sqrt{h\sum_{i=1}^{m-1}\left(u_i^k(h, \tau) - u_i^{2k}\left(h, \frac{\tau}{2}\right)\right)^2},$$

$$G(h, \tau) = \max_{0 \leqslant k \leqslant n} \sqrt{h\sum_{i=1}^{m-1}\left(u_i^k(h, \tau) - u_{2i}^k\left(\frac{h}{2}, \tau\right)\right)^2}.$$

Table 9.1 (Example 9.1) The posterior errors of numerical solutions with different spatial step sizes ($\tau = 1/2^{10}$)

h	$G(h, \tau)$	$G(2h, \tau)/G(h, \tau)$
1/32	1.275285e$-$03	
1/64	6.466659e$-$04	1.9721
1/128	3.259361e$-$04	1.9840
1/256	1.636640e$-$04	1.9915

Table 9.2 (Example 9.1) The posterior errors of numerical solutions with different temporal step sizes ($h = 1/2^{17}$)

τ	$F(h, \tau)$	$F(h, 2\tau)/F(h, \tau)$
1/32	2.243459e$-$02	
1/64	5.647851e$-$03	3.9722
1/128	1.849730e$-$03	3.0533
1/256	4.039513e$-$04	4.5791

Fig. 9.1 (Example 9.1) The curves of energy E^k for the difference scheme (9.13)

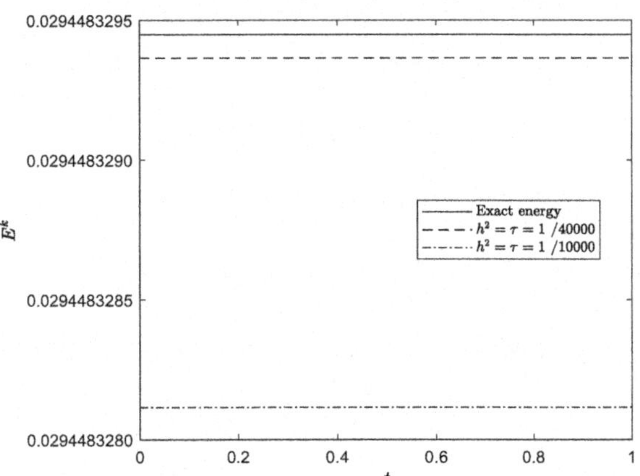

From Table 9.1, it can be concluded that when the step size h is halved, the resulting error is also halved. Similarly, from Table 9.2, it is observed that when the step size τ is halved, the error decreases to a quarter of its original value. Figure 9.1 illustrates the curves of the invariant E^k defined in Theorem 9.3.

9.3 Second-Order Difference Scheme in Space

Let

$$v = u_x,$$

then the problem (9.1) is equivalent to

$$
\begin{cases}
u_t + \gamma u u_x + v_{xx} = 0, & 0 < x < L, \quad 0 < t \leqslant T, & \text{(9.22a)} \\[2mm]
v = u_x, & 0 < x < L, \quad 0 \leqslant t \leqslant T, & \text{(9.22b)} \\[2mm]
u(x,0) = \varphi(x), & 0 < x < L, & \text{(9.22c)} \\[2mm]
u(0,t) = 0, \quad u(L,t) = 0, & v(L,t) = 0, \quad 0 \leqslant t \leqslant T. & \text{(9.22d)}
\end{cases}
$$

9.3.1 Derivation of the Difference Scheme

Define the grid functions U and V as follows:

$$
U_i^k = u(x_i, t_k), \quad V_i^k = v(x_i, t_k), \quad 0 \leqslant i \leqslant m, \quad 0 \leqslant k \leqslant n.
$$

Considering Eq. (9.22a) at the point $(x_i, t_{k+\frac{1}{2}})$ and Eq. (9.22b) at the point $(x_{i+\frac{1}{2}}, t_k)$, we have

$$
u_t(x_i, t_{k+\frac{1}{2}}) + \gamma u(x_i, t_{k+\frac{1}{2}}) u_x(x_i, t_{k+\frac{1}{2}}) + v_{xx}(x_i, t_{k+\frac{1}{2}}) = 0,
$$
$$
1 \leqslant i \leqslant m-1, \quad 0 \leqslant k \leqslant n-1,
$$
$$
v(x_{i+\frac{1}{2}}, t_k) = u_x(x_{i+\frac{1}{2}}, t_k), \quad 0 \leqslant i \leqslant m-1, \quad 0 \leqslant k \leqslant n.
$$

It follows by applying the numerical differential formula (Lemma 1.2) that

$$
\begin{cases}
\delta_t U_i^{k+\frac{1}{2}} + \gamma \psi(U^{k+\frac{1}{2}}, U^{k+\frac{1}{2}})_i + \delta_x^2 V_i^{k+\frac{1}{2}} = P_i^k, \\[2mm]
\qquad\qquad 1 \leqslant i \leqslant m-1, \quad 0 \leqslant k \leqslant n-1, & \text{(9.23)} \\[2mm]
V_{i+\frac{1}{2}}^k = \delta_x U_{i+\frac{1}{2}}^k + Q_{i+\frac{1}{2}}^k, \quad 0 \leqslant i \leqslant m-1, \quad 0 \leqslant k \leqslant n,
\end{cases}
$$

where there is a constant c_3 such that

$$
\begin{cases}
|P_i^k| \leqslant c_3(h^2 + \tau^2), & 1 \leqslant i \leqslant m-1, \quad 0 \leqslant k \leqslant n-1, & \text{(9.24a)} \\[2mm]
|Q_{i+\frac{1}{2}}^k| \leqslant c_3 h^2, & 0 \leqslant i \leqslant m-1, \quad 0 \leqslant k \leqslant n. & \text{(9.24b)}
\end{cases}
$$

In view of

$$
v(x_{i+\frac{1}{2}}, t_k) = \frac{1}{2}\left[v(x_i, t_k) + v(x_{i+1}, t_k) \right]
$$
$$
- \frac{h^2}{8} \int_0^1 \left[v_{xx}\left(x_{i+\frac{1}{2}} + \frac{h}{2}s, t_k\right) + v_{xx}\left(x_{i+\frac{1}{2}} - \frac{h}{2}s, t_k\right) \right](1-s)\,ds,
$$

$$u_x(x_{i+\frac{1}{2}}, t_k) = \frac{u(x_{i+1}, t_k) - u(x_i, t_k)}{h}$$

$$- \frac{h^2}{16} \int_0^1 \left[u_{xxx}\left(x_{i+\frac{1}{2}} + \frac{h}{2}s, t_k\right) + u_{xxx}\left(x_{i+\frac{1}{2}} - \frac{h}{2}s, t_k\right) \right](1-s)^2 ds,$$

we have

$$Q^k_{i+\frac{1}{2}} = \frac{h^2}{16} \int_0^1 (1-s^2)\left[u_{xxx}\left(x_{i+\frac{1}{2}} + \frac{sh}{2}, t_k\right) + u_{xxx}\left(x_{i+\frac{1}{2}} - \frac{sh}{2}, t_k\right) \right] ds.$$

About the estimate of the truncation error $Q^k_{i+\frac{1}{2}}$, there is the following further result.

Lemma 9.2 *For any fixed $t \in [0, T]$, denote $g(x) = u(x, t)$, $x \in [0, L]$ and suppose $g \in C^6([0, L])$. Let*

$$S_i^k = \frac{1}{h}\left(Q^k_{i+\frac{1}{2}} - Q^k_{i-\frac{1}{2}} \right), \quad 1 \leqslant i \leqslant m-1, \quad 0 \leqslant k \leqslant n,$$

$$R^k_{m-1} = 0, \quad R^k_j = \sum_{i=j+1}^{m-1} (-1)^{i-j-1} S_i^k, \quad j = m-2, m-3, \ldots, 0, \quad 0 \leqslant k \leqslant n,$$

then there is a constant c_4 such that

$$\left| S_i^k \right| \leqslant c_4 h^2, \quad 1 \leqslant i \leqslant m-1, \quad 0 \leqslant k \leqslant n,$$

$$\left| R_j^k \right| \leqslant c_4 h^2, \quad 0 \leqslant j \leqslant m-2, \quad 0 \leqslant k \leqslant n,$$

$$\left| \delta_x R^k_{j+\frac{1}{2}} \right| \leqslant c_4 h^2, \quad 0 \leqslant j \leqslant m-2, \quad 0 \leqslant k \leqslant n.$$

The proof for this lemma is given in Sect. 9.3.6.

Omitting the small terms P_i^k and $Q^k_{i+\frac{1}{2}}$ in (9.23) and noticing the initial-boundary value conditions

$$\begin{cases} U_i^0 = \varphi(x_i), & 1 \leqslant i \leqslant m-1, \\ U_0^k = 0, \quad U_m^k = 0, \quad V_m^k = 0, & 0 \leqslant k \leqslant n, \end{cases} \tag{9.25}$$

a difference scheme for solving (9.22) reads

$$
\begin{cases}
\delta_t u_i^{k+\frac{1}{2}} + \gamma \psi(u^{k+\frac{1}{2}}, u^{k+\frac{1}{2}})_i + \delta_x^2 v_i^{k+\frac{1}{2}} = 0, \\
\qquad\qquad 1 \leqslant i \leqslant m-1, \quad 0 \leqslant k \leqslant n-1, & \text{(9.26a)} \\[4pt]
v_{i+\frac{1}{2}}^k = \delta_x u_{i+\frac{1}{2}}^k, \quad 0 \leqslant i \leqslant m-1, \quad 0 \leqslant k \leqslant n, & \text{(9.26b)} \\[4pt]
u_i^0 = \varphi(x_i), \quad 1 \leqslant i \leqslant m-1, & \text{(9.26c)} \\[4pt]
u_0^k = 0, \quad u_m^k = 0, \quad 0 \leqslant k \leqslant n, & \text{(9.26d)} \\[4pt]
v_m^k = 0, \quad 0 \leqslant k \leqslant n. & \text{(9.26e)}
\end{cases}
$$

It is known from (9.24) that the local truncation error of the difference equations (9.26a)–(9.26b) is second-order accurate in both time and space.

Theorem 9.5 *The difference scheme (9.26) is equivalent to*

$$
\begin{cases}
\delta_t u_{i+\frac{1}{2}}^{k+\frac{1}{2}} + \dfrac{\gamma}{2}[\psi(u^{k+\frac{1}{2}}, u^{k+\frac{1}{2}})_i + \psi(u^{k+\frac{1}{2}}, u^{k+\frac{1}{2}})_{i+1}] + \delta_x^2\left(\delta_x u_{i+\frac{1}{2}}^{k+\frac{1}{2}}\right) = 0, \\
\qquad\qquad 1 \leqslant i \leqslant m-2, \quad 0 \leqslant k \leqslant n-1, & \text{(9.27a)} \\[4pt]
\delta_t u_{m-1}^{k+\frac{1}{2}} + \gamma \psi(u^{k+\frac{1}{2}}, u^{k+\frac{1}{2}})_{m-1} + \dfrac{2}{h}\left(-\dfrac{2}{h}\delta_x u_{m-\frac{1}{2}}^{k+\frac{1}{2}} - \delta_x^2 u_{m-1}^{k+\frac{1}{2}}\right) = 0, \\
\qquad\qquad 0 \leqslant k \leqslant n-1, & \text{(9.27b)} \\[4pt]
u_i^0 = \varphi(x_i), \quad 1 \leqslant i \leqslant m-1, & \text{(9.27c)} \\[4pt]
u_0^k = 0, \quad u_m^k = 0, \quad 0 \leqslant k \leqslant n, & \text{(9.27d)} \\[4pt]
v_m^k = 0, \quad 0 \leqslant k \leqslant n, & \text{(9.27e)} \\[4pt]
v_i^k = 2\delta_x u_{i+\frac{1}{2}}^k - v_{i+1}^k, \quad i = m-1, m-2, \ldots, 1, 0, \quad 0 \leqslant k \leqslant n. & \text{(9.27f)}
\end{cases}
$$

Proof Equation (9.26b) is equivalent to

$$
\begin{cases}
v_{i+\frac{1}{2}}^0 = \delta_x u_{i+\frac{1}{2}}^0, \quad 0 \leqslant i \leqslant m-1, & \text{(9.28a)} \\[6pt]
v_{i+\frac{1}{2}}^{k+\frac{1}{2}} = \delta_x u_{i+\frac{1}{2}}^{k+\frac{1}{2}}, \quad 0 \leqslant i \leqslant m-1, \quad 0 \leqslant k \leqslant n-1. & \text{(9.28b)}
\end{cases}
$$

Equation (9.26e) is equivalent to

$$
\begin{cases}
v_m^0 = 0, & \text{(9.29a)} \\[6pt]
v_m^{k+\frac{1}{2}} = 0, \quad 0 \leqslant k \leqslant n-1. & \text{(9.29b)}
\end{cases}
$$

Equation (9.26a) is equivalent to

$$\begin{cases} \delta_t u_{i+\frac{1}{2}}^{k+\frac{1}{2}} + \frac{\gamma}{2}\left[\psi(u^{k+\frac{1}{2}}, u^{k+\frac{1}{2}})_i + \psi(u^{k+\frac{1}{2}}, u^{k+\frac{1}{2}})_{i+1}\right] + \delta_x^2\left(\frac{v_i^{k+\frac{1}{2}} + v_{i+1}^{k+\frac{1}{2}}}{2}\right) = 0, \\[2mm] \qquad\qquad\qquad\qquad 1 \leqslant i \leqslant m-2, \quad 0 \leqslant k \leqslant n-1, \\[2mm] \delta_t u_{m-1}^{k+\frac{1}{2}} + \gamma\psi(u^{k+\frac{1}{2}}, u^{k+\frac{1}{2}})_{m-1} + \delta_x^2 v_{m-1}^{k+\frac{1}{2}} = 0, \quad 0 \leqslant k \leqslant n-1. \end{cases}$$

In view of (9.28b) and (9.29b) and noticing

$$\begin{aligned} \delta_x^2 v_{m-1}^{k+\frac{1}{2}} &= \frac{1}{h^2}\left(v_m^{k+\frac{1}{2}} - 2v_{m-1}^{k+\frac{1}{2}} + v_{m-2}^{k+\frac{1}{2}}\right) = \frac{1}{h^2}\left(-3v_m^{k+\frac{1}{2}} - 3v_{m-1}^{k+\frac{1}{2}} + v_{m-1}^{k+\frac{1}{2}} + v_{m-2}^{k+\frac{1}{2}}\right) \\[2mm] &= \frac{2}{h^2}\left(v_{m-\frac{3}{2}}^{k+\frac{1}{2}} - 3v_{m-\frac{1}{2}}^{k+\frac{1}{2}}\right) = \frac{2}{h}\left(-\frac{2}{h}v_{m-\frac{1}{2}}^{k+\frac{1}{2}} - \frac{v_{m-\frac{1}{2}}^{k+\frac{1}{2}} - v_{m-\frac{3}{2}}^{k+\frac{1}{2}}}{h}\right) \\[2mm] &= \frac{2}{h}\left(-\frac{2}{h}\delta_x u_{m-\frac{1}{2}}^{k+\frac{1}{2}} - \frac{\delta_x u_{m-\frac{1}{2}}^{k+\frac{1}{2}} - \delta_x u_{m-\frac{3}{2}}^{k+\frac{1}{2}}}{h}\right) = \frac{2}{h}\left(-\frac{2}{h}\delta_x u_{m-\frac{1}{2}}^{k+\frac{1}{2}} - \delta_x^2 u_{m-1}^{k+\frac{1}{2}}\right), \end{aligned}$$

it can be seen that (9.26a) is equivalent to

$$\begin{cases} \delta_t u_{i+\frac{1}{2}}^{k+\frac{1}{2}} + \frac{\gamma}{2}\left[\psi(u^{k+\frac{1}{2}}, u^{k+\frac{1}{2}})_i + \psi(u^{k+\frac{1}{2}}, u^{k+\frac{1}{2}})_{i+1}\right] + \delta_x^2\left(\delta_x u_{i+\frac{1}{2}}^{k+\frac{1}{2}}\right) = 0, \\[2mm] \qquad\qquad\qquad\qquad 1 \leqslant i \leqslant m-2, \quad 0 \leqslant k \leqslant n-1, \\[2mm] \delta_t u_{m-1}^{k+\frac{1}{2}} + \gamma\psi(u^{k+\frac{1}{2}}, u^{k+\frac{1}{2}})_{m-1} + \frac{2}{h}\left(-\frac{2}{h}\delta_x u_{m-\frac{1}{2}}^{k+\frac{1}{2}} - \delta_x^2 u_{m-1}^{k+\frac{1}{2}}\right) = 0, \\[2mm] \qquad\qquad\qquad\qquad 0 \leqslant k \leqslant n-1. \end{cases}$$

In addition, (9.27e) is precisely (9.26e); (9.27c) is precisely (9.26c); and (9.26b) can be rewritten as (9.27f) in view of (9.26e). $\qquad\square$

From Theorem 9.5, we establish a difference scheme (9.27a)–(9.27d) for solving the problem (9.1).

It can be seen from the symmetry that the truncation error of the difference equation (9.27a) is $O(\tau^2 + h^2)$.

In view of $u(x_m, t) = 0$, $u_x(x_m, t) = 0$, $u_{xxx}(x_m, t) = 0$, and $u_{xxxx}(x_m, t) = 0$, it follows from Taylor expansion and Lemma 1.2 that

$$u_{xxx}(x_{m-1}, t) = u_{xxx}(x_m, t) - h u_{xxxx}(x_m, t) + O(h^2) = O(h^2)$$

and

$$u_{xxx}(x_{m-1}, t)$$

$$= \frac{1}{h}[u_{xx}(x_m, t) - u_{xx}(x_{m-1}, t)] - \frac{h}{2}u_{xxx}(x_m, t) + O(h^2)$$

$$= \frac{1}{h}\left\{\frac{2}{h}\left[u_x(x_m, t) - \frac{u(x_m, t) - u(x_{m-1}, t)}{h}\right] + \frac{h}{3}u_{xxx}(x_m, t)\right.$$

$$- \frac{h^2}{12}u_{xxxx}(x_m, t) + O(h^3)$$

$$\left. - \left[\frac{u(x_m, t) - 2u(x_{m-1}, t) + u(x_{m-2}, t)}{h^2} - \frac{h^2}{12}u_{xxxx}(x_m, t) + O(h^4)\right]\right\}$$

$$- \frac{h}{2}u_{xxx}(x_m, t) + O(h^2)$$

$$= \frac{1}{h}\left[-\frac{2}{h}\frac{u(x_m, t) - u(x_{m-1}, t)}{h} - \frac{u(x_m, t) - 2u(x_{m-1}, t) + u(x_{m-2}, t)}{h^2}\right]$$

$$+ O(h^2).$$

Therefore,

$$\frac{1}{h}\left[-\frac{2}{h} \cdot \frac{u(x_m, t) - u(x_{m-1}, t)}{h} - \frac{u(x_{m-2}, t) - 2u(x_{m-1}, t) + u(x_m, t)}{h^2}\right] = O(h^2)$$

and for any constant α, we have

$$u_{xxx}(x_{m-1}, t)$$

$$= \alpha \cdot \frac{1}{h}\left[-\frac{2}{h} \cdot \frac{u(x_m, t) - u(x_{m-1}, t)}{h} - \frac{u(x_{m-2}, t) - 2u(x_{m-1}, t) + u(x_m, t)}{h^2}\right]$$

$$+ O(h^2).$$

Then the truncation error of the difference equation (9.27b) is also $O(\tau^2 + h^2)$.

By examining (9.27b) and (9.13b), we observe that the third terms on the left-hand side differ only by a constant. The difference equation (9.13b) aligns with (9.13a) at $i = m - 2$, and the difference equation (9.27b) matches (9.27a) at $i = m - 2$, where "match" refers to ensuring the conservation, boundedness, and convergence of the difference solution.

9.3.2 Existence of the Difference Solution

Theorem 9.6 *The difference scheme (9.26) has a solution.*

Proof From (9.26c) and (9.26d), we have $\{u_i^0 \mid 0 \leqslant i \leqslant m\}$.

Equation (9.26b) can be reformulated as the following equivalent form:

$$
\begin{cases}
v_{i+\frac{1}{2}}^0 = \delta_x u_{i+\frac{1}{2}}^0, & 0 \leqslant i \leqslant m-1, \\
v_{i+\frac{1}{2}}^{k+\frac{1}{2}} = \delta_x u_{i+\frac{1}{2}}^{k+\frac{1}{2}}, & 0 \leqslant i \leqslant m-1, \quad 0 \leqslant k \leqslant n-1.
\end{cases}
$$

Suppose now the value of $\{u^k, v^k\}$ at the k-th time level has been obtained. Denote

$$
w_i = u_i^{k+\frac{1}{2}}, \quad z_i = v_i^{k+\frac{1}{2}}, \quad 0 \leqslant i \leqslant m,
$$

then it is easy to see that

$$
u_i^{k+1} = 2w_i - u_i^k, \quad v_i^{k+1} = 2z_i - v_i^k, \quad 0 \leqslant i \leqslant m.
$$

From the difference equations (9.26a)–(9.26b) and (9.26d)–(9.26e), we can get the system of nonlinear equations in $\{w, z\}$ as follows:

$$
\begin{cases}
\dfrac{2}{\tau}(w_i - u_i^k) + \gamma \psi(w, w)_i + \delta_x^2 z_i = 0, & 1 \leqslant i \leqslant m-1, & (9.30\text{a}) \\
z_{i+\frac{1}{2}} = \delta_x w_{i+\frac{1}{2}}, & 0 \leqslant i \leqslant m-1, & (9.30\text{b}) \\
w_0 = 0, \quad w_m = 0, & & (9.30\text{c}) \\
z_m = 0. & & (9.30\text{d})
\end{cases}
$$

We can rewrite (9.30b) and (9.30d) as follows:

$$
z_m = 0; \quad z_i = 2\delta_x w_{i+\frac{1}{2}} - z_{i+1}, \quad i = m-1, m-2, \ldots, 0. \tag{9.31}
$$

Define an operator $\Pi : \mathring{\mathcal{U}}_h \to \mathring{\mathcal{U}}_h$ by

$$
\Pi(w)_i =
\begin{cases}
\dfrac{2}{\tau}(w_i - u_i^k) + \gamma \psi(w, w)_i + \delta_x^2 z_i, & 1 \leqslant i \leqslant m-1, \\
0, & i = 0, \, m,
\end{cases}
$$

where z_i $(i = m, m-1, \ldots, 0)$ is determined by (9.30b) and (9.30d).

Some calculations produce

$$
(\Pi(w), w) = \frac{2}{\tau}[(w, w) - (u^k, w)] + \gamma(\psi(w, w), w) + h \sum_{i=1}^{m-1} (\delta_x^2 z_i) w_i
$$

$$= \frac{2}{\tau}[(w, w) - (u^k, w)] - h \sum_{i=0}^{m-1} (\delta_x z_{i+\frac{1}{2}})(\delta_x w_{i+\frac{1}{2}})$$

$$= \frac{2}{\tau}[(w, w) - (u^k, w)] - h \sum_{i=0}^{m-1} (\delta_x z_{i+\frac{1}{2}}) z_{i+\frac{1}{2}}$$

$$= \frac{2}{\tau}[(w, w) - (u^k, w)] - \frac{1}{2}(z_m^2 - z_0^2)$$

$$\geqslant \frac{2}{\tau}\left(\|w\|^2 - \|u^k\| \cdot \|w\|\right) + \frac{1}{2} z_0^2$$

$$\geqslant \frac{2}{\tau}(\|w\| - \|u^k\|)\|w\|.$$

When $\|w\| = \|u^k\|$, it follows that $(\Pi(w), w) \geqslant 0$. By Theorem 7.4 (Browder theorem), there is a solution $w \in \overset{\circ}{\mathcal{U}}_h$ of the system (9.30) satisfying $\|w\| \leqslant \|u^k\|$ and $\Pi(w) = 0$. Once w is obtained, then z is determined by (9.31). $\square$

Remark 9.1 Combining Theorem 9.5 with Theorem 9.6, we have indirectly proved that the difference scheme (9.27a)–(9.27d) has a solution, that is, we have proved the existence of the solution to the following nonlinear system:

$$\begin{cases} \frac{2}{\tau}\left(w_{i+\frac{1}{2}} - u^k_{i+\frac{1}{2}}\right) + \frac{\gamma}{2}[\psi(w, w)_i + \psi(w, w)_{i+1}] + \delta_x^2(\delta_x w_{i+\frac{1}{2}}) = 0, \\ \qquad\qquad\qquad\qquad 1 \leqslant i \leqslant m - 2, \\ \frac{2}{\tau}\left(w_{m-1} - u^k_{m-1}\right) + \gamma \psi(w, w)_{m-1} + \frac{2}{h}\left(-\frac{2}{h}\delta_x w_{m-\frac{1}{2}} - \delta_x^2 w_{m-1}\right) = 0, \\ w_0 = 0, \quad w_m = 0. \end{cases}$$

$$(9.32)$$

9.3.3 Conservation and Boundedness of the Difference Solution

Theorem 9.7 *Let $\{u_i^k, v_i^k \mid 0 \leqslant i \leqslant m, 0 \leqslant k \leqslant n\}$ be the solution of the difference scheme (9.26). Denote*

$$E^k = \|u^{k+1}\|^2 + \tau \sum_{l=0}^{k} \left(v_0^{l+\frac{1}{2}}\right)^2,$$

then

$$E^k = \|u^0\|^2, \quad 0 \leqslant k \leqslant n - 1. \tag{9.33}$$

Proof Taking the inner product of (9.26a) with $u^{k+\frac{1}{2}}$ on both sides gives

$$\frac{1}{2\tau}(\|u^{k+1}\|^2 - \|u^k\|^2) + \gamma(\psi(u^{k+\frac{1}{2}}, u^{k+\frac{1}{2}}), u^{k+\frac{1}{2}}) + h\sum_{i=1}^{m-1}(\delta_x^2 v_i^{k+\frac{1}{2}})u_i^{k+\frac{1}{2}} = 0.$$

Noticing

$$(\psi(u^{k+\frac{1}{2}}, u^{k+\frac{1}{2}}), u^{k+\frac{1}{2}}) = 0,$$

$$h\sum_{i=1}^{m-1}\left(\delta_x^2 v_i^{k+\frac{1}{2}}\right)u_i^{k+\frac{1}{2}} = -h\sum_{i=0}^{m-1}\left(\delta_x v_{i+\frac{1}{2}}^{k+\frac{1}{2}}\right)\delta_x u_{i+\frac{1}{2}}^{k+\frac{1}{2}}$$

$$= -h\sum_{i=0}^{m-1}\left(\delta_x v_{i+\frac{1}{2}}^{k+\frac{1}{2}}\right)\left(v_{i+\frac{1}{2}}^{k+\frac{1}{2}}\right) = -\frac{1}{2}\left[(v_m^{k+\frac{1}{2}})^2 - (v_0^{k+\frac{1}{2}})^2\right],$$

we have

$$\frac{1}{2\tau}\left(\|u^{k+1}\|^2 - \|u^k\|^2\right) + \frac{1}{2}\left(v_0^{k+\frac{1}{2}}\right)^2 = 0, \quad 0 \leqslant k \leqslant n-1.$$

Hence,

$$\|u^{k+1}\|^2 + \tau\sum_{l=0}^{k}\left(v_0^{l+\frac{1}{2}}\right)^2 = \|u^0\|^2, \quad 0 \leqslant k \leqslant n-1.$$

$\square$

It easily follows from Theorem 9.7 that

$$\|u^k\| \leqslant \|u^0\|, \quad 0 \leqslant k \leqslant n.$$

9.3.4   Convergence of the Difference Solution

Now we focus on the convergence of the difference scheme.

Theorem 9.8 *Suppose* $\{u(x,t), v(x,t)\}$ *is the solution of the problem (9.22) and* $\{u_i^k, v_i^k \mid 0 \leqslant i \leqslant m, 0 \leqslant k \leqslant n\}$ *is the solution of the difference scheme (9.26). Denote*

$$e_i^k = u(x_i, t_k) - u_i^k, \quad f_i^k = v(x_i, t_k) - v_i^k, \quad 0 \leqslant i \leqslant m, \quad 0 \leqslant k \leqslant n.$$

Then there is a constant c_5 such that

$$\|e^k\| \leqslant c_5(h^2 + \tau^2), \quad 0 \leqslant k \leqslant n.$$

Proof Subtracting (9.26) from (9.23) and (9.25), the system of error equations is produced as

$$
\begin{cases}
\delta_t e_i^{k+\frac{1}{2}} + \gamma\left[\psi(U^{k+\frac{1}{2}}, U^{k+\frac{1}{2}})_i - \psi(u^{k+\frac{1}{2}}, u^{k+\frac{1}{2}})_i\right] + \delta_x^2 f_i^{k+\frac{1}{2}} = P_i^k, \\
\qquad\qquad\qquad\qquad 1 \leqslant i \leqslant m-1, \quad 0 \leqslant k \leqslant n-1, \quad (9.34a) \\
f_{i+\frac{1}{2}}^k = \delta_x e_{i+\frac{1}{2}}^k + Q_{i+\frac{1}{2}}^k, \quad 0 \leqslant i \leqslant m-1, \quad 0 \leqslant k \leqslant n, \qquad (9.34b) \\
e_i^0 = 0, \quad 1 \leqslant i \leqslant m-1, \qquad\qquad\qquad\qquad\qquad (9.34c) \\
e_0^k = 0, \quad e_m^k = 0, \quad f_m^k = 0, \quad 0 \leqslant k \leqslant n. \qquad\qquad (9.34d)
\end{cases}
$$

Taking the average of (9.34b) and (9.34d), respectively, with superscripts k and $k+1$, we obtain

$$f_{i+\frac{1}{2}}^{k+\frac{1}{2}} = \delta_x e_{i+\frac{1}{2}}^{k+\frac{1}{2}} + Q_{i+\frac{1}{2}}^{k+\frac{1}{2}}, \quad 0 \leqslant i \leqslant m-1, \quad 0 \leqslant k \leqslant n-1, \qquad (9.35)$$

$$e_0^{k+\frac{1}{2}} = 0, \quad e_m^{k+\frac{1}{2}} = 0, \quad f_m^{k+\frac{1}{2}} = 0, \quad 0 \leqslant k \leqslant n-1. \qquad (9.36)$$

Taking the inner product of (9.34a) with $e^{k+\frac{1}{2}}$ on both sides gives

$$\left(\delta_t e^{k+\frac{1}{2}}, e^{k+\frac{1}{2}}\right) + \gamma\left(\psi(U^{k+\frac{1}{2}}, U^{k+\frac{1}{2}}) - \psi(u^{k+\frac{1}{2}}, u^{k+\frac{1}{2}}), e^{k+\frac{1}{2}}\right)$$

$$+\left(\delta_x^2 f^{k+\frac{1}{2}}, e^{k+\frac{1}{2}}\right) = \left(P^k, e^{k+\frac{1}{2}}\right), \quad 0 \leqslant k \leqslant n-1. \qquad (9.37)$$

For the first term on the left-hand side of (9.37), we have

$$\left(\delta_t e^{k+\frac{1}{2}}, e^{k+\frac{1}{2}}\right) = \frac{1}{2\tau}\left(\|e^{k+1}\|^2 - \|e^k\|^2\right).$$

In view of Lemma 8.1, some calculations yield

$$\left(\psi(U^{k+\frac{1}{2}}, U^{k+\frac{1}{2}}) - \psi(u^{k+\frac{1}{2}}, u^{k+\frac{1}{2}}), e^{k+\frac{1}{2}}\right)$$

$$=\left(\psi(U^{k+\frac{1}{2}}, U^{k+\frac{1}{2}}) - \psi(U^{k+\frac{1}{2}} - e^{k+\frac{1}{2}}, U^{k+\frac{1}{2}} - e^{k+\frac{1}{2}}), e^{k+\frac{1}{2}}\right)$$

$$=\left(\psi(e^{k+\frac{1}{2}}, U^{k+\frac{1}{2}}), e^{k+\frac{1}{2}}\right).$$

Thus,

$$\left| \left(\psi(U^{k+\frac{1}{2}}, U^{k+\frac{1}{2}}) - \psi(u^{k+\frac{1}{2}}, u^{k+\frac{1}{2}}), e^{k+\frac{1}{2}} \right) \right|$$

$$= \left| \frac{h}{3} \sum_{i=1}^{m-1} \left[e_i^{k+\frac{1}{2}} \Delta_x U_i^{k+\frac{1}{2}} + \Delta_x (e^{k+\frac{1}{2}} U^{k+\frac{1}{2}})_i \right] \cdot e_i^{k+\frac{1}{2}} \right|$$

$$= \left| \frac{h}{3} \sum_{i=1}^{m-1} (e_i^{k+\frac{1}{2}})^2 \Delta_x U_i^{k+\frac{1}{2}} + \frac{h}{3} \sum_{i=1}^{m-1} e_i^{k+\frac{1}{2}} \frac{U_{i+1}^{k+\frac{1}{2}} e_{i+1}^{k+\frac{1}{2}} - U_{i-1}^{k+\frac{1}{2}} e_{i-1}^{k+\frac{1}{2}}}{2h} \right|$$

$$= \left| \frac{h}{3} \sum_{i=1}^{m-1} (e_i^{k+\frac{1}{2}})^2 \Delta_x U_i^{k+\frac{1}{2}} + \frac{h}{6} \sum_{i=0}^{m-1} e_i^{k+\frac{1}{2}} e_{i+1}^{k+\frac{1}{2}} (\delta_x U_{i+\frac{1}{2}}^{k+\frac{1}{2}}) \right|$$

$$\leq \frac{\hat{c}_1}{2} \| e^{k+\frac{1}{2}} \|^2, \tag{9.38}$$

where

$$\hat{c}_1 = \max_{0 \leq x \leq L, 0 \leq t \leq T} |u_x(x, t)|.$$

The following borrows notations S_i^k and R_j^k from Lemma 9.2. According to (9.35) and (9.36), we get

$$-\left(\delta_x^2 f^{k+\frac{1}{2}}, e^{k+\frac{1}{2}} \right)$$

$$= \left(\delta_x f^{k+\frac{1}{2}}, \delta_x e^{k+\frac{1}{2}} \right)$$

$$= h \sum_{i=0}^{m-1} \left(\delta_x f_{i+\frac{1}{2}}^{k+\frac{1}{2}} \right) \cdot \left(f_{i+\frac{1}{2}}^{k+\frac{1}{2}} - Q_{i+\frac{1}{2}}^{k+\frac{1}{2}} \right)$$

$$= h \sum_{i=0}^{m-1} \left(\delta_x f_{i+\frac{1}{2}}^{k+\frac{1}{2}} \right) \cdot f_{i+\frac{1}{2}}^{k+\frac{1}{2}} - h \sum_{i=0}^{m-1} \left(\delta_x f_{i+\frac{1}{2}}^{k+\frac{1}{2}} \right) \cdot Q_{i+\frac{1}{2}}^{k+\frac{1}{2}}$$

$$= \frac{1}{2} \sum_{i=0}^{m-1} \left[(f_{i+1}^{k+\frac{1}{2}})^2 - (f_i^{k+\frac{1}{2}})^2 \right] - \sum_{i=0}^{m-1} \left(f_{i+1}^{k+\frac{1}{2}} - f_i^{k+\frac{1}{2}} \right) \cdot Q_{i+\frac{1}{2}}^{k+\frac{1}{2}}$$

$$= -\frac{1}{2} (f_0^{k+\frac{1}{2}})^2 + h \sum_{i=1}^{m-1} f_i^{k+\frac{1}{2}} S_i^{k+\frac{1}{2}} + f_0^{k+\frac{1}{2}} Q_{\frac{1}{2}}^{k+\frac{1}{2}}. \tag{9.39}$$

Rewrite f_i^k as

$$f_i^k = \left(f_i^k + f_{i-1}^k\right) - \left(f_{i-1}^k + f_{i-2}^k\right) + \cdots + (-1)^{i-1}\left(f_1^k + f_0^k\right) + (-1)^i f_0^k$$

$$= 2\sum_{j=0}^{i-1}(-1)^{i-j-1} f_{j+\frac{1}{2}}^k + (-1)^i f_0^k.$$

It follows from the definition of R_j^k and $\delta_x R_{j+\frac{1}{2}}^k$ that

$$h\sum_{i=1}^{m-1} f_i^{k+\frac{1}{2}} S_i^{k+\frac{1}{2}}$$

$$= h\sum_{i=1}^{m-1}\left[2\sum_{j=0}^{i-1}(-1)^{i-j-1} f_{j+\frac{1}{2}}^{k+\frac{1}{2}} + (-1)^i f_0^{k+\frac{1}{2}}\right] S_i^{k+\frac{1}{2}}$$

$$= 2h\sum_{j=0}^{m-2} f_{j+\frac{1}{2}}^{k+\frac{1}{2}}\sum_{i=j+1}^{m-1}(-1)^{i-j-1} S_i^{k+\frac{1}{2}} + h\sum_{i=1}^{m-1}(-1)^i f_0^{k+\frac{1}{2}} S_i^{k+\frac{1}{2}}$$

$$= 2h\sum_{j=0}^{m-2} f_{j+\frac{1}{2}}^{k+\frac{1}{2}} R_j^{k+\frac{1}{2}} + f_0^{k+\frac{1}{2}}\left[h\sum_{i=1}^{m-1}(-1)^i S_i^{k+\frac{1}{2}}\right]$$

$$= 2h\sum_{j=0}^{m-2}\left(\delta_x e_{j+\frac{1}{2}}^{k+\frac{1}{2}} + Q_{j+\frac{1}{2}}^{k+\frac{1}{2}}\right) R_j^{k+\frac{1}{2}} + f_0^{k+\frac{1}{2}}\left(-hR_0^{k+\frac{1}{2}}\right)$$

$$= 2h\sum_{j=0}^{m-2} Q_{j+\frac{1}{2}}^{k+\frac{1}{2}} R_j^{k+\frac{1}{2}} - 2h\sum_{j=1}^{m-1} e_j^{k+\frac{1}{2}}\left(\delta_x R_{j-\frac{1}{2}}^{k+\frac{1}{2}}\right) - hf_0^{k+\frac{1}{2}} R_0^{k+\frac{1}{2}}.$$

Inserting the above result into (9.39) and combining (9.24b) with Lemma 9.2, we arrive at

$$-\left(\delta_x^2 f^{k+\frac{1}{2}}, e^{k+\frac{1}{2}}\right)$$

$$= -\frac{1}{2}\left(f_0^{k+\frac{1}{2}}\right)^2 + f_0^{k+\frac{1}{2}} Q_{\frac{1}{2}}^{k+\frac{1}{2}} - hf_0^{k+\frac{1}{2}} R_0^{k+\frac{1}{2}}$$

$$+ 2h\sum_{j=0}^{m-2} Q_{j+\frac{1}{2}}^{k+\frac{1}{2}} R_j^{k+\frac{1}{2}} - 2h\sum_{j=1}^{m-1} e_j^{k+\frac{1}{2}}\left(\delta_x R_{j-\frac{1}{2}}^{k+\frac{1}{2}}\right)$$

$$\leqslant -\frac{1}{2}\left(f_0^{k+\frac{1}{2}}\right)^2 + \left[\frac{1}{4}\left(f_0^{k+\frac{1}{2}}\right)^2 + \left(Q_{\frac{1}{2}}^{k+\frac{1}{2}}\right)^2\right] + \left[\frac{1}{4}\left(f_0^{k+\frac{1}{2}}\right)^2 + h^2\left(R_0^{k+\frac{1}{2}}\right)^2\right]$$

$$+2h \sum_{j=0}^{m-2} \left| Q_{j+\frac{1}{2}}^{k+\frac{1}{2}} \right| \cdot \left| R_j^{k+\frac{1}{2}} \right| + \left\| e^{k+\frac{1}{2}} \right\|^2 + h \sum_{j=1}^{m-1} \left(\delta_x R_{j-\frac{1}{2}}^{k+\frac{1}{2}} \right)^2$$

$$\leqslant \left\| e^{k+\frac{1}{2}} \right\|^2 + (c_3^2 + c_4^2 + 2Lc_3c_4 + Lc_4^2)h^4, \quad 0 \leqslant k \leqslant n-1. \tag{9.40}$$

Substituting (9.38) and (9.40) into (9.37), it follows by using (9.24a) that

$$\frac{1}{2\tau} \left(\left\| e^{k+1} \right\|^2 - \left\| e^k \right\|^2 \right)$$

$$\leqslant \frac{\hat{c}_1|\gamma|}{2} \left\| e^{k+\frac{1}{2}} \right\|^2 + \left\| e^{k+\frac{1}{2}} \right\|^2 + (c_3^2 + c_4^2 + 2Lc_3c_4 + Lc_4^2)h^4 + \left(P^k, e^{k+\frac{1}{2}} \right)$$

$$\leqslant \frac{\hat{c}_1|\gamma|}{2} \left\| e^{k+\frac{1}{2}} \right\|^2 + \left\| e^{k+\frac{1}{2}} \right\|^2 + (c_3^2 + c_4^2 + 2Lc_3c_4 + Lc_4^2)h^4 + \left\| e^{k+\frac{1}{2}} \right\|^2 + \frac{1}{4} \left\| P^k \right\|^2$$

$$\leqslant \left(2 + \frac{\hat{c}_1|\gamma|}{2} \right) \left\| e^{k+\frac{1}{2}} \right\|^2 + \left(c_3^2 + c_4^2 + 2Lc_3c_4 + Lc_4^2 + \frac{1}{4}Lc_3^2 \right)(\tau^2 + h^2)^2$$

$$\leqslant \left(1 + \frac{\hat{c}_1|\gamma|}{4} \right) \left(\left\| e^k \right\|^2 + \left\| e^{k+1} \right\|^2 \right) + \left(c_3^2 + c_4^2 + 2Lc_3c_4 + Lc_4^2 + \frac{1}{4}Lc_3^2 \right)(\tau^2 + h^2)^2,$$

$$0 \leqslant k \leqslant n-1,$$

that is,

$$\left[1 - 2\left(1 + \frac{\hat{c}_1|\gamma|}{4} \right)\tau \right] \left\| e^{k+1} \right\|^2$$

$$\leqslant \left[1 + 2\left(1 + \frac{\hat{c}_1|\gamma|}{4} \right)\tau \right] \left\| e^k \right\|^2 + 2\left(c_3^2 + c_4^2 + 2Lc_3c_4 + Lc_4^2 + \frac{1}{4}Lc_3^2 \right)\tau(\tau^2 + h^2)^2,$$

$$0 \leqslant k \leqslant n-1.$$

When $(2 + \hat{c}_1|\gamma|/2)\tau \leqslant 1/3$, we have

$$\left\| e^{k+1} \right\|^2 \leqslant \left[1 + 6\left(1 + \frac{\hat{c}_1|\gamma|}{4} \right)\tau \right] \left\| e^k \right\|^2$$

$$+ 3\left(c_3^2 + c_4^2 + 2Lc_3c_4 + Lc_4^2 + \frac{1}{4}Lc_3^2 \right)\tau(\tau^2 + h^2)^2,$$

$$0 \leqslant k \leqslant n-1.$$

With the help of Lemma 3.3 (Gronwall inequality), it is easy to get

$$\|e^k\|^2 \leqslant \exp\left\{6\left(1 + \frac{\hat{c}_1|\gamma|}{4}\right)T\right\} \cdot \frac{c_3^2 + c_4^2 + 2Lc_3c_4 + Lc_4^2 + \frac{1}{4}Lc_3^2}{2\left(1 + \frac{\hat{c}_1|\gamma|}{4}\right)}\left(\tau^2 + h^2\right)^2$$

$$\equiv c_5^2\left(\tau^2 + h^2\right)^2, \quad 1 \leqslant k \leqslant n.$$

$\square$

Remark 9.2 The key step in the above convergence proof is the analysis of $(\delta_x^2 f^{k+\frac{1}{2}}, e^{k+\frac{1}{2}})$. By performing summation by parts several times and using (9.35), the information of $f^{k+\frac{1}{2}}$ is transferred to that of $e^{k+\frac{1}{2}}$.

9.3.5 Numerical Examples

Example 9.2 Apply the difference scheme (9.27a)–(9.27d) to compute the initial-boundary value problem (9.21).

From the proof process of Theorem 9.6 and Remark 9.1, we know that the system of nonlinear equations (9.32) needs to be solved. The Newton iteration method is employed to solve it. Once $\{w_i \mid 1 \leqslant i \leqslant m - 1\}$ is obtained, the value of u_i^{k+1} is determined by

$$u_i^{k+1} = 2w_i - u_i^k, \quad 1 \leqslant i \leqslant m - 1.$$

For different values of step sizes, Tables 9.3 and 9.4 record the posterior errors

$$F(h, \tau) = \max_{0 \leqslant k \leqslant n} \sqrt{h \sum_{i=1}^{m-1} \left(u_i^k(h, \tau) - u_i^{2k}\left(h, \frac{\tau}{2}\right)\right)^2},$$

$$G(h, \tau) = \max_{0 \leqslant k \leqslant n} \sqrt{h \sum_{i=1}^{m-1} \left(u_i^k(h, \tau) - u_{2i}^k\left(\frac{h}{2}, \tau\right)\right)^2}$$

of the numerical solutions obtained using the difference scheme (9.27a)–(9.27d).

From Table 9.3, it can be observed that when the step size h is halved, the posterior error is reduced to a quarter of its original value. Similarly, from Table 9.4, it is evident that when the step size τ is halved, the posterior error is also reduced to a quarter of the original. Figure 9.2 presents the curves of the invariant E^k defined in Theorem 9.7.

Table 9.3 (Example 9.2) The posterior errors of numerical solutions with different spatial step sizes ($\tau = 1/5120$)

h	$G(h, \tau)$	$G(2h, \tau)/G(h, \tau)$
1/80	8.250003e−07	
1/160	2.062497e−07	4.0000
1/320	5.156299e−08	4.0000
1/640	1.289139e−08	3.9998
1/1280	3.234988e−09	3.9850

Table 9.4 (Example 9.2) The posterior errors of numerical solutions with different temporal step sizes ($h = 1/5120$)

τ	$F(h, \tau)$	$F(h, 2\tau)/F(h, \tau)$
1/80	4.445484e−03	
1/160	1.030212e−03	4.3151
1/320	2.507906e−04	4.1079
1/640	7.133053e−05	3.5159
1/1280	1.842227e−05	3.8720

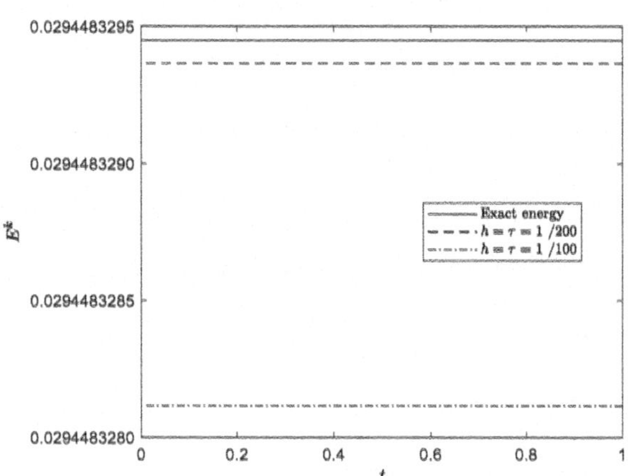

Fig. 9.2 (Example 9.2) The curves of energy E^k for the difference scheme (9.27)

9.3.6 Proof for Lemma 9.2

(I) By the differential mean value theorem, there exists an $\xi_i \in (0, 1)$ such that

$$
\begin{aligned}
S_i^k =& \frac{1}{h}\left(Q_{i+\frac{1}{2}}^k - Q_{i-\frac{1}{2}}^k\right) \\
=& \frac{h}{16}\int_0^1 (1 - s^2)\left\{\left[u_{xxx}\left(x_{i+\frac{1}{2}} + \frac{sh}{2}, t_k\right) + u_{xxx}\left(x_{i+\frac{1}{2}} - \frac{sh}{2}, t_k\right)\right]\right. \\
& \left. - \left[u_{xxx}\left(x_{i-\frac{1}{2}} + \frac{sh}{2}, t_k\right) + u_{xxx}\left(x_{i-\frac{1}{2}} - \frac{sh}{2}, t_k\right)\right]\right\}\mathrm{d}s
\end{aligned}
$$

$$= \frac{h^2}{16} \int_0^1 (1-s^2) \left[u_{xxxx}\left(x_{i-\frac{1}{2}} + \xi_i h + \frac{sh}{2}, t_k\right) \right.$$
$$\left. + u_{xxxx}\left(x_{i-\frac{1}{2}} + \xi_i h - \frac{sh}{2}, t_k\right) \right] ds.$$

Thus, we have

$$|S_i^k| \leqslant \frac{h^2}{16} \cdot 2\hat{c}_4 \int_0^1 (1-s^2) ds = \frac{\hat{c}_4}{12} h^2, \quad 1 \leqslant i \leqslant m-1, \quad 0 \leqslant k \leqslant n,$$

$$(9.41)$$

where

$$\hat{c}_4 = \max_{0 \leqslant x \leqslant L, 0 \leqslant t \leqslant T} |u_{xxxx}(x, t)|.$$

(II) By the differential mean value theorem, there exists an $\eta_i \in (-1, 1)$ such that

$$\delta_x S_{i+\frac{1}{2}}^k = \frac{1}{h^2}\left(Q_{i+\frac{3}{2}}^k - 2Q_{i+\frac{1}{2}}^k + Q_{i-\frac{1}{2}}^k\right)$$
$$= \frac{1}{16} \int_0^1 (1-s^2) \left\{ \left[u_{xxx}\left(x_{i+\frac{3}{2}} + \frac{sh}{2}, t_k\right) + u_{xxx}\left(x_{i+\frac{3}{2}} - \frac{sh}{2}, t_k\right) \right] \right.$$
$$- 2\left[u_{xxx}\left(x_{i+\frac{1}{2}} + \frac{sh}{2}, t_k\right) + u_{xxx}\left(x_{i+\frac{1}{2}} - \frac{sh}{2}, t_k\right) \right]$$
$$\left. + \left[u_{xxx}\left(x_{i-\frac{1}{2}} + \frac{sh}{2}, t_k\right) + u_{xxx}\left(x_{i-\frac{1}{2}} - \frac{sh}{2}, t_k\right) \right] \right\} ds$$
$$= \frac{h^2}{16} \int_0^1 (1-s^2) \left[u_{xxxxx}\left(x_{i+\frac{1}{2}} + \eta_i h + \frac{sh}{2}, t_k\right) \right.$$
$$\left. + u_{xxxxx}\left(x_{i+\frac{1}{2}} + \eta_i h - \frac{sh}{2}, t_k\right) \right] ds.$$

Thus, we have

$$|\delta_x S_{i+\frac{1}{2}}^k| \leqslant \frac{h^2}{16} \cdot 2\hat{c}_5 \int_0^1 (1-s^2) ds = \frac{\hat{c}_5}{12} h^2, \quad 1 \leqslant i \leqslant m-2, \quad 0 \leqslant k \leqslant n,$$

$$(9.42)$$

where

$$\hat{c}_5 = \max_{0 \leqslant x \leqslant L, 0 \leqslant t \leqslant T} |u_{xxxxx}(x, t)|.$$

(III) By the differential mean value theorem, there exists a $\theta_i \in (-1, 2)$ such that

$$
\begin{aligned}
\delta_x^2 S_i^k &= \frac{1}{h^3}\left(Q_{i+\frac{3}{2}}^k - 3Q_{i+\frac{1}{2}}^k + 3Q_{i-\frac{1}{2}}^k - Q_{i-\frac{3}{2}}^k\right) \\
&= \frac{1}{16h}\int_0^1 (1 - s^2)\left\{\left[u_{xxx}\left(x_{i+\frac{3}{2}} + \frac{sh}{2}, t_k\right) + u_{xxx}\left(x_{i+\frac{3}{2}} - \frac{sh}{2}, t_k\right)\right]\right. \\
&\quad - 3\left[u_{xxx}\left(x_{i+\frac{1}{2}} + \frac{sh}{2}, t_k\right) + u_{xxx}\left(x_{i+\frac{1}{2}} - \frac{sh}{2}, t_k\right)\right] \\
&\quad + 3\left[u_{xxx}\left(x_{i-\frac{1}{2}} + \frac{sh}{2}, t_k\right) + u_{xxx}\left(x_{i-\frac{1}{2}} - \frac{sh}{2}, t_k\right)\right] \\
&\quad \left. - \left[u_{xxx}\left(x_{i-\frac{3}{2}} + \frac{sh}{2}, t_k\right) + u_{xxx}\left(x_{i-\frac{3}{2}} - \frac{sh}{2}, t_k\right)\right]\right\} ds \\
&= \frac{h^2}{16}\int_0^1 (1 - s^2)\left[u_{xxxxxx}\left(x_{i-\frac{1}{2}} + \theta_i h + \frac{sh}{2}, t_k\right)\right. \\
&\quad \left. + u_{xxxxxx}\left(x_{i-\frac{1}{2}} + \theta_i h - \frac{sh}{2}, t_k\right)\right] ds.
\end{aligned}
$$

Hence, we have

$$
|\delta_x^2 S_i^k| \leqslant \frac{h^2}{16} \cdot 2\hat{c}_6 \int_0^1 (1 - s^2)ds = \frac{\hat{c}_6}{12}h^2, \quad 2 \leqslant i \leqslant m - 2, \quad 0 \leqslant k \leqslant n,
$$

$$
\tag{9.43}
$$

where

$$
\hat{c}_6 = \max_{0 \leqslant x \leqslant L, 0 \leqslant t \leqslant T} |u_{xxxxxx}(x, t)|.
$$

(IV) It follows from $R_{m-2}^k = S_{m-1}^k$ and (9.41) that

$$
|R_{m-2}^k| = |S_{m-1}^k| \leqslant \frac{\hat{c}_4}{12}h^2, \quad 0 \leqslant k \leqslant n. \tag{9.44}
$$

When $0 \leqslant j \leqslant m - 3$, $0 \leqslant k \leqslant n$, we have

$$
\begin{aligned}
R_j^k &= \sum_{i=j+1}^{m-1} (-1)^{i-j-1} S_i^k = \sum_{i=1}^{m-j-1} (-1)^{i-1} S_{j+i}^k \\
&= \sum_{l=1}^{\lfloor \frac{m-j-1}{2} \rfloor} \left[(-1)^{2l-1-1} S_{j+2l-1}^k + (-1)^{2l-1} S_{j+2l}^k\right]
\end{aligned}
$$

$$+ \frac{1-(-1)^{m-j-1}}{2} \cdot (-1)^{m-j-2} S_{m-1}^k$$

$$= h \sum_{l=1}^{\lfloor \frac{m-j-1}{2} \rfloor} (-1)^{2l-1} \delta_x S_{j+2l-\frac{1}{2}}^k + \frac{1-(-1)^{m-j-1}}{2} \cdot (-1)^{m-j-2} R_{m-2}^k.$$

Combining (9.42) with (9.44) leads to

$$|R_j^k| \leqslant h \sum_{l=1}^{\lfloor \frac{m-j-1}{2} \rfloor} |\delta_x S_{j+2l-\frac{1}{2}}^k| + \frac{1-(-1)^{m-j-1}}{2} \cdot |R_{m-2}^k|$$

$$\leqslant \frac{L}{2} \cdot \frac{\hat{c}_5}{12} h^2 + \frac{\hat{c}_4}{12} h^2 = \frac{1}{12}\left(\hat{c}_4 + \frac{L}{2}\hat{c}_5\right) h^2, \quad 0 \leqslant j \leqslant m-3, \ 0 \leqslant k \leqslant n.$$

(V) Differentiating Eq. (9.1a) once with respect to x yields

$$u_{xt} + \gamma u u_{xx} + \gamma (u_x)^2 + u_{xxxx} = 0.$$

Noticing (9.1c), we have

$$u_{xxxx}(L,t) = 0.$$

From (I), there is a $\xi_{m-1} \in (0,1)$ such that

$$S_{m-1}^k = \frac{h^2}{16} \int_0^1 (1-s^2)\left[u_{xxxx}\left(x_{m-\frac{3}{2}} + \xi_{m-1}h + \frac{sh}{2}, t_k\right) \right.$$

$$\left. + u_{xxxx}\left(x_{m-\frac{3}{2}} + \xi_{m-1}h - \frac{sh}{2}, t_k\right) \right] ds$$

$$= \frac{h^2}{16} \int_0^1 (1-s^2)\left\{ \left[u_{xxxx}\left(x_{m-\frac{3}{2}} + \xi_{m-1}h + \frac{sh}{2}, t_k\right) - u_{xxxx}(x_m, t_k) \right] \right.$$

$$\left. + \left[u_{xxxx}\left(x_{m-\frac{3}{2}} + \xi_{m-1}h - \frac{sh}{2}, t_k\right) - u_{xxxx}(x_m, t_k) \right] \right\} ds.$$

It follows from the differential mean value theorem that

$$|S_{m-1}^k| \leqslant \frac{h^2}{16}\hat{c}_5 \int_0^1 (1-s^2)\left\{ \left[x_m - \left(x_{m-\frac{3}{2}} + \xi_{m-1}h + \frac{sh}{2}\right) \right] \right.$$

$$\left. + \left[x_m - \left(x_{m-\frac{3}{2}} + \xi_{m-1}h - \frac{sh}{2}\right) \right] \right\} ds$$

$$\leqslant \frac{\hat{c}_5}{8} h^3, \quad 0 \leqslant k \leqslant n.$$

In view of $R_{m-2}^k = S_{m-1}^k$, we have

$$|R_{m-2}^k| = |S_{m-1}^k| \leqslant \frac{\hat{c}_5}{8}h^3, \quad 0 \leqslant k \leqslant n. \tag{9.45}$$

(VI) When $1 \leqslant j \leqslant m - 2$, $0 \leqslant k \leqslant n$, we have

$$
\begin{aligned}
\delta_x R_{j+\frac{1}{2}}^k &= \frac{1}{h}\left[\sum_{i=1}^{m-(j+1)-1} (-1)^{i-1} S_{j+1+i}^k - \sum_{i=1}^{m-j-1} (-1)^{i-1} S_{j+i}^k \right] \\
&= \frac{1}{h}\left[\sum_{i=1}^{m-j-2} (-1)^{i-1}\left(S_{j+1+i}^k - S_{j+i}^k\right) - (-1)^{m-j-2} S_{m-1}^k \right] \\
&= \left[h \sum_{l=1}^{\lfloor \frac{m-j-2}{2} \rfloor} (-1)^{2l-1} \delta_x^2 S_{j+2l}^k + \frac{1-(-1)^{m-j-2}}{2}\cdot(-1)^{m-j-3}\delta_x S_{m-\frac{3}{2}}^k \right. \\
&\quad \left. + \frac{1}{h}(-1)^{m-j-2} R_{m-2}^k \right].
\end{aligned}
$$

Combining (9.42) and (9.43) with (9.45), we arrive at

$$
\begin{aligned}
|\delta_x R_{j+\frac{1}{2}}^k| &\leqslant h \sum_{l=1}^{\lfloor \frac{m-j-2}{2} \rfloor} |\delta_x^2 S_{j+2l}^k| + |\delta_x S_{m-\frac{3}{2}}^k| + \frac{1}{h}|R_{m-2}^k| \\
&\leqslant \frac{L}{2}\cdot\frac{\hat{c}_6}{12}h^2 + \frac{\hat{c}_5}{12}h^2 + \frac{1}{h}\cdot\frac{\hat{c}_5}{8}h^3 \\
&= \frac{1}{24}(5\hat{c}_5 + L\hat{c}_6)h^2, \quad 1 \leqslant j \leqslant m - 2, \quad 0 \leqslant k \leqslant n.
\end{aligned}
$$

9.4   Summary and Extension

The KdV equation is a third-order equation in space. The solution to the initial-boundary value problem of the KdV equation (9.1) satisfies the conservation law (9.2). In this chapter, two difference schemes are derived. The first scheme, originating from [1], is first-order accurate in space, while the second scheme, originating from [3], is second-order accurate in space. In [2], a second-order convergent and linearized difference scheme is derived and analyzed. For the periodic initial-boundary value problem of the KdV equation, which possesses infinite conservation laws, the derivation of the first four conservation laws based on the energy method is referred to [4].

By introducing a new variable $v = u_x,$, the original problem (9.1) can be rewritten as an equivalent problem (9.22), where (9.22a) is a second-order equation in space, and (9.22b) is a first-order equation in space. We then derive a difference scheme (9.26) for solving (9.22), which is a natural approach for deriving such a scheme. The introduced intermediate variable $\{v_i^k\}$ does not need to actively participate in the calculation. By separating the variables for the difference scheme (9.26), we obtain a difference scheme (9.27a)–(9.27d) that only involves the variable $\{u_i^k\}$. We have indirectly proved the existence, boundedness, and convergence of the solution to the difference scheme (9.27a)–(9.27d).

9.5 Exercise

9.1 Let $\{u(x,t) \mid 0 \leqslant x \leqslant L, 0 \leqslant t \leqslant T\}$ be the solution of the initial-boundary value problem of the Rosenau-KdV equation

$$
\begin{cases}
u_t + u_x + u_{xxx} + u_{xxxxt} + uu_x = 0, & 0 < x < L, \quad 0 < t \leqslant T, \\
u(x,0) = \varphi(x), & 0 < x < L, \\
u(0,t) = 0, \quad u(L,t) = 0, \quad u_x(0,t) = 0, \quad u_x(L,t) = 0, & 0 < t \leqslant T,
\end{cases}
\tag{9.46}
$$

where $\varphi(0) = \varphi(L) = \varphi'(0) = \varphi'(L) = 0$. Denote

$$
E(t) = \int_0^L \left[u^2(x,t) + u_{xx}^2(x,t) \right] dx.
$$

Try to prove

$$
E(t) = E(0), \quad 0 \leqslant t \leqslant T.
$$

9.2 Suppose $v \in \mathcal{U}_h$. Define

$$
\delta_x^2 v_0 = \frac{2}{h} \delta_x v_{\frac{1}{2}}, \quad \delta_x^2 v_m = \frac{2}{h}\left(-\delta_x v_{m-\frac{1}{2}} \right),
$$

$$
\|\delta_x^2 v\|^2 = h\left[\frac{1}{2}\left(\delta_x^2 v_0\right)^2 + \sum_{i=1}^{m-1} \left(\delta_x^2 v_i\right)^2 + \frac{1}{2}\left(\delta_x^2 v_m\right)^2 \right],
$$

$$
\delta_x^4 v_i = \delta_x^2\left(\delta_x^2 v_i\right), \quad 1 \leqslant i \leqslant m-1.
$$

For the problem (9.46), a difference scheme is established as follows:

$$\begin{cases} \delta_t u_i^{k+\frac{1}{2}} + \Delta_x u_i^{k+\frac{1}{2}} + \Delta_x \delta_x^2 u_i^{k+\frac{1}{2}} + \delta_x^4 \delta_t u_i^{k+\frac{1}{2}} + \psi(u^{k+\frac{1}{2}}, u^{k+\frac{1}{2}})_i = 0, \\ \qquad\qquad\qquad\qquad\qquad 1 \leqslant i \leqslant m-1, \quad 0 \leqslant k \leqslant n-1, \\ u_i^0 = \varphi(x_i), \quad 1 \leqslant i \leqslant m-1, \\ u_0^k = 0, \quad u_m^k = 0, \quad 0 \leqslant k \leqslant n. \end{cases}$$

(1) Show that the difference scheme satisfies certain conservation law.
(2) Analyze the local truncation error of the difference scheme.
(3) Show the existence of the difference solution.
(4) Prove the convergence of the difference solution.

References

1. Shen, J.Y., Wang, X.P., Sun, Z.Z.: The conservation and convergence of two finite difference schemes for Korteweg-de Vries equations with the initial and boundary value conditions. Numer. Math. Theor. Meth. Appl. **13**(1), 253–280 (2020)
2. Wang, X.P., Sun, Z.Z.: A second order convergent and linearized difference scheme for the initial-boundary value problem of KdV equation. J. Southeast Univ. (English Ed.) **38**(2), 203–212 (2022)
3. Wang, X.P., Sun, Z.Z.: A second order convergent difference scheme for the initial-boundary value problem of Korteweg-de Vries equation. Numer. Methods Partial Differ. Equations **37**, 2873–2894 (2021)
4. Zhang, Q.F., Yan, T., Gao, G.H.: The energy method for high-order invariants in shallow water wave equations. Appl. Math. Lett. **142**, 108626 (2023)

Index

Symbols

H^1-norm, 19
H^2-norm, 19
h, 11, 98, 330, 362
$\delta_x^2 v_{ij}^k$, 254
$\|\delta_x^2 u\|$, 18
$(\delta_x u, \delta_x v)$, 18, 362
$(\delta_x \delta_y v, \delta_x \delta_y w)$, 69
$(\delta_x^2 u, \delta_x^2 v)$, 18
$(\Delta_h v, \Delta_h w)$, 69
(u, v), 18, 362
$D_t v_i^k$, 99
$D_x v_i^k$, 99
$D_{\bar{t}} v_i^k$, 99
$D_{\bar{x}} v_i^k$, 99
Γ, 50, 254
Ω, 50, 254
Ω_h, 11, 99, 196, 330, 362
Ω_τ, 99, 196, 254, 330, 362
$\Omega_{h\tau}$, 99, 330
$\bar{\omega}$, 55, 254
$\delta_t^2 v_i^k$, 196
$\delta_t v_i^{k+\frac{1}{2}}$, 99
$\delta_t v_{ij}^{k+\frac{1}{2}}$, 254
$\delta_x^2 v_i$, 11
$\delta_x^2 v_i^k$, 99
$\delta_x^2 v_{ij}$, 55
$\delta_x v_{i-\frac{1}{2},j}$, 55, 254
$\delta_x v_{i-\frac{1}{2}}$, 11
$\delta_x v_{i+\frac{1}{2}}^k$, 99
$\delta_y^2 v_{ij}$, 55

$\delta_y^2 v_{ij}^k$, 254
$\delta_y v_{i,j-\frac{1}{2}}$, 55, 254
γ, 55, 254
$\mathcal{A}$, 33
$\mathcal{U}_h$, 18, 100, 330
$\mathcal{V}_h$, 55, 254
$\mathcal{W}_h$, 362
ω, 55, 254
τ, 99, 254, 330, 362
$\Delta_h v_{ij}$, 55, 254
$\Delta_t v_i^k$, 99
$\Delta_x u_i$, 400
$\Delta_t v_{ij}^k$, 254
$|u|_1$, 18, 362
$|u|_2$, 18
$\|u\|$, 18, 362
$\|u\|_\infty$, 18, 362
$\|u\|_p$, 362
$\|\delta_x u\|$, 18, 362
$\|\delta_x \delta_y v\|$, 69
$\|\Delta_h v\|$, 69
$a_l^{(\alpha)}$, 324
$b_l^{(\gamma)}$, 328, 337
h_1, 54, 254
h_2, 54, 254
$v_i^{\bar{k}}$, 99
$v_i^{k+\frac{1}{2}}$, 99
$v_{i-\frac{1}{2}}$, 11
$v_{ij}^{k+\frac{1}{2}}$, 254
Γ_h, 55
$\overset{\circ}{\Omega}_h$, 55

© Science Press 2026

Z.-Z. Sun et al., *Numerical Solutions to Partial Differential Equations
with Finite Difference Methods*, Springer Asia Pacific Mathematics Series 9,
https://doi.org/10.1007/978-981-95-5563-5

The manufacturer's authorised representative in the EU is Springer
Nature Customer Service Centre GmbH, Europaplatz 3, 69115 Heidelberg,
Germany. If you have any concerns regarding our products, please
contact ProductSafety@springernature.com

Printed and bound by CPI Group (UK) Ltd, Croydon, CR0 4YY
08/06/2026
02126691-0001